Parameter Estimation in Reliability and Life Span Models

STATISTICS: Textbooks and Monographs

A Series Edited by

Vol. 1: The Generalized Jackknife Statistic, *H. L. Gray and W. R. Schucany*
Vol. 2: Multivariate Analysis, *Anant M. Kshirsagar*
Vol. 3: Statistics and Society, *Walter T. Federer*
Vol. 4: Multivariate Analysis: A Selected and Abstracted Bibliography, 1957-1972, *Kocherlakota Subrahmaniam and Kathleen Subrahmaniam* (out of print)
Vol. 5: Design of Experiments: A Realistic Approach, *Virgil L. Anderson and Robert A. McLean*
Vol. 6: Statistical and Mathematical Aspects of Pollution Problems, *John W. Pratt*
Vol. 7: Introduction to Probability and Statistics (in two parts), Part I: Probability; Part II: Statistics, *Narayan C. Giri*
Vol. 8: Statistical Theory of the Analysis of Experimental Designs, *J. Ogawa*
Vol. 9: Statistical Techniques in Simulation (in two parts), *Jack P. C. Kleijnen*
Vol. 10: Data Quality Control and Editing, *Joseph I. Naus* (out of print)
Vol. 11: Cost of Living Index Numbers: Practice, Precision, and Theory, *Kali S. Banerjee*
Vol. 12: Weighing Designs: For Chemistry, Medicine, Economics, Operations Research, Statistics, *Kali S. Banerjee*
Vol. 13: The Search for Oil: Some Statistical Methods and Techniques, *edited by D. B. Owen*
Vol. 14: Sample Size Choice: Charts for Experiments with Linear Models, *Robert E. Odeh and Martin Fox*
Vol. 15: Statistical Methods for Engineers and Scientists, *Robert M. Bethea, Benjamin S. Duran, and Thomas L. Boullion*
Vol. 16: Statistical Quality Control Methods, *Irving W. Burr*
Vol. 17: On the History of Statistics and Probability, *edited by D. B. Owen*
Vol. 18: Econometrics, *Peter Schmidt*
Vol. 19: Sufficient Statistics: Selected Contributions, *Vasant S. Huzurbazar (edited by Anant M. Kshirsagar)*
Vol. 20: Handbook of Statistical Distributions, *Jagdish K. Patel, C. H. Kapadia, and D. B. Owen*
Vol. 21: Case Studies in Sample Design, *A. C. Rosander*
Vol. 22: Pocket Book of Statistical Tables, *compiled by R. E. Odeh, D. B. Owen, Z. W. Birnbaum, and L. Fisher*
Vol. 23: The Information in Contingency Tables, *D. V. Gokhale and Solomon Kullback*

Vol. 24: Statistical Analysis of Reliability and Life-Testing Models: Theory and Methods, *Lee J. Bain*
Vol. 25: Elementary Statistical Quality Control, *Irving W. Burr*
Vol. 26: An Introduction to Probability and Statistics Using BASIC, *Richard A. Groeneveld*
Vol. 27: Basic Applied Statistics, *B. L. Raktoe and J. J. Hubert*
Vol. 28: A Primer in Probability, *Kathleen Subrahmaniam*
Vol. 29: Random Processes: A First Look, *R. Syski*
Vol. 30: Regression Methods: A Tool for Data Analysis, *Rudolf J. Freund and Paul D. Minton*
Vol. 31: Randomization Tests, *Eugene S. Edgington*
Vol. 32: Tables for Normal Tolerance Limits, Sampling Plans, and Screening, *Robert E. Odeh and D. B. Owen*
Vol. 33: Statistical Computing, *William J. Kennedy, Jr. and James E. Gentle*
Vol. 34: Regression Analysis and Its Application: A Data-Oriented Approach, *Richard F. Gunst and Robert L. Mason*
Vol. 35: Scientific Strategies to Save Your Life, *I. D. J. Bross*
Vol. 36: Statistics in the Pharmaceutical Industry, *edited by C. Ralph Buncher and Jia-Yeong Tsay*
Vol. 37: Sampling from a Finite Population, *J. Hajek*
Vol. 38: Statistical Modeling Techniques, *S. S. Shapiro*
Vol. 39: Statistical Theory and Inference in Research, *T. A. Bancroft and C.-P. Han*
Vol. 40: Handbook of the Normal Distribution, *Jagdish K. Patel and Campbell B. Read*
Vol. 41: Recent Advances in Regression Methods, *Hrishikesh D. Vinod and Aman Ullah*
Vol. 42: Acceptance Sampling in Quality Control, *Edward G. Schilling*
Vol. 43: The Randomized Clinical Trial and Therapeutic Decisions, *edited by Niels Tygstrup, John M. Lachin, and Erik Juhl*
Vol. 44: Regression Analysis of Survival Data in Cancer Chemotherapy, *Walter H. Carter, Jr., Galen L. Wampler, and Donald M. Stablein*
Vol. 45: A Course in Linear Models, *Anant M. Kshirsagar*
Vol. 46: Clinical Trials: Issues and Approaches, *edited by Stanley H. Shapiro and Thomas H. Louis*
Vol. 47: Statistical Analysis of DNA Sequence Data, *edited by B. S. Weir*
Vol. 48: Nonlinear Regression Modeling: A Unified Practical Approach, *David A. Ratkowsky*
Vol. 49: Attribute Sampling Plans, Tables of Tests and Confidence Limits for Proportions, *Robert E. Odeh and D. B. Owen*
Vol. 50: Experimental Design, Statistical Models, and Genetic Statistics, *edited by Klaus Hinkelmann*
Vol. 51: Statistical Methods for Cancer Studies, *edited by Richard G. Cornell*
Vol. 52: Practical Statistical Sampling for Auditors, *Arthur J. Wilburn*
Vol. 53: Statistical Signal Processing, *edited by Edward J. Wegman and James G. Smith*
Vol. 54: Self-Organizing Methods in Modeling: GMDH Type Algorithms, *edited by Stanley J. Farlow*
Vol. 55: Applied Factorial and Fractional Designs, *Robert A. McLean and Virgil L. Anderson*
Vol. 56: Design of Experiments: Ranking and Selection, *edited by Thomas J. Santner and Ajit C. Tamhane*
Vol. 57: Statistical Methods for Engineers and Scientists. Second Edition, Revised and Expanded, *Robert M. Bethea, Benjamin S. Duran, and Thomas L. Boullion*
Vol. 58: Ensemble Modeling: Inference from Small-Scale Properties to Large-Scale Systems, *Alan E. Gelfand and Crayton C. Walker*
Vol. 59: Computer Modeling for Business and Industry, *Bruce L. Bowerman and Richard T. O'Connell*
Vol. 60: Bayesian Analysis of Linear Models, *Lyle D. Broemeling*

Vol. 61: Methodological Issues for Health Care Surveys, *Brenda Cox and Steven Cohen*

Vol. 62: Applied Regression Analysis and Experimental Design, *Richard J. Brook and Gregory C. Arnold*

Vol. 63: Statpal: A Statistical Package for Microcomputers – PC-DOS Version for the IBM PC and Compatibles, *Bruce J. Chalmer and David G. Whitmore*

Vol. 64: Statpal: A Statistical Package for Microcomputers – Apple Version for the II, II+, and IIe, *David G. Whitmore and Bruce J. Chalmer*

Vol. 65: Nonparametric Statistical Inference, Second Edition, Revised and Expanded, *Jean Dickinson Gibbons*

Vol. 66: Design and Analysis of Experiments, *Roger G. Petersen*

Vol. 67: Statistical Methods for Pharmaceutical Research Planning, *Sten W. Bergman and John C. Gittins*

Vol. 68: Goodness-of-Fit Techniques, *edited by Ralph B. D'Agostino and Michael A. Stephens*

Vol. 69: Statistical Methods in Discrimination Litigation, *edited by D. H. Kaye and Mikel Aickin*

Vol. 70: Truncated and Censored Samples from Normal Populations, *Helmut Schneider*

Vol. 71: Robust Inference, *M. L. Tiku, W. Y. Tan, and N. Balakrishnan*

Vol. 72: Statistical Image Processing and Graphics, *edited by Edward J. Wegman and Douglas J. DePriest*

Vol. 73: Assignment Methods in Combinatorial Data Analysis, *Lawrence J. Hubert*

Vol. 74: Econometrics and Structural Change, *Lyle D. Broemeling and Hiroki Tsurumi*

Vol. 75: Multivariate Interpretation of Clinical Laboratory Data, *Adelin Albert and Eugene K. Harris*

Vol. 76: Statistical Tools for Simulation Practitioners, *Jack P. C. Kleijnen*

Vol. 77: Randomization Tests, Second Edition, *Eugene S. Edgington*

Vol. 78: A Folio of Distributions: A Collection of Theoretical Quantile-Quantile Plots, *Edward B. Fowlkes*

Vol. 79: Applied Categorical Data Analysis, *Daniel H. Freeman, Jr.*

Vol. 80: Seemingly Unrelated Regression Equations Models : Estimation and Inference, *Virendra K. Srivastava and David E. A. Giles*

Vol. 81: Response Surfaces: Designs and Analyses, *Andre I. Khuri and John A. Cornell*

Vol. 82: Nonlinear Parameter Estimation: An Integrated System in BASIC, *John C. Nash and Mary Walker-Smith*

Vol. 83: Cancer Modeling, *edited by James R. Thompson and Barry W. Brown*

Vol. 84: Mixture Models: Inference and Applications to Clustering, *Geoffrey J. McLachlan and Kaye E. Basford*

Vol. 85: Randomized Response: Theory and Techniques, *Arijit Chaudhuri and Rahul Mukerjee*

Vol. 86: Biopharmaceutical Statistics for Drug Development, *edited by Karl E. Peace*

Vol. 87: Parts per Million Values for Estimating Quality Levels, *Robert E. Odeh and D. B. Owen*

Vol. 88: Lognormal Distributions: Theory and Applications, *edited by Edwin L. Crow and Kunio Shimizu*

Vol. 89: Properties of Estimators for the Gamma Distribution, *K. O. Bowman and L. R. Shenton*

Vol. 90: Spline Smoothing and Nonparametric Regression, *Randall L. Eubank*

Vol. 91: Linear Least Squares Computations, *R. W. Farebrother*

Vol. 92: Exploring Statistics, *Damaraju Raghavarao*

Vol. 93: Applied Time Series Analysis for Business and Economic Forecasting, *Sufi M. Nazem*

Vol. 94: Bayesian Analysis of Time Series and Dynamic Models, *edited by James C. Spall*

Vol. 95: The Inverse Gaussian Distribution: Theory, Methodology, and Applications, *Raj S. Chhikara and J. Leroy Folks*

Vol. 96: Parameter Estimation in Reliability and Life Span Models, *A. Clifford Cohen and Betty Jones Whitten*

Vol. 97: Pooled Cross-Sectional and Time Series Data Analysis, *Terry E. Dielman*

ADDITIONAL VOLUMES IN PREPARATION

Parameter Estimation in Reliability and Life Span Models

A. Clifford Cohen
Betty Jones Whitten
University of Georgia
Athens, Georgia

CRC Press is an imprint of the
Taylor & Francis Group, an **informa** business

CRC Press
Taylor & Francis Group
6000 Broken Sound Parkway NW, Suite 300
Boca Raton, FL 33487-2742

First issued in paperback 2019

ISBN-13: 978-0-8247-7980-1 (hbk)
ISBN-13: 978-0-367-40334-8 (pbk)

Visit the Taylor & Francis Web site at
http://www.taylorandfrancis.com

and the CRC Press Web site at
http://www.crcpress.com

to Dorothy, Judy, Susan, and Debbie — A.C.C.

and Ken, Andy, and Kathryn — B.J.W.

Library of Congress Cataloging-in-Publication Data

Cohen, A. Clifford
Parameter estimation in reliability and life span models.

(Statistics, textbooks and monographs ; v. 96)
Bibliography: p.
Includes index.
1. Parameter estimation. 2. Distribution (Probability theory) I. Whitten, Betty Jones, II. Title. III. Series.
QA276.8.C63 1988 519.5 88-20342
ISBN 0-8247-7980-0

Preface

Skewed distributions play an important role in the analysis of sample data originating from life span, reaction time, reliability, survivor, and related studies. They are also important in descriptive studies involving measurements of weight, volume, and various other physical measurements. In the field of engineering, skewed distributions are employed as models for life tests and for the distribution of characteristics such as dimension, strength, and hardness of materials and products. They are also useful as models for distributions of voltage, amperage, capacitance, resistance, and other characteristics of interest in electric and electronic devices. In medicine and biology, skewed distributions play a role in survival studies and in distributions of characteristics such as blood pressure, other measures of vital functions, cholesterol level, and various other measures of body chemistry. In economics, they serve as models for distributions of income, sales volume, tax collection, insurance premiums and claims, and for other items of interest in economic and financial studies. Accordingly, these distributions are important in business, manufacturing, engineering, quality control, medical and biological sciences, management sciences at all levels, and in all areas of the physical sciences. They are of particular importance in the area of research and development.

Parameter estimation from both complete and censored samples is a major aspect of any analysis of skewed data; it is also the principal concern of this book. Consideration is given to estimation in the Weibull, exponential, lognormal, inverse Gaussian, gamma, Rayleigh, Pareto, generalized gamma, and extreme value distributions.

Modified moment estimators (MME) for parameters of these distributions are featured throughout this volume. Various tabular and graphical aids are provided, which simplify the calculations of estimates in practical applications.

Since maximum likelihood and moment estimators are usually considered to be the standard or traditional estimators for parameters of the distributions considered here, these estimators are also presented for comparison with the MME. When the origin is known, both maximum likelihood and moment estimation in the resulting distributions are relatively free from computational problems. However, introduction of the threshold (origin) parameter creates complications. Maximum likelihood estimators are sometimes difficult to calculate, and for some parameter combinations they break down and fail to exist. Although moment estimators are usually easy to calculate, their sampling errors are often unacceptably large as a result of using higher order moments. For these reasons, the MME provide an attractive alternative to both maximum likelihood and moment estimators. They retain most of the desirable, while eliminating most of the undesirable, properties of maximum likelihood and moment estimators. The MME are unbiased with respect to distribution means and variances. They are applicable over the entire parameter space, and, with the aid of tables and charts, they are easy to calculate from sample data. In the three-parameter distributions, the third moment is replaced as an estimator in the MME by the first-order statistic. This statistic contains more information about the threshold parameter than any of the other sample observations and often more than all the other sample observations combined. In the Weibull distribution, the replacement is $E(X_1) = x_1$; in the lognormal, it is $E[\ln(X_1 - \gamma)] = \ln(x_1 - \gamma)$; and in the inverse Gaussian and gamma distributions, it is $E[F(X_1)] = F(x_1)$. The first-order statistic (a random variable whose expected value is a function of the distribution parameters) is denoted by X_1, and x_1 denotes the corresponding sample value. $E(\cdot)$ signifies expectation and $F(\cdot)$ signifies the cumulative distribution function (cdf).

To a considerable extent, this volume is based on the authors' work over the past several decades. Unified notations have been introduced, and estimators are presented in a simplified format that will enable users to calculate estimates and otherwise to analyze sample data with a minimum of computational effort. Emphasis is placed on graphical techniques. Through their use, estimates can be quickly and easily calculated. Although their precision is limited, graphical estimates are sufficiently accurate for most practical applications, and when greater accuracy is required they provide good first approximations which can be improved by iteration.

This volume is intended primarily as a reference book and as a handbook for practitioners. It is also intended as a supplemental text for advanced coures in reliability and survivor analysis such as might be included in a Ph.D. program in statistics. It should be useful to quality control and quality assurance personnel; to design engineers, statisticians, biometricians, medical researchers, economists, reliability engineers, marketing research workers, operations research practitioners, physicists; and to scientists in all disciplines where complete or censored samples from skewed distributions are likely to arise. It should also appeal to graduate students and consultants with interests in these areas.

We are indebted to Dr. Ralph A. Bradley and to Dr. H. K. Lam for reading and critically evaluating portions of the draft copies of our manuscript. Thanks are also extended to anonymous referees who made valuable suggestions for numerous improvements in our presentation. We express appreciation to Dr. Don Owen, the coordinating editor of this series, and to Vickie Kearn, Maria Allegra, Brian Black, and other staff members of Marcel Dekker, Inc., for their encouragement and patient guidance during the writing of this book.

We extend thanks to Dr. Lynne Billard, Head, Department of Statistics, and to Dr. Roscoe Davis, Head, Department of Management Science and Information Technology, of the University of Georgia for encouragement and support. We express appreciation to Dawn Tolbert, Sandra Roberts, and Molly Rema for typing and word processing.

Thanks are given to the American Society for Quality Control, the American Statistical Association, Marcel Dekker, Inc., and the Biometrika Trustees for permission to reproduce previously published material by the authors. We are also grateful to Dr. M. S. Arora, Dr. Paul Hsieh, and Dr. Vani Sundaraiyer for permission to use material from their dissertations, and Dr. Wayne Nelson for permission to use the data in Table 7.1 for an illustrative example.

A. Clifford Cohen
Betty Jones Whitten

Contents

List of Illustrations

List of Tables

List of Programs

1

Introduction

1.1 INTRODUCTORY REMARKS

This book is primarily concerned with parameter estimation in skewed distributions. More specifically, it is concerned with estimation in the Weibull, exponential, lognormal, inverse Gaussian (IG), gamma, Rayleigh, Pareto, generalized gamma, and extreme value distributions. Emphasis is placed on estimators that are efficient, low in bias, easy to calculate, and free from restrictions imposed as a consequence of regularity conditions. Modified estimators that employ the first-order statistic are featured, but the traditional moment and maximum likelihood estimators are also presented when they are applicable. Since parameter estimation is a major aspect of the analysis of lifetime and reaction-time data as well as of the analysis of descriptive data of various types, this volume might serve as a handbook and reference for all who are involved with data analyses of these types. The following examples illustrate some of the situations that give rise to the need for estimators presented here.

Example 1.1.1. A manufacturer of dry cell batteries has conducted a life test on a random sample consisting of n cells of a new design. The test was continued until all sample units failed. The manufacturer wished to know the mean life and the probability of failure prior to expiration of a proposed minimum life warranty. This example involves selection of a model and the estimation of its parameters from a complete uncensored sample.

Example 1.1.2. A purchasing agent who buys large quantities of electric light bulbs wishes to know the mean life of bulbs offered by a certain supplier. A life test conducted on N randomly selected bulbs is continued until n failures occur. The test is terminated at time T with c surviving bulbs. This example involves the selection of a model and the estimation of its parameters from a sample that is singly censored on the right.

Example 1.1.3. In medical studies dealing with a specified fatal disease, interest is focused on the survival time of individuals measured from the date of diagnosis, or from some other appropriate starting point, until death. In order to evaluate treatment effectiveness, it is necessary to estimate mean survival time of patients who receive different treatments. Observed samples thus might be complete if observation is continued until all patients die, or they might be censored if observation is discontinued with some surviving patients. Accordingly, this example involves model selection and parameter estimation both from complete and from censored samples.

Example 1.1.4. In investigations of carcinogenic substances, it is standard practice to subject laboratory animals to measured doses of the substance and then observe them until tumors develop. The principal variable of interest is the elapsed time measured from the time of administering the dosage until the appearance of a tumor or until death of the animal. This example involves model selection and parameter estimation both from complete and from censored samples.

Example 1.1.5. The director of physical education at a major university has collected a large sample consisting of the recorded weights of entering male students. He wishes to summarize his data in terms of an appropriate statistical model. This is a descriptive example which involves model selection and parameter estimation.

Example 1.1.6. A metallurgist has conducted hardness tests on a number of specimens of an alloy, and wishes to summarize the data by fitting them to an appropriate distribution. This is a descriptive example that involves both model selection and parameter estimation.

Example 1.1.7. A reliability engineer has collected data on the time between failures of a complex electronic system. The engineer wants to know the hazard function for the system and to estimate its parameters. This example involves model selection, estimation of instantaneous failure rates, and parameter estimation.

These examples illustrate a few of the numerous situations for which the methods of this book are applicable. Each example involves the selection of an appropriate skewed statistical model and the estimation of its parameters.

1.2 AN OVERVIEW OF SKEWED DISTRIBUTIONS

Statistical models of interest in this volume are positively skewed distributions—i.e., distributions that are skewed to the right. They originate at a finite threshold on the left, which may or may not be known, and they tail off to zero on the right as the random variable approaches infinity. Brief introductory descriptions of these distributions follow in this chapter. More detailed accounts are given in subsequent chapters.

1.2.1 Weibull Distribution

The probability density function (pdf) of the three-parameter Weibull distribution is

$$f(x;\gamma,\delta,\beta) = (\delta/\beta^{\delta})(x-\gamma)^{\delta-1}\exp\left\{-\left[\frac{x-\gamma}{\beta}\right]^{\delta}\right\} \qquad (1.2.1)$$

$$\gamma < x < \infty,\ \delta > 0,\ \beta > 0$$

$$= 0 \quad \text{otherwise}$$

and the cumulative distribution function (cdf) is

$$F(x;\gamma,\delta,\beta) = 1 - \exp\left\{-\left[\frac{x-\gamma}{\beta}\right]^{\delta}\right\} \qquad (1.2.2)$$

In the notation employed here, γ is the location or threshold parameter, δ is the shape parameter, and β is a scale parameter.

The frequency curve for this distribution can be either reverse J-shaped or bell-shaped, depending on the shape parameter. Although skewness as measured by the third standard moment, α_3, can be either positive, zero, or negative, our primary interest here is in the case where skewness is positive. The exponential distribution with $\alpha_3 = 2$ ($\delta = 1$) is a special case of the Weibull distribution. The frequency curve is reverse J-shaped when $\alpha_3 \geq 2$ ($\delta \leq 1$). When $\alpha_3 < 2$ ($\delta > 1$), it is bell-shaped. When $\alpha_3 = 0$, the frequency curve is almost normal in shape. The cdf is a simple exponential function and is quite easy to calculate. The Weibull distribution is widely employed as a model for the distribution of lifetime and reaction-time data. In many applications the threshold (origin) is known to equal zero, in which case only two parameters are retained.

1.2.2 Exponential Distribution

As previously noted, the exponential distribution is a special case of the Weibull distribution in which $\delta = 1$ and thus $\alpha_3 = 2$. It is also a special case of the gamma distribution. The pdf of the two-parameter exponential distribution is

$$f(x;\gamma,\beta) = \frac{1}{\beta}\exp\left[\frac{-(x-\gamma)}{\beta}\right]; \quad \gamma < x < \infty,\ \beta > 0 \tag{1.2.3}$$

$$= 0 \quad \text{otherwise}$$

and the cdf is

$$F(x;\gamma,\beta) = 1 - \exp\left[-\frac{(x-\gamma)}{\beta}\right] \tag{1.2.4}$$

This distribution is also used extensively in the analysis of certain types of lifetime data. Its frequency curve is reverse J-shaped.

1.2.3 Lognormal Distribution

The random variable X is said to have a three-parameter lognormal distribution if the random variable $Y = \ln(X - \gamma)$, where $X > \gamma$, is distributed normally (μ, σ^2), $\sigma > 0$. Accordingly, the pdf of X is

$$f(x;\gamma,\mu,\sigma^2) = \frac{1}{\sqrt{2\pi\sigma^2}\,(x-\gamma)}\exp\left\{-\frac{[\ln(x-\gamma)-\mu]^2}{2\sigma^2}\right\} \tag{1.2.5}$$

$$\gamma < x < \infty,\ \sigma > 0$$

$$= 0 \quad \text{otherwise}$$

where γ is the threshold parameter, σ is a shape parameter, and μ is a scale parameter. The cdf for this distribution is

$$F(x;\gamma,\mu,\sigma^2) = \Phi\left[\frac{\ln(x-\gamma)-\mu}{\sigma}\right] \tag{1.2.6}$$

where $\Phi(\cdot)$ is the cdf of the standard normal distribution (0, 1). This distribution is appropriate as a model when data are highly skewed; i.e., when α_3 is large. Its frequency curve rises from an origin (threshold) on the left to a discernible mode before tailing off to zero as $x \to \infty$. The frequency curve of this distribution can never actually be reverse J-shaped. However, when α_3 is sufficiently large, the rise from the origin to the mode is sometimes so steep as to give the appearance at a first glance of being almost reverse J-shaped.

1.2.4 Inverse Gaussian Distribution

The IG distribution is bell-shaped for all values of α_3. It is a long-tailed distribution that more nearly resembles the lognormal than it does the

Weibull or gamma distributions. It is most likely to be employed as a model when α_3 is large. In the notation of Chan et al. (1983), the pdf of this distribution is

$$f(x;\gamma,\mu,\sigma) = \frac{1}{\sigma\sqrt{2\pi}}\left(\frac{\mu}{x-\gamma}\right)^{3/2} \exp\left\{-\frac{1}{2}\left(\frac{\mu}{x-\gamma}\right)\left[\frac{(x-\gamma)-\mu}{\sigma}\right]^2\right\} \quad (1.2.7)$$

$$x > \gamma,\ \mu > 0,\ \sigma > 0$$

$$= 0 \quad \text{elsewhere}$$

1.2.5 Gamma Distribution

In appearance, the frequency curve of the gamma distribution more nearly resembles that of the Weibull distribution than that of the lognormal distribution. Depending on α_3, the frequency curve can be either reverse J-shaped or bell-shaped. If $\alpha_3 < 2$, it is bell-shaped. If $\alpha_3 \geq 2$, it is reverse J-shaped. When $\alpha_3 = 2$, the gamma distribution becomes the exponential distribution, as does the Weibull distribution. The pdf of the gamma distribution is

$$f(x;\gamma,\rho,\beta) = \frac{\beta^{-\rho}}{\Gamma(\rho)}(x-\gamma)^{\rho-1} \exp\left[-\frac{(x-\gamma)}{\beta}\right] \quad (1.2.8)$$

$$\gamma < x < \infty,\ \rho > 0,\ \beta > 0$$

$$= 0 \quad \text{otherwise}$$

where $\Gamma(\,)$ is the gamma function.

1.2.6 Rayleigh Distribution

The most widely recognized version of the Rayleigh distribution has the pdf

$$f(x;\sigma) = \left(\frac{x}{\sigma^2}\right) \exp\left\{\frac{-1}{2}\left(\frac{x}{\sigma}\right)^2\right\}, \quad 0 < x < \infty \quad (1.2.9)$$

$$= 0 \quad \text{otherwise}$$

This is a special case of the two-parameter Weibull distribution (β, δ) in which $\delta = 2$ and $\beta^2 = 2\sigma^2$. It is positively skewed with $\alpha_3 = 0.631110$.

In its most general form, the Rayleigh distribution is the distribution of the distance X from the origin to a point $(Y_1, Y_2, \ldots, Y_p)$ in a p-dimensional space, where components Y_j, $j = 1, 2, \ldots, p$, are independent random variables, each of which is normally distributed $(0, \sigma^2)$. The pdf of (1.2.9) is for the two-dimensional special case; i.e., $p = 2$. The

Rayleigh distribution is frequently employed by engineers and physicists as a model for the analysis of data resulting from wave propagation, radiation, and related phenomena.

1.2.7 Pareto Distribution

The Pareto distribution is a reverse J-shaped positively skewed distribution that is of special interest to economists as a model for the distribution of income data. The pdf is

$$f(x;\gamma,\alpha) = \alpha\gamma^{\alpha}x^{(-\alpha+1)}, \quad \gamma < x < \infty, \ \alpha > 0 \qquad (1.2.10)$$

$$= 0 \quad \text{elsewhere}$$

The cdf is

$$F(x;\gamma,\alpha) = 1 - \left(\frac{\gamma}{x}\right)^{\alpha} \qquad (1.2.11)$$

1.2.8 Generalized Gamma Distribution

The four-parameter generalized gamma distribution results from the addition of a Weibull shape parameter to the gamma distribution. With two shape parameters, it is more flexible than either the Weibull or the gamma distribution alone. The pdf of this distribution is

$$f(x;\gamma,\beta,\rho,\delta) = \frac{\delta}{\beta^{\rho\delta}\Gamma(\rho)}(x-\gamma)^{\rho\delta-1}\exp\left\{-\left[\frac{x-\gamma}{\beta}\right]^{\delta}\right\} \qquad (1.2.12)$$

$$\gamma < x < \infty$$

$$= 0 \quad \text{elsewhere}$$

where γ is the threshold parameter, ρ and δ are shape parameters, and β is the scale parameter. This distribution provides an excellent alternative to the four-parameter Pearson distributions as a model for the graduation of large samples.

1.2.9 Extreme Value Distribution

The extreme value distribution is sometimes referred to as the Gumbel distribution, after E. J. Gumbel, who pioneered its use in the analysis of lifetime data (Gumbel 1958).

The pdf and the cdf of this distribution are

$$f(x;\mu,\alpha) = \frac{1}{\alpha}\exp\left[\frac{x-\mu}{\alpha}\right]\exp\left\{-\exp\left[\frac{x-\mu}{\alpha}\right]\right\} \qquad (1.2.13)$$

$$-\infty < x < \infty, \ \alpha > 0$$

$$F(x;\mu,\alpha) = 1 - \exp\left\{-\exp\left[\frac{x-\mu}{\alpha}\right]\right\}$$

where μ is a location parameter and α is a scale parameter.

This distribution could appropriately be labeled as a log-Weibull distribution. Its relation to the Weibull distribution is the same as that of the lognormal to the normal distribution. If Y has a Weibull distribution with pdf (1.2.1) with $\gamma = 0$, then $X = \ln Y$ has an extreme value distribution with pdf (1.2.13), where $\alpha = \delta^{-1}$ and $\mu = \ln \beta$. Although δ is the shape parameter of the Weibull distribution, its reciprocal becomes the scale parameter of the extreme value distribution.

1.3 PARAMETER ESTIMATION

Maximum likelihood estimators (MLEs) and moment estimators (MEs) are the traditional or standard estimators of choice. Moment estimators are easy to calculate, provided that samples are not censored or truncated. However, estimate variances are sometimes quite large. Maximum likelihood estimators are usually characterized by smaller estimate variances, but they are often difficult to calculate. Maximum likelihood estimators for Weibull and gamma parameters are badly behaved in the vicinity of the transition point, i.e., the point ($\alpha_3 = 2$) where the shape changes from bell-shaped to reverse J-shaped. In the three-parameter lognormal distribution, it is necessary to settle for local MLEs, because global estimators lead to inadmissible estimates.

As a compromise between the ME and the MLE, the authors have proposed a modification of the moment estimators (MME). In the three-parameter distributions, MME employ the first two sample moments plus a function of the first-order statistic. The first-order statistic was chosen for this role since it contains more information about the threshold parameter than any of the other sample observations, often more than all the other observations combined. The MME are unbiased with respect to distribution mean and variance, the parameters which often are of primary interest to investigators. Furthermore, simulation studies have shown that these estimators compare quite favorably with the MLE in respect to estimate variances. They are easy to calculate, and they are applicable to complete and censored samples. With regard to censored samples, we note that the first-order statistic is available even in samples that are right censored.

In the following chapters, MMEs along with MEs and MLEs are presented for each of the distributions under consideration. Tables and charts which simplify calculation of the MME are included.

1.4 SOME COMPARISONS

Comparisons of the Weibull, lognormal, inverse Gaussian, and gamma distributions with respect to modes and thresholds are presented in Tables 1.1 and 1.2, with the skewness α_3 as the argument, where $\alpha_3 = E\{[X - E(X)]/\sigma\}^3$.

The normal distribution is the limiting distribution of the lognormal, IG, and gamma distributions as $\alpha_3 \to 0$, but not of the Weibull distribution. However, the standard Weibull distribution with $\alpha_3 = 0$ is "almost" normal with a left terminus at -3.2431 standard units.

Further comparisons of the Weibull, lognormal, IG, and gamma distributions are provided by additional tables and figures. These various comparisons should be helpful in selecting a model for a specific application. Table 1.3 contains entries for α_4, the lower limit, the mode, and various percentiles as functions of α_3. Figures 1.1, 1.2, and 1.3 contain density curves for $\alpha_3 = 1.0$, 3.0, and 5.0, respectively. Figure 1.4 is a plot of the β_1-β_2 curves for the Weibull, lognormal, and IG distributions superimposed on the Pearson β_1-β_2 plane. The gamma (Type III) distribution is

TABLE 1.1 Lower Limit or Threshold (Terminal) Values in Standard Units, E(Z) = 0, V(Z) = 1

α_3	Lognormal	IG $\gamma = -3/\alpha_3$	Gamma $\gamma = -2/\alpha_3$	Weibull
0.0	$-\infty$	$-\infty$	$-\infty$	-3.2431
0.5	-6.0546	-6.0000	-4.0000	-2.0973
1.0	-3.1038	-3.0000	-2.0000	-1.5304
1.5	-2.1449	-2.0000	-1.3333	-1.2055
2.0	-1.6761	-1.5000	-1.0000	-1.0000
3.0	-1.2229	-1.0000	-0.6667	-0.7595
4.0	-1.0000	-0.7500	-0.5000	-0.6243
4.5	-0.9259	-0.6667	-0.4444	-0.5767
5.0	-0.8664	-0.6000	-0.4000	-0.5375
6.0	-0.7764	-0.5000	-0.3333	-0.4767

Source: Adapted from Chan, M., Cohen, A. C., and Whitten, B. J. (1983).

TABLE 1.2 Modes in Standard Units
$E(Z) = 0$, $V(Z) = 1$

α_3	Lognormal	IG	Gamma	Weibull
0.0	0	0	0	0.0452
0.5	-0.2396	-0.2448	-0.2500	-0.2912
1.0	-0.4274	-0.4586	-0.5000	-0.6433
1.5	-0.5480	-0.6140	-0.7500	-0.9018
2.0	-0.6144	-0.6972	J-Shaped	J-Shaped
3.0	-0.6556	-0.6972	J-Shaped	J-Shaped
4.0	-0.6464	-0.6140	J-Shaped	J-Shaped
4.5	-0.6355	-0.5700	J-Shaped	J-Shaped
5.0	-0.6231	-0.5290	J-Shaped	J-Shaped
6.0	-0.5973	-0.4586	J-Shaped	J-Shaped

Source: Adapted from Chan, M., Cohen, A. C., and Whitten, B. J. (1983).

included as a member of the Pearson system. Tables of the cumulative distribution functions are included in an appendix.

An examination of the frequency curves of Figures 1.1–1.3 discloses that the right tails of the four distributions almost coincide. When $\alpha_3 \leq 1$, discrepancies are small except in the vicinity of the origins. For large values of α_3, the differences are more pronounced. However, even for large values of α_3, the right tails of these distributions tend to coincide. Distinguishing characteristics which might be important in model selection are the threshold values, the modes, and the shapes. The Weibull and the gamma distributions are reverse J-shaped when $\alpha_3 \geq 2$, whereas the lognormal and the IG distributions are bell-shaped. All are bell-shaped when $\alpha_3 < 2$.

TABLE 1.3 Characteristics of Standardized Distributions

	α_3	α_4	Lower Limit	Mode	Percentiles								
					.5	2.5	5	10	50	90	95	97.5	99.5
GAMMA	.10	3.01500	-20.00000	-.05000	-2.48209	-1.91223	-1.61593	-1.27033	-.01666	1.29173	1.67276	2.00691	2.66988
LOGNORMAL	.10	3.01778	-30.01110	-.04991	-2.48301	-1.91242	-1.61592	-1.27018	-.01665	1.29156	1.67270	2.00703	2.67068
WEIBULL	.10	2.71299	-2.93340	-.01599	-2.30061	-1.88733	-1.63104	-1.30502	-.01138	1.30798	1.66893	1.97605	2.55994
I.G.	.10	3.16667	-30.00000	-.04996	-2.48256	-1.91232	-1.61592	-1.27026	-.01666	1.29165	1.67274	2.00698	2.67027
GAMMA	.20	3.06000	-10.00000	-.10000	-2.38795	-1.86360	-1.58607	-1.25824	-.03331	1.30105	1.69971	2.05289	2.76321
LOGNORMAL	.20	3.07120	-15.02216	-.09930	-2.39264	-1.86471	-1.58611	-1.25757	-.03317	1.30014	1.69925	2.05331	2.76710
WEIBULL	.20	2.74246	-2.67306	-.08082	-2.18890	-1.82752	-1.59487	-1.29167	-.03084	1.32142	1.70020	2.02505	2.64870
I.G.	.20	3.33333	-15.00000	-.09967	-2.39082	-1.86424	-1.58606	-1.25781	-.03325	1.30053	1.69947	2.05318	2.76551
GAMMA	.30	3.13500	-6.66667	-.15000	-2.29423	-1.81427	-1.55527	-1.24516	-.04993	1.30936	1.72562	2.09795	2.85636
LOGNORMAL	.30	3.16043	-10.03311	-.14767	-2.30512	-1.81713	-1.55565	-1.24384	-.04947	1.30721	1.72432	2.09851	2.86456
WEIBULL	.30	2.80492	-2.45174	-.14879	-2.08125	-1.76713	-1.55699	-1.27623	-.05059	1.33354	1.73069	2.07410	2.74004
I.G.	.30	3.50000	-10.00000	-.14888	-2.30096	-1.81591	-1.55540	-1.24428	-.04971	1.30815	1.72496	2.09842	2.86120
GAMMA	.40	3.24000	-5.00000	-.20000	-2.20093	-1.76426	-1.52357	-1.23114	-.06651	1.31671	1.75048	2.14202	2.94900
LOGNORMAL	.40	3.28581	-7.54393	-.19456	-2.22072	-1.76997	-1.52478	-1.22914	-.06542	1.31276	1.74777	2.14235	2.96255
WEIBULL	.40	2.90014	-2.26176	-.21920	-1.97827	-1.70668	-1.51777	-1.25884	-.07046	1.34418	1.76016	2.12290	2.83358
I.G.	.40	3.66667	-7.50000	-.19733	-2.21329	-1.76754	-1.52407	-1.22974	-.06598	1.31450	1.74914	2.14254	2.95700
GAMMA	.50	3.37500	-4.00000	-.25000	-2.10825	-1.71365	-1.49101	-1.21618	-.08302	1.32309	1.77428	2.18505	3.04101
LOGNORMAL	.50	3.44776	-6.05456	-.23956	-2.13971	-1.72350	-1.49371	-1.21361	-.08093	1.31678	1.76948	2.18458	3.06053
WEIBULL	.50	3.02800	-2.09732	-.29124	-1.88038	-1.64672	-1.47760	-1.23970	-.09026	1.35323	1.78839	2.17112	2.92888
I.G.	.50	3.83333	-6.00000	-.24479	-2.12810	-1.71933	-1.49222	-1.21425	-.08201	1.31959	1.77195	2.18542	3.05257
GAMMA	.60	3.54000	-3.33333	-.30000	-2.01644	-1.66253	-1.45762	-1.20028	-.09945	1.32850	1.79701	2.22702	3.13232
LOGNORMAL	.60	3.64684	-5.06497	-.28235	-2.06223	-1.67793	-1.46265	-1.19742	-.09592	1.31929	1.78935	2.22500	3.15800
WEIBULL	.60	3.18850	-1.95394	-.36402	-1.78788	-1.58769	-1.43689	-1.21906	-.10984	1.36060	1.81517	2.21845	3.02550
I.G.	.60	4.00000	-5.00000	-.29101	-2.04564	-1.67146	-1.45999	-1.19792	-.09773	1.32343	1.79335	2.22695	3.14764
GAMMA	.70	3.73500	-2.85714	-.35000	-1.92580	-1.61099	-1.42345	-1.18347	-.11578	1.33294	1.81864	2.26790	3.22281
LOGNORMAL	.70	3.88372	-4.36084	-.32267	-1.98840	-1.63346	-1.43180	-1.18072	-.11032	1.32035	1.80735	2.26343	3.25448
WEIBULL	.70	3.38168	-1.82810	-.43657	-1.70090	-1.53001	-1.39599	-1.19716	-.12904	1.36625	1.84033	2.26462	3.12297
I.G.	.70	4.16667	-4.28571	-.33573	-1.96610	-1.62413	-1.42753	-1.18083	-.11310	1.32603	1.81331	2.26706	3.24193
GAMMA	.80	3.96000	-2.50000	-.40000	-1.83660	-1.55914	-1.38855	-1.16574	-.13199	1.33640	1.83916	2.30764	3.31243
LOGNORMAL	.80	4.15916	-3.83499	-.36034	-1.91825	-1.59027	-1.40131	-1.16364	-.12409	1.32002	1.82345	2.29975	3.34951
WEIBULL	.80	3.60770	-1.71700	-.50793	-1.61947	-1.47402	-1.35526	-1.17426	-.14772	1.37014	1.86371	2.30937	3.22083
I.G.	.80	4.33333	-3.75000	-.37873	-1.88962	-1.57749	-1.39498	-1.16308	-.12808	1.32742	1.83180	2.30565	3.33522

	α_3	α_4	Lower Limit	Mode	Percentiles								
					.5	2.5	5	10	50	90	95	97.5	99.5
GAMMA	.90	4.21500	-2.22222	-.45000	-1.74919	-1.50712	-1.35299	-1.14712	-.14807	1.33889	1.85856	2.34623	3.40110
LOGNORMAL	.90	4.47404	-3.42789	-.39525	-1.85177	-1.54845	-1.37135	-1.14634	-.13717	1.31839	1.83767	2.33387	3.44270
WEIBULL	.90	3.86676	-1.61839	-.57714	-1.54349	-1.41998	-1.31500	-1.15061	-.16576	1.37230	1.88521	2.35248	3.31866
I.G.	.90	4.50000	-3.33333	-.41976	-1.81631	-1.53169	-1.36248	-1.14477	-.14261	1.32764	1.84883	2.34269	3.42730
GAMMA	1.00	4.50000	-2.00000	-.50000	-1.66390	-1.45507	-1.31684	-1.12762	-.16397	1.34039	1.87683	2.38364	3.48874
LOGNORMAL	1.00	4.82931	-3.10380	-.42737	-1.78891	-1.50812	-1.34202	-1.12893	-.14955	1.31555	1.85005	2.36573	3.53371
WEIBULL	1.00	4.15914	-1.53042	-.64326	-1.47279	-1.36809	-1.27547	-1.12645	-.18306	1.37275	1.90476	2.39375	3.41602
I.G.	1.00	4.66667	-3.00000	-.45862	-1.74621	-1.48687	-1.33015	-1.12600	-.15667	1.32672	1.86441	2.37814	3.51801
GAMMA	1.25	5.34375	-1.60000	-.62500	-1.46234	-1.32590	-1.22448	-1.07520	-.20282	1.33988	1.91747	2.47184	3.70300
LOGNORMAL	1.25	5.90119	-2.52543	-.49564	-1.64687	-1.41408	-1.27206	-1.08565	-.17738	1.30387	1.87338	2.43550	3.74996
WEIBULL	1.25	5.03798	-1.34778	-.78982	-1.31753	-1.24852	-1.18117	-1.06526	-.22259	1.36684	1.94488	2.48803	3.65496
I.G.	1.25	5.08333	-2.40000	-.54495	-1.58501	-1.37976	-1.25080	-1.07770	-.18956	1.31974	1.89709	2.45967	3.73792
GAMMA	1.50	6.37500	-1.33333	-.75000	-1.28167	-1.20059	-1.13075	-1.01810	-.23996	1.33330	1.95083	2.55222	3.90973
LOGNORMAL	1.50	7.25033	-2.14491	-.54800	-1.52464	-1.32983	-1.20758	-1.04371	-.20090	1.28699	1.88698	2.49170	3.94794
WEIBULL	1.50	6.13302	-1.20546	-.90183	-1.18925	-1.14373	-1.09488	-1.00511	-.25636	1.35203	1.97273	2.56864	3.88370
I.G.	1.50	5.50000	-2.00000	-.61400	-1.44315	-1.28054	-1.17468	-1.02851	-.21901	1.30668	1.92119	2.53107	3.94690
GAMMA	1.75	7.59375	-1.14286	-.87500	-1.12562	-1.08248	-1.03795	-.95740	-.27485	1.32079	1.97690	2.62459	4.10831
LOGNORMAL	1.75	8.89708	-1.87656	-.58674	-1.41968	-1.25495	-1.14888	-1.00394	-.22047	1.26646	1.89248	2.53559	4.12650
WEIBULL	1.75	7.45120	-1.09208	-.97321	-1.08318	-1.05295	-1.01744	-.94789	-.28431	1.33004	1.98921	2.63538	4.09882
I.G.	1.75	5.91667	-1.71429	-.66460	-1.31903	-1.18970	-1.10279	-.97961	-.24486	1.28833	1.93730	2.59262	4.14418
GAMMA	2.00	9.00000	-1.00000	J-SHAPED	-.99499	-.97468	-.94871	-.89464	-.30685	1.30259	1.99573	2.68888	4.29832
LOGNORMAL	2.00	10.86346	-1.67765	-.61437	-1.32942	-1.18861	-1.09581	-.96677	-.23659	1.24359	1.89150	2.56876	4.28582
WEIBULL	2.00	9.00000	-1.00000	-1.00000	-.99499	-.97468	-.94871	-.89464	-.30685	1.30259	1.99573	2.68888	4.29832
I.G.	2.00	6.33333	-1.50000	-.69722	-1.21066	-1.10718	-1.03568	-.93189	-.26715	1.26547	1.94609	2.64479	4.32950
GAMMA	2.50	12.37500	-.80000	J-SHAPED	-.79973	-.79667	-.79015	-.77062	-.35992	1.25039	2.01247	2.79345	4.65176
LOGNORMAL	2.50	15.84874	-1.40316	-.64538	-1.18389	-1.07777	-1.00495	-.90056	-.26049	1.19469	1.87576	2.60953	4.55139
WEIBULL	2.50	12.82031	-.86019	J-SHAPED	-.85846	-.84891	-.83463	-.80133	-.33819	1.23737	1.98504	2.76089	4.64797
I.G.	2.50	7 16667	-1.20000	-.71723	-1.03312	-.96535	-.91636	-.84231	-.30159	1.20919	1.94456	2.72335	4.66498
GAMMA	3.00	16.50000	-.66667	J-SHAPED	-.66666	-.66638	-.66532	-.66023	-.39554	1.18006	2.00335	2.86735	4.96959
LOGNORMAL	3.00	22.39997	-1.22290	-.65558	-1.07302	-.99003	-.93113	-.84451	-.27622	1.14556	1.84873	2.62562	4.75687
WEIBULL	3.00	17.65516	-.75947	J-SHAPED	-.75881	-.75403	-.74583	-.72466	-.35566	1.16599	1.95213	2.79545	4.93443
I.G.	3.00	8.00000	-1.00000	-.69722	-.89607	-.85020	-.81589	-.76238	-.32416	1.14303	1.92208	2.77184	4.95630

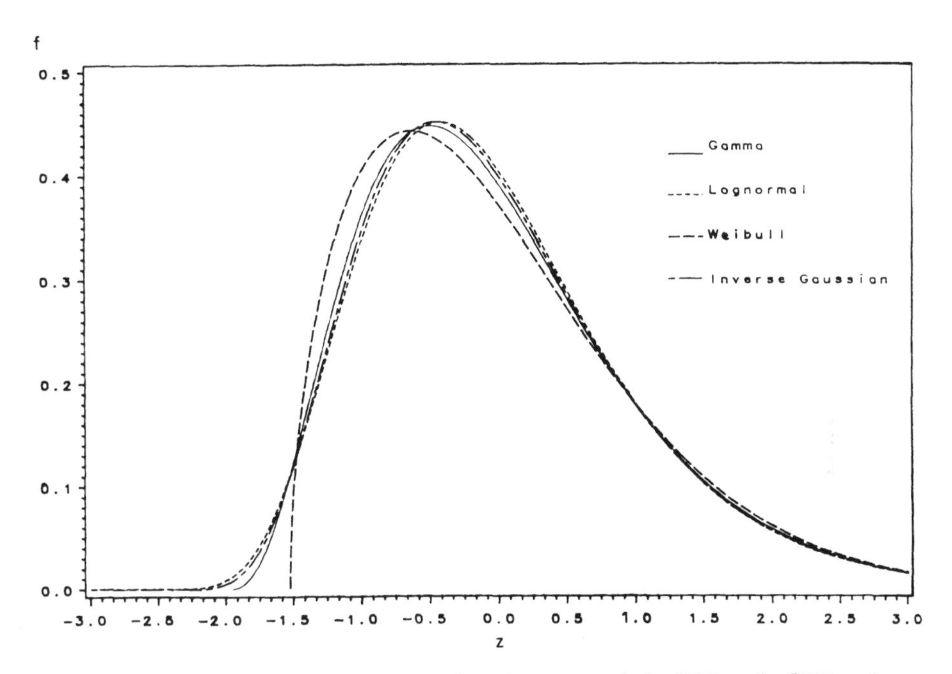

FIGURE 1.1 Standardized density functions $\alpha_3 = 1.0$, $E(Z) = 0$, $V(Z) = 1$. Adapted with permission from Sundaraiyer (1986).

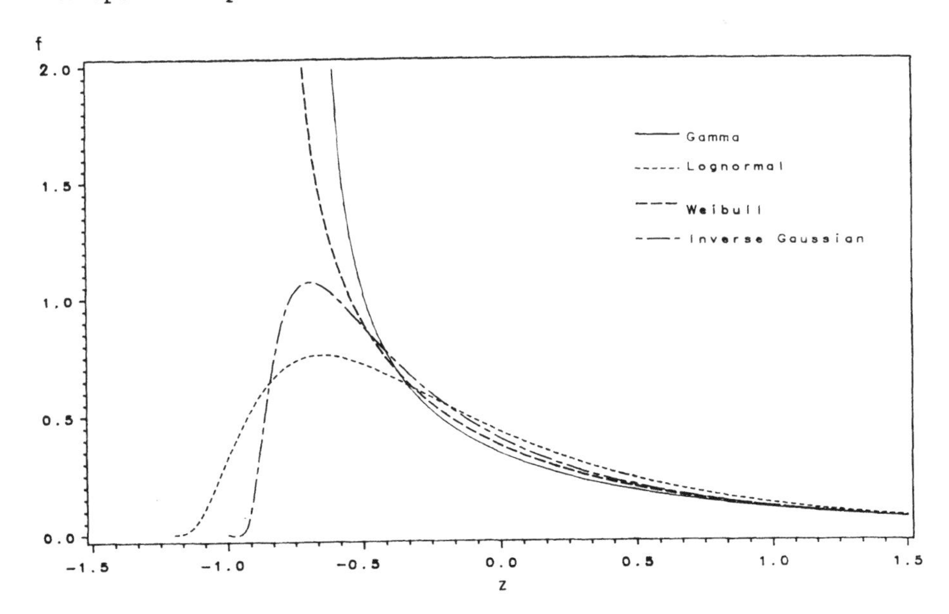

FIGURE 1.2 Standardized density functions $\alpha_3 = 3.0$, $E(Z) = 0$, $V(Z) = 1$. Adapted with permission from Sundaraiyer (1986).

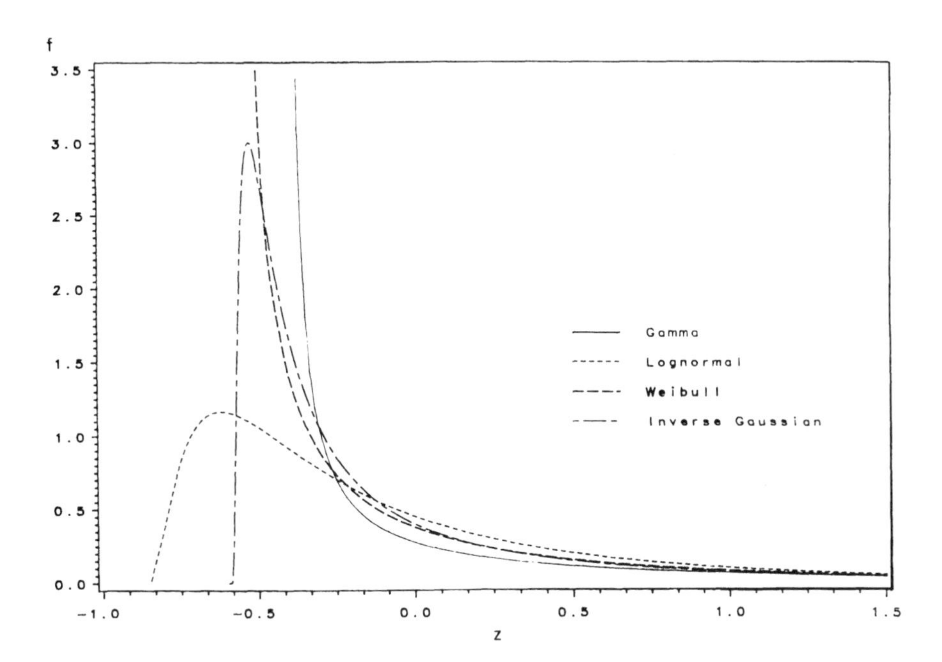

FIGURE 1.3 Standardized density functions $\alpha_3 = 5.0$, $E(Z) = 0$, $V(Z) = 1$. Adapted with permission from Sundaraiyer (1986).

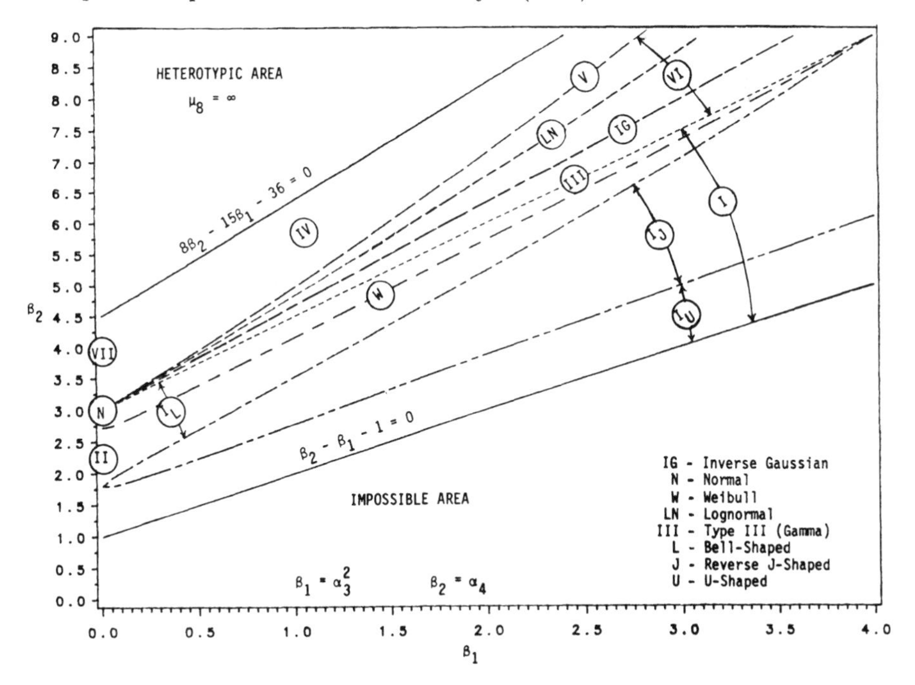

FIGURE 1.4 Pearson β_1/β_2 curves. Adapted with permission from Sundaraiyer (1986).

2

Basic Concepts for the Analysis of Reliability and Life Distribution Data

2.1 LIFE DISTRIBUTIONS AND RELIABILITY

Reliability might be defined as the ability of a product or system to perform over a period of time according to design specifications or according to user expectations. Longer life is identified with greater reliability. In measuring reliability, we are therefore observing life-spans of products or systems of interest. Analysis of the resulting sample data involves the selection of an appropriate model and the estimation of its parameters. Various factors must be considered in the selection of models. The instantaneous failure rate is a factor of considerable importance. An appropriate model should account for the observed data and therefore provide valid predictions about future performance of the items that are being evaluated. Eventually the analysis of lifetime data leads to the estimation of parameters of skewed distributions such as the Weibull, lognormal, inverse Gaussian (IG), gamma, exponential, and related distributions. Before proceeding further, we need to examine some basic concepts.

2.1.1 The Reliability Function

Let T be a continuous nonnegative random variable that represents lifetimes of individual items in a specified population. Let $f(t; \theta)$ denote the probability density function (pdf) of T, where θ is an unknown parameter. There may be more than a single parameter. Let the cumulative distribution (cdf) be

$$F(t; \theta) = \Pr(T \leq t) = \int_0^t f(x; \theta)\, dx \tag{2.1.1}$$

Unless otherwise specified, all functions are defined over the interval $[0, \infty)$. In some circumstances, they might be defined over the interval $[\gamma, \infty)$.

The probability that an individual item will survive until at least time t is given by the reliability function

$$R(t; \theta) = \Pr(T \geq t) = \int_t^\infty f(x; \theta)\, dx \tag{2.1.2}$$

In certain contexts, particularly in the life sciences, $R(t; \theta)$ is referred to as the survivor function where it is written as $S(t; \theta)$. Of course, $R(t; \theta) \equiv S(t; \theta)$. Observe that $R(t; \theta)$ is a monotone decreasing continuous function with $R(0; \theta) = 1$ and $R(\infty; \theta) = \lim_{t \to \infty} R(t; \theta) = 0$.

The 100pth percentile of the distribution of T is the value of t_p such that $\Pr(T \leq t_p = p$; i.e., $t_p = F^{-1}(p)$.

2.1.2 The Hazard Function

The instantaneous failure rate, often referred to as the hazard function, $h(t; \theta)$, is another useful concept in the analysis of life distribution data. We define this function as

$$h(t; \theta) = \lim_{\Delta t \to 0} \frac{\Pr(t \leq T \leq t + \Delta t \mid T \geq t)}{\Delta t} = \frac{f(t; \theta)}{R(t; \theta)} \tag{2.1.3}$$

The hazard function specifies the instantaneous failure or death rate at time t given that an item has survived until t. Thus, $h(t; \theta)\Delta t$ approximates the probability of failure of death in the time interval $[t, t+\Delta t]$, given survival until t. The hazard function is sometimes designated as the hazard rate and sometimes as the age-specific failure rate; in the actuarial field it is called the force of mortality, and in economics its inverse is known as Mill's ratio.

Attention is invited to the fact that $f(t; \theta)$, $F(t; \theta)$, $R(t; \theta)$, and $h(t; \theta)$ are mathematically equivalent specifications of the distribution of T. Accordingly, models for the distribution of T can be specified in terms of any of these functions. Relationships that exist among these functions are relatively easy to derive. Since $f(t; \theta) = F'(t; \theta) = -R'(t; \theta)$ and since $h(t; \theta) = f(t; \theta)/R(t; \theta)$, it follows that

$$h(x; \theta) = -\frac{d \ln R(x; \theta)}{dx} \tag{2.1.4}$$

Accordingly,

$$\ln R(x; \theta) \Big|_0^t = -\int_0^t h(x; \theta)\, dx \tag{2.1.5}$$

and since $R(0; \theta) = 1$, it follows that

$$R(t; \theta) = \exp\left(-\int_0^t h(x; \theta)\, dx\right) \tag{2.1.6}$$

2.1.3 The Cumulative Hazard Function

For some purposes it is useful to employ the cumulative hazard function, defined as

$$H(t; \theta) = \int_0^t h(x; \theta)\, dx \tag{2.1.7}$$

which is related to the reliability function by

$$R(t; \theta) = \exp[-H(t; \theta)] \tag{2.1.8}$$

Since $R(\infty; \theta) = 0$, then $H(\infty; \theta) = \lim_{t\to\infty} H(t) = \infty$. Thus for a continuous lifetime distribution, $h(t; \theta)$ possesses the properties

$$h(t; \theta) \geq 0, \qquad \int_0^\infty h(t; \theta)\, dt = \infty$$

Finally it follows that

$$f(t; \theta) = h(t; \theta) \exp\left(-\int_0^t h(x)\, dx\right) \tag{2.1.9}$$

2.1.4 An Illustrative Example

As an illustrative example, let us consider T as being distributed in accordance with a two-parameter Weibull distribution with pdf

$$f(t; \beta, \delta) = (\delta/\beta^\delta)t^{\delta-1}e^{-(t/\beta)^\delta}, \qquad 0 < t < \infty$$

$$= 0 \qquad \text{elsewhere}$$

The hazard function for this distribution becomes

$$h(t; \beta, \delta) = (\delta/\beta^\delta)t^{\delta-1}$$

Note that the behavior of this function varies according to the value of δ, the shape parameter of the Weibull distribution. We consider three cases: (i) $\delta < 1$; (ii) $\delta = 1$; (iii) $\delta > 1$. When $\delta = 1$, then $\alpha_3 = 2$; when $\delta < 1$, then $\alpha_3 > 2$; and when $\delta > 1$, then $\alpha_3 < 2$.

Case (i): $\delta < 1$, $\alpha_3 > 2$. In this case $h(t; \beta, \delta)$ is a decreasing function; and it might be an appropriate model for use during the development stages of a product or system when the elimination of problem sources results in increased reliability with the passage of time.

Case (ii): $\delta = 1$, $\alpha_3 = 2$. The pdf in this case is the exponential distribution, which is a special case of the Weibull distribution. The hazard function and, thus, the instantaneous failure rate are constant. This case is applicable when used items are equal in reliability to new ones.

Case (iii): $\delta > 1$, $\alpha_3 < 2$. In this case the hazard function is an increasing function, and it might be an appropriate model when, due to wearout, products become more susceptible to failure with the passage of time.

2.2 DISCRETE MODELS

2.2.1 Fundamentals

When lifetime data are grouped into discrete grouping intervals or when life is measured in terms of a discrete number of cycles of some type, it may be desirable to treat T as a discrete random variable. In this case, suppose that T can assume the values $t_1, t_2, \ldots$ such that $0 \leq t_1 < \cdots < t_k$. Let the probability function (pf) be

$$p(t_j; \theta) = \Pr(T = t_j), \quad j = 1, 2, \ldots, k \qquad (2.2.1)$$

The reliability (survivor) function is then

$$R(t; \theta) = \Pr(T \geq t) = \sum_{j; t_j \geq t} p(t_j; \theta) \qquad (2.2.2)$$

In the discrete case, as in the continuous case, $R(t: \theta)$ is a monotone decreasing left-continuous function, with $R(0; \theta) = 1$ and $R(\infty; \theta) = 0$. The hazard function is now defined as

$$h(t_j; \theta) = \Pr(T = t_j \mid T \geq t_j) = \frac{p(t_j; \theta)}{R(t_j; \theta)} \qquad (2.2.3)$$

As in the continuous case, the probability, reliability, and hazard functions provide equivalent specifications of the distribution of T. Since $p(t_j; \theta) = R(t_j; \theta) - R(t_{j+1}; \theta)$, it follows that the hazard function can be expressed as

$$h(t_j; \theta) = 1 - \frac{R(t_{j+1}; \theta)}{R(t_j; \theta)}, \quad j = 1, 2, \ldots \tag{2.2.4}$$

and then

$$R(t: \theta) = \prod_{j:t_j<t} [1 - h(t_j: \theta)] \tag{2.2.5}$$

An analog to the cumulative hazard function of the continuous case, which is applicable in the discrete case, can be defined as

$$H(t; \theta) = -\ln R(t; \theta) \tag{2.2.6}$$

where $R(t; \theta)$ is given by (2.2.5). In general, in the discrete case,

$$H(t; \theta) \neq \sum_{j;t_j<t} h(t_j) \tag{2.2.7}$$

For this reason, we might wish to label the analog function as something other than cumulative hazard function. Perhaps it might be labeled simply as the discrete analog hazard function.

2.2.2 Equivalent Representations

Since, in addition to the density, the distribution, the hazard, and the reliability functions, each provides equivalent specifications for the distribution of T, they present alternatives to the sole use of the density function as a vehicle for the estimation of distribution parameters. Furthermore, the hazard function is a valuable tool to be used in selecting a model for a particular life distribution. Requirements for specific applications frequently dictate whether the hazard function must be decreasing, increasing, constant, or perhaps even bathtub shaped. Knowledge of this function provides guidance to insure selection of an appropriate model.

Nelson (1982), Lawless (1982), and various other writers have employed the hazard function and the reliability function as vehicles for the estimation of distribution parameters. Hazard and reliability function plots have been widely employed to provide simple graphical procedures for parameter estimation. Although sometimes lacking in precision, these estimates are often sufficiently accurate for practical applications. When greater accuracy is required, these estimates provide good first approximations for use in iterative procedures for calculating improved estimates. The standard or usual estimates are based on the probability density function and utilize maximum likelihood or moment estimating procedures. As previously stated, emphasis in this book is focused on modified moment estimators which employ the first-order statistic, but, at the same time, full consideration is given to maximum likelihood and moment estimators.

As aids in the selection of an appropriate model, where the hazard function is of primary concern, Table 2.1 and Figure 2.1 have been included. The hazard plot, which is discussed in Chapter 7, provides a convenient procedure for employing the hazard function in the selection process.

Table 2.1 contains entries of standardized hazard functions of the gamma, Weibull, lognormal, and IG distributions for $\alpha_3 = 0$, 0.5, 1.5, 2.0, 2.5, 3.0, 4.0, and 4.5. This table clearly shows that for $\alpha_3 > 2$, the hazard function of the gamma and the Weibull ascend almost instantaneously from zero to a maximum and then decrease as the random variable increases. For $\alpha_3 < 2$, the hazard functions for these two distributions are increasing functions. For $\alpha_3 = 2$, the hazard function is constant. For $\alpha_3 = 0$ and $\alpha_3 = 0.5$, the lognormal and the IG hazard functions are increasing functions over the range of values that are tabulated. For $\alpha_3 = 1.0$, the IG hazard function is increasing over the range of tabulated values. The lognormal reaches a maximum and then slowly decreases. For $\alpha_3 > 1.0$, the hazard functions for both the lognormal and the IG distributions increase to a maximum and then begin to decrease. This behavior may or may not conform to specified requirements for a model in a specific application. Figure 2.1 contains graphs of the standardized hazard function of the lognormal distribution. A more extensive tabulation of the lognormal hazard function is contained in Table 4.4.

TABLE 2.1 Hazard Functions of the Standard Weibull, Lognormal, Inverse Gaussian, and Gamma Distributions: E(Z) = 0, V(Z) = 1

	$\alpha_3 = 0.5$					$\alpha_3 = 1.0$			
z	Gamma	Weibull	Lognormal	IG	z	Gamma	Weibull	Lognormal	IG
-3.0	.000060	.000000	.000187	.000139	-2.0	.000000	.000000	.008513	.005140
-2.0	.036404	.019322	.037185	.036812	-1.5	.125000	.094872	.122862	.121780
-1.5	.145988	.175357	.143164	.144090	-1.0	.421053	.475604	.391463	.402720
-1.0	.342917	.367276	.338699	.340485	-.5	.692308	.691633	.685467	.691739
-.5	.591151	.579704	.591159	.591556	.0	.901408	.864479	.923234	.916644
.0	.850293	.807194	.856585	.854301	.5	1.059322	1.013886	1.094279	1.078648
.5	1.096455	1.046789	1.106184	1.101968	1.0	1.180328	1.147898	1.211126	1.194034
1.0	1.319988	1.296596	1.327364	1.323527	1.5	1.275093	1.270760	1.288271	1.277034
1.5	1.518949	1.555284	1.517212	1.516843	2.0	1.350923	1.385047	1.337226	1.337701
2.0	1.694735	1.821865	1.677332	1.683716	2.5	1.412791	1.492461	1.366296	1.382795
2.5	1.849888	2.095569	1.811011	1.827295	3.0	1.464129	1.594201	1.381310	1.416845
3.0	1.987128	2.375778	1.921885	1.950917	3.5	1.507361	1.691149	1.386349	1.442921
3.5	2.108977	2.661985	2.013389	2.057648	4.0	1.544236	1.783977	1.384279	1.463138
4.0	2.217637	2.953764	2.088553	2.150149	5.0	1.603741	1.959270	1.366392	1.491509
5.0	2.402599	3.552639	2.199823	2.301055	6.0	1.649617	2.123187	1.338013	1.509496
6.0	2.553711	4.170040	2.271846	2.417401					

	$\alpha_3 = 1.5$					$\alpha_3 = 2.0$			
z	Gamma	Weibull	Lognormal	IG	z	Gamma	Weibull	Lognormal	IG
-1.5	.000000	.000000	.063558	.035533	-1.0	.000000	.000000	.457834	.495002
-1.0	.551536	.640309	.435999	.467240	-.5	1.000000	1.000000	.893765	.963141
-.5	.822456	.830805	.787575	.813680	.0	1.000000	1.000000	1.060811	1.054048
.0	.951281	.930301	.992592	.983611	.5	1.000000	1.000000	1.105868	1.058804
.5	1.027607	1.001008	1.096421	1.065520	1.0	1.000000	1.000000	1.100661	1.042849
1.0	1.078300	1.056845	1.142386	1.105922	1.5	1.000000	1.000000	1.074620	1.022637
1.5	1.114481	1.103435	1.156000	1.125747	2.0	1.000000	1.000000	1.040358	1.002829
2.0	1.141627	1.143658	1.151423	1.134814	2.5	1.000000	1.000000	1.003555	.984722
2.5	1.162757	1.179198	1.136594	1.138023	3.0	1.000000	1.000000	.966839	.968548
3.0	1.179677	1.211135	1.116075	1.137960	3.5	1.000000	1.000000	.931421	.954199
3.5	1.193535	1.240203	1.092548	1.136052	4.0	1.000000	1.000000	.897819	.941470
4.0	1.205094	1.266929	1.067617	1.133115	5.0	1.000000	1.000000	.836600	.920037
5.0	1.223282	1.314814	1.017050	1.125886	6.0	1.000000	1.000000	.783018	.902795
6.0	1.236945	1.356949	.968382	1.118246					

(continued)

TABLE 2.1 (continued)

	$\alpha_3 = 2.5$					$\alpha_3 = 3.0$			
z	Gamma	Weibull	Lognormal	IG	z	Gamma	Weibull	Lognormal	IG
-1.0	.000000	.000000	.434568	.295621	-1.0	.000000	.000000	.331599	.000000
-.5	1.264534	1.205857	1.001777	1.148768	-.5	1.722986	1.461496	1.110500	1.383856
.0	1.047660	1.070452	1.125875	1.127061	.0	1.094352	1.139922	1.187043	1.202003
.5	.975620	1.005399	1.118573	1.056076	.5	.953842	1.014017	1.132389	1.055814
1.0	.937691	.963243	1.074407	.995704	1.0	.886711	.938533	1.057268	.959164
1.5	.913843	.932373	1.021428	.948101	1.5	.846379	.885759	.984732	.891708
2.0	.897313	.908180	.968753	.910401	2.0	.819148	.845721	.919604	.842055
2.5	.885119	.888379	.919355	.880019	2.5	.799394	.813754	.862207	.803952
3.0	.875722	.871676	.874031	.855079	3.0	.784346	.787321	.811743	.773752
3.5	.868244	.857270	.832768	.834261	3.5	.772470	.764899	.767226	.749197
4.0	.862141	.844630	.795268	.816627	4.0	.762840	.745505	.727742	.728820
5.0	.852765	.823282	.730041	.788380	5.0	.748138	.713324	.660902	.696902
6.0	.845885	.805724	.675469	.766742	6.0	.737414	.687382	.606486	.672998

	$\alpha_3 = 4.0$					$\alpha_3 = 4.5$			
z	Gamma	Weibull	Lognormal	IG	z	Gamma	Weibull	Lognormal	IG
-.5	.000000	2.248120	1.328730	2.099739	-.5	.000000	2.954674	1.438252	2.669660
.0	1.185153	1.273041	1.297708	1.355935	.0	1.229401	1.336265	1.347764	1.434345
.5	.916363	1.034700	1.159675	1.059214	.5	.900042	1.045315	1.172443	1.061884
1.0	.807786	.908887	1.037680	.906070	1.0	.776441	.899690	1.031989	.886074
1.5	.746947	.826876	.938328	.811788	1.5	.709051	.807315	.922910	.782303
2.0	.707403	.767525	.857380	.747453	2.0	.665873	.741641	.836574	.713209
2.5	.679389	.721778	.790487	.700519	2.5	.635549	.691672	.766628	.663604
3.0	.658387	.684991	.734335	.664637	3.0	.612945	.651896	.708757	.626104
3.5	.641996	.654496	.686511	.636238	3.5	.595374	.619196	.660016	.596671
4.0	.628814	.628631	.645256	.613156	4.0	.581283	.591653	.618341	.572898
5.0	.608858	.586726	.577571	.577813	5.0	.560013	.547408	.550651	.536745
6.0	.594413	.553848	.524212	.551934	6.0	.544656	.513024	.497842	.510451

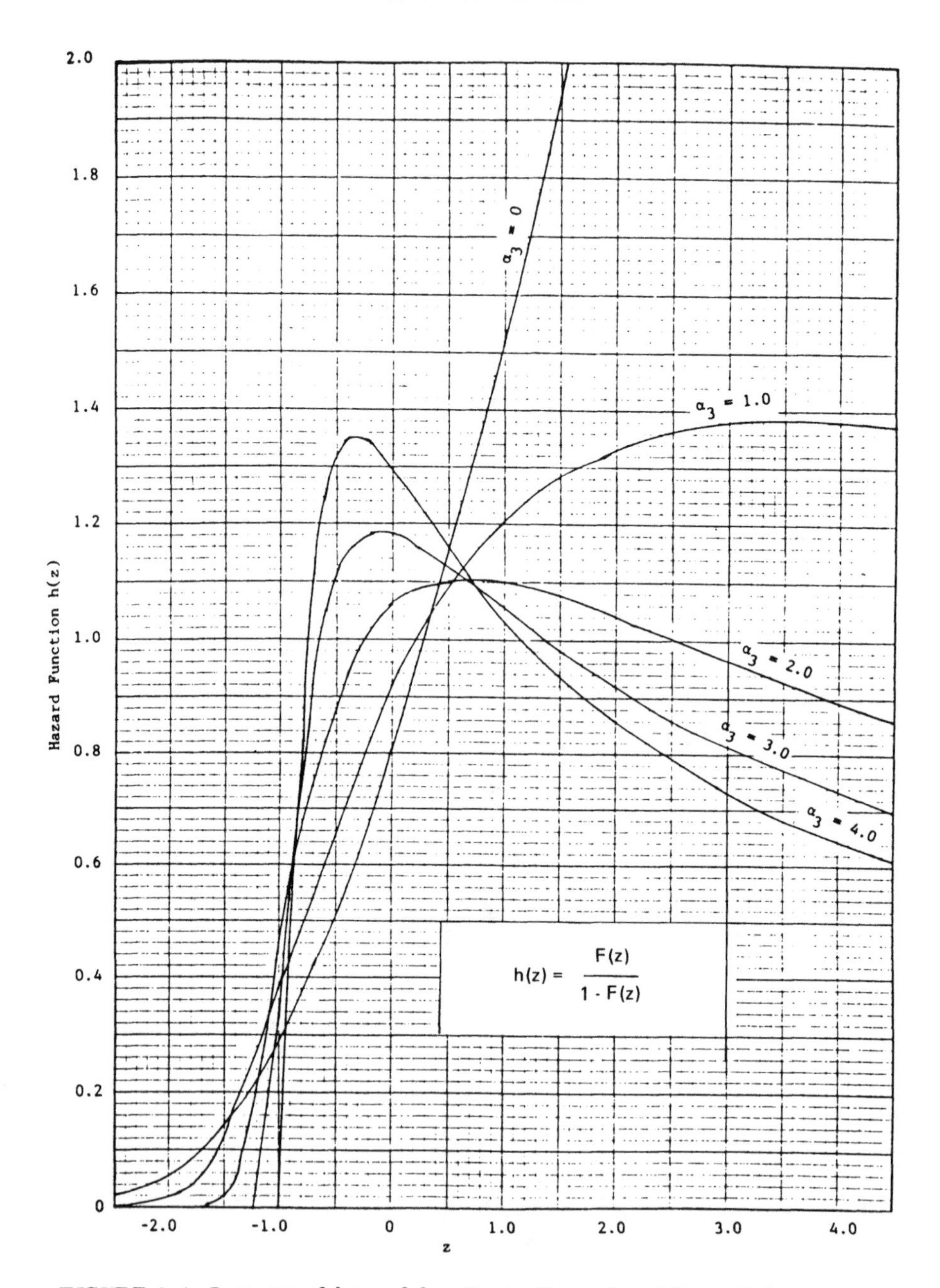

FIGURE 2.1 Lognormal hazard functions. Reproduced from Cohen, A. C. (1986).

3

The Weibull, Exponential, and Extreme Value Distributions

3.1 BACKGROUND

The Weibull distribution is widely employed as a model for distributions of life-spans and reaction times. From a computational point of view, it is particularly appealing, since its cumulative distribution function (cdf) can be expressed explicitly as a simple function of the random variable. Both two- and three-parameter Weibull distributions have been studied extensively, by numerous writers, including Cohen (1965, 1973, 1975), Dubey (1966), Harter and Moore (1965, 1967), Bain (1978), Mann (1968), Mann and Fertig (1975a, b), Lemon (1975), Wingo (1973), Ringer and Sprinkle (1972), Rockette et al. (1974), Zanakis (1977, 1979a, b), Zanakis and Mann (1981), Wycoff et al. (1980), and Königer (1981).

Moment estimators (ME) and maximum likelihood estimators (MLE) have been considered by most of these writers to be the traditional or standard estimators for parameters of this distribution. It is well known, however, that for some combinations of parameter values, MLE in the three-parameter Weibull fail to satisfy the usual regularity conditions. Consequently, for these combinations, the MLE sometimes lead to inconsistent estimates, and the usual asymptotic properties do not apply. For some samples, MLE simply fail to exist. When any of these conditions prevail, modified estimators, which utilize information contained in the first-order statistic, provide alternatives which often enjoy advantages over both ME and MLE. In most instances, modified estimates are easier to calculate and frequently exhibit smaller biases and variances than MLE. Modified estimators presented here for the Weibull parameters are similar to those previously proposed by Cohen and Whitten (1980, 1981, 1982, 1985, 1986),

Cohen et al. (1984, 1985), and Chan et al. (1984) for parameters of the Weibull, lognormal, gamma, and inverse Gaussian distributions.

3.2 CHARACTERISTICS OF THE WEIBULL DISTRIBUTION

Let X designate a random variable that is Weibull (γ, δ, β) with probability density function

$$f(x;\gamma,\delta,\beta) = \frac{\delta}{\beta^{\delta}}(x-\gamma)^{\delta-1}\exp\left\{-\left[\frac{x-\gamma}{\beta}\right]^{\delta}\right\} \tag{3.2.1}$$

for $\gamma < x < \infty$, $\delta > 0$, $\beta > 0$

$$= 0 \quad \text{otherwise}$$

The corresponding cdf is

$$F(x;\gamma,\delta,\beta) = 1 - \exp\left\{-\left[\frac{x-\gamma}{\beta}\right]^{\delta}\right\} \tag{3.2.2}$$

In the notation employed here, γ is the location or threshold parameter, δ is the shape parameter, and β is the scale parameter. The expected value E(X), variance V(X), median Me(X), mode Mo(X), third standard moment α_3(X), and fourth standard moment α_4(X) are

$$E(X) = \gamma + \beta\Gamma_1, \qquad V(X) = \beta^2[\Gamma_2 - \Gamma_1^2]$$

$$Me(X) = \gamma + \beta(\ln 2)^{1/\delta}, \qquad \alpha_3(X) = \frac{\Gamma_3 - 3\Gamma_2\Gamma_1 + 2\Gamma_1^3}{[\Gamma_2 - \Gamma_1^2]^{3/2}} \tag{3.2.3}$$

$$Mo(X) = \gamma + \beta\left[\frac{\delta-1}{\delta}\right]^{1/\delta}, \qquad \alpha_4(X) = \frac{\Gamma_4 - 4\Gamma_3\Gamma_1 + 6\Gamma_2\Gamma_1^2 - 3\Gamma_1^4}{[\Gamma_2 - \Gamma_1^2]^2}$$

where

$$\Gamma_k = \Gamma\left(1 + \frac{k}{\delta}\right) \tag{3.2.4}$$

and $\Gamma(\ )$ is the gamma function defined by the definite integral

$$\Gamma(z) = \int_0^{\infty} t^{z-1} e^{-t}\, dt \tag{3.2.5}$$

It is sometimes expedient to substitute θ for the scale parameter β, where

$$\theta = \beta^{\delta} \quad \text{and} \quad \beta = \theta^{1/\delta} \tag{3.2.6}$$

Although the Weibull distribution is usually perceived to be positively skewed, it becomes negatively skewed when $\delta > \delta_0 = 3.6023494257197$ (Cohen 1973). It is positively skewed when $\delta < \delta_0$. When $\delta = \delta_0$, it is "almost" normal with $\alpha_3 = 0$. In applications concerning life-spans and reaction times, values of δ in excess of 3.22 ($\alpha_3 < 0.10$) seldom occur. However, in other applications, larger values of δ are sometimes encountered. Excellent expository accounts of this distribution and its applications have been given by Johnson and Kotz (1970) and by Mann et al. (1974).

3.3 MAXIMUM LIKELIHOOD ESTIMATION

The likelihood function of an ordered random sample $\{x_i\}$, $i = 1, 2, \ldots, n$, from a distribution with probability distribution function (pdf) (3.2.1) is

$$L(x_1, x_2, \ldots, x_n; \gamma, \delta, \beta) = \left(\frac{\delta}{\beta^{\delta}}\right)^{n} \prod_{i=1}^{n} (x_i - \gamma)^{\delta - 1} \exp\left\{-\sum_{1}^{n}\left(\frac{x_i - \gamma}{\beta}\right)^{\delta}\right\} \tag{3.3.1}$$

When $\delta < 1$, the distribution is reverse J-shaped and the likelihood function becomes infinite as $\gamma \to x_1$, the smallest sample observation. Accordingly, in this situation the MLE of γ would be x_1, but estimates of β and δ would not exist. When $\delta = 1$, the two-parameter exponential, a reverse J-shaped distribution, emerges from (3.2.1) as a special case.

The Weibull distribution is bell-shaped when $\delta > 1$, and MLE in that case can be found by simultaneously solving the system of equations obtained by equating to zero the partial derivatives of the loglikelihood function with respect to the parameters. Computational problems are likely to be encountered when δ is near to 1, even though δ actually exceeds 1. Furthermore, the usual asymptotic properties of MLE do not hold unless $\delta > 2$. As a practical matter, it is not considered advisable to employ the MLE for parameters of the three-parameter Weibull distribution unless there is reason to expect that $\delta > 2.2$.

Taking the logarithm of (3.3.1), differentiating, and equating partial derivatives to zero, we obtain

$$\frac{\partial \ln L}{\partial \gamma} = \frac{\delta}{\theta} \sum_1^n (x_i - \gamma)^{\delta - 1} - (\delta - 1) \sum_1^n (x_i - \gamma)^{-1} = 0$$

$$\frac{\partial \ln L}{\partial \delta} = \frac{n}{\delta} + \sum_1^n \ln (x_i - \gamma) - \frac{1}{\theta} \sum_1^n (\gamma_i - \gamma)^{\delta} \ln (x_i - \gamma) = 0 \tag{3.3.2}$$

$$\frac{\partial \ln L}{\partial \theta} = -\frac{n}{\theta} + \frac{1}{\theta^2} \sum_1^n (x_i - \gamma)^{\delta} = 0$$

When δ is sufficiently large, as indicated previously, the MLE $\hat{\gamma}$, $\hat{\delta}$, $\hat{\theta}$ follow as the simultaneous solution of (3.3.2). If an estimate of β is desired, it follows from (3.2.6) as $\hat{\beta} = \hat{\theta}^{1/\hat{\delta}}$

The three equations of (3.3.2) do not yield explicit solutions for the estimates. However, as shown by Cohen (1965), θ can be eliminated from the last two equations to give

$$\left[\frac{\sum_1^n (x_i - \gamma)^{\delta} \ln (x_i - \gamma)}{\sum_1^n (x_i - \gamma)^{\delta}} - \frac{1}{\delta} \right] - \frac{1}{n} \sum_1^n \ln (x_i - \gamma) = 0 \tag{3.3.3}$$

When γ is known, it is quite easy to solve this equation iteratively for $\hat{\delta}$. The chart of δ as a function of the sample coefficient of variation, given by Cohen (1965) and reproduced here as Figure 3.1, may be useful in finding a first approximation to $\hat{\delta}$ for use in the iterative process. An alternate first approximation might be found by interpolating in Table 3.1 with $\alpha_3 = a_3$, where

$$a_3 = \frac{1}{n} \sum_1^n (x_i - \bar{x})^3 \Big/ \left[\frac{1}{n} \sum_1^n (x_i - \bar{x})^2 \right]^{3/2}$$

and

$$s^2 = \frac{1}{n - 1} \sum_1^n (x_i - \bar{x})^2$$

Subsequently, $\hat{\theta}$ follows from the third equation of (3.3.2) as

$$\hat{\theta} = \frac{1}{n} \sum_1^n (x_i - \hat{\gamma})^{\hat{\delta}} \tag{3.3.4}$$

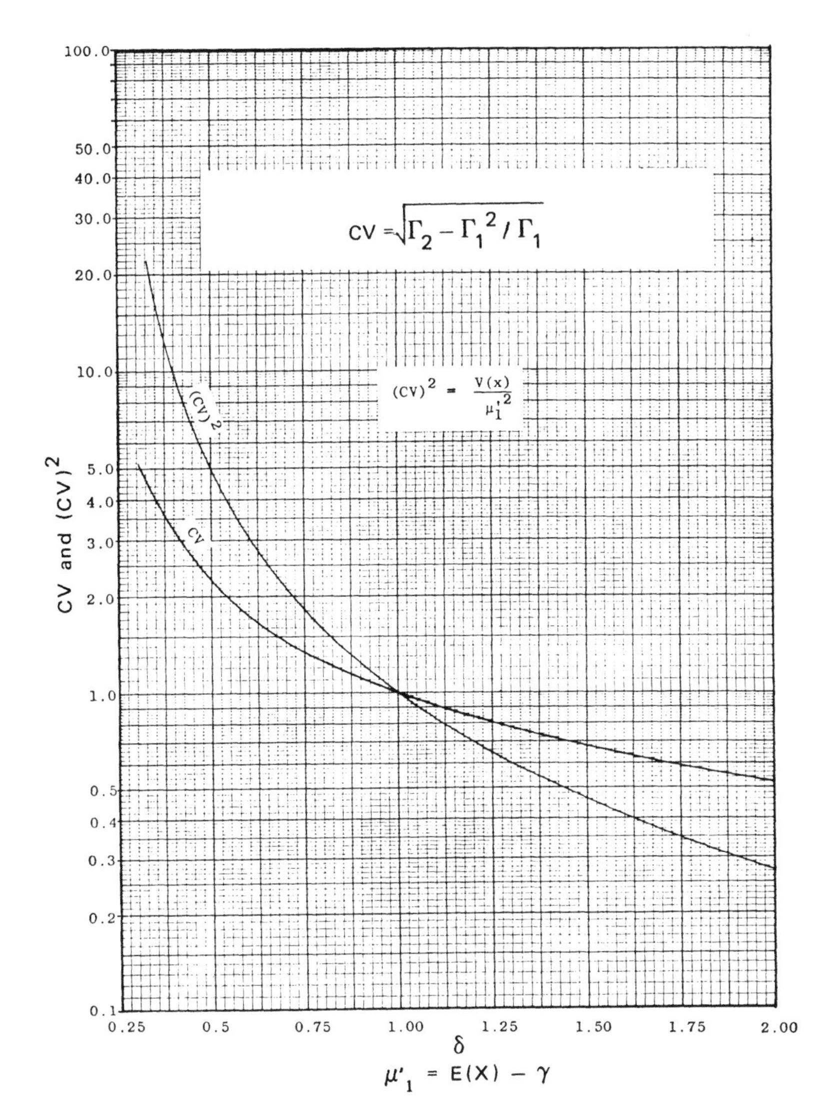

FIGURE 3.1 The Weibull coefficient of variation and its square as functions of the shape parameter. Adapted from Cohen, A. C. (1965).

TABLE 3.1 Values of C, D, α_3, α_4, Mode, Median, and Coefficient of Variation for the Weibull Distribution

δ	C	D	α_3	α_4	Mo	Me	C.V.
.50	.22361	.44721	6.61876	87.72000	J-SHAPE	-.33978	2.23607
.55	.29893	.50890	5.43068	57.39817	J-SHAPE	-.35538	1.96502
.60	.37805	.56881	4.59341	40.48166	J-SHAPE	-.36357	1.75807
.65	.45895	.62706	3.97420	30.20718	J-SHAPE	-.36591	1.59475
.70	.54020	.68380	3.49837	23.54202	J-SHAPE	-.36379	1.46242
.75	.62082	.73917	3.12124	18.98700	J-SHAPE	-.35834	1.35286
.80	.70020	.79333	2.81465	15.74074	J-SHAPE	-.35048	1.26051
.85	.77796	.84638	2.56009	13.34657	J-SHAPE	-.34092	1.18150
.90	.85389	.89845	2.34496	11.53005	J-SHAPE	-.33020	1.11303
.95	.92791	.94963	2.16040	10.11872	J-SHAPE	-.31874	1.05305
1.00	1.00000	1.00000	2.00000	9.00000	J-SHAPE	-.30685	1.00000
1.05	1.07020	1.04965	1.85904	8.09795	-.99074	-.29478	.95270
1.10	1.13859	1.09864	1.73397	7.35985	-.96992	-.28269	.91022
1.15	1.20525	1.14703	1.62204	6.74819	-.94199	-.27071	.87181
1.20	1.27027	1.19488	1.52113	6.23571	-.90950	-.25894	.83690
1.25	1.33375	1.24223	1.42955	5.80215	-.87419	-.24744	.80500
1.30	1.39580	1.28913	1.34593	5.43226	-.83731	-.23624	.77572
1.35	1.45651	1.33560	1.26920	5.11432	-.79975	-.22539	.74873
1.40	1.51597	1.38169	1.19844	4.83923	-.76215	-.21490	.72375
1.45	1.57427	1.42742	1.13291	4.59983	-.72495	-.20477	.70056
1.50	1.63149	1.47282	1.07199	4.39040	-.68848	-.19500	.67897
1.55	1.68771	1.51792	1.01515	4.20636	-.65296	-.18561	.65880
1.60	1.74300	1.56273	.96196	4.04396	-.61852	-.17657	.63991
1.65	1.79743	1.60728	.91202	3.90015	-.58527	-.16788	.62217
1.70	1.85104	1.65158	.86502	3.77238	-.55324	-.15953	.60548
1.75	1.90391	1.69566	.82068	3.65855	-.52245	-.15151	.58974
1.80	1.95608	1.73952	.77874	3.55688	-.49291	-.14380	.57487
1.85	2.00760	1.78317	.73899	3.46588	-.46459	-.13639	.56080
1.90	2.05850	1.82664	.70124	3.38428	-.43747	-.12927	.54745
1.95	2.10885	1.86993	.66533	3.31100	-.41150	-.12243	.53478
2.00	2.15866	1.91306	.63111	3.24509	-.38666	-.11586	.52272
2.25	2.40084	2.12650	.48121	3.00148	-.27762	-.08655	.47026
2.50	2.63389	2.33696	.35863	2.85678	-.18983	-.06224	.42791
2.75	2.86013	2.54511	.25589	2.77332	-.11846	-.04186	.39291
3.00	3.08119	2.75144	.16810	2.72946	-.05977	-.02460	.36345
3.25	3.29822	2.95630	.09196	2.71207	-.01093	-.00983	.33826
3.50	3.51206	3.15997	.02511	2.71273	.03018	.00292	.31646
3.75	3.72336	3.36265	-.03419	2.72591	.06515	.01402	.29738
4.00	3.93258	3.56450	-.08724	2.74783	.09518	.02376	.28054
4.25	4.14008	3.76564	-.13504	2.77585	.12119	.03237	.26556
4.50	4.34616	3.96619	-.17838	2.80811	.14390	.04002	.25213
5.00	4.75490	4.36580	-.25411	2.88029	.18156	.05302	.22905
6.00	5.56274	5.16066	-.37326	3.03546	.23559	.07245	.19377
7.00	6.36237	5.95160	-.46319	3.18718	.27219	.08621	.16802
8.00	7.15690	6.73996	-.53373	3.32768	.29847	.09645	.14837

When γ is unknown, and must therefore be estimated from the sample data, we select a first approximation $\gamma_1 < x_1$ and employ a trial-and-error procedure to calculate the required estimates. Any one of various standard iterative procedures could be employed, but the trial-and-error technique is simple in concept and easy to apply in practice. Its use is therefore recommended. With γ_1 fixed, (3.3.3) is solved for δ_1, and θ_1 follows from (3.3.4). We then calculate $(\partial \ln L/\partial\gamma)_1$ by substituting γ_1, δ_1, and θ_1 into the first equation of (3.3.2). If $(\partial \ln L/\partial\gamma)_1 = 0$, then $\hat{\gamma} = \gamma_1$, $\hat{\delta} = \delta_1$, $\hat{\theta} = \theta_1$, and our task is completed. Otherwise, we repeat the cycle of computations with a new approximation γ_2 and continue until we find a pair of values (γ_i, γ_j) such that $|\gamma_i - \gamma_j|$ is sufficiently small such that $(\partial \ln L/\partial\gamma)_i \lessgtr 0 \lessgtr (\partial \ln L/\partial\gamma)_j$. We then interpolate linearly for the required estimates. When desired, we calculate $\hat{\beta} = \hat{\theta}^{1/\hat{\delta}}$.

3.4 MOMENT ESTIMATORS

On equating the first three sample moments to corresponding distribution moments as given by (3.2.3), the ME become

$$\hat{\gamma} + \hat{\beta}\hat{\Gamma}_1 = \bar{x}, \quad \hat{\beta}^2[\hat{\Gamma}_2 - \hat{\Gamma}_1^2] = s^2, \quad \frac{\hat{\Gamma}_3 - 3\hat{\Gamma}_2\hat{\Gamma}_1 + 2\hat{\Gamma}_1^3}{[\hat{\Gamma}_2 - \hat{\Gamma}_1^2]^{3/2}} = a_3 \qquad (3.4.1)$$

where Γ_k is defined by (3.2.4).

The third equation of (3.4.1) is a function of the shape parameter of δ only, and can easily be solved for the required estimate $\hat{\delta}$ through use of standard iterative procedures or even by use of the previously described trial-and-error scheme. A first approximation to $\hat{\delta}$ for use in the iterative process might be obtained by interpolation in Table 3.1 from either the δ-versus-α_3 or the δ-versus-coefficient-of-variation relationship. With $\hat{\delta}$ thus calculated, $\hat{\beta}$ follows from the second equation of (3.4.1) as

$$\hat{\beta} = \frac{s}{\sqrt{\hat{\Gamma}_2 - \hat{\Gamma}_1^2}} \qquad (3.4.2)$$

and $\hat{\gamma}$ then follows from the first equation of (3.4.1) as

$$\hat{\gamma} = \bar{x} - \hat{\beta}\hat{\Gamma}_1 \qquad (3.4.3)$$

Although calculation of ME requires considerably less computational effort than MLE, it should be remembered that estimate variances of the MLE are smaller than corresponding variances of the ME. However, ME are applicable over the entire parameter space, whereas computational problems arise with MLE when $\delta < 1$. Unless $\delta > 2$, the asymptotic variance-covariance matrix is singular. Because of computational problems due to the matrix singularity or near singularity, Johnson and Kotz (1970) suggest that asymptotic variances and covariances should not be calculated unless $\delta > 2.2$ (approximately). The authors' experience confirms the justification for this recommendation.

3.5 MODIFIED MOMENT ESTIMATORS

In an effort to alleviate problems encountered with the ME and the MLE, various modifications of these estimators have from time to time been proposed. Kao (1959), Dubey (1966), Wycoff, Bain, and Engelhardt (1980), Zanakis (1977, 1979a, b), and Zanakis and Mann (1981) proposed estimators based on percentiles. Cohen and Whitten (1982) proposed several modifications of MLE and ME which employ the first-order statistic. Although several of the various modified estimators were reasonably satisfactory, primary consideration here is limited to a modification of the moment estimators (MME) in which the equation involving the third moment (i.e., the third equation of (3.4.1) is replaced by one in which the sample first-order statistic is equated to its expectation). A detailed account of this estimator, along with tables and graphs (reproduced here) which greatly reduce the computational effort otherwise required, was given by Cohen et al. (1984).

The MME are applicable over the full range of shape parameter values, although they might not perform quite as well as the MLE when α_3 (i.e., the skewness) is near zero. The MME are unbiased with respect to both mean and variance. Their performance is particularly good when skewness is large, in contrast to the MLE, which are most unsatisfactory in this case.

It is well known that the first-order statistic X_1 in a random sample of size n from a Weibull distribution (γ, δ, β) is Weibull (γ, δ, β'), where $\beta' = \beta/n^{1/\delta}$. Accordingly $E(X_1)$ and $V(X_1)$ follow from (3.2.3) when β is replaced by β'. As estimating equations, the MME employ $E(X) = \bar{x}$, $V(X) = s^2$, and $E(X_1) = x_1$. Accordingly, as estimating equations, we have

$$\hat{\gamma} + \hat{\beta}\hat{\Gamma}_1 = \bar{x} \qquad \hat{\beta}^2[\hat{\Gamma}_2 - \hat{\Gamma}_1^2] = s^2 \qquad \hat{\gamma} + \left(\frac{\hat{\beta}}{n^{1/\hat{\delta}}}\right)\hat{\Gamma}_1 = x_1 \tag{3.5.1}$$

Following a few simple algebraic manipulations, the three equations of (3.5.1) are reduced to

$$\frac{s^2}{(\bar{x} - x_1)^2} = \frac{\hat{\Gamma}_2 - \hat{\Gamma}_1^2}{[(1 - n^{-1/\hat{\delta}})\hat{\Gamma}_1]^2} = W(\hat{\delta}, n)$$

$$\hat{\beta} = \frac{s}{\sqrt{\hat{\Gamma}_2 - \hat{\Gamma}_1^2}} \tag{3.5.2}$$

$$\hat{\gamma} = \bar{x} - \hat{\beta}\hat{\Gamma}_1$$

Equivalent expressions for $\hat{\beta}$ and $\hat{\gamma}$, which are often more convenient for routine calculations, are

$$\hat{\beta} = s \cdot C(\hat{\delta}), \qquad \hat{\gamma} = \bar{x} - s \cdot D(\hat{\delta}) \tag{3.5.3}$$

where

$$C(\delta) = [\Gamma_2 - \Gamma_1^2]^{-\frac{1}{2}}, \qquad D(\delta) = \Gamma_1 \cdot [\Gamma_2 - \Gamma_1^2]^{-\frac{1}{2}} \tag{3.5.4}$$

The first equation of (3.5.2) involves $\hat{\delta}$ as the only unknown quantity. As aids for use in solving this equation, Table 3.2 and Figure 3.2 are included here. Figure 3.2 has been reproduced from Cohen et al. (1984) with permission of the American Society for Quality Control. With $s^2/(\bar{x} - x_1)^2$ and n available from sample data, an approximation to $\hat{\delta}$ that is sufficiently accurate for most practical purposes can be read from the graphs of Figure 3.2 or obtained by interpolation from Table 3.2. When greater accuracy is required, standard iterative procedures can be employed to improve this approximation. With $\hat{\delta}$ thus determined, $C(\hat{\delta})$ and $D(\hat{\delta})$ can be evaluated from (3.5.4). Estimates $\hat{\beta}$ and $\hat{\gamma}$ then follow from (3.5.3). To aid in the calculation of $C(\hat{\delta})$ and $D(\hat{\delta})$, a table and a graph of these functions are reproduced here with permission of the American Society for Quality Control from Cohen and Whitten (1982) as Table 3.1 and as Figure 3.3. When additional values of $W(\delta, n)$ in (3.5.2) are required, values of the gamma function needed for their calculation can be obtained from tables such as those of Abramowitz and Stegun (1964). A FORTRAN program for calculating the MME is included in Appendix A.4.

Since $\hat{\gamma} < x_1$, the inequality $[(\hat{\Gamma}_2 - \hat{\Gamma}_1^2)/\hat{\Gamma}_1^2] < [s/(\bar{x} - x_1)]^2$ follows from the first two equations of (3.5.1). Thus we have an upper bound on a function of our estimate of the shape parameter δ. The following table presents this function for selected values of $\hat{\delta}$ and of α_3 as an aid which can be used in solving the first equation of (3.5.2) for $\hat{\delta}$.

α_3	δ	$(\Gamma_2 - \Gamma_1^2)/\Gamma_1^2$
0.1	3.22197	0.11621
0.5	2.21560	0.22734
1.0	1.56391	0.42695
1.5	1.21112	0.68817
2.0	1.00000	1.00000
2.5	0.86317	1.35149
3.0	0.76862	1.73372

TABLE 3.2 The Weibull Estimating Function

$$W(n, \delta) = [\Gamma_2 - \Gamma_1^2]/[(1 - n^{-1/\delta})\Gamma_1]^2$$

δ \ n	5	10	15	20	25	30	35	40
.40	10.22762	9.92767	9.88766	9.87602	9.87129	9.86898	9.86770	9.86693
.50	5.42535	5.10152	5.04474	5.02509	5.01604	5.01113	5.00817	5.00626
.54	4.49833	4.16963	4.10751	4.08493	4.07414	4.06811	4.06439	4.06193
.56	4.13789	3.80743	3.74293	3.71898	3.70734	3.70076	3.69665	3.69390
.58	3.82868	3.49684	3.43014	3.40487	3.39241	3.38528	3.38079	3.37775
.60	3.56132	3.22841	3.15968	3.13318	3.11992	3.11225	3.10737	3.10405
.62	3.32847	2.99476	2.92418	2.89650	2.88248	2.87428	2.86902	2.86542
.64	3.12434	2.79008	2.71779	2.68899	2.67424	2.66553	2.65990	2.65601
.66	2.94431	2.60970	2.53584	2.50600	2.49055	2.48134	2.47534	2.47118
.68	2.78466	2.44986	2.37457	2.34375	2.32762	2.31794	2.31159	2.30715
.70	2.64235	2.30751	2.23090	2.19917	2.18241	2.17227	2.16557	2.16086
.72	2.51489	2.18014	2.10234	2.06975	2.05238	2.04180	2.03476	2.02980
.74	2.40025	2.06568	1.98679	1.95340	1.93546	1.92445	1.91709	1.91187
.76	2.29670	1.96241	1.88253	1.84838	1.82990	1.81849	1.81082	1.80536
.78	2.20282	1.86888	1.78809	1.75325	1.73426	1.72246	1.71450	1.70880
.80	2.11742	1.78388	1.70227	1.66678	1.64730	1.63514	1.62689	1.62096
.82	2.03945	1.70638	1.62402	1.58792	1.56799	1.55548	1.54695	1.54080
.84	1.96807	1.63549	1.55245	1.51579	1.49543	1.48259	1.47380	1.46744
.86	1.90251	1.57046	1.48680	1.44962	1.42886	1.41571	1.40667	1.40011
.88	1.84215	1.51065	1.42644	1.38877	1.36763	1.35418	1.34491	1.33815
.90	1.78642	1.45550	1.37078	1.33266	1.31116	1.29743	1.28794	1.28099
.92	1.73486	1.40452	1.31934	1.28080	1.25897	1.24498	1.23527	1.22815
.94	1.68702	1.35728	1.27169	1.23276	1.21062	1.19638	1.18646	1.17917
.96	1.64256	1.31342	1.22745	1.18817	1.16574	1.15125	1.14115	1.13370
.98	1.60115	1.27261	1.18631	1.14669	1.12398	1.10928	1.09899	1.09138
1.00	1.56250	1.23457	1.14796	1.10803	1.08507	1.07015	1.05969	1.05194
1.02	1.52636	1.19904	1.11215	1.07194	1.04874	1.03362	1.02299	1.01511
1.04	1.49252	1.16579	1.07866	1.03819	1.01476	.99946	.98867	.98065
1.06	1.46077	1.13463	1.04729	1.00656	.98292	.96745	.95651	.94837
1.08	1.43093	1.10538	1.01784	.97689	.95306	.93742	.92634	.91808
1.10	1.40286	1.07788	.99017	.94901	.92499	.90919	.89799	.88961
1.12	1.37640	1.05199	.96412	.92277	.89858	.88263	.87130	.86282
1.14	1.35143	1.02758	.93957	.89804	.87369	.85761	.84616	.83757
1.16	1.32783	1.00453	.91640	.87471	.85020	.83399	.82243	.81375
1.18	1.30551	.98274	.89451	.85266	.82802	.81168	.80002	.79124
1.20	1.28436	.96212	.87379	.83180	.80703	.79058	.77882	.76995
1.22	1.26430	.94257	.85417	.81205	.78715	.77060	.75874	.74979
1.24	1.24526	.92403	.83556	.79332	.76831	.75165	.73971	.73068
1.26	1.22716	.90642	.81789	.77555	.75043	.73368	.72164	.71255
1.28	1.20994	.88968	.80110	.75865	.73344	.71660	.70448	.69532
1.30	1.19354	.87374	.78512	.74259	.71728	.70035	.68817	.67894
1.32	1.17791	.85856	.76991	.72729	.70190	.68489	.67264	.66334
1.34	1.16299	.84409	.75541	.71272	.68724	.67016	.65784	.64849
1.36	1.14875	.83028	.74158	.69881	.67326	.65612	.64373	.63432
1.38	1.13514	.81708	.72837	.68554	.65992	.64271	.63027	.62080

TABLE 3.2 Continued

$$W(n,\delta) = [\Gamma_2 - \Gamma_1^2]/[(1 - n^{-1/\delta})\Gamma_1]^2$$

δ \ n	5	10	15	20	25	30	35	40
1.40	1.12212	.80447	.71575	.67286	.64718	.62990	.61740	.60789
1.42	1.10965	.79240	.70368	.66073	.63499	.61766	.60511	.59555
1.44	1.09771	.78084	.69212	.64913	.62333	.60595	.59335	.58374
1.46	1.08626	.76976	.68105	.63802	.61217	.59473	.58208	.57244
1.48	1.07528	.75914	.67044	.62737	.60147	.58399	.57130	.56161
1.50	1.06473	.74895	.66026	.61715	.59121	.57368	.56095	.55123
1.52	1.05460	.73916	.65049	.60735	.58136	.56380	.55103	.54127
1.54	1.04487	.72975	.64110	.59793	.57191	.55430	.54150	.53170
1.56	1.03550	.72070	.63208	.58888	.56282	.54518	.53234	.52252
1.58	1.02648	.71200	.62340	.58018	.55409	.53642	.52354	.51369
1.60	1.01780	.70362	.61505	.57181	.54569	.52798	.51508	.50520
1.62	1.00944	.69555	.60700	.56374	.53760	.51986	.50693	.49702
1.64	1.00137	.68777	.59925	.55598	.52981	.51204	.49909	.48915
1.66	.99359	.68026	.59177	.54849	.52230	.50451	.49153	.48157
1.68	.98608	.67302	.58457	.54127	.51506	.49725	.48424	.47426
1.70	.97883	.66603	.57761	.53431	.50807	.49024	.47721	.46721
1.72	.97183	.65927	.57089	.52758	.50133	.48348	.47043	.46041
1.74	.96506	.65275	.56439	.52108	.49482	.47694	.46388	.45384
1.76	.95851	.64643	.55812	.51480	.48852	.47064	.45755	.44749
1.78	.95218	.64033	.55205	.50873	.48244	.46454	.45144	.44136
1.80	.94605	.63442	.54618	.50286	.47656	.45864	.44552	.43543
1.82	.94011	.62870	.54049	.49718	.47086	.45293	.43980	.42970
1.84	.93436	.62316	.53499	.49168	.46535	.44741	.43427	.42415
1.86	.92879	.61779	.52966	.48635	.46002	.44206	.42891	.41878
1.88	.92339	.61258	.52449	.48118	.45485	.43688	.42371	.41357
1.90	.91816	.60753	.51948	.47618	.44984	.43186	.41868	.40853
1.92	.91307	.60263	.51462	.47132	.44498	.42699	.41380	.40364
1.94	.90814	.59787	.50990	.46661	.44026	.42227	.40907	.39890
1.96	.90336	.59326	.50532	.46204	.43569	.41769	.40449	.39431
1.98	.89871	.58877	.50088	.45760	.43125	.41325	.40003	.38984
2.00	.89419	.58441	.49656	.45329	.42694	.40893	.39571	.38551
2.02	.88980	.58018	.49236	.44911	.42275	.40473	.39151	.38130
2.04	.88553	.57606	.48828	.44504	.41868	.40066	.38743	.37722
2.06	.88138	.57205	.48431	.44108	.41472	.39670	.38346	.37325
2.08	.87734	.56815	.48045	.43723	.41087	.39285	.37961	.36938
2.10	.87341	.56436	.47669	.43348	.40713	.38910	.37586	.36563
2.12	.86959	.56067	.47304	.42984	.40349	.38546	.37221	.36198
2.14	.86587	.55707	.46948	.42629	.39994	.38191	.36866	.35842
2.16	.86224	.55356	.46601	.42283	.39649	.37846	.36520	.35496
2.18	.85871	.55015	.46263	.41947	.39313	.37510	.36184	.35159
2.20	.85526	.54682	.45934	.41619	.38985	.37182	.35856	.34831
2.22	.85191	.54357	.45613	.41299	.38666	.36863	.35536	.34512
2.24	.84864	.54041	.45300	.40988	.38355	.36552	.35225	.34200
2.26	.84545	.53732	.44994	.40684	.38051	.36248	.34922	.33896
2.28	.84234	.53431	.44697	.40387	.37755	.35953	.34626	.33600

TABLE 3.2 Continued

$$W(n,\delta) = [\Gamma_2 - \Gamma_1^2]/[(1 - n^{-1/\delta})\Gamma_1]^2$$

δ \ n	5	10	15	20	25	30	35	40
2.30	.83930	.53136	.44406	.40098	.37467	.35664	.34337	.33312
2.32	.83634	.52849	.44122	.39816	.37185	.35382	.34056	.33030
2.34	.83344	.52569	.43845	.39540	.36910	.35108	.33781	.32755
2.36	.83062	.52295	.43575	.39271	.36641	.34839	.33513	.32487
2.38	.82786	.52027	.43310	.39008	.36379	.34577	.33251	.32225
2.40	.82517	.51766	.43052	.38751	.36123	.34321	.32995	.31969
2.42	.82254	.51510	.42800	.38500	.35873	.34072	.32745	.31719
2.44	.81996	.51260	.42553	.38255	.35628	.33827	.32501	.31475
2.46	.81745	.51016	.42312	.38015	.35389	.33589	.32262	.31236
2.48	.81499	.50777	.42076	.37781	.35155	.33355	.32029	.31003
2.50	.81259	.50543	.41845	.37551	.34927	.33127	.31801	.30775
2.52	.81024	.50314	.41619	.37327	.34703	.32904	.31578	.30552
2.54	.80794	.50090	.41398	.37108	.34484	.32685	.31360	.30334
2.56	.80569	.49871	.41182	.36893	.34270	.32472	.31147	.30121
2.58	.80349	.49656	.40970	.36682	.34061	.32263	.30938	.29912
2.60	.80133	.49446	.40763	.36477	.33856	.32058	.30733	.29708
2.62	.79922	.49240	.40560	.36275	.33655	.31858	.30533	.29508
2.64	.79716	.49039	.40361	.36078	.33458	.31662	.30337	.29312
2.66	.79513	.48841	.40166	.35884	.33266	.31469	.30145	.29120
2.68	.79315	.48647	.39975	.35695	.33077	.31281	.29957	.28933
2.70	.79121	.48457	.39788	.35509	.32892	.31097	.29773	.28749
2.72	.78931	.48271	.39605	.35327	.32711	.30916	.29593	.28568
2.74	.78745	.48089	.39425	.35148	.32533	.30739	.29416	.28392
2.76	.78562	.47910	.39249	.34973	.32359	.30565	.29243	.28219
2.78	.78383	.47734	.39076	.34802	.32188	.30395	.29073	.28049
2.80	.78207	.47562	.38906	.34634	.32021	.30228	.28906	.27883
2.82	.78035	.47393	.38739	.34468	.31856	.30064	.28743	.27719
2.84	.77867	.47228	.38576	.34306	.31695	.29904	.28583	.27559
2.86	.77701	.47065	.38416	.34148	.31537	.29746	.28425	.27402
2.88	.77539	.46905	.38258	.33992	.31382	.29591	.28271	.27248
2.90	.77379	.46748	.38104	.33838	.31230	.29440	.28120	.27097
2.92	.77223	.46594	.37952	.33688	.31080	.29291	.27971	.26949
2.94	.77069	.46443	.37803	.33540	.30933	.29144	.27825	.26803
2.96	.76919	.46295	.37657	.33395	.30789	.29001	.27682	.26660
2.98	.76771	.46149	.37513	.33253	.30648	.28860	.27541	.26520
3.00	.76626	.46005	.37372	.33113	.30509	.28721	.27403	.26382
3.50	.73703	.43097	.34508	.30278	.27693	.25920	.24612	.23598
4.00	.71724	.41090	.32532	.28325	.25757	.23995	.22697	.21691
4.50	.70312	.39629	.31092	.26904	.24349	.22599	.21309	.20310
5.00	.69264	.38523	.30000	.25826	.23283	.21542	.20260	.19267
5.50	.68462	.37659	.29146	.24983	.22450	.20717	.19441	.18454
6.00	.67832	.36966	.28460	.24306	.21781	.20055	.18785	.17802
6.50	.67327	.36400	.27898	.23752	.21234	.19513	.18248	.17269
7.00	.66915	.35929	.27430	.23290	.20777	.19061	.17800	.16826
8.00	.66286	.35192	.26694	.22564	.20061	.18353	.17099	.16131

TABLE 3.2 Continued

$$W(n,\delta) = [\Gamma_2 - \Gamma_1^2]/[(1 - n^{-1/\delta})\Gamma_1]^2$$

δ \ n	45	50	55	60	65	70	75	80
.40	9.86643	9.86609	9.86586	9.86569	9.86556	9.86546	9.86538	9.86532
.50	5.00494	5.00400	5.00331	5.00278	5.00237	5.00204	5.00178	5.00156
.54	4.06021	4.05896	4.05802	4.05729	4.05672	4.05627	4.05589	4.05559
.56	3.69197	3.69055	3.68949	3.68866	3.68800	3.68747	3.68704	3.68668
.58	3.37561	3.37402	3.37282	3.37188	3.37113	3.37052	3.37003	3.36961
.60	3.10168	3.09993	3.09858	3.09753	3.09669	3.09600	3.09544	3.09496
.62	2.86283	2.86089	2.85941	2.85824	2.85730	2.85653	2.85590	2.85536
.64	2.65320	2.65110	2.64947	2.64818	2.64714	2.64629	2.64558	2.64498
.66	2.46815	2.46587	2.46410	2.46269	2.46155	2.46061	2.45983	2.45917
.68	2.30391	2.30145	2.29953	2.29801	2.29676	2.29574	2.29488	2.29415
.70	2.15740	2.15477	2.15272	2.15107	2.14973	2.14861	2.14768	2.14688
.72	2.02613	2.02333	2.02113	2.01936	2.01792	2.01672	2.01571	2.01485
.74	1.90801	1.90504	1.90270	1.90081	1.89927	1.89798	1.89689	1.89596
.76	1.80129	1.79816	1.79568	1.79368	1.79204	1.79066	1.78950	1.78850
.78	1.70454	1.70125	1.69863	1.69652	1.69477	1.69331	1.69207	1.69101
.80	1.61651	1.61307	1.61032	1.60809	1.60625	1.60470	1.60339	1.60226
.82	1.53618	1.53258	1.52971	1.52737	1.52543	1.52380	1.52241	1.52121
.84	1.46264	1.45890	1.45590	1.45346	1.45142	1.44971	1.44825	1.44699
.86	1.39514	1.39126	1.38814	1.38559	1.38347	1.38167	1.38014	1.37881
.88	1.33302	1.32900	1.32577	1.32311	1.32090	1.31903	1.31743	1.31604
.90	1.27571	1.27155	1.26821	1.26546	1.26316	1.26121	1.25954	1.25809
.92	1.22271	1.21843	1.21498	1.21213	1.20975	1.20772	1.20599	1.20448
.94	1.17359	1.16919	1.16563	1.16269	1.16023	1.15814	1.15633	1.15477
.96	1.12798	1.12346	1.11980	1.11677	1.11423	1.11207	1.11020	1.10858
.98	1.08554	1.08091	1.07715	1.07404	1.07142	1.06919	1.06726	1.06558
1.00	1.04597	1.04123	1.03738	1.03419	1.03149	1.02920	1.02721	1.02548
1.02	1.00902	1.00418	1.00024	.99696	.99420	.99184	.98979	.98801
1.04	.97445	.96951	.96548	.96213	.95930	.95688	.95478	.95295
1.06	.94206	.93703	.93292	.92949	.92660	.92412	.92196	.92008
1.08	.91166	.90654	.90235	.89886	.89590	.89336	.89116	.88923
1.10	.88310	.87789	.87362	.87006	.86705	.86445	.86220	.86022
1.12	.85621	.85092	.84659	.84296	.83988	.83724	.83494	.83292
1.14	.83088	.82551	.82110	.81742	.81429	.81159	.80924	.80718
1.16	.80697	.80153	.79706	.79331	.79013	.78738	.78499	.78289
1.18	.78439	.77887	.77434	.77053	.76730	.76451	.76208	.75993
1.20	.76302	.75744	.75285	.74899	.74570	.74287	.74040	.73822
1.22	.74279	.73715	.73249	.72858	.72525	.72238	.71986	.71765
1.24	.72361	.71791	.71320	.70924	.70586	.70295	.70040	.69815
1.26	.70541	.69965	.69489	.69088	.68746	.68450	.68192	.67964
1.28	.68812	.68230	.67749	.67344	.66998	.66699	.66437	.66205
1.30	.67168	.66581	.66095	.65686	.65336	.65033	.64767	.64533
1.32	.65603	.65011	.64520	.64107	.63753	.63447	.63178	.62941
1.34	.64112	.63515	.63021	.62603	.62246	.61936	.61665	.61425
1.36	.62691	.62089	.61590	.61169	.60809	.60496	.60221	.59978
1.38	.61334	.60728	.60226	.59801	.59437	.59121	.58844	.58598

TABLE 3.2 Continued

$$W(n,\delta) = [\Gamma_2 - \Gamma_1^2]/[(1 - n^{-1/\delta})\Gamma_1]^2$$

δ \ n	45	50	55	60	65	70	75	80
1.40	.60038	.59428	.58922	.58494	.58127	.57808	.57528	.57281
1.42	.58800	.58186	.57676	.57245	.56875	.56553	.56271	.56021
1.44	.57615	.56998	.56484	.56050	.55677	.55353	.55068	.54816
1.46	.56481	.55860	.55344	.54906	.54531	.54204	.53917	.53662
1.48	.55394	.54770	.54251	.53811	.53432	.53103	.52814	.52557
1.50	.54353	.53725	.53203	.52760	.52380	.52048	.51757	.51498
1.52	.53353	.52723	.52198	.51753	.51370	.51036	.50742	.50482
1.54	.52394	.51761	.51233	.50785	.50400	.50065	.49769	.49507
1.56	.51472	.50837	.50306	.49856	.49469	.49131	.48834	.48570
1.58	.50587	.49948	.49416	.48964	.48574	.48235	.47936	.47670
1.60	.49735	.49094	.48559	.48105	.47714	.47372	.47071	.46804
1.62	.48915	.48272	.47735	.47279	.46886	.46543	.46240	.45971
1.64	.48126	.47480	.46941	.46483	.46088	.45744	.45440	.45169
1.66	.47365	.46718	.46177	.45717	.45320	.44974	.44669	.44397
1.68	.46632	.45983	.45440	.44979	.44580	.44233	.43926	.43653
1.70	.45925	.45274	.44730	.44266	.43867	.43517	.43209	.42935
1.72	.45243	.44590	.44044	.43579	.43178	.42827	.42518	.42242
1.74	.44584	.43930	.43382	.42916	.42513	.42161	.41851	.41574
1.76	.43948	.43292	.42743	.42275	.41871	.41518	.41206	.40929
1.78	.43333	.42676	.42125	.41656	.41251	.40897	.40584	.40305
1.80	.42739	.42080	.41528	.41058	.40652	.40296	.39982	.39702
1.82	.42164	.41504	.40951	.40479	.40072	.39715	.39400	.39120
1.84	.41608	.40946	.40392	.39920	.39511	.39154	.38837	.38556
1.86	.41069	.40407	.39851	.39378	.38968	.38610	.38293	.38010
1.88	.40548	.39884	.39327	.38853	.38442	.38083	.37765	.37482
1.90	.40043	.39378	.38820	.38344	.37933	.37573	.37254	.36970
1.92	.39553	.38887	.38328	.37852	.37440	.37078	.36759	.36474
1.94	.39078	.38411	.37852	.37374	.36961	.36599	.36279	.35993
1.96	.38617	.37949	.37389	.36911	.36497	.36134	.35813	.35527
1.98	.38170	.37501	.36940	.36461	.36047	.35683	.35362	.35075
2.00	.37736	.37067	.36505	.36025	.35610	.35246	.34924	.34636
2.02	.37315	.36644	.36082	.35602	.35186	.34821	.34498	.34210
2.04	.36905	.36234	.35671	.35190	.34774	.34408	.34085	.33796
2.06	.36507	.35836	.35272	.34791	.34373	.34007	.33683	.33394
2.08	.36121	.35449	.34884	.34402	.33984	.33618	.33293	.33003
2.10	.35745	.35072	.34507	.34024	.33606	.33239	.32914	.32624
2.12	.35379	.34706	.34140	.33657	.33238	.32871	.32545	.32254
2.14	.35023	.34349	.33783	.33300	.32880	.32512	.32186	.31895
2.16	.34677	.34002	.33436	.32952	.32532	.32164	.31837	.31545
2.18	.34339	.33665	.33098	.32613	.32193	.31824	.31497	.31205
2.20	.34011	.33336	.32769	.32284	.31863	.31494	.31167	.30874
2.22	.33691	.33016	.32448	.31962	.31541	.31172	.30844	.30551
2.24	.33379	.32703	.32135	.31650	.31228	.30858	.30530	.30237
2.26	.33075	.32399	.31831	.31345	.30923	.30553	.30225	.29931
2.28	.32779	.32103	.31534	.31047	.30625	.30255	.29926	.29633

TABLE 3.2 Continued

$$W(n,\delta) = [\Gamma_2 - \Gamma_1^2]/[1 - n^{-1/\delta})\Gamma_1]^2$$

δ \ n	45	50	55	60	65	70	75	80
2.30	.32490	.31813	.31244	.30758	.30335	.29965	.29636	.29342
2.32	.32208	.31531	.30962	.30475	.30052	.29681	.29352	.29058
2.34	.31933	.31256	.30686	.30199	.29776	.29405	.29076	.28781
2.36	.31664	.30987	.30418	.29930	.29507	.29135	.28806	.28511
2.38	.31402	.30725	.30155	.29667	.29244	.28872	.28543	.28248
2.40	.31146	.30469	.29899	.29411	.28987	.28616	.28286	.27990
2.42	.30896	.30219	.29649	.29161	.28737	.28365	.28035	.27739
2.44	.30652	.29974	.29404	.28916	.28492	.28120	.27789	.27494
2.46	.30414	.29736	.29165	.28677	.28253	.27880	.27550	.27254
2.48	.30180	.29502	.28932	.28443	.28019	.27647	.27316	.27020
2.50	.29952	.29274	.28704	.28215	.27791	.27418	.27087	.26791
2.52	.29729	.29051	.28481	.27992	.27567	.27195	.26864	.26567
2.54	.29511	.28833	.28262	.27774	.27349	.26976	.26645	.26349
2.56	.29298	.28620	.28049	.27560	.27136	.26762	.26431	.26135
2.58	.29089	.28411	.27840	.27351	.26927	.26553	.26222	.25925
2.60	.28885	.28207	.27636	.27147	.26722	.26349	.26018	.25721
2.62	.28685	.28007	.27436	.26947	.26522	.26149	.25817	.25520
2.64	.28489	.27811	.27240	.26751	.26326	.25953	.25621	.25324
2.66	.28298	.27620	.27049	.26560	.26135	.25761	.25430	.25133
2.68	.28110	.27432	.26861	.26372	.25947	.25573	.25242	.24945
2.70	.27926	.27248	.26677	.26188	.25763	.25390	.25058	.24761
2.72	.27746	.27068	.26497	.26008	.25583	.25210	.24878	.24581
2.74	.27570	.26892	.26321	.25832	.25407	.25033	.24701	.24404
2.76	.27397	.26719	.26148	.25659	.25234	.24860	.24528	.24231
2.78	.27227	.26550	.25979	.25490	.25065	.24691	.24359	.24062
2.80	.27061	.26383	.25813	.25323	.24898	.24525	.24193	.23895
2.82	.26898	.26220	.25650	.25161	.24736	.24362	.24030	.23733
2.84	.26738	.26061	.25490	.25001	.24576	.24202	.23870	.23573
2.86	.26581	.25904	.25333	.24844	.24419	.24046	.23714	.23416
2.88	.26427	.25750	.25180	.24691	.24266	.23892	.23560	.23263
2.90	.26276	.25599	.25029	.24540	.24115	.23741	.23409	.23112
2.92	.26128	.25451	.24881	.24392	.23967	.23594	.23262	.22964
2.94	.25983	.25306	.24736	.24247	.23822	.23449	.23117	.22819
2.96	.25840	.25163	.24593	.24105	.23680	.23306	.22974	.22677
2.98	.25700	.25023	.24453	.23965	.23540	.23166	.22835	.22537
3.00	.25562	.24886	.24316	.23827	.23403	.23029	.22697	.22400
3.50	.22784	.22112	.21546	.21061	.20638	.20267	.19937	.19641
4.00	.20884	.20217	.19655	.19173	.18754	.18386	.18058	.17764
4.50	.19509	.18847	.18289	.17811	.17396	.17030	.16704	.16413
5.00	.18471	.17814	.17260	.16786	.16373	.16010	.15687	.15398
5.50	.17662	.17009	.16459	.15987	.15577	.15217	.14896	.14609
6.00	.17014	.16365	.15818	.15349	.14942	.14584	.14265	.13980
6.50	.16485	.15839	.15295	.14829	.14423	.14067	.13751	.13467
7.00	.16045	.15401	.14860	.14396	.13993	.13638	.13323	.13041
8.00	.15355	.14716	.14179	.13719	.13319	.12967	.12655	.12376

TABLE 3.2 Continued

$$W(n,\delta) = [\Gamma_2 - \Gamma_1^2]/[(1 - n^{-1/\delta})\Gamma_1]^2$$

δ \ n	90	100	150	200	250	300	500	1000
.40	9.86523	9.86517	9.86505	9.86501	9.86500	9.86499	9.86498	9.86498
.50	5.00123	5.00100	5.00044	5.00025	5.00016	5.00011	5.00004	5.00001
.54	4.05511	4.05476	4.05392	4.05360	4.05345	4.05337	4.05324	4.05318
.56	3.68612	3.68571	3.68469	3.68430	3.68412	3.68401	3.68384	3.68376
.58	3.36896	3.36848	3.36728	3.36681	3.36658	3.36644	3.36623	3.36613
.60	3.09422	3.09367	3.09226	3.09170	3.09142	3.09126	3.09099	3.09086
.62	2.85452	2.85389	2.85226	2.85161	2.85127	2.85107	2.85075	2.85058
.64	2.64404	2.64333	2.64147	2.64071	2.64031	2.64008	2.63968	2.63947
.66	2.45811	2.45732	2.45522	2.45434	2.45388	2.45360	2.45314	2.45288
.68	2.29299	2.29211	2.28975	2.28876	2.28823	2.28791	2.28736	2.28704
.70	2.14561	2.14464	2.14202	2.14090	2.14029	2.13992	2.13928	2.13891
.72	2.01346	2.01240	2.00951	2.00825	2.00757	2.00715	2.00641	2.00597
.74	1.89446	1.89331	1.89014	1.88874	1.88798	1.88751	1.88666	1.88614
.76	1.78689	1.78564	1.78219	1.78065	1.77980	1.77927	1.77831	1.77771
.78	1.68928	1.68794	1.68421	1.68252	1.68158	1.68099	1.67991	1.67922
.80	1.60042	1.59899	1.59496	1.59313	1.59209	1.59144	1.59024	1.58946
.82	1.51926	1.51774	1.51343	1.51144	1.51031	1.50959	1.50826	1.50738
.84	1.44492	1.44331	1.43871	1.43656	1.43534	1.43455	1.43308	1.43210
.86	1.37664	1.37494	1.37004	1.36774	1.36642	1.36557	1.36396	1.36286
.88	1.31376	1.31196	1.30678	1.30432	1.30290	1.30198	1.30023	1.29901
.90	1.25570	1.25382	1.24836	1.24574	1.24422	1.24323	1.24132	1.23999
.92	1.20199	1.20002	1.19427	1.19150	1.18988	1.18882	1.18676	1.18529
.94	1.15217	1.15012	1.14410	1.14117	1.13944	1.13831	1.13610	1.13450
.96	1.10589	1.10375	1.09745	1.09437	1.09254	1.09134	1.08897	1.08724
.98	1.06279	1.06058	1.05401	1.05077	1.04884	1.04756	1.04504	1.04317
1.00	1.02260	1.02030	1.01347	1.01008	1.00805	1.00670	1.00401	1.00200
1.02	.98504	.98267	.97557	.97203	.96990	.96848	.96563	.96348
1.04	.94989	.94744	.94010	.93640	.93418	.93269	.92967	.92738
1.06	.91694	.91442	.90682	.90299	.90066	.89910	.89593	.89348
1.08	.88600	.88342	.87558	.87160	.86917	.86754	.86421	.86161
1.10	.85692	.85426	.84619	.84207	.83955	.83785	.83435	.83161
1.12	.82954	.82682	.81852	.81425	.81164	.80987	.80622	.80332
1.14	.80373	.80094	.79242	.78802	.78532	.78348	.77966	.77661
1.16	.77936	.77651	.76778	.76325	.76045	.75854	.75458	.75137
1.18	.75634	.75343	.74448	.73983	.73694	.73497	.73084	.72748
1.20	.73455	.73159	.72244	.71766	.71468	.71264	.70837	.70486
1.22	.71392	.71090	.70155	.69665	.69359	.69149	.68706	.68340
1.24	.69436	.69128	.68175	.67672	.67358	.67141	.66683	.66302
1.26	.67579	.67266	.66294	.65780	.65458	.65235	.64762	.64366
1.28	.65815	.65497	.64507	.63982	.63652	.63423	.62935	.62524
1.30	.64137	.63814	.62807	.62271	.61933	.61698	.61197	.60771
1.32	.62540	.62213	.61189	.60642	.60297	.60056	.59541	.59100
1.34	.61018	.60687	.59647	.59090	.58737	.58491	.57962	.57507
1.36	.59567	.59232	.58176	.57609	.57249	.56998	.56455	.55986
1.38	.58183	.57843	.56773	.56196	.55829	.55572	.55016	.54533

TABLE 3.2 Continued

$$W(n,\delta) = [\Gamma_2 - \Gamma_1^2]/[(1 - n^{-1/\delta})\Gamma_1]^2$$

δ \ n	90	100	150	200	250	300	500	1000
1.40	.56860	.56517	.55432	.54846	.54472	.54210	.53641	.53144
1.42	.55596	.55249	.54150	.53555	.53174	.52908	.52326	.51815
1.44	.54387	.54036	.52924	.52320	.51933	.51661	.51067	.50543
1.46	.53230	.52875	.51750	.51138	.50745	.50468	.49862	.49325
1.48	.52121	.51763	.50626	.50005	.49606	.49325	.48707	.48156
1.50	.51058	.50697	.49548	.48919	.48514	.48228	.47599	.47036
1.52	.50038	.49674	.48514	.47877	.47467	.47176	.46536	.45960
1.54	.49060	.48693	.47521	.46877	.46461	.46167	.45515	.44927
1.56	.48120	.47750	.46568	.45917	.45495	.45197	.44535	.43935
1.58	.47217	.46844	.45652	.44993	.44567	.44264	.43592	.42980
1.60	.46348	.45973	.44771	.44106	.43674	.43368	.42686	.42062
1.62	.45513	.45135	.43923	.43252	.42815	.42505	.41813	.41178
1.64	.44708	.44328	.43107	.42429	.41989	.41675	.40973	.40327
1.66	.43933	.43551	.42321	.41637	.41192	.40875	.40164	.39507
1.68	.43186	.42802	.41564	.40874	.40425	.40104	.39384	.38717
1.70	.42466	.42080	.40834	.40138	.39685	.39360	.38632	.37954
1.72	.41772	.41383	.40129	.39429	.38971	.38643	.37906	.37219
1.74	.41101	.40710	.39449	.38744	.38282	.37951	.37206	.36508
1.76	.40454	.40061	.38793	.38082	.37616	.37283	.36529	.35822
1.78	.39828	.39434	.38159	.37443	.36974	.36637	.35876	.35160
1.80	.39224	.38828	.37546	.36826	.36353	.36013	.35245	.34519
1.82	.38639	.38241	.36953	.36229	.35752	.35410	.34634	.33900
1.84	.38073	.37674	.36380	.35651	.35172	.34827	.34043	.33301
1.86	.37526	.37126	.35826	.35092	.34610	.34262	.33472	.32721
1.88	.36996	.36594	.35289	.34551	.34066	.33716	.32919	.32159
1.90	.36483	.36080	.34769	.34028	.33539	.33186	.32383	.31616
1.92	.35986	.35581	.34265	.33520	.33028	.32674	.31864	.31089
1.94	.35504	.35098	.33777	.33028	.32534	.32177	.31361	.30578
1.96	.35036	.34629	.33303	.32552	.32054	.31695	.30874	.30083
1.98	.34583	.34174	.32844	.32089	.31589	.31228	.30401	.29603
2.00	.34142	.33733	.32399	.31640	.31138	.30775	.29942	.29138
2.02	.33715	.33305	.31966	.31205	.30700	.30335	.29497	.28686
2.04	.33300	.32889	.31546	.30782	.30275	.29908	.29065	.28247
2.06	.32897	.32485	.31139	.30372	.29862	.29494	.28645	.27821
2.08	.32506	.32093	.30742	.29973	.29461	.29091	.28237	.27407
2.10	.32125	.31711	.30357	.29585	.29072	.28699	.27841	.27005
2.12	.31755	.31340	.29983	.29208	.28693	.28319	.27456	.26614
2.14	.31395	.30979	.29619	.28842	.28324	.27949	.27082	.26233
2.16	.31045	.30628	.29265	.28485	.27966	.27589	.26718	.25864
2.18	.30704	.30287	.28920	.28138	.27617	.27239	.26364	.25504
2.20	.30372	.29954	.28585	.27801	.27278	.26898	.26019	.25154
2.22	.30048	.29630	.28258	.27472	.26948	.26567	.25683	.24814
2.24	.29734	.29315	.27940	.27152	.26626	.26244	.25357	.24482
2.26	.29427	.29007	.27630	.26840	.26313	.25929	.25039	.24159
2.28	.29128	.28708	.27329	.26537	.26008	.25623	.24729	.23844

TABLE 3.2 Continued

$$W(n,\delta) = [\Gamma_2 - \Gamma_1^2]/[(1 - n^{-1/\delta})\Gamma_1]^2$$

δ \ n	90	100	150	200	250	300	500	1000
2.30	.28836	.28416	.27034	.26241	.25710	.25324	.24427	.23538
2.32	.28552	.28131	.26747	.25952	.25420	.25033	.24132	.23239
2.34	.28275	.27853	.26468	.25671	.25138	.24749	.23845	.22948
2.36	.28004	.27582	.26195	.25396	.24862	.24472	.23565	.22664
2.38	.27740	.27318	.25928	.25129	.24593	.24202	.23292	.22387
2.40	.27482	.27060	.25669	.24867	.24331	.23939	.23026	.22117
2.42	.27231	.26808	.25415	.24612	.24074	.23682	.22766	.21853
2.44	.26985	.26562	.25167	.24363	.23824	.23431	.22513	.21596
2.46	.26745	.26321	.24925	.24120	.23580	.23186	.22265	.21344
2.48	.26511	.26087	.24689	.23883	.23342	.22947	.22023	.21099
2.50	.26282	.25857	.24458	.23651	.23109	.22713	.21787	.20860
2.52	.26058	.25633	.24233	.23424	.22881	.22484	.21557	.20626
2.54	.25839	.25414	.24012	.23203	.22659	.22261	.21331	.20397
2.56	.25624	.25199	.23797	.22986	.22441	.22043	.21111	.20174
2.58	.25415	.24990	.23586	.22774	.22229	.21830	.20896	.19956
2.60	.25210	.24784	.23380	.22567	.22021	.21621	.20685	.19742
2.62	.25010	.24584	.23178	.22365	.21818	.21418	.20479	.19534
2.64	.24813	.24387	.22981	.22167	.21619	.21218	.20278	.19330
2.66	.24621	.24195	.22788	.21973	.21424	.21023	.20081	.19130
2.68	.24433	.24007	.22599	.21783	.21234	.20832	.19888	.18935
2.70	.24249	.23823	.22414	.21597	.21048	.20645	.19700	.18744
2.72	.24069	.23642	.22232	.21415	.20865	.20462	.19515	.18557
2.74	.23892	.23465	.22055	.21237	.20686	.20283	.19334	.18374
2.76	.23719	.23292	.21881	.21063	.20512	.20107	.19157	.18195
2.78	.23550	.23123	.21711	.20892	.20340	.19936	.18984	.18019
2.80	.23383	.22956	.21544	.20724	.20172	.19767	.18814	.17847
2.82	.23220	.22793	.21380	.20560	.20008	.19602	.18648	.17679
2.84	.23061	.22633	.21220	.20400	.19846	.19441	.18485	.17514
2.86	.22904	.22477	.21063	.20242	.19688	.19282	.18326	.17353
2.88	.22750	.22323	.20909	.20088	.19534	.19127	.18169	.17194
2.90	.22600	.22172	.20758	.19936	.19382	.18975	.18016	.17039
2.92	.22452	.22024	.20610	.19787	.19233	.18825	.17865	.16887
2.94	.22307	.21879	.20464	.19642	.19087	.18679	.17718	.16738
2.96	.22165	.21737	.20322	.19499	.18943	.18535	.17573	.16592
2.98	.22025	.21597	.20182	.19358	.18803	.18394	.17431	.16448
3.00	.21888	.21460	.20044	.19221	.18665	.18256	.17292	.16308
3.50	.19130	.18704	.17289	.16464	.15904	.15492	.14515	.13508
4.00	.17257	.16834	.15428	.14605	.14048	.13636	.12658	.11643
4.50	.15910	.15490	.14095	.13278	.12724	.12315	.11341	.10328
5.00	.14899	.14482	.13098	.12288	.11738	.11332	.10365	.09357
5.50	.14114	.13700	.12327	.11523	.10978	.10575	.09616	.08615
6.00	.13488	.13077	.11714	.10916	.10375	.09976	.09024	.08031
6.50	.12978	.12570	.11215	.10424	.09887	.09491	.08546	.07561
7.00	.12555	.12149	.10803	.10017	.09484	.09090	.08153	.07176
8.00	.11894	.11493	.10161	.09384	.08858	.08470	.07545	.06582

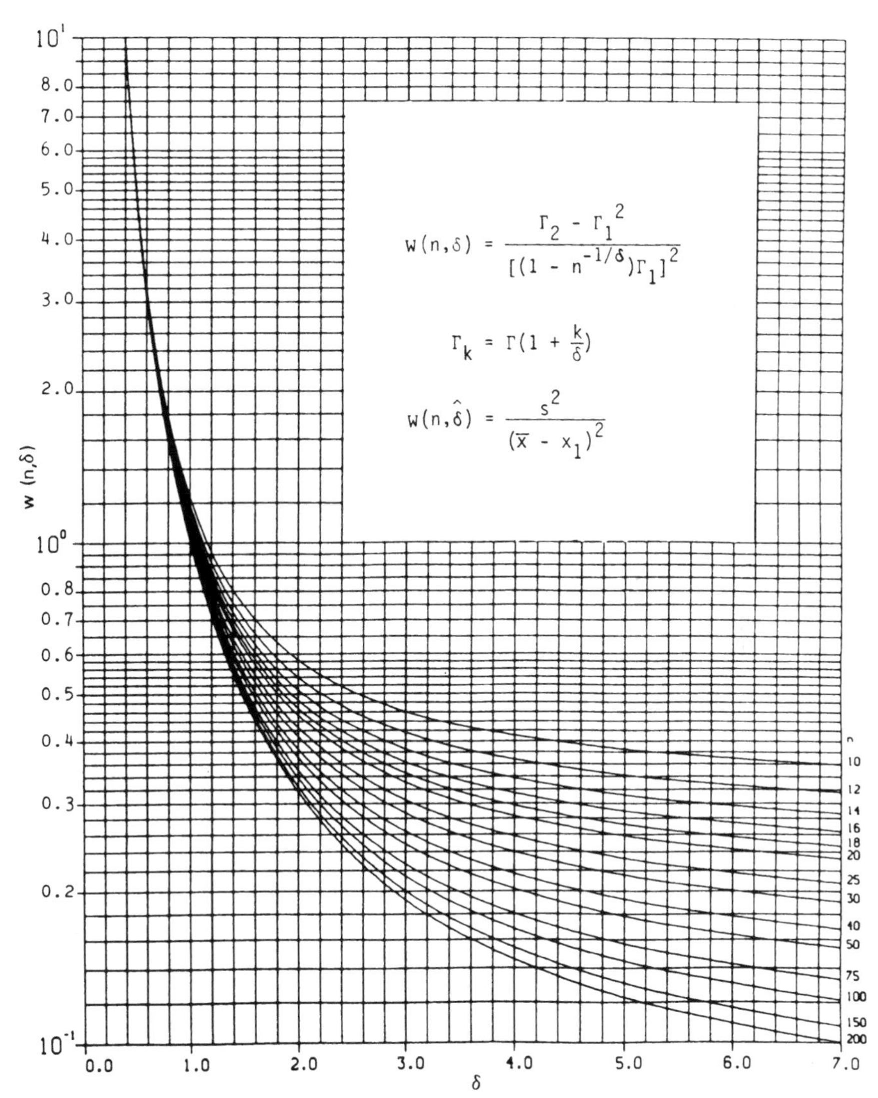

FIGURE 3.2 Graphs of the Weibull estimating function. Reproduced from Cohen, A. C., Whitten, B. J., and Ding, Y. (1984) with permission of the American Society for Quality Control.

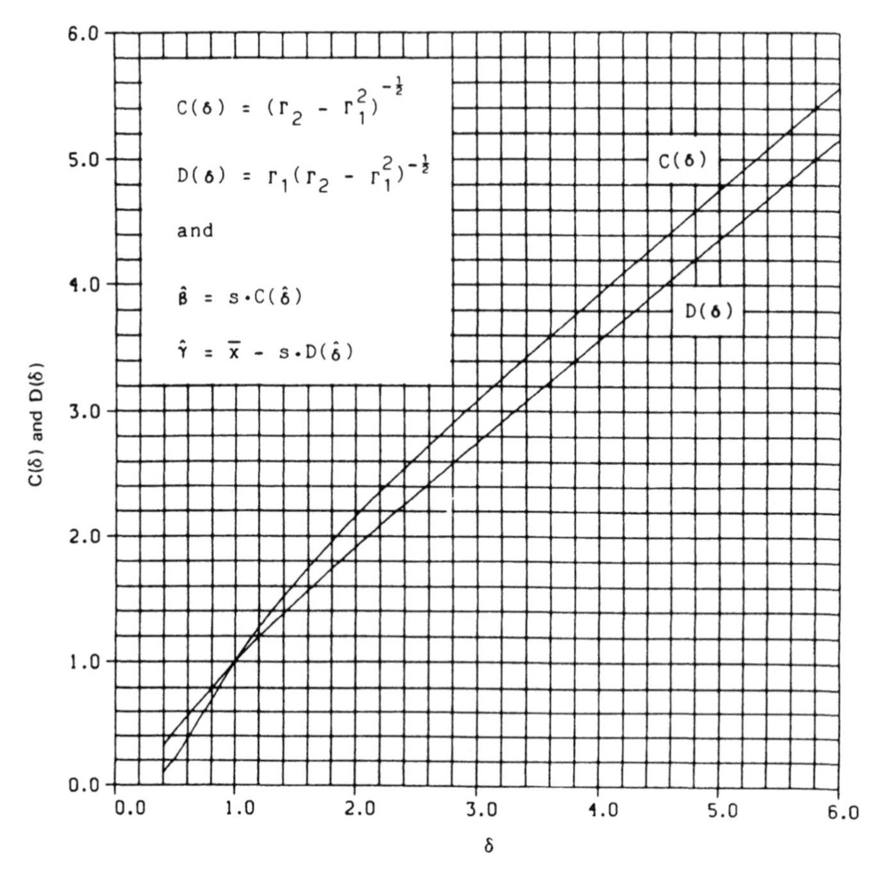

FIGURE 3.3 Graphs of C(δ) and D(δ) for the Weibull distribution. Reproduced from Cohen, A. C., Whitten, B. J., and Ding, Y. (1984) with permission of the American Society for Quality Control.

Although MME are applicable over the entire parameter space, they are most useful when skewness is large. When skewness is small, the MLE or the ME might be preferred. For very small values of α_3, it might be better to replace the Weibull distribution as a model with the normal distribution, in which case the MLE and the ME are identical.

3.6 WYCOFF, BAIN, ENGELHARDT, AND ZANAKIS ESTIMATORS

The Wycoff, Bain, and Engelhardt (1980) recommended estimators (WBE) are

$$\delta_0 = \frac{2.989}{\ln\{(x_{[np_k+1]} - x_1)/(x_{[np_i+1]} - x_1)\}}$$

$$\frac{x_1 - n^{1/\delta_0}}{1 - n^{1/\delta_0}} \tag{3.6.1}$$

$$\delta^* = nk_n \left[-\sum_1^s \ln(x_i - \gamma^*) + \frac{s}{n-s} \sum_{s+1}^n \ln(x_i - \gamma^*) \right]^{-1}$$

$$\beta^* = \exp\left[\frac{0.5772}{\delta^*} + \frac{1}{n} \sum_1^n \ln(x_i - \gamma^*) \right]$$

where [w] denotes the largest integer less than w, $p_i = 0.16731$, $p_k = 0.97366$, $s = [0.84n]$, and the constants k_n are tabulated by Engelhardt and Bain (1977). For sample sizes 10 and 20, $k_{10} = 1.3644$ and $k_{20} = 1.4192$. The first of equations (3.6.1) is employed to provide an initial value of δ, which is then used in the second equation to calculate γ^*. The last two equations are subsequently used to calculate estimates δ^* and β^*.

The Zanakis (1979a) recommended estimators are

$$\tilde{\gamma} = \frac{x_1 x_n - x_2^2}{x_1 + x_n - 2x_2}$$

$$\tilde{\delta} = \frac{2.989}{\ln\{(x_{[np_k+1]} - \tilde{\gamma})/(x_{[np_i+1]} - \tilde{\gamma})\}} \tag{3.6.2}$$

$$\tilde{\beta} = -\tilde{\gamma} + x_{[0.63n+1]}$$

Note that the Zanakis estimator $\tilde{\delta}$ differs from the WBE initial estimate δ_0 only in that $\tilde{\gamma}$ has replaced x_1.

Comparisons reported by Cohen et al. (1984) indicate that when both bias and MSE are considered, the MME appear to be at least as good as the WBE and Zanakis estimators. The MSE of the three estimators are approximately equal with respect to estimates of each of the three parameters. Biases of the three estimators are approximately equal with respect to estimates of β and δ, and the MME appears to exhibit a slightly smaller bias in estimating γ than the WBE or Zanakis estimators.

A possible disadvantage in using percentile estimators when sample sizes are large and when sample observations are not naturally ordered with respect to magnitude is the additional computing effort involved in the ordering process. The MME require only that the smallest sample observation be identified. With the aid of the graphs provided here, calculation of the MME is quite simple.

3.7 SPECIAL CASES: SHAPE PARAMETER δ KNOWN

When δ is known, expected values of the expressions in (3.5.2) for $\hat{\beta}$ and $\hat{\gamma}$ reduce to

$$E(\hat{\beta}) = \beta, \qquad E(\hat{\gamma}) = \gamma \tag{3.7.1}$$

Therefore, $\hat{\beta}$ and $\hat{\gamma}$ in this case are unbiased estimators of β and γ.

3.8 ERRORS OF ESTIMATES

The asymptotic variance-covariance matrix for MLE is obtained by inverting the Fisher information matrix in which elements are negatives of expected values of second partial derivatives of the loglikelihood function. In the general case, when all three parameters (γ, δ, β) must be estimated, we thereby obtain

$$V(\hat{\gamma}, \hat{\delta}, \hat{\beta}) = \|a_{ij}\|^{-1}, \quad i, j = 1, 2, 3 \tag{3.8.1}$$

where

$$a_{11} = \frac{n\delta^2}{\beta^2} C, \quad a_{22} = \frac{n}{\delta^2} K, \quad a_{33} = \frac{n\delta^2}{\beta^2}$$
$$a_{12} = a_{21} = \frac{n}{\beta} J, \quad a_{13} = a_{31} = \frac{n\delta^2}{\beta^2} \Gamma\left(2 - \frac{1}{\delta}\right), \quad a_{23} = a_{32} = -\frac{n}{\beta}\psi(2) \tag{3.8.2}$$

where

$$A = 1 + \psi\left(2 - \frac{1}{\delta}\right)$$
$$C = \left[\Gamma\left(1 - \frac{2}{\delta}\right) + \delta\Gamma\left(2 - \frac{2}{\delta}\right)\right] \frac{(\delta - 1)}{\delta^2}$$
$$J = \Gamma\left(1 - \frac{1}{\delta}\right) - A\Gamma\left(2 - \frac{1}{\delta}\right) \tag{3.8.3}$$
$$K = \psi'(1) + \psi^2(2)$$

where $\psi(\cdot)$ is the digamma function and $\psi'(\ )$ is the trigamma function (i.e., the first derivative of the digamma function (cf. Abramowitz and Stegun 1964).

When the matrix $V(\hat{\gamma}, \hat{\delta}, \hat{\beta})$ is expanded, we obtain variances and covariances which we write as

$$V(\hat{\gamma}) = \frac{\beta^2}{n}\phi_{11}, \quad V(\hat{\delta}) = \frac{\delta^2}{n}\phi_{22}$$
$$V(\hat{\beta}) = \frac{\beta^2}{n}\phi_{33}, \quad \mathrm{Cov}(\hat{\gamma}, \hat{\delta}) = \frac{\beta}{n}\phi_{12} \tag{3.8.4}$$
$$\mathrm{Cov}(\hat{\gamma}, \hat{\beta}) = \frac{\beta^2}{n}\phi_{13}, \quad \mathrm{Cov}(\hat{\delta}, \hat{\beta}) = \frac{\beta}{n}\phi_{23}$$

where

$$\phi_{11} = \frac{\psi'(1)}{\delta^2 M}, \quad \phi_{22} = \frac{C - \Gamma^2(2 - 1/\delta)}{M}$$
$$\phi_{33} = \frac{KC - J^2}{\delta^2 M}, \quad \phi_{12} = \frac{J + \psi(2)\Gamma(2 - 1/\delta)}{M} \tag{3.8.5}$$
$$\phi_{13} = -\frac{J\psi(2) + K\Gamma(2 - 1/\delta)}{\delta^2 M}, \quad \phi_{23} = \frac{C\psi(2) - J\Gamma(2 - 1/\delta)}{M}$$

where

$$M = KC - 2J\psi(2)\Gamma\left(2 - \frac{1}{\delta}\right) - C\psi^2(2) - K\Gamma^2\left(2 - \frac{1}{\delta}\right) - J^2 \tag{3.8.6}$$

In the notation employed here $\Gamma^2(\ ) = [\Gamma(\ )]^2$, and $\psi^2(\ ) = [\psi(\ )]^2$, etc.

The ϕ_{ij} are thus functions of the shape parameter δ alone, and since α_3 is a function of δ alone the ϕ_{ij} can be considered as functions of α_3 rather than of δ. In order to facilitate the calculation of estimate variances and covariances, Table 3.3, in which entries of the ϕ_{ij} are given as functions of α_3, has been included.

Unfortunately, variances and covariances given here are valid only if $\delta > 2$, and they are strictly applicable only for the MLE. As previously mentioned, computational difficulties might be encountered unless δ is greater than approximately 2.2. Although these results do not strictly apply to the MME, simulation investigations by Cohen and Whitten (1982) and by Cohen et al. (1984) indicate that the variances and covariances of (3.5.4) provide close approximations to those of the MME.

TABLE 3.3 Variance-Covariance Factors for Maximum Likelihood Estimates of Weibull Parameters[a]

α_3	δ	ϕ_{11}	ϕ_{22}	ϕ_{33}	ϕ_{12}	ϕ_{13}	ϕ_{23}
0.05	3.40325	1.81053	3.11756	2.11502	-7.25440	1.91206	3.64075
0.06	3.36564	1.73452	3.02575	2.03759	-6.89240	1.83425	3.36154
0.07	3.32873	1.66133	2.93695	1.96301	-6.54777	1.75926	3.10390
0.08	3.29249	1.59085	2.85105	1.89116	-6.21964	1.68696	2.86620
0.09	3.25691	1.52298	2.76792	1.82193	-5.90715	1.61727	2.64695
0.10	3.22197	1.45760	2.68748	1.75523	-5.60953	1.55009	2.44475
0.11	3.18766	1.39464	2.60962	1.69097	-5.32601	1.48531	2.25832
0.12	3.15397	1.33400	2.53424	1.62905	-5.05590	1.42284	2.08648
0.13	3.12087	1.27559	2.46125	1.56939	-4.79851	1.36261	1.92813
0.14	3.08836	1.21932	2.39057	1.51190	-4.55323	1.30452	1.78226
0.15	3.05642	1.16513	2.32212	1.45651	-4.31945	1.24849	1.64794
0.16	3.02503	1.11292	2.25580	1.40313	-4.09661	1.19446	1.52430
0.17	2.99419	1.06263	2.19156	1.35170	-3.88417	1.14235	1.41054
0.18	2.96388	1.01419	2.12930	1.30215	-3.68162	1.09208	1.30591
0.19	2.93409	0.96753	2.06897	1.25440	-3.48850	1.04359	1.20974
0.20	2.90481	0.92259	2.01049	1.20840	-3.30433	0.99681	1.12138
0.21	2.87603	0.87930	1.95381	1.16409	-3.12869	0.95169	1.04024
0.22	2.84773	0.83760	1.89884	1.12140	-2.96118	0.90817	0.96579
0.23	2.81990	0.79744	1.84555	1.08027	-2.80141	0.86618	0.89752
0.24	2.79254	0.75876	1.79387	1.04066	-2.64900	0.82567	0.83495
0.25	2.76563	0.72150	1.74374	1.00251	-2.50361	0.78660	0.77766
0.26	2.73917	0.68563	1.69511	0.96578	-2.36491	0.74890	0.72525
0.27	2.71314	0.65109	1.64794	0.93040	-2.23260	0.71254	0.67734
0.28	2.68753	0.61783	1.60216	0.89635	-2.10636	0.67746	0.63359
0.29	2.66234	0.58581	1.55774	0.86357	-1.98591	0.64363	0.59367
0.30	2.63756	0.55498	1.51463	0.83201	-1.87100	0.61099	0.55730
0.31	2.61317	0.52530	1.47279	0.80165	-1.76136	0.57950	0.52420
0.32	2.58918	0.49674	1.43217	0.77244	-1.65675	0.54914	0.49412
0.33	2.56557	0.46926	1.39273	0.74434	-1.55694	0.51985	0.46681
0.34	2.54233	0.44281	1.35444	0.71732	-1.46171	0.49161	0.44206
0.35	2.51946	0.41736	1.31726	0.69133	-1.37085	0.46437	0.41967
0.36	2.49694	0.39289	1.28115	0.66636	-1.28417	0.43811	0.39946
0.37	2.47478	0.36934	1.24608	0.64236	-1.20147	0.41279	0.38123
0.38	2.45296	0.34670	1.21201	0.61930	-1.12258	0.38837	0.36485
0.39	2.43148	0.32494	1.17891	0.59716	-1.04732	0.36483	0.35014
0.40	2.41032	0.30401	1.14675	0.57590	-0.97554	0.34215	0.33699
0.41	2.38950	0.28390	1.11551	0.55550	-0.90707	0.32028	0.32525
0.42	2.36899	0.26458	1.08514	0.53593	-0.84178	0.29921	0.31482
0.43	2.34879	0.24601	1.05564	0.51716	-0.77951	0.27891	0.30557
0.44	2.32889	0.22819	1.02695	0.49917	-0.72014	0.25935	0.29741
0.45	2.30929	0.21107	0.99907	0.48193	-0.66353	0.24050	0.29025
0.50	2.21560	0.13533	0.87087	0.40630	-0.41794	0.15627	0.26648
0.55	2.12856	0.07410	0.75933	0.34639	-0.22546	0.08681	0.25688
0.60	2.04757	0.02514	0.66209	0.30002	-0.07555	0.02991	0.25552
0.63	2.00166	0.00083	0.60978	0.27791	-0.00249	0.00100	0.25695

*Valid only if $\delta>2$.

3.8.1 Variances and Covariances in the Two-Parameter Special Case

In the two-parameter case with γ known, asymptotic variances for the MLE $\hat{\delta}$ and $\hat{\beta}$ are valid for $\delta > 1$. These results, as given by Cohen and Whitten (1982), are

$$\begin{aligned} V(\hat{\delta}) &= 0.607927\left(\frac{\delta^2}{n}\right) \\ \mathrm{Cov}(\hat{\delta}, \hat{\beta}) &= \frac{0.257022\beta}{n} \\ V(\hat{\beta}) &= 1.108665\left(\frac{\beta^2}{n\delta^2}\right) \end{aligned} \qquad (3.8.7)$$

The variances of (3.8.7) also give close approximations to corresponding variances of the MME.

3.9 THE EXPONENTIAL DISTRIBUTION

When $\delta = 1$, and thus $\alpha_3 = 2$, the exponential distribution emerges as a special case of the Weibull distribution. In this case, the pdf of (3.2.1) becomes

$$\begin{aligned} f(x; \gamma, \beta) &= \frac{1}{\beta} \exp\left\{-\frac{x-\gamma}{\beta}\right\}, \qquad \gamma < x < \infty \\ &= 0 \qquad \text{elsewhere} \end{aligned} \qquad (3.9.1)$$

and the cdf becomes

$$F(x; \gamma, \beta) = 1 - \exp\left\{-\frac{x-\gamma}{\beta}\right\} \qquad (3.9.2)$$

The exponential distribution is reverse J-shaped, with $f(\gamma) = 1/\beta$ and $\lim_{x \to \infty} f(x) = 0$. Important characteristics are

$$\begin{aligned} &E(X) = \gamma + \beta, \qquad V(X) = \beta^2 \\ &Me(X) = \gamma + \beta \ln 2, \qquad \alpha_3 = 2, \text{ and } \alpha_4 = 9 \end{aligned} \qquad (3.9.3)$$

As mentioned in Chapter 2, this distribution has constant hazard function $h(x) = 1/\beta$. It is therefore a suitable model for lifetime data where used items are to be considered as good as new ones.

3.9.1 Estimators

Moment, maximum likelihood, and modified moment estimators for the parameters of this distribution are given next.

Moment Estimators

$$\hat{\beta} = s, \quad \hat{\gamma} = \bar{x} - s \tag{3.9.4}$$

Maximum Likelihood Estimators

$$\hat{\gamma} = x_1, \quad \hat{\beta} = \bar{x} - x_1 \tag{3.9.5}$$

Modified Moment Estimators. The estimating equations are $E(X) = \bar{x}$ and $E(X_1) = x_1$. It is easy to show that the pdf of the first-order statistic X_1 in a sample of size n from a distribution that is exponential (γ, β) with pdf (3.9.1) is exponential $[\gamma, \beta/n]$. Thus

$$\begin{aligned} f(x_1; \gamma, \beta, n) &= \frac{n}{\beta} \exp\left\{-\frac{n(x_1 - \gamma)}{\beta}\right\} \\ &= 0 \quad \text{elsewhere} \end{aligned} \tag{3.9.6}$$

and

$$E(X_1) = \gamma + \frac{\beta}{n} \tag{3.9.7}$$

The estimating equations may now be written as

$$\gamma + \beta = \bar{x}, \quad \gamma + \frac{\beta}{n} = x_1 \tag{3.9.8}$$

The MME follow from (3.9.8) as

$$\hat{\beta} = \frac{n(\bar{x} - x_1)}{n - 1}, \quad \hat{\gamma} = \frac{nx_1 - \bar{x}}{n - 1} \tag{3.9.9}$$

where $\bar{x} = \Sigma_1^n x_i/n$.

As shown by Cohen and Helm (1973) and Sarhan and Greenberg (1956), these estimators are both best linear unbiased (BLUE) and minimum variance unbiased (MVUE). They are therefore the preferred estimators.

3.9.2 Estimate Variances and Covariances

The exact variances and covariance of the MME of (3.9.9) were shown by Cohen and Helm (1973) to be

$$V(\hat{\beta}) = \frac{\beta^2}{n - 1}$$

$$V(\hat{\gamma}) = \frac{\beta^2}{n(n - 1)} \tag{3.9.10}$$

$$\mathrm{Cov}(\hat{\beta}, \hat{\gamma}) = -\frac{\beta^2}{n(n - 1)}$$

Additionally, it was pointed out by David (1970) that, in this case, $2(n - 1)(\hat{\beta}/\beta)$ is distributed as chi-square with $2(n - 1)$ degrees of freedom. This result permits the calculation of exact confidence limits on β. Accordingly, a $(100 - \alpha)\%$ confidence interval can be written as

$$\frac{2(n - 1)\hat{\beta}}{\chi^2_{(1-\alpha/2);2(n-1)}} < \beta < \frac{2(n - 1)\hat{\beta}}{\chi^2_{(\alpha/2);2(n-1)}} \tag{3.9.11}$$

Of course, when skewness is small and sample sizes are large, estimate distributions will approach normality, and the usual normal theory can be employed to calculate approximate confidence intervals on parameters of interest.

3.10 THE EXTREME VALUE DISTRIBUTION

This distribution is sometimes referred to as the Gumbel distribution, in recognition of the pioneering efforts of E. J. Gumbel (1958) to promote its use in the analysis of lifetime data. It has received attention from Mann, Schafer, and Singpurwalla (1974), Lawless (1982), Nelson (1982), Meeker and Nelson (1974), Johnson and Kotz (1980), Dumonceaux and Antle (1973), and many others.

There are three types of extreme value distributions with cdfs as follows [cf. Johnson and Kotz (1970)].

Type I:

$$F(x) = \Pr(X \leq x) = \exp\left\{-\exp\left[-\frac{x - \mu}{\alpha}\right]\right\}, \qquad -\infty < x < \infty \tag{3.10.1}$$

Type II:

$$F(x) = \Pr(X \leq x) = \exp\left\{-\left[\frac{x-\mu}{\alpha}\right]^{-\delta}\right\} \qquad x \geq \mu \tag{3.10.2}$$

$$= 0 \quad \text{otherwise}$$

Type III:

$$F(x) = \Pr(X \leq x) = \exp\left\{-\left[\frac{\mu-x}{\alpha}\right]^{\delta}\right\} \qquad x \leq \mu \tag{3.10.3}$$

$$= 1 \quad \text{otherwise}$$

Type I is the distribution of the smallest extreme value, and it is the one that is most often employed as a model in the analysis of lifetime data. It is the one most often associated with Gumbel's name, and when no qualifying restrictions are mentioned it is understood to be "the" extreme value distribution. The pdf of this distribution is

$$f(x;\mu,\alpha) = \frac{1}{\alpha}\exp\left[\frac{x-\mu}{\alpha}\right]\exp\left\{-\exp\left[\frac{x-\mu}{\alpha}\right]\right\}, \qquad -\infty < x < \infty,\ \alpha > 0 \tag{3.10.4}$$

where μ is a location parameter and α is a scale parameter. This distribution is an appropriate model for descriptions of certain extreme phenomena such as rainfall minima, rainfall during droughts, electrical strength of materials, and certain types of life data, for example, mortality of the aged. It is also used as a model in "weakest link" situations. If a unit consists of a number of identical components with identical life distributions, and if the unit fails when the first component fails, then the smallest extreme value distribution might be a suitable model for describing the life of the population of units.

To a considerable extent, interest in the extreme value distribution is due to its relation to the Weibull distribution. The Weibull distribution is to the extreme value distribution as the normal distribution is to the lognormal distribution. If Y has a Weibull distribution with pdf (3.2.1), it can be shown that $X = \ln(Y - \gamma)$ has an extreme value distribution with pdf (3.10.4) and cdf (3.10.1) [cf. Mann et al. (1974)]. Our principal concern in this connection is with the two-parameter Weibull distribution for which $\gamma = 0$. In view of this relationship, the extreme value distribution might be thought of as a log-Weibull distribution. In some analyses, rather than work with original sample data from a Weibull distribution, it might be simpler to transform the data by taking logarithms (to base e) and then to treat the transformed sample as one from an extreme value distribution.

3.10.1 Some Important Characteristics

The extreme value distribution is unimodal with inflection points at $\mu \pm 0.96242\alpha$. As given by Johnson and Kotz (1970), the moment generating function of the Type I distribution is

$$M_X(t) = E[e^{tX}] = e^{\mu t}\Gamma(1 - \alpha t), \quad \alpha|t| < 1 \tag{3.10.5}$$

and the cumulant generating function is

$$\Psi(t) = \mu t + \ln \Gamma(1 - \alpha t) \tag{3.10.6}$$

The cumulants of X follow from (3.10.6) as

$$\kappa_1(X) = \mu - \alpha\psi(1) = \mu + 0.57722\alpha$$

$$\kappa_r(X) = (-\alpha)^r \psi^{(r-1)}(1), \quad r \geq 2 \tag{3.10.7}$$

where $\psi(\,)$ is the digamma function. It follows that

$$E(X) = \mu + 0.57722\alpha$$

where 0.57722 is Euler's constant,

$$V(X) = \left(\frac{\pi^2}{6}\right)\alpha^2 = 1.64493\alpha^2$$

$$\alpha_3^2(X) = \beta_1(X) = 1.29857 \tag{3.10.8}$$

$$\alpha_4(X) = \beta_2(X) = 5.4$$

$$Mo(X) = \mu$$

$$Me(X) = \mu - \alpha \ln(\ln 2) = \mu + 0.36651\alpha$$

3.10.2 Parameter Estimation

Johnson and Kotz (1970, Vol. 1, Chapter 21) contains an excellent discussion of MLE for the parameters of this distribution. Unfortunately the calculations are somewhat complicated and time consuming.

Moment Estimators. Although they are subject to larger sampling errors, ME of extreme value parameters are much simpler to calculate. We simply equate the sample mean, $\bar{x}$, and sample variance, s^2, to corresponding distribution values as given in (3.10.8), and simplify to obtain

$$\alpha^* = \left(\frac{\sqrt{6}}{\pi}\right)s = 0.779697s, \qquad \mu^* = \bar{x} - 0.57722\alpha^* \tag{3.10.9}$$

Tiago de Oliveria (1963) derived approximate variances for β^* and μ^* as

$$V(\mu^*) = \frac{1.1678\alpha^2}{n}, \qquad V(\alpha^*) = \frac{1.1\alpha^2}{n} \tag{3.10.10}$$

These might be useful in calculating approximate confidence intervals on μ and α.

3.11 ILLUSTRATIVE EXAMPLES

Practical application of the MME presented here for the Weibull distribution is demonstrated with three illustrative examples. Although chosen from real-life situations, these examples are intended only as illustrations for computing techniques. Each is assumed to be from a Weibull population.

Example 3.11.1. The maximum flood levels in millions of cubic feet per second for the Susquehanna River at Harrisburg, Pennsylvania, over 20 four-year periods from 1890 to 1969 are

0.654	0.613	0.315	0.449	0.297
0.402	0.379	0.423	0.379	0.3235
0.269	0.740	0.418	0.412	0.494
0.416	0.338	0.392	0.484	0.265

These data were previously used as an illustrative example for the extreme value distribution by Dumonceaux and Antle (1973). Here, we consider this sample to be from a Weibull distribution. In summary, $n = 20$, $\bar{x} = 0.423125$, $s^2 = 0.0156948$, $s = 0.1252789$, $a_3 = 1.0673243$, $x_1 = 0.265$, and $s^2/(\bar{x} - x_1)^2 = 0.627704$.

To calculate MME, we go to Figure 3.2, with $n = 20$ and $s^2/(\bar{x} - x_1)^2 = 0.628$, and read $\delta_1 = 1.48$. Then from Figure 3.3, with $\delta_1 = 1.48$, we read $C(1.48) = 1.62$ and $D(1.48) = 1.45$. Equations (3.5.3) then yield $\beta_1 = s \cdot C(1.48) = (0.1253)(1.62) = 0.203$, $\gamma_1 = \bar{x} - s \cdot D(1.48) = 0.423125 - (0.1253 \times 1.45) = 0.241$. Alternatively, we interpolate in Table 3.2 between $\delta = 1.46$ and $\delta = 1.48$ as shown below.

δ	$W(n, \delta)$
1.4600	0.63802
1.4794	0.62770
1.4800	0.62737

TABLE 3.4 Estimates for Example 3.11.1, Weibull Distribution

Estimator	$\hat{\gamma}$	$\hat{\delta}$	$\hat{\beta}$	$\hat{E}(X)$	$\sqrt{\hat{V}(X)}$	$\hat{\alpha}_3(X)$
ME	0.2382	1.5040	0.2050	0.4231	0.1253	1.0673
MLE	0.2611	1.2445	0.1727	0.4222	0.1302	1.4393
MME	0.2410	1.4794	0.2014	0.4231	0.1253	1.0966
Zanakis	0.2650	1.3280	0.1530	0.4057	0.1070	1.3021
WBE	0.2465	1.3963	0.2011	0.4299	0.1330	1.2034

Asymptotic variances of (3.8.4) are not applicable since $\delta < 2$. From the central limit theorem $\sigma_{\bar{X}} = 0.028$.

We thus have $\hat{\delta} = 1.479$, and we subsequently use the last two equations of (3.5.2) to calculate $\hat{\gamma} = 0.241$ and $\hat{\beta} = 0.201$. These same results also follow from (3.5.3).

Estimates that might be more accurate, calculated by use of the authors' (1982) FORTRAN program on the University of Georgia Cyber 70/74 computer, are given in Table 3.4. In addition to the MME, moment estimates, Zanakis (1979a) percentile estimates, and Wycoff, Bain, Engelhardt estimates (WBE) are also included for comparison.

As previously mentioned, Dumonceaux and Antle (1973) considered Example 3.11.1 to be a sample from an extreme value distribution and gave MLE for μ and α as $\hat{\mu} = 0.3688$ and $\hat{\alpha} = 0.089775$. Moment estimates for this distribution, calculated from (3.10.10), are $\mu^* = 0.3667$ and $\alpha^* = 0.097680$.

Example 3.11.2. This example is from McCool (1974). The data consist of fatigue life in hours of 10 bearings of a certain type. Listed in increasing order, these data are

152.7	204.7
172.0	216.5
172.5	234.9
173.3	262.6
193.0	422.6

In summary, $n = 10$, $\bar{x} = 220.48$, $s^2 = 6147.444$, $s = 78.405638$, $a_3 = 1.8635835$, $x_1 = 152.7$, and $s^2/(\bar{x} - x_1)^2 = 1.338109$.

To calculate MME, we go to Figure 3.2, with $n = 10$ and $s^2/(\bar{x} - x_1)^2 = 1.34$ (rounded off), and read $\delta_1 = 0.95$. From Figure 3.3, we read

TABLE 3.5 Estimates for Example 3.11.2, Weibull Distribution

Estimator	$\hat{\gamma}$	$\hat{\delta}$	$\hat{\beta}$	$\hat{E}(X)$	$\sqrt{\hat{V}(X)}$	$\hat{\alpha}_3(X)$
ME	138.3138	1.0483	83.7243	220.4800	78.4056	1.8636
MLE[a]						
MME	146.1385	0.9486	72.5884	220.4800	78.4056	2.1653
Zanakis	151.0896	1.1659	65.4104	213.0975	53.3503	1.5889
WBE	142.4775	1.1565	88.5735	226.6351	72.9700	1.6083
BLUE/MVUE[b] (2-parameter)	145.1689	1	75.3111	220.4800	75.3111	2

Asymptotic variances of (3.8.4) are not applicable since $\delta < 2$. From the central limit theorem $\sigma_{\bar{X}} = 24.79$.
[a]Not calculated.
[b]Calculated by using (3.9.9) and assuming $\delta = 1$. Standard deviations of these estimates calculated from (3.9.10): $\sigma_{\hat{\gamma}} = 7.938$, $\sigma_{\hat{\beta}} = 25.104$, $\mathrm{Cov}(\hat{\gamma}, \hat{\beta}) = -63.019$.

$C(0.95) = 0.93$ and $D(0.95) = 0.96$. On substituting in (3.5.3), we calculate $\beta_1 = 72.92$ and $\gamma_1 = 145.21$. Computer calculations based on the authors' (1982) FORTRAN program yield the results displayed in Table 3.5. Again, ME, Zanakis, and WBE estimates have been included for comparison. In view of the small sample size and since estimates of δ are close to 1, it seems appropriate to consider the possibility that indeed $\delta = 1$ and that our sample is thus from a two-parameter exponential distribution. Under this hypothesis we employ (3.9.9) and calculate BLUE and MVUE as

$$\hat{\beta} = 75.3111, \quad \hat{\gamma} = 145.1689$$

Using (3.9.10) with $\hat{\beta} = 75.3111$ substituted for β, we calculate

$$V(\hat{\beta}) = 630.19408, \quad \sqrt{V(\hat{\beta})} = 25.104$$
$$V(\hat{\gamma}) = 63.0194, \quad \sqrt{V(\hat{\gamma})} = 7.938$$

Since $2(n-1)\hat{\beta}/\beta \sim \chi^2_{2(n-1)}$ when $\delta = 1$, we compute a 0.95 confidence interval on β under this hypothesis from (3.9.11) as $43.0 < \beta < 164.7$.

These same two examples are also employed in subsequent chapters to illustrate estimation in the lognormal, inverse Gaussian, and gamma distributions. Use of the same examples permits comparisons to be made between estimates of skewness and of threshold parameters for different models. These are often important considerations in the choice of a model.

Example 3.11.3. In order to illustrate application of the modified moment estimators when sample data are grouped, we have selected a sample consisting of 1000 observations of University of Michigan first-year women students measured to the nearest 1/10 pound. These data were originally compiled by Lillian Shook (1930), who tabulated them as follows:

Weight in pounds	Observed frequency	Weight in pounds	Observed frequency	Weight in pounds	Observed frequency
70-79.9	2	120-129.9	196	170-179.9	7
80-89.9	16	130-139.9	122	180-189.9	1
90-99.9	82	140-149.9	63	190-199.9	2
100-109.9	231	150-159.9	23	200-209.9	1
110-119.9	248	160-169.9	5	210-219.9	1

After applying Sheppard's corrections for grouping errors, these data are summarized as $n = 1000$, $\bar{x} = 118.74$, $s = 16.9175$, $a_3 = 0.976424$, $a_4 = 5.3206$, and $69.95 < x_1 < 79.95$.

Unfortunately, as frequently happens with grouped data, x_1 is not known explicitly and must be approximated from the information that it lies in the interval (B_0, B_1), where B_j, $j = 0, 1, \ldots, k$, are the class boundaries and f_j, $j = 1, 2, \ldots, k$, are the class frequencies. Let the class width be $W = B_{j+1} - B_j$. An approximation to x_1 that has proven satisfactory in most practical applications is

$$x_1 = B_1 - \frac{f_1}{f_2} W \tag{3.11.1}$$

This approximation is obtained when we project a straight line between the two points (B_1, f_1) and $[B_2, f_1 + f_2]$ of the cdf plot until it intersects the abscissa at $(x_1, 0)$. For this example, substitution into (3.11.1) gives $x_1 = 79.95 - (2/16)10 = 78.70$, and this is the value which we use in calculating MME.

Miss Shook treated this sample as being from a Pearson Type III (gamma) distribution. At this time, however, we entertain the possibility that it might be from a Weibull population, and we employ the MME of (3.5.2) and (3.5.3) to estimate the Weibull parameters γ, β, and δ. For comparison, this same sample is employed in subsequent chapters to calculate corresponding estimates based in turn on the assumption of lognormal, inverse Gaussian, and gamma populations. Comparisons of estimates of γ and of α_3 for each of the four distributions is helpful in choosing an appropriate model to best describe a particular sample.

In order to estimate δ, we calculate

$$W(n, \hat{\delta}) = \frac{s^2}{(\bar{x} - x_1)^2} = \left(\frac{16.9175}{118.74 - 78.70}\right)^2 = 0.17852$$

We enter Table 3.2 with $n = 1000$ and $W = 0.17852$, and interpolate to find $\hat{\delta} = 2.7994$. Subsequent interpolation in Table 3.1 yields $\hat{\alpha}_3 = 0.23854$, $C(\hat{\delta}) = 2.90381$, and $D(\hat{\delta}) = 2.58588$. Substitution in (3.5.3) gives

$$\hat{\beta} = 16.9175(2.90381) = 49.125$$

$$\hat{\gamma} = 118.74 - 16.9175(2.58588) = 74.993$$

$$\hat{\theta} = \hat{\beta}^{\hat{\delta}} = 54279$$

The MME for the Weibull parameters are thus $\hat{\delta} = 2.799$, $\hat{\beta} = 49.125$, $\hat{\gamma} = 74.993$, $\hat{\alpha}_3 = 0.23854$, and $\hat{\alpha}_4 = 2.765$.

3.12 REFLECTIONS

The MLE are of limited applicability in estimating parameters of the three-parameter Weibull distribution from sample data, whereas the MME are applicable over the entire parameter space and are clearly superior to ME. The MLE exist only if $\delta > 1$. The asymptotic variance-covariance matrix for the MLE exists only if $\delta > 2$ ($\alpha_3 < 0.63$). Furthermore, as previously noted, computational difficulties are likely to be encountered unless $\delta > 2.2$ (approximately). Like the ME, the MME are unbiased with respect to the distribution mean and variance. In view of these considerations, use of the MME therefore is recommended over both the ME and MLE unless there is reason to expect that α_3 is quite small, and, in that case, use of the Weibull as a model does not appear to be justified.

4

The Lognormal Distribution

4.1 INTRODUCTION

The lognormal distribution has been studied extensively by Yuan (1933), Cohen (1951, 1987), Aitchison and Brown (1957), Hill (1963), Harter and Moore (1966), Johnson and Kotz (1970), Giesbrecht and Kempthorne (1976), Kane (1978, 1982), Cohen et al. (1985), Wingo (1975, 1976), Munro and Wixley (1970), Stedinger (1980), Rukhin (1984), and many others. It is a "long-tailed," positively skewed distribution that is an appropriate model in life-span and reaction-time studies where data are often highly skewed.

4.2 SOME FUNDAMENTALS

The name of this distribution is derived from the relation that exists between random variables X and $Y = \ln(X - \gamma)$. If Y is distributed normally (μ, σ^2), then X is lognormal (γ, μ, σ^2). When γ is known, it is a simple matter to make the transformation from X to Y. Subsequent analyses, including parameter estimation, can then be made by the well-known theory of normal distributions. When the threshold parameter γ is unknown, estimation procedures become more complex. The probability density function (pdf) of the three-parameter lognormal distribution follows from the definition as

$$f(x; \gamma, \mu, \sigma^2) = \frac{1}{\sigma\sqrt{2\pi}(x-\gamma)} \exp\left\{-\frac{[\ln(x-\gamma)-\mu]^2}{2\sigma^2}\right\}, \quad \gamma < x < \infty, \ \sigma^2 > 0 \tag{4.2.1}$$

$$= 0 \quad \text{otherwise}$$

In the notation employed here, σ^2 and μ are the variance and mean of Y, but they become the shape and scale parameters, respectively, of X, and γ is the threshold (location) parameter. Sometimes it is more convenient to employ $\beta = \exp(\mu)$ as the scale parameter and $\omega = \exp(\sigma^2)$ as the shape parameter of the lognormal distribution.

The cumulative distribution function (cdf) of X may conveniently be expressed as

$$F(x; \gamma, \mu, \sigma) = \Phi\left[\frac{\ln(x - \gamma) - \mu}{\sigma}\right] \tag{4.2.2}$$

where $\Phi(\cdot)$ is the cumulative standard normal distribution function.

The expected value, median, mode, variance, coefficient of variation, β_1 and β_2 (Pearson's betas) for this distribution [cf. Yuan (1933)] are

$$\begin{aligned}
E(X) &= \gamma + \beta\sqrt{\omega} \\
Me(X) &= \gamma + \beta, \qquad Mo(X) = \gamma + \frac{\beta}{\omega} \\
V(X) &= \beta^2\omega(\omega - 1), \qquad v(X) = \sqrt{\omega - 1} \\
\beta_1 &= \alpha_3^2 = (\omega + 2)^2(\omega - 1) \\
\beta_2 &= \alpha_4 = \omega^4 + 2\omega^3 + 3\omega^2 - 3
\end{aligned} \tag{4.2.3}$$

where $E(\cdot)$ is the expected value symbol, $V(\cdot)$ is the variance, $v(\cdot)$ is the coefficient of variation, and where α_3 and α_4 denote the third and fourth standard moments. The coefficient of variation is defined as

$$v(X) = \frac{\sqrt{V(X)}}{E(X) - \gamma}$$

4.2.1 The Standard Distribution

If we make the standardizing transformation

$$Z = \frac{X - E(X)}{\sqrt{V(X)}} \tag{4.2.4}$$

the pdf of the standard lognormal distribution with mean zero, unit variance, and shape parameter ω becomes

$$g(z; 0, 1, \omega) = \frac{\sqrt{\omega - 1}}{\sqrt{2\pi \ln \omega}\,[1 + z\sqrt{\omega - 1}]} \exp\left\{\frac{-1}{2(\ln \omega)}\right\}[\ln\{\sqrt{\omega} + z\sqrt{\omega(\omega - 1)}\}]^2 \tag{4.2.5}$$

$$-(\omega - 1)^{-\frac{1}{2}} < z, \quad \omega > 1$$

$$= 0 \quad \text{otherwise}$$

The pdf of (4.2.5) can also be expressed in terms of the standard normal pdf $\phi(\cdot)$ as

$$g(z; 0, 1, \omega) = \frac{1}{\sqrt{\ln \omega}}\left(\frac{\sqrt{\omega - 1}}{1 + z\sqrt{\omega - 1}}\right) \phi\left[\frac{\ln \{\sqrt{\omega} + z\sqrt{\omega(\omega - 1)}\}}{\sqrt{\ln \omega}}\right] \qquad (4.2.6)$$

The standard cdf becomes

$$G(z; 0, 1, \omega) = \Phi\left[\frac{\ln [\sqrt{\omega} + z\sqrt{\omega(\omega - 1)}]}{\sqrt{\ln \omega}}\right] \qquad (4.2.7)$$

4.3 MOMENT ESTIMATORS

On equating the first three distribution moments as given in (4.2.3) to corresponding sample moments, we have

$$\hat{\gamma} + \hat{\beta}\sqrt{\hat{\omega}} = \bar{x}, \quad \hat{\beta}^2\hat{\omega}(\hat{\omega} - 1) = s^2, \quad (\hat{\omega} + 2)\sqrt{\hat{\omega} - 1} = a_3 \qquad (4.3.1)$$

where

$$\bar{x} = \frac{1}{n}\sum_1^n x_i, \quad s^2 = \frac{1}{n-1}\sum_1^n (x_i - \bar{x})^2$$

$$a_3 = \frac{1}{n}\sum_1^n (x_i - \bar{x})^3 \Big/ \left[\frac{1}{n}\sum_1^n (x_i - \bar{x})^2\right]^{3/2} \qquad (4.3.2)$$

On squaring both sides and collecting terms, we find that the third equation of (4.3.1) is reduced to the following cubic equation in

$$\omega^3 + 3\omega^2 - (4 + a_3^2) = 0 \qquad (4.3.3)$$

This equation has a single positive real root $\hat{\omega} > 1$, which is the required estimate and which, according to Abramowitz and Stegun (1964), can be expressed explicitly as

$$\hat{\omega} = \sqrt[3]{1 + \frac{a_3}{2}[a_3 + \sqrt{a_3^2 + 4}]} + \sqrt[3]{1 + \frac{a_3}{2}[a_3 - \sqrt{a_3^2 + 4}]} - 1 \qquad (4.3.4)$$

An equivalent cubic equation with the coefficient of variation v as the variable might be preferred since it has a simpler solution.

Since $v = \sqrt{\omega - 1}$, it follows that $\omega = 1 + v^2$, $\alpha_3 = v(v^2 + 3)$, and $\sigma = \sqrt{\ln(1 + v^2)}$. On equating α_3 to a_3, we obtain the cubic equation

$$v^3 + v - a_3 = 0 \qquad (4.3.5)$$

$$v^3 + v - a_3 = 0 \tag{4.3.5}$$

This equation has one positive root and a pair of conjugate imaginary roots. Accordingly, the required estimate $\hat{v}$ is the positive root, which can be expressed as

$$\hat{v} = \sqrt[3]{\tfrac{1}{2}[a_3 + \sqrt{4 + a_3^2}]} + \sqrt[3]{\tfrac{1}{2}[a_3 - \sqrt{4 + a_3^2}]} \tag{4.3.6}$$

It follows that $\hat{\omega} = 1 + \hat{v}^2$, and estimates $\hat{\beta}$ and $\hat{\gamma}$ then follow from the first two equations of (4.3.1) as

$$\hat{\beta} = \frac{s}{\sqrt{\hat{\omega}(\hat{\omega} - 1)}}, \qquad \hat{\gamma} = \bar{x} - \hat{\beta}\sqrt{\hat{\omega}} \tag{4.3.7}$$

If required, $\hat{\mu}$ can be calculated as $\hat{\mu} = \ln \hat{\beta}$.

The principal objection to moment estimation (ME) centers about the large sampling errors due to the third moment, plus the fact that, as shown by Heyde (1963), the three parameter lognormal distribution is not uniquely determined by its moments. We are therefore led to consider maximum likelihood estimation (MLE).

4.4 MAXIMUM LIKELIHOOD ESTIMATORS

We seek estimates that will maximize the likelihood function of a random sample consisting of observations $\{x_i\}$, $i = 1, \ldots, n$. Without any loss of generality it will be assumed that the sample is ordered so that x_1 is the smallest sample observation. The likelihood function may be written as

$$L(x_1, \ldots, x_n; \gamma, \mu, \sigma) = \prod_{i=1}^{n} f(x_i; \gamma, \mu, \sigma)$$

It is immediately obvious that L( ) approaches infinity as $\gamma \to x_1$. It would thus appear that we should take $\hat{\gamma} = x_1$ as our estimate. However, Hill (1963) demonstrated the existence of paths along which the likelihood function of any ordered sample $x_1, \ldots, x_n$ tends to ∞ as (γ, μ, σ^2) approach $(x_1, -\infty, \infty)$. This global maximum thereby leads to the inadmissible estimates $\hat{\mu} = -\infty$ and $\hat{\sigma}^2 = \infty$ regardless of the sample.

As an alternative, Cohen (1951), Cohen and Whitten (1980), and Harter and Moore (1966) equated partial derivatives of the loglikelihood function to zero and solved the resulting equations to obtain local maximum likelihood estimators (LMLE), which in most cases would be considered reasonable in comparison with corresponding moment estimators. Harter and Moore and, later, Calitz (1973) noted that these LMLE appear to possess most of the desirable properties ordinarily associated with MLE.

On differentiating the loglikelihood function and equating to zero, we obtain the LMLE estimating equations

$$\frac{\partial \ln L}{\partial \mu} = \frac{1}{\sigma^2}\sum_1^n [\ln(x_i - \gamma) - \mu] = 0$$

$$\frac{\partial \ln L}{\partial \sigma} = -\frac{n}{\sigma} + \frac{1}{\sigma^3}\sum_1^n [\ln(x_i - \gamma) - \mu]^2 = 0 \tag{4.4.1}$$

$$\frac{\partial \ln L}{\partial \gamma} = \frac{1}{\sigma^2}\sum_1^n \frac{\ln(x_i - \gamma) - \mu}{x_i - \gamma} + \sum_1^n (x_i - \gamma)^{-1} = 0$$

When σ^2 and μ are eliminated from these equations, as was done by Cohen (1951), the resulting equation in γ becomes

$$\lambda(\hat{\gamma}) = \sum_1^n (x_i - \hat{\gamma})^{-1}$$

$$\left[\sum_1^n \ln(x_i - \hat{\gamma}) - \sum_1^n \ln^2(x_i - \hat{\gamma}) + \frac{1}{n}\left\{\sum_1^n \ln(x_i - \hat{\gamma})\right\}^2\right] - n\sum_1^n \frac{\ln(x_i - \hat{\gamma})}{x_i - \hat{\gamma}} = 0 \tag{4.4.2}$$

Equation (4.4.2) may be solved iteratively for $\hat{\gamma}$. It then follows from the first two equations of (4.4.1) that

$$\hat{\mu} = \frac{1}{n}\sum_1^n \ln(x_i - \hat{\gamma})$$

$$\hat{\sigma}^2 = \frac{1}{n}\sum_1^n \ln^2(x_i - \hat{\gamma}) - \left[\frac{1}{n}\sum_1^n \ln(x_i - \hat{\gamma})\right]^2 \tag{4.4.3}$$

In solving (4.4.2) for $\hat{\gamma}$, we accept only admissible roots for which $\gamma < x_1$. Usually, only a single admissible root will be found. In the event that multiple admissible roots occur, we choose as our estimate the root which results in closest agreement between $\bar{x}$ and $\hat{E}(X)$.

Standard iterative procedures such as the Newton-Raphson method are satisfactory for solving (4.2.2), but in many instances it is more convenient to use the trial-and-error technique with linear interpolation. We begin with a first approximation $\gamma_1 < x_1$ and evaluate $\lambda(\gamma_1)$. If this value is zero, then no further calculations are required. Otherwise we continue until we find a pair of values γ_i and γ_j in a sufficiently narrow interval such that $\lambda(\gamma_i) \gtrless 0 \gtrless \lambda(\gamma_j)$, and interpolate for the final estimate $\hat{\gamma}$.

Wilson and Worcester (1945) and Lambert (1964) attempted to solve the three equations of (4.4.1) simultaneously without simplifying the form given in (4.4.2), but they encountered convergence problems. Calitz (1973) examined the convergence problem further and concluded that the simplification of Cohen (1951) using (4.4.2) led to fewer convergence problems and was therefore to be recommended.

4.5 ASYMPTOTIC VARIANCES AND COVARIANCES

In the absence of regularity restrictions, asymptotic variances and covariances of MLE of distribution parameters can be obtained by inverting the Fisher information matrix in which elements are negatives of expected values of second partial derivatives of the likelihood function with respect to the parameters. Unfortunately, the lognormal distribution is subject to regularity problems as previously mentioned, and this raises questions about the validity of asymptotic variances and covariances thus obtained as they might apply to the local MLE. Nevertheless, they have been found to be in reasonably close agreement with estimate variances, obtained in simulation studies by Cohen and Whitten (1980), Harter and Moore (1966), and Cohen et al. (1984). They are therefore offered here as possible useful approximations to the applicable variances and covariances. Based on the information matrix for $\hat{\gamma}$, $\hat{\beta}$, $\hat{\sigma}$ given by Cohen (1951) or that for $\hat{\gamma}$, $\hat{\mu}$, $\hat{\sigma}^2$ given by Hill (1963), we have

$$
\begin{aligned}
V(\hat{\gamma}) &\doteq \frac{\sigma^2}{n}\frac{\beta^2}{\omega}H, & \mathrm{Cov}(\hat{\gamma}, \hat{\beta}) &\doteq \frac{-\sigma^3}{n}\frac{\beta^2}{\sqrt{\omega}}H \\
V(\hat{\beta}) &\doteq \frac{\sigma^2}{n}\beta^2[1 + H], & \mathrm{Cov}(\hat{\gamma}, \hat{\sigma}) &\doteq \frac{\sigma^3}{n}\frac{\beta^2}{\sqrt{\omega}}H \\
V(\hat{\sigma}) &\doteq \frac{\sigma^2}{2n}[1 + 2\sigma^2 H], & \mathrm{Cov}(\hat{\beta}, \hat{\sigma}) &\doteq \frac{-\sigma^3}{n}\beta^2 H \\
V(\hat{\mu}) &\doteq \frac{\sigma^2}{n}[1 + H], & V(\hat{\sigma}^2) &\doteq \frac{2\sigma^4}{n}[1 + 2\sigma^2 H]
\end{aligned}
\tag{4.5.1}
$$

where

$$H = [\omega(1 + \sigma^2) - (1 + 2\sigma^2)]^{-1}$$

For large samples, the central limit theorem can be employed to approximate the variance of the estimate $\hat{m}$ of the distribution mean $[m = E(X)]$ as

$$V(\hat{m}) = \frac{V(X)}{n}, \quad \text{where } V(X) \text{ is given in (4.2.3)}$$

The variances and covariances of (4.5.1) can be expressed in a simpler format as

TABLE 4.1 Variance-Covariance Factors for Maximum Likelihood Estimates of Lognormal Parameters

α_3	ω	ϕ_{11}	ϕ_{22}	ϕ_{33}	ϕ_{12}	ϕ_{23}
0.50	1.02728	885.23263	910.38125	49.95010	-147.19413	-149.18831
0.55	1.03289	607.65422	628.63763	41.61692	-111.08790	-112.89975
0.60	1.03898	431.38171	449.19715	35.27801	-85.98520	-87.64504
0.65	1.04555	315.03444	330.38447	30.34403	-67.98669	-69.51786
0.70	1.05258	235.69195	249.08577	26.42824	-54.74136	-56.16220
0.75	1.06007	180.05004	191.86570	23.26838	-44.77401	-46.09919
0.80	1.06799	140.07179	150.59584	20.68148	-37.12694	-38.36839
0.85	1.07634	110.73199	120.18559	18.53674	-31.15990	-32.32744
0.90	1.08510	88.79275	97.34930	16.73867	-26.43369	-27.53552
0.95	1.09426	72.11193	79.90929	15.21622	-22.64022	-23.68324
1.00	1.10380	59.23866	66.38784	13.91565	-19.55896	-20.54904
1.05	1.11372	49.16952	55.76087	12.79571	-17.02923	-17.97141
1.10	1.12398	41.19733	47.30516	11.82430	-14.93205	-15.83068
1.15	1.13460	34.81515	40.50113	10.97615	-13.17803	-14.03690
1.20	1.14554	29.65392	34.96965	10.23114	-11.69914	-12.52156
1.25	1.15679	25.44114	30.43014	9.57311	-10.44293	-11.23182
1.30	1.16835	21.97300	26.67221	8.98892	-9.36858	-10.12653
1.35	1.18020	19.09523	23.53622	8.46783	-8.44393	-9.17324
1.40	1.19233	16.68980	20.89971	8.00100	-7.64347	-8.34620
1.45	1.20472	14.66545	18.66773	7.58107	-6.94674	-7.62472
1.50	1.21736	12.95095	16.76599	7.20190	-6.33721	-6.99210
1.55	1.23025	11.49026	15.13585	6.85830	-5.80144	-6.43475
1.60	1.24336	10.23888	13.73062	6.54590	-5.32841	-5.94150
1.65	1.25669	9.16122	12.51284	6.26099	-4.90903	-5.50313
1.70	1.27023	8.22860	11.45225	6.00038	-4.53575	-5.11201
1.75	1.28397	7.41778	10.52422	5.76133	-4.20229	-4.76172
1.80	1.29790	6.70977	9.70861	5.54149	-3.90336	-4.44691
1.85	1.31200	6.08899	8.98879	5.33881	-3.63449	-4.16304
1.90	1.32628	5.54257	8.35100	5.15152	-3.39191	-3.90627
1.95	1.34071	5.05983	7.78379	4.97805	-3.17240	-3.67330
2.00	1.35530	4.63185	7.27756	4.81705	-2.97322	-3.46134
2.25	1.43024	3.08881	5.41775	4.16174	-2.20975	-2.64270
2.50	1.50791	2.16869	4.27019	3.68628	-1.70671	-2.09578
2.75	1.58758	1.58683	3.51921	3.32880	-1.35930	-1.71271
3.00	1.66869	1.20085	3.00385	3.05208	-1.11001	-1.43389
3.25	1.75079	0.93447	2.63607	2.83263	-.92535	-1.22440
3.50	1.83355	0.74441	2.36492	2.65498	-.78485	-1.06276
3.75	1.91669	0.60491	2.15943	2.50865	-.67550	-.93519
4.00	2.00000	0.50000	2.00000	2.38629	-.58871	-.83255
4.25	2.08331	0.41942	1.87378	2.28263	-.51863	-.74858
4.50	2.16650	0.35637	1.77207	2.19380	-.46121	-.67886
5.00	2.33211	0.26578	1.61983	2.04972	-.37349	-.57037
6.00	2.65871	0.16333	1.43425	1.84925	-.26335	-.42941
7.00	2.97764	0.11031	1.32847	1.71682	-.19884	-.34311
8.00	3.28840	0.07956	1.26164	1.62290	-.15742	-.28546

$\phi_{13} = -\phi_{12}$

$$
\begin{aligned}
V(\hat{\gamma}) &= \frac{\sigma^2}{n}\beta^2\phi_{11}, & V(\hat{\sigma}^2) &= \frac{2\sigma^4}{n}\phi_{33} \\
V(\hat{\beta}) &= \frac{\sigma^2}{n}\beta^2\phi_{22}, & \text{Cov}(\hat{\gamma},\hat{\beta}) &= \frac{\sigma^2}{n}\beta^2\phi_{12} \\
V(\hat{\sigma}) &= \frac{\sigma^2}{2n}\phi_{33}, & \text{Cov}(\hat{\gamma},\hat{\sigma}) &= \frac{\sigma^2}{n}\beta^2\phi_{13} \\
V(\hat{\mu}) &= \frac{\sigma^2}{n}\phi_{22}, & \text{Cov}(\hat{\beta},\hat{\sigma}) &= \frac{\sigma^2}{n}\beta^2\phi_{23}
\end{aligned}
\tag{4.5.2}
$$

where

$$
\begin{aligned}
\phi_{11} &= \frac{H}{\omega}, & \phi_{22} &= 1 + H, & \phi_{33} &= 1 + 2\sigma^2 H \\
\phi_{12} &= -\frac{\sigma}{\sqrt{\omega}}H, & \phi_{13} &= -\phi_{12}, & \phi_{23} &= -\sigma H
\end{aligned}
\tag{4.5.3}
$$

The ϕ_{ij} of (4.5.3) are functions of σ alone. A table of these factors which will facilitate the calculation of variances and covariances is included as Table 4.1 with α_3 as the argument. This choice of the argument is possible since α_3 and ω are functions of σ alone.

4.6 MODIFIED MOMENT ESTIMATORS

In modified moment estimators (MME) the third moment, which is subject to somewhat large sampling errors, is replaced by a function of the first-order statistic. This replacement was selected since the first-order statistic contains more information about the threshold parameter than do any of the other sample observations and often more than all the other observations combined. Cohen and Whitten (1980) first proposed the MME under consideration here. Cohen et al. (1985) further investigated these estimators and presented tabular and graphical aids which greatly simplify calculations in practical applications. The presentation which follows is based on the 1985 paper.

For an ordered random sample of size n, the estimating equations are

$$E(X) = \bar{x}, \quad V(X) = s^2, \quad E[\ln(X_1 - \gamma)] = \ln(x_1 - \gamma) \tag{4.6.1}$$

where $\bar{x}$ and s^2 are the sample mean and variance (unbiased). X_1 is the first-order statistic (a random variable) in a random sample of size n and x_1 is the corresponding sample value. The symbol $E(\cdot)$ designates expected value.

The third equation of (4.6.1) reduces to

$$\gamma + \beta \exp[\sqrt{\ln\omega}\, E(Z_{1,n})] = x_1$$

TABLE 4.2 Expected Values of the First-Order Statistic from the Standard Normal Distribution (0, 1)

n	$E(Z_{1,n})$	n	$E(Z_{1,n})$	n	$E(Z_{1,n})$
5	-1.16296	36	-2.11812	100	-2.50759
10	-1.53875	38	-2.14009	125	-2.58634
12	-1.62923	40	-2.16078	150	-2.64925
14	-1.70338	45	-2.20772	175	-2.70148
16	-1.75699	50	-2.24907	200	-2.74604
18	-1.82003	55	-2.28598	225	-2.78485
20	-1.86748	60	-2.31928	250	-2.81918
22	-1.90969	65	-2.34958	280	-2.85572
24	-1.94767	70	-2.37736	300	-2.87777
26	-1.98216	75	-2.40299	315	-2.89327
28	-2.01371	80	-2.42677	350	-2.92651
30	-2.04276	85	-2.44894	375	-2.94810
32	-2.06967	90	-2.46970	400	-2.96818
34	-2.09471	95	-2.48920	1000[a]	-3.09053

Source: Extracted from Harter's (1961) tables.
[a]Entry approximated as $Z_{1,1000} \doteq \Phi^{-1}(1/1001)$.

where $E(Z_{1,n})$ is the expected value of the first-order statistic in a random sample of size n from a standard normal distribution (0, 1). Values of $E(Z_{1,n})$ can be obtained from tables compiled by Harter (1961, 1969). Selected values from this source are reproduced here as Table 4.2. Linear interpolation in this table will usually provide sufficient accuracy for most practical applications.

Appropriate substitutions from (4.2.3) into (4.6.1) enable us to write the modified estimating equations as

$$\hat{\gamma} + \hat{\beta}\hat{\omega}^{\frac{1}{2}} = \bar{x}, \quad \hat{\beta}^2\hat{\omega}(\hat{\omega} - 1) = s^2, \quad \hat{\gamma} + \hat{\beta}\exp[\sqrt{\ln \hat{\omega}}E(Z_{1,n})] = x_1 \tag{4.6.2}$$

where $\sqrt{\ln \omega} = \sigma$. After a few simple algebraic manipulations, the three equations of (4.6.2) are reduced to

$$\frac{s^2}{(\bar{x} - x_1)^2} = \frac{\hat{\omega}(\hat{\omega} - 1)}{[\sqrt{\hat{\omega}} - \exp\{\sqrt{\ln \hat{\omega}}\, E(Z_{1,n})\}]^2} = J(n, \hat{\omega})$$

$$\hat{\beta} = s[\hat{\omega}(\hat{\omega} - 1)]^{-\frac{1}{2}}, \qquad \hat{\gamma} = \bar{x} - s(\hat{\omega} - 1)^{-\frac{1}{2}} \tag{4.6.3}$$

With $\bar{x}$, s^2, and x_1 available from sample data, we calculate $s^2/(\bar{x} - x_1)^2$ and solve the first equation of (4.6.3) for $\hat{\omega}$. We subsequently calculate $\hat{\beta}$ and $\hat{\gamma}$ from the second and third equations of (4.6.3). Estimates $\hat{\mu}$ and $\hat{\sigma}$ follow as

$$\hat{\mu} = \ln \hat{\beta} \quad \text{and} \quad \hat{\sigma} = \sqrt{\ln \hat{\omega}} \tag{4.6.4}$$

Since $\hat{\gamma} < x_1$, the inequality $\hat{\omega} < 1 + s^2/(\bar{x} - x_1)^2$ follows from the third equation of (4.6.3). Thus we have an upper bound on $\hat{\omega}$ which can facilitate solution of the first equation of (4.6.3) for this estimate. In the special case where ω and thus σ are known, we calculate $\hat{\beta}$ and $\hat{\gamma}$ from (4.6.3) with ω substituted for $\hat{\omega}$.

As with ME and LMLE, we need to solve a nonlinear estimating equation for a single unknown. In this instance the first equation of (4.6.3) involves only the single unknown $\hat{\omega}$. Again, the Newton-Raphson technique is satisfactory for this purpose. Various other iterative schemes might serve as well. A simple trial-and-error scheme with linear interpolation will often suffice. We need only find two values ω_i and ω_j in a sufficiently narrow interval such that

$$J(n, \omega_i) \gtrless \frac{s^2}{(\bar{x} - x_1)^2} \gtrless J(n, \omega_j)$$

and then interpolate for the final estimate $\hat{\omega}$, where $J(n, \omega)$ is defined in (4.6.3) above. The only problem likely to be encountered is that of finding a good (close) first approximation. To meet this requirement, Cohen et al. (1985) presented a table and a chart of the function $J(n, \omega)$ for values of σ and n that are most likely to be encountered in practice. These aids to computation are reproduced here as Table 4.3 and Figure 4.1.

In using these aids, we enter the table or the chart with the sample values $s^2/(\bar{x} - x_1)^2$ and n. Inverse interpolation in the table or a direct reading from the chart yields a first approximation σ_1 that is sufficiently accurate to serve as the final estimate in many applications. When greater accuracy is required, this approximation is close enough to the required solution to produce rapid convergence of the iterative process to the final estimate. With $\hat{\sigma}$ and/or $\hat{\omega}$ thus calculated, we calculate $\hat{\beta}$ and $\hat{\gamma}$ from the last two equations of (4.6.3).

TABLE 4.3 The Lognormal Estimating Function

$$J(n,\sigma) = \omega(\omega-1)/[\sqrt{\omega} - \exp\{\sqrt{\ln \omega}\, E(Z_{1,n})\}]^2;\ \sigma = \sqrt{\ln \omega}$$

σ \ n	10	15	20	25	30	35	40	45
.01	.42615	.33575	.29061	.26272	.24341	.22906	.21786	.2088
.02	.43007	.33975	.29457	.26664	.24727	.23288	.22164	.2125
.03	.43411	.34386	.29864	.27065	.25123	.23679	.22551	.2163
.04	.43826	.34807	.30281	.27476	.25529	.24080	.22947	.2203
.05	.44253	.35239	.30708	.27897	.25945	.24491	.23354	.2243
.10	.46574	.37572	.33011	.30168	.28186	.26706	.25546	.2460
.15	.49230	.40216	.35617	.32736	.30722	.29214	.28030	.2706
.20	.52257	.43212	.38565	.35643	.33593	.32055	.30845	.2986
.25	.55700	.46603	.41901	.38933	.36845	.35275	.34038	.3303
.30	.59609	.50444	.45678	.42660	.40532	.38928	.37663	.3663
.35	.64046	.54793	.49958	.46885	.44715	.43077	.41783	.4072
.40	.69079	.59723	.54811	.51682	.49467	.47793	.46470	.4538
.45	.74792	.65317	.60322	.57132	.54871	.53161	.51807	.5070
.50	.81279	.71670	.66587	.63334	.61026	.59278	.57894	.5676
.55	.88655	.78897	.73719	.70402	.68045	.66259	.64844	.6368
.60	.97050	.87129	.81852	.78467	.76061	.74237	.72792	.7160
.65	1.06620	.96522	.91142	.87688	.85231	.83369	.81894	.8068
.70	1.17548	1.07261	1.01773	.98248	.95741	.93840	.92335	.9110
.75	1.30051	1.19562	1.13962	1.10364	1.07806	1.05868	1.04333	1.0307
.80	1.44387	1.33683	1.27966	1.24295	1.21686	1.19710	1.18145	1.1686
.85	1.60860	1.49928	1.44091	1.40345	1.37685	1.35672	1.34079	1.3277
.90	1.79834	1.68661	1.62700	1.58879	1.56168	1.54119	1.52499	1.5117
.94	1.97105	1.85730	1.79667	1.75785	1.73034	1.70955	1.69314	1.6797
.95	2.01746	1.90319	1.84230	1.80333	1.77571	1.75486	1.73840	1.7249
.96	2.06524	1.95045	1.88931	1.85018	1.82247	1.80154	1.78502	1.7715
.97	2.11446	1.99914	1.93774	1.89845	1.87064	1.84964	1.83307	1.8195
.98	2.16515	2.04929	1.98763	1.94819	1.92027	1.89921	1.88259	1.8690
.99	2.21736	2.10097	2.03904	1.99945	1.97143	1.95029	1.93362	1.9200
1.00	2.27116	2.15422	2.09202	2.05228	2.02416	2.00295	1.98623	1.9726
1.05	2.56573	2.44598	2.38244	2.34191	2.31329	2.29174	2.27477	2.2609
1.10	2.90875	2.78605	2.72112	2.67980	2.65068	2.62879	2.61158	2.5976
1.15	3.30941	3.18361	3.11725	3.07513	3.04551	3.02328	3.00584	2.9916
1.20	3.77887	3.64980	3.58196	3.53903	3.50891	3.48635	3.46868	3.4543
1.25	4.33073	4.19823	4.12886	4.08509	4.05447	4.03159	4.01369	3.9992
1.30	4.98165	4.84552	4.77456	4.72995	4.69882	4.67561	4.65750	4.6428
1.35	5.75206	5.61211	5.53950	5.49401	5.46236	5.43883	5.42050	5.4057
1.40	6.66717	6.52316	6.44882	6.40244	6.37026	6.34639	6.32785	6.3129
1.45	7.75811	7.60980	7.53366	7.48634	7.45362	7.42941	7.41064	7.3955
1.50	9.06354	8.91066	8.83263	8.78433	8.75104	8.72649	8.70750	8.6922
1.55	10.63161	10.47387	10.39383	10.34451	10.31064	10.28573	10.26650	10.2511
1.60	12.52251	12.35959	12.27743	12.22703	12.19254	12.16725	12.14778	12.1322
1.65	14.81178	14.64330	14.55889	14.50735	14.47221	14.44652	14.42680	14.4111
1.70	17.59453	17.42010	17.33329	17.28054	17.24471	17.21860	17.19860	17.1827
1.75	20.99102	20.81020	20.72082	20.66678	20.63021	20.60365	20.58335	20.5672
2.00	54.27287	54.05161	53.94619	53.88409	53.84292	53.81349	53.79132	53.7739

TABLE 4.3 The Lognormal Estimating Function (continued)

σ \ n	50	55	60	65	70	75	80	85
.01	.20131	.19494	.18946	.18467	.18044	.17666	.17327	.1701
.02	.20501	.19862	.19311	.18829	.18404	.18024	.17682	.1737
.03	.20881	.20238	.19685	.19201	.18773	.18391	.18047	.1773
.04	.21270	.20625	.20068	.19581	.19151	.18767	.18421	.1810
.05	.21669	.21021	.20461	.19972	.19539	.19153	.18805	.1849
.10	.23823	.23159	.22584	.22082	.21637	.21239	.20881	.2055
.15	.26267	.25585	.24995	.24479	.24021	.23612	.23243	.2290
.20	.29039	.28340	.27735	.27204	.26733	.26312	.25932	.2558
.25	.32189	.31471	.30850	.30304	.29821	.29387	.28996	.2864
.30	.35769	.35033	.34395	.33835	.33338	.32892	.32489	.3212
.35	.39843	.39088	.38434	.37858	.37348	.36890	.36476	.3609
.40	.44483	.43709	.43038	.42448	.41924	.41454	.41029	.4064
.45	.49774	.48981	.48293	.47688	.47151	.46669	.46233	.4583
.50	.55813	.55002	.54298	.53678	.53127	.52634	.52187	.5178
.55	.62717	.61887	.61166	.60533	.59969	.59464	.59007	.5859
.60	.70618	.69770	.69034	.68386	.67810	.67294	.66827	.6640
.65	.79674	.78809	.78057	.77396	.76809	.76282	.75805	.7537
.70	.90070	.89188	.88422	.87748	.87150	.86613	.86128	.8568
.75	1.02025	1.01126	1.00346	.99661	.99052	.98506	.98012	.9756
.80	1.15796	1.14882	1.14089	1.13392	1.12773	1.12218	1.11718	1.1126
.85	1.31690	1.30761	1.29955	1.29248	1.28620	1.28058	1.27550	1.2708
.90	1.50071	1.49128	1.48311	1.47594	1.46958	1.46389	1.45875	1.4540
.94	1.66857	1.65904	1.65078	1.64354	1.63712	1.63138	1.62620	1.6214
.95	1.71375	1.70419	1.69592	1.68866	1.68223	1.67647	1.67128	1.6665
.96	1.76031	1.75073	1.74243	1.73516	1.72871	1.72294	1.71774	1.7130
.97	1.80829	1.79868	1.79036	1.78307	1.77661	1.77083	1.76562	1.7608
.98	1.85774	1.84810	1.83977	1.83246	1.82599	1.82020	1.81498	1.8102
.99	1.90870	1.89905	1.89069	1.88337	1.87688	1.87108	1.86585	1.8611
1.00	1.96124	1.95156	1.94319	1.93585	1.92935	1.92354	1.91830	1.9135
1.05	2.24945	2.23966	2.23119	2.22379	2.21723	2.21137	2.20610	2.2013
1.10	2.58595	2.57606	2.56751	2.56004	2.55343	2.54754	2.54223	2.5374
1.15	2.97992	2.96993	2.96131	2.95378	2.94713	2.94120	2.93587	2.9310
1.20	3.44248	3.43240	3.42372	3.41614	3.40946	3.40350	3.39815	3.3933
1.25	3.98722	3.97707	3.96833	3.96071	3.95399	3.94801	3.94264	3.9377
1.30	4.63077	4.62054	4.61174	4.60408	4.59734	4.59134	4.58596	4.5811
1.35	5.39352	5.38321	5.37437	5.36668	5.35991	5.35390	5.34852	5.3436
1.40	6.30062	6.29025	6.28136	6.27364	6.26685	6.26083	6.25544	6.2505
1.45	7.38317	7.37273	7.36380	7.35605	7.34924	7.34321	7.33782	7.3329
1.50	8.67978	8.66927	8.66029	8.65251	8.64569	8.63966	8.63426	8.6294
1.55	10.23852	10.22795	10.21892	10.21111	10.20428	10.19823	10.19284	10.1879
1.60	12.11953	12.10888	12.09981	12.09198	12.08512	12.07907	12.07368	12.0688
1.65	14.39827	14.38754	14.37842	14.37055	14.36367	14.35761	14.35221	14.3473
1.70	17.16976	17.15895	17.14977	17.14187	17.13497	17.12889	17.12349	17.1186
1.75	20.55419	20.54329	20.53405	20.52610	20.51918	20.51308	20.50767	20.5028
2.00	53.75998	53.74844	53.73874	53.73045	53.72329	53.71703	53.71150	53.7065

σ \ n	90	95	100	125	150	200	400	1000[a]
.01	.16738	.16480	.16243	.15283	.14576	.13582	.11654	.1076
.02	.17090	.16831	.16592	.15625	.14913	.13911	.11965	.1106
.03	.17451	.17190	.16950	.15976	.15259	.14249	.12286	.1137
.04	.17822	.17559	.17317	.16337	.15614	.14596	.12616	.1169
.05	.18202	.17938	.17694	.16707	.15979	.14953	.12956	.1203
.10	.20259	.19986	.19734	.18713	.17959	.16892	.14807	.1383
.15	.22601	.22320	.22060	.21004	.20223	.19115	.16942	.1592
.20	.25270	.24980	.24712	.23621	.22812	.21664	.19403	.1834
.25	.28314	.28015	.27739	.26613	.25776	.24588	.22240	.2113
.30	.31788	.31480	.31195	.30034	.29170	.27942	.25510	.2436
.35	.35755	.35437	.35144	.33948	.33058	.31791	.29278	.2809
.40	.40287	.39962	.39660	.38430	.37515	.36209	.33620	.3239
.45	.45472	.45138	.44828	.43566	.42625	.41284	.38622	.3736
.50	.51408	.51065	.50748	.49454	.48489	.47115	.44387	.4309
.55	.58210	.57859	.57534	.56210	.55224	.53818	.51030	.4971
.60	.66013	.65654	.65323	.63971	.62964	.61529	.58689	.5735
.65	.74975	.74609	.74272	.72894	.71868	.70408	.67523	.6616
.70	.85282	.84910	.84566	.83164	.82122	.80640	.77719	.7635
.75	.97153	.96775	.96425	.95003	.93946	.92445	.89496	.8812
.80	1.10845	1.10462	1.10108	1.08667	1.07598	1.06082	1.03113	1.0173
.85	1.26667	1.26279	1.25920	1.24463	1.23384	1.21857	1.18876	1.1749
.90	1.44982	1.44590	1.44228	1.42758	1.41671	1.40135	1.37152	1.3577
.94	1.61719	1.61324	1.60960	1.59481	1.58389	1.56850	1.53870	1.5250
.95	1.66226	1.65830	1.65465	1.63985	1.62892	1.61352	1.58374	1.5701
.96	1.70870	1.70474	1.70109	1.68627	1.67533	1.65993	1.63016	1.6165
.97	1.75657	1.75261	1.74895	1.73411	1.72316	1.70776	1.67801	1.6644
.98	1.80591	1.80194	1.79828	1.78342	1.77247	1.75706	1.72734	1.7137
.99	1.85677	1.85280	1.84913	1.83426	1.82330	1.80789	1.77820	1.7646
1.00	1.90921	1.90523	1.90155	1.88667	1.87571	1.86030	1.83064	1.8171
1.05	2.19694	2.19294	2.18925	2.17431	2.16334	2.14795	2.11848	2.1051
1.10	2.53303	2.52901	2.52531	2.51034	2.49937	2.48403	2.45482	2.4417
1.15	2.92664	2.92260	2.91889	2.90392	2.89298	2.87771	2.84882	2.8359
1.20	3.38889	3.38485	3.38114	3.36618	3.35528	3.34011	3.31158	3.2989
1.25	3.93337	3.92933	3.92562	3.91070	3.89984	3.88480	3.85667	3.8443
1.30	4.57669	4.57265	4.56894	4.55407	4.54328	4.52837	4.50068	4.4886
1.35	5.33924	5.33521	5.33151	5.31669	5.30598	5.29123	5.26400	5.2522
1.40	6.24617	6.24215	6.23845	6.22371	6.21308	6.19849	6.17175	6.1602
1.45	7.32856	7.32455	7.32087	7.30620	7.29566	7.28124	7.25499	7.2438
1.50	8.62501	8.62101	8.61734	8.60276	8.59230	8.57806	8.55232	8.5414
1.55	10.18361	10.17961	10.17596	10.16145	10.15109	10.13703	10.11179	10.1012
1.60	12.06445	12.06047	12.05683	12.04241	12.03214	12.01825	11.99350	11.9832
1.65	14.34300	14.33903	14.33540	14.32105	14.31087	14.29716	14.27289	14.2628
1.70	17.11428	17.11032	17.10670	17.09243	17.08233	17.06878	17.04499	17.0352
1.75	20.49846	20.49450	20.49089	20.47669	20.46667	20.45328	20.42994	20.4204
2.00	53.70216	53.69817	53.69456	53.68049	53.67075	53.65796	53.63649	53.6281

[a] For entries in this column, $E(Z_{1:n})$ was approximated as $E(Z_{1:1000}) \doteq \Phi^{-1}[1/1001]$.

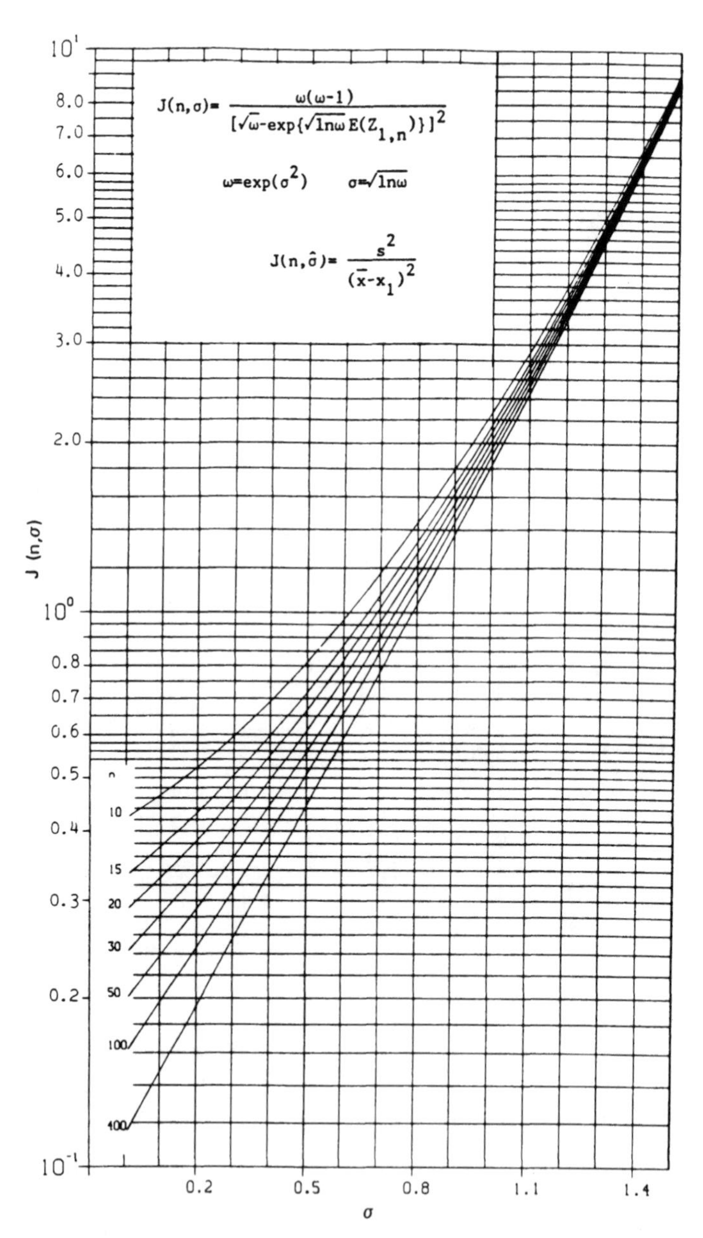

FIGURE 4.1 Graphs of the lognormal estimating function, $J(n, \sigma) = [\omega(\omega - 1)]/[\sqrt{\omega} - \exp\{\sqrt{\ln \omega}\ E(z_{1,n})\}]^2$; $\sigma = \sqrt{\ln \omega}$. Reproduced from Cohen, A. C., Whitten, B. J., and Ding, Y. (1985) with permission of the American Society for Quality Control.

TABLE 4.4 The Lognormal Hazard Function

z \ α_3	0	0.5	1.0	1.5	2.0	2.5	3.0	4.0	4.5
-3.00	.00444	.00019	.00000						
-2.80	.00794	.00079	.00000						
-2.60	.01365	.00268	.00000						
-2.40	.02258	.00751	.00005						
-2.20	.03597	.01791	.00116						
-2.00	.05525	.03719	.00851	.00000					
-1.80	.08189	.06860	.03306	.00130					
-1.60	.11735	.11449	.08528	.02708	.00001				
-1.40	.16288	.17561	.16744	.11823	.03017	.00000			
-1.20	.21944	.25109	.27302	.26742	.20861	.08143	.00003		
-1.00	.28760	.33870	.39146	.43600	.45783	.43457	.33160	.00000	
-.80	.36756	.43550	.51284	.59435	.67442	.74692	.80380	.81072	.69935
-.60	.45915	.53837	.62995	.72960	.83367	.93981	1.04659	1.25863	1.36208
-.40	.56188	.64440	.73851	.83934	.94299	1.04712	1.15064	1.35462	1.45522
-.20	.67507	.75112	.83646	.92579	1.01510	1.10212	1.18591	1.34316	1.41696
.00	.79788	.85658	.92323	.99259	1.06081	1.12588	1.18704	1.29771	1.34776
.20	.92942	.95932	.99913	1.04339	1.08803	1.13087	1.17099	1.24264	1.27449
.40	1.06876	1.05832	1.06490	1.08138	1.10232	1.12452	1.14639	1.18677	1.20501
.60	1.21503	1.15289	1.12149	1.10921	1.10753	1.11132	1.11777	1.13340	1.14141
.80	1.36740	1.24263	1.16991	1.12898	1.10635	1.09403	1.08761	1.08363	1.08389
1.00	1.52514	1.32736	1.21113	1.14239	1.10066	1.07441	1.05727	1.03768	1.03199
1.20	1.68755	1.40703	1.24603	1.15073	1.09181	1.05354	1.02750	.99542	.98512
1.40	1.85406	1.48171	1.27542	1.15507	1.08075	1.03216	.99870	.95657	.94266
1.60	2.02413	1.55152	1.30001	1.15622	1.06818	1.01072	.97107	.92082	.90406
1.80	2.19731	1.61666	1.32043	1.15483	1.05459	.98952	.94470	.88785	.86884
2.00	2.37322	1.67733	1.33723	1.15142	1.04036	.96875	.91960	.85738	.83657
2.20	2.55150	1.73377	1.35087	1.14641	1.02575	.94853	.89575	.82914	.80691
2.40	2.73186	1.78621	1.36176	1.14013	1.01097	.92892	.87310	.80291	.77954
2.60	2.91406	1.83490	1.37027	1.13284	.99615	.90996	.85159	.77849	.75420
2.80	3.09787	1.88004	1.37670	1.12476	.98142	.89166	.83116	.75568	.73067
3.00	3.28310	1.92189	1.38131	1.11608	.96684	.87403	.81174	.73433	.70876
3.20	3.46959	1.96063	1.38434	1.10692	.95248	.85705	.79328	.71432	.68830
3.40	3.65720	1.99649	1.38599	1.09741	.93837	.84071	.77570	.69550	.66914
3.60	3.84581	2.02964	1.38643	1.08764	.92455	.82498	.75895	.67778	.65117
3.80	4.03531	2.06027	1.38581	1.07769	.91103	.80984	.74298	.66106	.63427
4.00	4.22561	2.08855	1.38428	1.06762	.89782	.79527	.72774	.64526	.61834
4.20	4.41662	2.11464	1.38194	1.05748	.88493	.78124	.71318	.63029	.60330
4.40	4.60827	2.13868	1.37890	1.04733	.87237	.76772	.69925	.61610	.58908
4.60	4.80051	2.16080	1.37524	1.03719	.86013	.75470	.68592	.60261	.57560
4.80	4.99327	2.18115	1.37105	1.02709	.84821	.74215	.67315	.58979	.56281
5.00	5.18650	2.19982	1.36639	1.01705	.83660	.73004	.66090	.57757	.55065
5.20	5.38017	2.21695	1.36133	1.00710	.82530	.71836	.64915	.56592	.53908
5.40	5.57424	2.23262	1.35592	.99724	.81430	.70708	.63785	.55479	.52805
5.60	5.76867	2.24694	1.35020	.98750	.80359	.69618	.62699	.54415	.51752
5.80	5.96342	2.25998	1.34422	.97788	.79317	.68565	.61655	.53397	.50746
6.00	6.15848	2.27185	1.33801	.96838	.78302	.67547	.60649	.52421	.49784
L.L.	$-\infty$	-6.05456	-3.10380	-2.14491	.67765	-1.40316	-1.22290	-1.00000	-.92589

Source: Reproduced from Cohen, A. C. (1987).

In many practical applications, MME are the preferred estimators. These estimators are unbiased with respect to the population mean and variance. They are easy to calculate, and their variances are minimal or at least near minimal. They do not suffer from regularity problems. However, instances will arise in which an investigator might prefer MLE or LMLE. In those cases the MME will at least provide excellent first approximations from which to begin iterations toward the LMLE. A table of the hazard function (more complete than the abridged table included in Chapter 2) of the standardized lognormal distribution $(\alpha_3; 0, 1)$, which is of interest in reliability applications, is included here as Table 4.4.

4.6.1 Sampling Errors

Simulation results reported by Cohen and Whitten (1980) and by Cohen et al. (1985) disclose that asymptotic estimate variances and covariances calculated using (4.5.2) are reasonably close to corresponding simulated variances and covariances of the MME. These results may therefore be used in practical applications to calculate approximate variances and covariances of the MME.

4.7 ILLUSTRATIVE EXAMPLES

As illustrative examples, we choose the same examples that were employed in Chapter 3 to illustrate estimation in the Weibull distribution. The data are recorded in Chapter 3 and, except for summaries, will not be repeated here.

Example 4.7.1. For this example, $n = 20$, $\bar{x} = 0.423125$, $s^2 = 0.0156948$, $s = 0.1252789$, $a_3 = 1.0673243$, $x_1 = 0.265$, and $s^2/(\bar{x} - x_1)^2 = 0.627704$.

This time, we assume our sample to be from a three-parameter lognormal distribution. Accordingly, we enter the graphs of Figure 4.1 with $n = 20$ and $s^2/(\bar{x} - x_1)^2 = 0.628$ (rounded off) and read $\sigma_1 = 0.47$. This same value is obtained by inverse linear interpolation in Table 4.3 between $\sigma = 0.45$ and $\sigma = 0.50$. For additional accuracy, we make further calculations of $J(n, \omega)$ and interpolate as follows:

σ	ω	$J(n, \omega)$
0.4710	1.24837	0.6286
0.4703	1.24755	0.6277
0.4700	1.24720	0.6273

As a final estimate, we have $\hat{\sigma} = 0.4703$, which is subsequently rounded off to 0.470. We use the second and third equations of (4.6.3) to calculate

TABLE 4.5 Estimates for Example 4.7.1: Lognormal Distribution

Estimator	$\hat{\gamma}$	$\hat{\sigma}$	$\hat{\omega}$	$\hat{\mu}$	$\hat{\beta}$	$\hat{E}(X)$	$\sqrt{\hat{V}(X)}$	$\hat{\alpha}_3(X)$
ME	0.0572	0.3330	1.1173	-1.0607	0.3462	0.4231	0.1253	1.0673
LMLE	0.1850	0.5073	1.2935	-1.5606	0.2100	0.4238	0.1294	1.7843
MME	0.1714	0.4703	1.2476	-1.4899	0.2254	0.4231	0.1253	1.6159

Asymptotic standard deviations calculated from (4.5.2): $\sigma_{\hat{\gamma}} = 0.063$, $\sigma_{\hat{\sigma}} = 0.190$, $\sigma_{\hat{\beta}} = 0.075$, $\sigma_{\hat{\mu}} = 0.358$. From the central limit theorem $\sigma_{\bar{X}} = 0.028$.

$\hat{\gamma} = 0.171$ and $\hat{\beta} = 0.226$. It follows from (4.6.4) that $\hat{\mu} = -1.487$. Of course, $\hat{\omega} = 1.247$. With $\hat{\sigma} = 0.4703$, the MME in Table 4.5 differ slightly from these values.

A FORTRAN program in Appendix A.4.2 was used to calculate MME that agree with those presented above. This program was also used to calculate ME and MLE. For comparisons, the three sets of estimates are listed in Table 4.5.

Asymptotic standard deviations of the estimates calculated using (4.5.2) with the LMLE estimate $\sigma = 0.507$ and $\beta = 0.210$ are approximately $\sigma_{\hat{\gamma}} = 0.063$, and $\sigma_{\hat{\mu}} = 0.358$, $\sigma_{\hat{\beta}} = 0.075$, and $\sigma_{\hat{\sigma}} = 0.190$. Observe that differences between the MME and LMLE are quite small. Comparable differences between these estimators have been observed by the authors in most applications where the MME were calculated.

Example 4.7.2. For this example, $n = 10$, $\bar{x} = 220.48$, $s^2 = 6147.444$, $s = 78.405638$, $a_3 = 1.8635835$, $x_1 = 152.7$, and $s^2/(\bar{x} - x_1{}^2) = 1.338109$. To calculate the MME, we enter the graphs of Figure 4.1 with $n = 10$ and $s^2/(\bar{x} - x_1)^2$ rounded off to 1.34 and read $\sigma_1 = 0.76$. Interpolation in Table 4.2 between entries for $\sigma = 0.75$ and $\sigma = 0.80$ also yields this same value. By calculating additional values of $J(n, \omega)$ for $\sigma = 0.760$ and 0.770, and then interpolating, as was done in Example 4.7.1, we obtain as a final

TABLE 4.6 Estimates for Example 4.7.2: Lognormal Distribution

Estimator	$\hat{\gamma}$	$\hat{\sigma}$	$\hat{\omega}$	$\hat{\mu}$	$\hat{\beta}$	$\hat{E}(X)$	$\sqrt{\hat{V}(X)}$	$\hat{\alpha}_3(X)$
ME	80.9732	0.5239	1.3159	4.8009	121.6157	220.4800	78.4056	1.8636
MME	132.3829	0.7638	1.7921	4.1866	65.8086	220.4800	78.4056	3.3749

Asymptotic standard deviations not applicable since estimating equations failed to yield MLE. From the central limit theorem, $\sigma_{\bar{X}} = 24.79$.

estimate $\hat{\sigma} = 0.764$. We then use the last two equations of (4.6.3) to calculate $\hat{\beta} = 65.8$ and $\hat{\gamma} = 132.4$. Computer calculations using the FORTRAN program differed but slightly from these values for the MME, and also produced ME. The LMLE failed to produce estimates from this sample. Failure of the LMLE is not an uncommon occurrence for small samples. Sometimes failures occur even for large samples. This limitation of the LMLE is one of the principal reasons for favoring the MME. Computer calculations of the ME and the MME from Example 4.7.2 are displayed in Table 4.6.

Example 4.7.3. The Shook data given for Example 3.11.3 is treated here as though it were a random sample from a lognormal distribution with pdf (4.2.1). In summary, for this sample, we have $n = 1000$, $\bar{x} = 118.74$, $s = 16.9175$, $a_3 = 0.976424$, $a_4 = 5.3206$, $x_1 = 78.70$, and $s^2/(\bar{x} - x_1)^2 = 0.17852$. To calculate MME for the lognormal parameters, we enter Table 4.3 with $n = 1000$ and $s^2/(\bar{x} - x_1)^2 = 0.17852$ and interpolate to obtain $\hat{\sigma} = 0.1899$. It then follows that $\hat{\omega} = 1.0367$, $\hat{\alpha}_3 = 0.5819$, $\hat{\alpha}_4 = 3.608$, $\hat{\gamma} = 30.455$, $\hat{\beta} = 86.686$, and $\hat{\mu} = \ln \hat{\beta} = 4.462$. A close approximation to the above value of $\hat{\sigma}$ might have been obtained from the graphs of Figure 4.1.

4.8 REFLECTIONS

The superiority of the MME over ME and MLE in estimating parameters of the three-parameter lognormal distribution has been demonstrated in numerous practical applications. Moment estimators have the disadvantage that they are not uniquely determined by their moments and that inherently large sampling errors of the third moment introduce correspondingly larger sampling errors into parameter estimates. Maximum likelihood estimators lead to inadmissible estimates, and LMLE lead to questionable variance-covariance matrices. The MME are unbiased with respect to distribution mean and variance. They are easy to calculate by using the aids provided here. They are applicable over the entire parameter space, and they avoid the problems associated with ME and MLE. Therefore, unless α_3 is quite small, use of the MME is recommended. If α_3 is small, then the normal rather than the lognormal distribution would be a better choice as a model.

5

The Inverse Gaussian Distribution

5.1 BACKGROUND

The inverse Gaussian (IG) is a "long-tailed" positively skewed distribution that has recently received considerable attention as a model in various reliability studies. Its shape is similar to that of the lognormal distribution. Comparisons with the lognormal, Weibull, and gamma distributions with respect to thresholds and modes are provided in Tables 1.1 and 1.2. This distribution was originally derived as the first passage of time distribution of Brownian motion with positive drift. In reliability and life-span applications it is primarily useful when there is substantial skewness. It is applicable in the physical, management, and biological sciences.

The IG distribution has been studied extensively by numerous investigators, beginning with Schrödinger (1915) and including Smoluchowsky (1915), Wald (1944), Tweedie (1956, 1957a, b), Wasan (1968), Wasan and Roy (1969), Chhikara and Folks (1974), Folks and Chhikara (1978), Padgett and Wei (1979), Cheng and Amin (1981), Chan et al. (1983, 1984), Cohen and Whitten (1985). Except for Padgett and Wei, Cheng and Amin, Chan et al., and Cohen and Whitten, who considered three-parameter distributions, the previous writers were concerned with one- or two-parameter distributions with zero as the origin. An excellent expository account of this distribution and its properties is given by Johnson and Kotz (1970).

This presentation is primarily concerned with parameter estimation in the three-parameter version of the IG distribution with parameters γ, μ, σ, where γ is the threshold or origin, $\mu + \gamma$ is the mean, and σ is the standard deviation. Although maximum likelihood estimators (MLE) and moment estimators (ME) are considered to be the traditional or standard estimators, emphasis here is placed on the modified moment estimators (MME) of Cohen and Whitten (1985).

5.2 THE PROBABILITY DENSITY FUNCTION

In the parameterization employed by Tweedie (1956), the two-parameter probability density function (pdf) is

$$g(y; \mu, \lambda) = \left(\frac{\lambda}{2\pi y^3}\right)^{\frac{1}{2}} \exp\left[\frac{-\lambda}{2\mu^2} \frac{(y-\mu)^2}{y}\right], \quad y > 0,\ \lambda > 0,\ \mu > 0$$
$$= 0 \quad \text{elsewhere} \tag{5.2.1}$$

The three-parameter pdf as written by Cheng and Amin (1981) follows from the transformation $Y = X - \gamma$ as

$$f(x; \gamma, \mu, \lambda) = \left(\frac{\lambda}{2\pi(x-\gamma)^3}\right)^{\frac{1}{2}} \exp\left[\frac{-\lambda}{2\mu^2} \frac{(x-\gamma-\mu)^2}{x-\gamma}\right], \quad x > \gamma,\ \lambda > 0,\ \mu > 0$$
$$= 0 \quad \text{elsewhere} \tag{5.2.2}$$

The parametrization employed here is that of Chan et al. (1983). When we set $\lambda = \mu^3/\sigma^2$, it follows from (5.2.2) that

$$f(x; \gamma, \mu, \sigma) = \frac{1}{\sigma\sqrt{2\pi}}\left(\frac{\mu}{x-\gamma}\right)^{3/2} \exp\left\{-\frac{1}{2}\left(\frac{\mu}{x-\gamma}\right)\left[\frac{(x-\gamma)-\mu}{\sigma}\right]^2\right\}$$
$$x > \gamma,\ \mu > 0,\ \sigma > 0$$
$$= 0 \quad \text{elsewhere} \tag{5.2.3}$$

The expected value (mean), variance, third standard moment (a measure of skewness), and fourth standard moment follow from (5.2.3) as

$$E(X) = \gamma + \mu, \qquad V(X) = \sigma^2$$
$$\alpha_3(X) = \sqrt{\beta_1} = \frac{3\sigma}{\mu} \tag{5.2.4}$$
$$\alpha_4(X) = \beta_2 = 3 + \frac{5}{3}\beta_1$$

Accordingly, γ is the threshold parameter, $\gamma + \mu$ is the mean, σ is the standard deviation, and $\alpha_3 = 3\sigma/\mu$ becomes a shape parameter. In the notation used here, α_4 is the fourth standard moment, and β_1 and β_2 are Pearson's notation for α_3^2 and α_4. Note that β_2 is a linear function of β_1. The

β_1-β_2 line for the IG distribution as shown in Figure 1.4 lies in the Type VI region of Pearson's β_1-β_2 plane, approximately midway between corresponding lines for the lognormal and gamma (Type III) distributions.

The standardized pdf of the IG distribution $(0, 1, \alpha_3)$ as derived by Chan et al. (1983), where $Z = [X - E(X)]/\sigma$, is

$$g(z; 0, 1, \alpha_3) = \frac{1}{\sqrt{2\pi}}\left(\frac{3}{3 + \alpha_3 z}\right)^{3/2} \exp\left\{-\frac{z^2}{2}\left(\frac{3}{3 + \alpha_3 z}\right)\right\}, \quad -\frac{3}{\alpha_3} < z < \infty$$

$$= 0 \quad \text{elsewhere} \tag{5.2.5}$$

We note that $g(0) = 1/\sqrt{2\pi}$ for all values of α_3. Furthermore, it is readily seen that the limiting form of the pdf of (5.2.5) as $\alpha_3 \to 0$ is

$$\lim_{\alpha_3 \to 0} g(z; 0, 1, \alpha_3) = \frac{1}{\sqrt{2\pi}} e^{-z^2/2}, \quad -\infty < z < \infty \tag{5.2.6}$$

which is the pdf of the standardized normal distribution (0, 1).

The mode of the standardized IG distribution can be obtained by setting the first derivative of the pdf of (5.2.5) equal to zero. We accordingly obtain

$$\text{Mo}(Z) = \frac{\sqrt{36 + \alpha_3^4} - \alpha_3^2 - 6}{2\alpha_3} \tag{5.2.7}$$

5.3 MAXIMUM LIKELIHOOD ESTIMATION

The likelihood function of a random sample consisting of observations $\{x_i\}$, $i = 1, 2, \ldots, n$, from a distribution with pdf (5.2.3) is

$$L(x_1, \ldots, x_n; \gamma, \mu, \sigma) = \left(\frac{1}{\sigma\sqrt{2\pi}}\right)^n \prod_{i=1}^{n}\left(\frac{\mu}{x_i - \gamma}\right)^{3/2} \exp\left[\frac{-\mu}{2\sigma^2}\sum_1^n \frac{(x_i - \gamma - \mu)^2}{x_i - \gamma}\right] \tag{5.3.1}$$

Without any loss of generality, the sample is assumed to be ordered in magnitude. On taking logarithms of (5.3.1), differentiating with respect to γ, μ, and σ in turn, and equating to zero, we obtain the MLE equations

$$\frac{\partial \ln L}{\partial \gamma} = \frac{3}{2}\sum_1^n (x_i - \gamma)^{-1} + \frac{n\mu}{2\sigma^2} - \frac{\mu^3}{2\sigma^2}\sum_1^n (x_i - \gamma)^{-2} = 0 \tag{5.3.2}$$

$$\frac{\partial \ln L}{\partial \mu} = \frac{3n}{2\mu} - \frac{1}{2\sigma^2}\sum_1^n \frac{(x_i - \gamma - \mu)^2}{x_i - \gamma} + \frac{n\mu}{\sigma^2} - \frac{\mu^2}{\sigma^2}\sum_1^n (x_i - \gamma)^{-1} = 0 \qquad (5.3.2)$$

(continued)

$$\frac{\partial \ln L}{\partial \sigma} = \frac{-n}{\sigma} + \frac{\mu}{\sigma^3}\sum_1^n \frac{(x_i - \gamma - \mu)^2}{x_i - \gamma} = 0$$

It is necessary to solve the three equations of (5.3.2) simultaneously for the estimates $\hat{\gamma}$, $\hat{\mu}$, and $\hat{\sigma}$. On eliminating μ and σ from these three equations, we obtain the following equation, which can be solved for $\hat{\gamma}$:

$$n + \frac{3(\bar{x} - \hat{\gamma})^2}{n}\left[\sum_1^n (x_i - \hat{\gamma})^{-1}\right]^2 - 3(\bar{x} - \hat{\gamma})\sum_1^n (x_i - \hat{\gamma})^{-1} - (\bar{x} - \hat{\gamma})^2 \sum_1^n (x_i - \hat{\gamma})^{-2} = 0 \qquad (5.3.3)$$

On eliminating σ from the last two equations of (5.3.2), we have

$$\hat{\mu} = \bar{x} - \hat{\gamma} \qquad (5.3.4)$$

In deriving (5.3.4) it has been demonstrated that when $\partial \ln L/\partial\mu = 0$ and $\partial \ln L/\partial\sigma = 0$, then $E(X) = \bar{x}$, and thus the MLE of the mean $\gamma + \mu$ is unbiased.

From the last equation of (5.3.2), $\hat{\sigma}$ follows as

$$\hat{\sigma} = (\bar{x} - \hat{\gamma})\left[\frac{\bar{x} - \hat{\gamma}}{n}\sum_1^n (x_i - \hat{\gamma})^{-1} - 1\right]^{\frac{1}{2}} \qquad (5.3.5)$$

Equation (5.3.3) in which $\hat{\gamma}$ is the only unknown and in which $\bar{x}$ is the sample mean can be solved by using standard iterative procedures or even by employing simple trial-and-error techniques coupled with linear interpolation. With $\hat{\gamma}$ thus determined, $\hat{\mu}$ follows from (5.3.4) and $\hat{\sigma}$ follows from (5.3.5).

An alternative form of (5.3.3), which might be easier to handle in some applications, is

$$n - (\bar{x} - \gamma)^2 S_1 + \frac{3S_2S_3}{n} = 0 \qquad (5.3.6)$$

where

$$S_1 = \sum_1^n (x_i - \gamma)^{-2}, \qquad S_2 = \sum_1^n (x_i - \gamma)^{-1}$$

$$S_3 = \sum_1^n \frac{(x_i - \bar{x})^2}{x_i - \gamma} = (\bar{x} - \gamma)^2 S_2 - n(\bar{x} - \gamma) \qquad (5.3.7)$$

Padgett and Wei (1979) derived equivalent MLE for μ and $\lambda = \mu^3/\sigma^2$ and proved their existence provided $a_3 > 0$. Their results therefore guarantee that a solution exists for (5.3.3). They further guarantee the existence of $\hat{\mu}$ and $\hat{\sigma}$ as given by (5.3.4) and (5.3.5) provided that $a_3 > 0$. It is possible to obtain samples in which $a_3 < 0$ even though $\alpha_3 > 0$ when α_3 and/or n are small. The primary usefulness of the IG distribution, however, is in situations where α_3 is large. Consequently, existence or nonexistence problems are unlikely to prove troublesome in most practical applications. The occurrence of a sample with negative skewness might suggest that the IG distribution is an inappropriate model.

5.4 ASYMPTOTIC VARIANCES AND COVARIANCES

Asymptotic variances and covariances of the MLE $\hat{\gamma}$, $\hat{\mu}$, and $\hat{\sigma}$ can be obtained by inverting the Fisher information matrix in which elements are negatives of expected values of second partials of the loglikelihood function. Thereby, we obtain the variance-covariance matrix

$$V(\hat{\gamma}, \hat{\mu}, \hat{\sigma}) = \|a_{ij}\|^{-1}, \quad i, j = 1, 2, 3 \tag{5.4.1}$$

where

$$\begin{aligned} a_{11} &= \frac{n}{\sigma^2}\left[\frac{7\alpha_3^4}{54}\left(\frac{\alpha_3^2}{9} + 1\right) + \frac{\alpha_3^2}{2} + 1\right] \\ a_{22} &= \frac{n}{\sigma^2}\left(\frac{\alpha_3^2}{2} + 1\right) \\ a_{33} &= \frac{2n}{\sigma^2} \\ a_{12} &= a_{21} = \frac{n}{\sigma^2}\left[\frac{\alpha_3^2}{2}\left(\frac{\alpha_3^2}{9} + 1\right) + 1\right] \\ a_{13} &= a_{31} = -\frac{n}{\sigma^2}\alpha_3\left(\frac{\alpha_3^2}{9} + 1\right) \\ a_{23} &= a_{32} = -\frac{n}{\sigma^2}\alpha_3 \end{aligned} \tag{5.4.2}$$

In the special case of the two-parameter distribution with $\gamma = 0$, the variance-covariance matrix reduces to

$$V(\hat{\mu}, \hat{\sigma}) = \begin{bmatrix} \frac{n}{\sigma^2}\left(\frac{\alpha_3^2}{2} + 1\right) & -\frac{n}{\sigma^2}\alpha_3 \\ -\frac{n}{\sigma^2}\alpha_3 & \frac{2n}{\sigma^2} \end{bmatrix}^{-1} \tag{5.4.3}$$

TABLE 5.1 Variance-Covariance Factors for Maximum Likelihood Estimates of Inverse Gaussian Parameters

α_3	ϕ_{11}	ϕ_{33}	ϕ_{13}	ϕ_{23}
0.50	777.60000	0.60000	5.40000	5.15000
0.55	520.18733	0.62007	4.80812	4.53312
0.60	359.19540	0.64172	4.31034	4.01034
0.65	254.68604	0.66491	3.88573	3.56073
0.70	184.68582	0.68956	3.51929	3.16929
0.75	136.53333	0.71563	3.20000	2.82500
0.80	102.64044	0.74304	2.91955	2.51955
0.85	78.30306	0.77177	2.67155	2.24655
0.90	60.51803	0.80176	2.45098	2.00098
0.95	47.31804	0.83298	2.25385	1.77885
1.00	37.38462	0.86538	2.07692	1.57692
1.05	29.81606	0.89895	1.91755	1.39255
1.10	23.98443	0.93364	1.77352	1.22352
1.15	19.44521	0.96945	1.64299	1.06799
1.20	15.87907	1.00634	1.52439	0.92439
1.25	13.05348	1.04431	1.41639	0.79139
1.30	10.79709	1.08335	1.31784	0.66784
1.35	8.98215	1.12344	1.22775	0.55275
1.40	7.51246	1.16458	1.14523	0.44523
1.45	6.31489	1.20677	1.06954	0.34454
1.50	5.33333	1.25000	1.00000	0.25000
1.55	4.52442	1.29427	0.93602	0.16102
1.60	3.85435	1.33958	0.87708	0.07708
1.65	3.29660	1.38594	0.82271	-0.00229
1.70	2.83020	1.43335	0.77249	-0.07751
1.75	2.43851	1.48180	0.72605	-0.14895
1.80	2.10821	1.53131	0.68306	-0.21694
1.85	1.82858	1.58188	0.64322	-0.28178
1.90	1.59098	1.63352	0.60625	-0.34375
1.95	1.38836	1.68622	0.57192	-0.40308
2.00	1.21500	1.74000	0.54000	-0.46000
2.25	0.64831	2.02524	0.41026	-0.71474
2.50	0.36593	2.33824	0.31765	-0.93235
2.75	0.21650	2.67964	0.25014	-1.12486
3.00	0.13333	3.05000	0.20000	-1.30000
3.25	0.08500	3.44977	0.16210	-1.46290
3.50	0.05584	3.87931	0.13300	-1.61700
3.75	0.03766	4.33890	0.11034	-1.76466
4.00	0.02601	4.82877	0.09247	-1.90753
4.25	0.01833	5.34909	0.07819	-2.04681
4.50	0.01317	5.90000	0.06667	-2.18333
5.00	0.00713	7.09404	0.04954	-2.45046
6.00	0.00245	9.85294	0.02941	-2.97059
7.00	0.00099	13.10854	0.01882	-3.48118
8.00	0.00045	16.86226	0.01274	-3.98726

$\phi_{22} = \phi_{11} + 1$ $\qquad$ $\phi_{12} = -\phi_{11}$

and it follows that

$$V(\hat{\mu}) = \frac{\sigma^2}{n}, \qquad V(\hat{\sigma}) = \frac{\sigma^2}{2n}\left(\frac{\alpha_3^2}{2} + 1\right) \qquad (5.4.4)$$

$$\text{Cov}(\hat{\mu}, \hat{\sigma}) = \frac{\sigma^2}{2n}\alpha_3$$

When $\alpha_3 = 0$ the well-known normal distribution estimate variances and covariance follow from (5.4.4) as

$$V(\hat{\mu}) = \frac{\sigma^2}{n}, \qquad V(\hat{\sigma}) = \frac{\sigma^2}{2n}, \qquad \text{Cov}(\hat{\mu}, \hat{\sigma}) = 0 \qquad (5.4.5)$$

When the three-parameter variance-covariance matrix (5.4.1) is expanded, the asymptotic variances and covariances in this general case can be written as (cf. Cohen and Whitten 1985)

$$V(\hat{\gamma}) = \frac{\sigma^2}{n}\phi_{11}, \qquad V(\hat{\mu}) = \frac{\sigma^2}{n}\phi_{22}, \qquad V(\hat{\sigma}) = \frac{\sigma^2}{n}\phi_{33}$$

$$\text{Cov}(\hat{\gamma}, \hat{\mu}) = \frac{\sigma^2}{n}\phi_{12}, \qquad \text{Cov}(\hat{\gamma}, \hat{\sigma}) = \frac{\sigma^2}{n}\phi_{13}, \qquad \text{Cov}(\hat{\mu}, \hat{\sigma}) = \frac{\sigma^2}{n}\phi_{23} \qquad (5.4.6)$$

where

$$\phi_{11} = \frac{2}{D}, \qquad \phi_{22} = \phi_{11} + 1, \qquad \phi_{33} = \frac{BC - E^2}{D}$$

$$\phi_{12} = -\phi_{11}, \qquad \phi_{13} = \frac{\alpha_3^3}{9D}, \qquad \phi_{23} = -\alpha_3\frac{(C - AE)}{D} \qquad (5.4.7)$$

and

$$A = \frac{\alpha_3^2}{9} + 1, \qquad B = \frac{\alpha_3^2}{2} + 1, \qquad C = \frac{7}{54}\alpha_3^4 A + B$$

$$E = \frac{\alpha_3^2}{2}A + 1, \qquad D = 2(C - 1) - \alpha_3^2 A^2 \qquad (5.4.8)$$

The ϕ_{ij} are thus functions of α_3 alone, and in order to facilitate the calculation of estimate variances and covariances, these functions are entered in Table 5.1 for selected values of α_3. It must be remembered that these asymptotic variances and covariances are strictly applicable only for the MLE. However, simulation results presented by Chan et al. (1984) and by Cohen and Whitten (1985) indicate that they closely approximate corresponding variances and covariances of the MME.

These results enable us to calculate asymptotic $(1 - \alpha)100\%$ confidence intervals on any parameter θ as $\hat{\theta} \pm z_{\alpha/2}\sqrt{\mathrm{Var}(\hat{\theta})}$, where $z_{\alpha/2}$ is the standard normal variate. For a 95% confidence interval, $z_{0.025} = -1.96$. The central limit theorem enables us to approximate a 95% confidence interval on the distribution mean $(\mu_X = \gamma + \mu)$ as

$$\bar{x} - 1.96\frac{\hat{\sigma}}{\sqrt{n}} < \mu_X < \bar{x} + 1.96\frac{\hat{\sigma}}{\sqrt{n}}$$

The variances and covariances given by (5.4.6) might not be reliable for small values of α_3. In particular, $V(\hat{\gamma})$ becomes quite large as $\alpha_3 \to 0$. This restriction, however, should not affect most practical applications, since the IG distribution is unlikely to be employed as a model unless $\alpha_3 >> 1$.

5.5 MOMENT ESTIMATORS

On equating the first three sample moments to corresponding distribution moments as given by (5.2.4), we obtain

$$\gamma + \mu = \bar{x}, \quad \sigma^2 = s^2, \quad \frac{3\sigma}{\mu} = a_3 \tag{5.5.1}$$

Moment estimators σ^*, μ^*, and γ^* follow as

$$\sigma^* = s, \quad \mu^* = \frac{3s}{a_3}, \quad \gamma^* = \bar{x} - \frac{3s}{a_3} \tag{5.5.2}$$

Since μ must be positive, we note that ME exist only if $a_3 > 0$. It has previously been noted that existence of MLE is guaranteed when $a_3 > 0$.

Although ME are quite easy to calculate, they are often unacceptable because of the large sampling errors which a_3 introduces into ME of μ and γ. However, the ME of γ might be useful as a first approximation in solving (5.3.3) for the MLE $\hat{\gamma}$.

5.6 MODIFIED MOMENT ESTIMATORS

In MME the third moment, because of its large sampling errors, is replaced by a function of the first-order statistic. The reason for choosing the first-order statistic is the same as that which prompted this choice in other skewed distributions. The first-order statistic contains more information concerning the threshold parameter γ than do any of the other sample observations, often more than all the other observations combined. Chan et al. (1984) first proposed the MME considered here. Cohen and Whitten (1985) investigated these estimators further and presented tabular and graphic aids which greatly simplify their calculation in practical applications. The presentation which follows is an elaboration based on the Cohen-Whitten paper. Estimating equations for the MME are

$$E(X) = \bar{x}, \quad V(X) = s^2, \quad E[F(X_1)] = F(x_1) \tag{5.6.1}$$

where $F(\cdot)$ is the cdf of the IG distribution, X_1 is the first-order statistic in a random sample of size n, x_1 is the smallest sample observation, $\bar{x}$ is the sample mean, and s^2 is the sample variance (unbiased). It is well known that $E[F(X_1)] = 1/(n+1)$, and the estimating equations may thus be written as

$$\hat{\gamma} + \hat{\mu} = \bar{x}, \quad \hat{\sigma}^2 = s^2, \quad G(z_1; 0, 1, \hat{\alpha}_3) = \frac{1}{n+1} \tag{5.6.2}$$

where

$$z_1 = \frac{x_1 - \bar{x}}{s} \tag{5.6.3}$$

and, as given by Chan et al. (1983),

$$G(z; 0, 1, \alpha_3) = \Phi\left[\frac{z}{\sqrt{1 + (\alpha_3/3)z}}\right] + \exp\left(\frac{18}{\alpha_3^2}\right) \cdot \Phi\left[\frac{-(z + 6/\alpha_3)}{\sqrt{1 + (\alpha_3/3)z}}\right] \tag{5.6.4}$$

where $G(\cdot)$ is the cdf of the standard IG distribution with mean 0 unit variance and with skewness α_3. The symbol $\Phi(\ )$ denotes the cdf of the standard normal distribution (0, 1). We note that $F(x_1) = G(z_1)$. Thus for any given values of n and z_1, the third equation of (5.6.1) can be solved for $\hat{\alpha}_3$. With $\hat{\alpha}_3$ thus determined estimating equations become

$$\hat{\sigma}^2 = s^2, \quad \hat{\mu} = \frac{3s}{\hat{\alpha}_3}, \quad \hat{\gamma} = \bar{x} - \frac{3s}{\hat{\alpha}_3} \tag{5.6.5}$$

The estimate $\hat{\mu}$ is derived from the relation $\alpha_3 = 3\sigma/\mu$.

Since $\hat{\gamma} < x_1$, the inequality $\hat{\alpha}_3 < 3s/(\bar{x} - x_1)$ follows from the third equation of (5.6.5). Thus we have an upper bound on $\hat{\alpha}_3$ which can facilitate solution of the third equation of (5.6.1) for this estimate.

5.6.1 Computational Procedures

In solving the third equation of (5.6.1) for $\hat{\alpha}_3$ we might resort to inverse interpolation in the cdf tables of Chan et al. (1984). However, tables and a

TABLE 5.2 α_3 as a Function of z_1 and n for the Inverse Gaussian Distribution

z_1 \ n	5	10	20	25	30	40	50	100	250	500	1000
-.30	9.51790	9.68114	9.76891	9.78860	9.80258	9.82152	9.83406	9.86431	9.89123	9.90564	9.91679
-.32	8.85896	9.03428	9.12825	9.14931	9.16426	9.18450	9.19789	9.23021	9.25895	9.27433	9.28623
-.34	8.27317	8.46083	8.56107	8.58351	8.59943	8.62098	8.63524	8.66962	8.70019	8.71655	8.72919
-.36	7.74824	7.94852	8.05511	8.07894	8.09584	8.11871	8.13383	8.17030	8.20270	8.22003	8.23343
-.38	7.27446	7.48767	7.60069	7.62592	7.64381	7.66801	7.68401	7.72256	7.75680	7.77511	7.78926
-.40	6.84405	7.07054	7.19007	7.21672	7.23560	7.26114	7.27802	7.31867	7.35475	7.37404	7.38895
-.42	6.45070	6.69083	6.81698	6.84506	6.86495	6.89183	6.90959	6.95235	6.99029	7.01056	7.02623
-.44	6.08924	6.34342	6.47629	6.50582	6.52672	6.55496	6.57361	6.61849	6.65829	6.67955	6.69597
-.46	5.75538	6.02407	6.16376	6.19475	6.21667	6.24628	6.26583	6.31285	6.35451	6.37676	6.39394
-.48	5.44553	5.72922	5.87586	5.90832	5.93128	5.96227	5.98273	6.03189	6.07543	6.09867	6.11661
-.50	5.15667	5.45591	5.60962	5.64358	5.66758	5.69997	5.72133	5.77266	5.81807	5.84231	5.86101
-.52	4.88623	5.20161	5.36252	5.39800	5.42306	5.45685	5.47914	5.53263	5.57994	5.60517	5.62464
-.54	4.63199	4.96416	5.13243	5.16944	5.19557	5.23079	5.25400	5.30968	5.35889	5.38512	5.40536
-.56	4.39203	4.74174	4.91750	4.95607	4.98329	5.01994	5.04409	5.10198	5.15309	5.18033	5.20134
-.58	4.16468	4.53274	4.71617	4.75632	4.78463	4.82273	4.84782	4.90794	4.96097	4.98922	5.01100
-.60	3.94846	4.33577	4.52704	4.56880	4.59822	4.63779	4.66384	4.72619	4.78115	4.81041	4.83296
-.62	3.74206	4.14963	4.34894	4.39232	4.42286	4.46393	4.49093	4.55554	4.61244	4.64272	4.66605
-.64	3.54428	3.97326	4.18080	4.22584	4.25752	4.30009	4.32807	4.39495	4.45380	4.48510	4.50921
-.66	3.35406	3.80573	4.02170	4.06843	4.10127	4.14536	4.17432	4.24349	4.30431	4.33663	4.36152
-.68	3.17038	3.64619	3.87084	3.91928	3.95330	3.99893	4.02889	4.10037	4.16317	4.19651	4.22219
-.70	2.99230	3.49392	3.72749	3.77767	3.81289	3.86009	3.89104	3.96487	4.02965	4.06403	4.09049
-.72	2.81891	3.34826	3.59101	3.64297	3.67940	3.72818	3.76016	3.83634	3.90311	3.93853	3.96579
-.74	2.64931	3.20860	3.46082	3.51460	3.55226	3.60265	3.63565	3.71421	3.78300	3.81947	3.84751
-.76	2.48257	3.07443	3.33642	3.39204	3.43096	3.48298	3.51703	3.59799	3.66880	3.70632	3.73516
-.78	2.31772	2.94524	3.21733	3.27484	3.31503	3.36872	3.40382	3.48721	3.56006	3.59863	3.62827
-.80	2.15369	2.82060	3.10314	3.16258	3.20408	3.25944	3.29561	3.38146	3.45636	3.49599	3.52643
-.82	1.98922	2.70010	2.99347	3.05489	3.09771	3.15479	3.19204	3.28036	3.35734	3.39803	3.42928
-.84	1.82277	2.58337	2.88799	2.95142	2.99559	3.05441	3.09277	3.18359	3.26265	3.30441	3.33647
-.86	1.65233	2.47007	2.78637	2.85187	2.89742	2.95801	2.99748	3.09084	3.17200	3.21484	3.24771
-.88	1.47504	2.35987	2.68834	2.75596	2.80292	2.86530	2.90591	3.00183	3.08511	3.12903	3.16271
-.90	1.28643	2.25246	2.59364	2.66342	2.71182	2.77604	2.81779	2.91631	3.00172	3.04673	3.08123
-.92	1.07847	2.14758	2.50202	2.57403	2.62390	2.68998	2.73290	2.83405	2.92161	2.96771	3.00304
-.94	.83324	2.04493	2.41327	2.48756	2.53894	2.60692	2.65103	2.75483	2.84456	2.89176	2.92792
-.96	.48283	1.94426	2.32718	2.40382	2.45673	2.52666	2.57197	2.67847	2.77039	2.81870	2.85568
-.98		1.84532	2.24357	2.32263	2.37711	2.44902	2.49556	2.60478	2.69891	2.74833	2.78615

z_1 \ n	5	10	20	25	30	40	50	100	250	500	1000
-1.00		1.74785	2.16226	2.24380	2.29991	2.37383	2.42162	2.53360	2.62996	2.68050	2.71916
-1.02		1.65161	2.08309	2.16720	2.22496	2.30095	2.35000	2.46479	2.56339	2.61506	2.65456
-1.04		1.55634	2.00592	2.09267	2.15212	2.23022	2.28056	2.39819	2.49906	2.55187	2.59221
-1.06		1.46179	1.93059	2.02007	2.08127	2.16152	2.21317	2.33369	2.43684	2.49079	2.53199
-1.08		1.36768	1.85699	1.94928	2.01227	2.09472	2.14772	2.27115	2.37662	2.43172	2.47378
-1.10		1.27373	1.78498	1.88018	1.94501	2.02972	2.08408	2.21048	2.31828	2.37454	2.41746
-1.12		1.17964	1.71444	1.81266	1.87938	1.96640	2.02215	2.15156	2.26172	2.31915	2.36293
-1.14		1.08505	1.64527	1.74661	1.81529	1.90467	1.96184	2.09431	2.20685	2.26545	2.31011
-1.16		.98960	1.57737	1.68194	1.75263	1.84443	1.90306	2.03863	2.15358	2.21336	2.25889
-1.18		.89282	1.51062	1.61856	1.69131	1.78560	1.84571	1.98443	2.10181	2.16279	2.20920
-1.20		.79421	1.44493	1.55636	1.63126	1.72810	1.78972	1.93165	2.05149	2.11367	2.16096
-1.22		.69312	1.38022	1.49529	1.57239	1.67185	1.73501	1.88020	2.00252	2.06591	2.11410
-1.24		.58875	1.31639	1.43524	1.51463	1.61678	1.68152	1.83003	1.95486	2.01947	2.06854
-1.26		.48007	1.25335	1.37615	1.45790	1.56282	1.62918	1.78105	1.90842	1.97426	2.02424
-1.28		.36568	1.19103	1.31795	1.40215	1.50990	1.57792	1.73322	1.86316	1.93024	1.98113
-1.30		.24360	1.12933	1.26057	1.34729	1.45798	1.52768	1.68648	1.81902	1.88734	1.93914
-1.32		.11083	1.06818	1.20393	1.29328	1.40698	1.47842	1.64077	1.77594	1.84552	1.89824
-1.34			1.00749	1.14797	1.24005	1.35686	1.43008	1.59605	1.73387	1.80473	1.85837
-1.36			.94719	1.09264	1.18755	1.30757	1.38260	1.55226	1.69278	1.76491	1.81948
-1.38			.88720	1.03787	1.13573	1.25905	1.33595	1.50937	1.65261	1.72603	1.78154
-1.40			.82743	.98359	1.08452	1.21126	1.29007	1.46732	1.61332	1.68805	1.74450
-1.42			.76779	.92976	1.03389	1.16415	1.24493	1.42609	1.57488	1.65092	1.70831
-1.44			.70822	.87631	.98377	1.11768	1.20048	1.38562	1.53724	1.61461	1.67296
-1.46			.64861	.82319	.93413	1.07182	1.15669	1.34589	1.50037	1.57908	1.63839
-1.48			.58889	.77033	.88491	1.02651	1.11351	1.30687	1.46425	1.54430	1.60458
-1.50			.52895	.71769	.83608	.98172	1.07092	1.26851	1.42883	1.51024	1.57150
-1.52			.46869	.66520	.78758	.93742	1.02888	1.23079	1.39409	1.47688	1.53911
-1.54			.40801	.61281	.73937	.89357	.98735	1.19368	1.36000	1.44417	1.50740
-1.56			.34681	.56045	.69142	.85014	.94631	1.15715	1.32654	1.41210	1.47633
-1.58			.28494	.50808	.64366	.80709	.90572	1.12118	1.29367	1.38065	1.44587
-1.60			.22229	.45562	.59607	.76439	.86556	1.08573	1.26137	1.34977	1.41601
-1.62			.15871	.40301	.54860	.72201	.82579	1.05080	1.22962	1.31946	1.38672
-1.64			.09404	.35020	.50120	.67991	.78640	1.01634	1.19840	1.28970	1.35798
-1.66			.02809	.29710	.45383	.63807	.74735	.98234	1.16769	1.26045	1.32976
-1.68				.24365	.40645	.59646	.70862	.94879	1.13746	1.23171	1.30206

TABLE 5.2 Continued

z_1 \ n	5	10	20	25	30	40	50	100	250	500	1000
-1.70				.18978	.35901	.55505	.67019	.91565	1.10770	1.20345	1.27485
-1.72				.13541	.31147	.51381	.63203	.88291	1.07840	1.17565	1.24810
-1.74				.08044	.26377	.47271	.59411	.85055	1.04952	1.14830	1.22182
-1.76				.02480	.21588	.43172	.55642	.81856	1.02106	1.12138	1.19597
-1.78					.16774	.39081	.51893	.78691	.99300	1.09487	1.17055
-1.80					.11930	.34996	.48163	.75558	.96532	1.06877	1.14554
-1.82					.07050	.30915	.44448	.72458	.93801	1.04305	1.12092
-1.84					.02129	.26833	.40746	.69387	.91106	1.01770	1.09668
-1.86						.22748	.37057	.66344	.88445	.99272	1.07281
-1.88						.18658	.33377	.63328	.85816	.96808	1.04929
-1.90						.14559	.29704	.60337	.83220	.94377	1.02612
-1.92						.10449	.26037	.57371	.80654	.91979	1.00328
-1.94						.06325	.22374	.54427	.78117	.89611	.98077
-1.96						.02183	.18711	.51506	.75609	.87274	.95856
-1.98							.15049	.48604	.73128	.84966	.93665
-2.00							.11384	.45722	.70673	.82686	.91503
-2.02							.07714	.42857	.68243	.80433	.89370
-2.04							.04038	.40010	.65837	.78207	.87263
-2.06							.00354	.37178	.63455	.76005	.85183
-2.08								.34361	.61095	.73828	.83129
-2.10								.31558	.58756	.71675	.81099
-2.12								.28768	.56439	.69545	.79093
-2.14								.25990	.54141	.67436	.77111
-2.16								.23222	.51863	.65349	.75150
-2.18								.20465	.49603	.63283	.73212
-2.20								.17716	.47361	.61237	.71295
-2.22								.14976	.45136	.59210	.69398
-2.24								.12242	.42927	.57202	.67521
-2.26								.09516	.40735	.55212	.65663
-2.28								.06795	.38557	.53239	.63824
-2.30								.04078	.36394	.51284	.62003
-2.40									.25776	.41740	.53152
-2.50									.15440	.32544	.44680
-2.60									.05328	.23641	.36534
-2.80										.06535	.21050

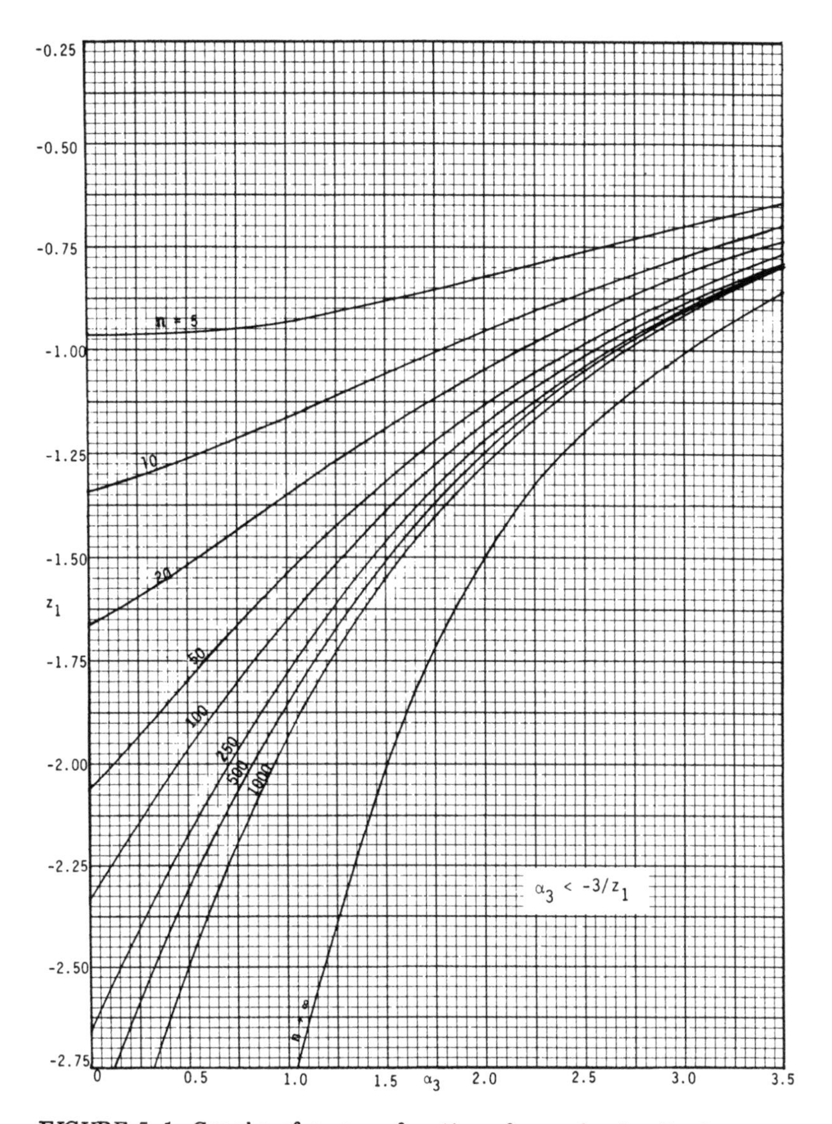

FIGURE 5.1 Graphs of α_3 as a function of z_1 and n for the inverse Gaussian distribution. Reproduced from Cohen and Whitten (1985) with permission of the American Society for Quality Control.

chart, given by Cohen and Whitten (1985) and reproduced here as Table 5.2 and Figure 5.1, in which $\hat{\alpha}_3$ is given as a function of n and z_1, greatly simplify the estimation procedure. For any given random sample of size n, we determine x_1, $\bar{x}$, s, and z_1. We then enter Table 5.2 or the graphs of

Figure 5.1 with z_1 and n and read $\hat{\alpha}_3$. We interpolate as necessary to improve the accuracy of $\hat{\alpha}_3$. Estimates $\hat{\sigma}$, $\hat{\mu}$, and $\hat{\gamma}$ then follow from (5.6.5). In many applications, these estimates will be sufficiently accurate to serve as final estimates. For greater accuracy, additional evaluations of the cdf as given by (5.6.4) might be made until two values $\alpha_{3(i)}$ and $\alpha_{3(j)}$ in a sufficiently narrow interval are found such that

$$G_{(i)} \lesseqgtr \frac{1}{n+1} \lesseqgtr G_{(j)}$$

In the event that MLE might be preferred, they can be calculated as described in Section 5.3 by using the MME as first approximations.

5.7 ILLUSTRATIVE EXAMPLES

Practical application of MME described here is illustrated with the same examples used in Chapters 3 and 4 to illustrate estimation for Weibull and lognormal parameters. The data are recorded in Chapter 3 and, except for summaries, will not be repeated here.

Example 5.7.1. For this example, $n = 20$, $\bar{x} = 0.423125$, $s^2 = 0.0156948$, $s = 0.1252789$, $a_3 = 1.0673243$, $x_1 = 0.265$, and $z_1 = (x_1 - \bar{x})/s = -1.262$.

From Figure 5.1, with $n = 20$ and $z_1 = -1.262$, we read $\hat{\alpha}_3 = 1.25$. This same value could be obtained by linear interpolation in Table 5.2. With $\hat{\alpha}_3 = 1.25$, we subsequently employ (5.6.5) to calculate $\hat{\sigma} = 0.1253$, $\hat{\mu} = 0.301$, and $\hat{\gamma} = 0.122$. These values should be sufficiently accurate for most purposes. They also provide close first approximations for an iterative solution of (5.6.2). For greater computational accuracy, the FORTRAN program of Chan et al. (1984) is available. For this example, computer calculations

TABLE 5.3 Estimates for Example 5.7.1: Inverse Gaussian Distribution

Estimator	$\hat{\gamma}$	$\hat{\mu}$	$\hat{E}(X)$	$\hat{\sigma}$	$\hat{\alpha}_3(X)$
ME	0.0710	0.3521	0.4231	0.1253	1.0673
MLE	0.1782	0.2449	0.4231	0.1268	1.5533
MME (approx)[a]	0.122	0.301	0.423	0.1253	1.25
MME (exact)[b]	0.1216	0.3015	0.4231	0.1253	1.2468

[a]Calculated by using graphs of Figure 5.1.
[b]Calculated by using authors' FORTRAN program. Asymptotic standard deviations calculated from (5.4.6): $\sigma_{\hat{\gamma}} = 0.060$, $\sigma_{\hat{\mu}} = 0.066$, $\sigma_{\hat{\sigma}} = 0.032$. From the central limit theorem, $\sigma_{\bar{X}} = 0.028$.

using that program gave the exact values displayed in Table 5.3. Note that these values differ only slightly from the approximations. For comparisons, MLE calculated using the authors' FORTRAN program are presented along with the MME and the ME. Asymptotic variances and covariances calculated from (5.4.6) are also given.

Example 5.7.2. For this example, $n = 10$, $\bar{x} = 220.48$, $s^2 = 6147.444$, $s = 78.405638$, $a_3 = 1.8635835$, $x_1 = 152.7$, and $z_1 = -0.864$.

From Figure 5.1 or by interpolation in Table 5.2 with $n = 10$ and $z_1 = -0.864$, we read $\hat{\alpha}_3 = 2.45$. From (5.6.5) we calculate $\hat{\sigma} = 78.406$, $\hat{\mu} = 96.0$, and $\hat{\gamma} = 124.5$. Computer computations using the FORTRAN program previously mentioned produced the exact values displayed in Table 5.4. Again these values differ only slightly from the approximations.

Maximum likelihood estimates are also presented in Table 5.4 along with the MME and ME, and the asymptotic variances and covariances from (5.4.6).

Example 5.7.3. The Shook data given for Example 3.11.3 will be considered here to be from an IG population. In summary, the sample data are $n = 1000$, $\bar{x} = 118.74$, $s = 16.9175$, $a_3 = 0.976424$, $a_4 = 5.3206$, $x_1 = 78.70$, and $z_1 = (x_1 - \bar{x})/s = -2.367$.

To calculate MME for the IG parameter, we enter Table 5.2 with $n = 1000$ and $z_1 = -2.367$ and interpolate to obtain $\hat{\alpha}_3 = 0.5607$. Since $\hat{\sigma} = s = 16.9175$, and since $\hat{\mu} = 3\hat{\sigma}/\hat{\alpha}_3$, it follows that $\hat{\mu} = 90.516$. It then follows that $\hat{\gamma} = \bar{x} - \hat{\mu} = 28.224$. We subsequently calculate $\hat{\alpha}_4 = 4.5890$.

A close approximation to the value obtained for $\hat{\alpha}_3$ might have been obtained from the graphs of Figure 5.1.

TABLE 5.4 Estimates for Example 5.7.2: Inverse Gaussian Distribution

Estimator	$\hat{\gamma}$	$\hat{\mu}$	$\hat{E}(X)$	$\hat{\sigma}$	$\hat{\alpha}_3(X)$
ME	94.2625	126.2175	220.4800	78.4056	1.8636
MLE	139.6586	80.8214	220.4800	77.3564	2.8714
MME (approx)[a]	124.5	96.0	220.5	78.4056	2.45
MME (exact)[b]	124.2820	96.1980	220.4800	78.4056	2.4451

[a]Calculated by using graphs of Figure 5.1.
[b]Calculated by using authors' FORTRAN program. Asymptotic standard deviations calculated from (5.4.6): $\sigma_{\hat{\gamma}} = 10.10$, $\sigma_{\hat{\mu}} = 26.46$, $\sigma_{\hat{\sigma}} = 41.34$. From the central limit theorem, $\sigma_{\bar{X}} = 24.79$.

5.8 REFLECTIONS

Moment estimates for the three-parameter IG distribution are easy to calculate, and their only major disadvantage is their rather large sampling error due to use of the third sample moment. Maximum likelihood estimates are only slightly more difficult to calculate than ME, and they enjoy the advantage of smaller sampling errors. Furthermore, the MLE are applicable over the entire parameter space. Asymptotic estimate variances and covariances, however, might not be reliable for small values of α_3. Modified moment estimators are easy to calculate, and in most instances they provide better estimates of the threshold parameter than either ME or MLE. The choice between the MME and the MLE with respect to scale and shape parameters is often a toss-up. Even in instances when the MLE might be preferred, the MME are useful as first approximations in iterative calculations of the MLE. Neither the MLE nor the MME are very good when α_3 is extremely small, but, as previously emphasized, the IG distribution is unlikely to be employed as a model unless $\alpha_3 >> 1$.

A table of the standardized IG cumulative distribution function has been included in Appendix A.3 as Table A.3.3.

6

The Gamma Distribution

6.1 BACKGROUND

The gamma distribution is positively skewed and along with the Weibull, lognormal, and inverse Gaussian (IG) distributions is also available as a model in reliability and life-span studies. In some respects it resembles the Weibull. When $\alpha_3 = 2$, the exponential is obtained as a special case from both of these distributions. Karl Pearson included the three-parameter gamma distribution as Type III of his generalized system of frequency distributions. Numerous writers have previously considered this distribution, but the list is much too long for inclusion here. Johnson and Kotz (1970) give an excellent expository account of the gamma distribution and its properties along with a list of 130 references at the end of Chapter 17, pp. 200-266. Bowman and Shenton (1988) also provide additional references.

Most previous writers have considered moment estimators (ME) and maximum likelihood estimators (MLE) to be the standard estimators for parameters of this distribution. Although ME are easy to calculate even in the three-parameter distribution, which is of primary concern here, large sampling errors of the third sample moment introduce corresponding large errors into ME of gamma parameters. Maximum likelihood estimators exhibit smaller sampling errors, but they are subject to regularity problems for some parameter values. Modified moment estimators (MME), which were presented in preceding chapters for the Weibull, lognormal, and IG distributions, provide alternatives which often enjoy advantages over both ME and MLE. The MME utilize information contained in the first-order statistic and thereby provide improved estimates of the threshold parameter in the gamma as well as in other skewed distributions.

6.2 THE DENSITY FUNCTION AND ITS CHARACTERISTICS

The probability density function (pdf) of the three-parameter gamma distribution may be written as

$$f(x;\gamma,\rho,\beta) = \frac{\beta^{-\rho}}{\Gamma(\rho)}(x-\gamma)^{\rho-1}\exp\left(\frac{-(x-\gamma)}{\beta}\right), \qquad \gamma < x < \infty,\ \rho > 0$$

$$= 0 \quad \text{otherwise} \tag{6.2.1}$$

The expected value (mean), variance, third standard moment, fourth standard moment, and mode (when $\rho > 1$) are

$$E(X) = \gamma + \rho\beta, \qquad V(X) = \rho\beta^2$$

$$\alpha_3(X) = \frac{2}{\sqrt{\rho}}, \qquad M_o(X) = \gamma + \beta(\rho - 1) \tag{6.2.2}$$

$$\alpha_4(X) = 3 + \frac{3}{2}\alpha_3^2$$

In this notation, γ is the threshold or location parameter, ρ is the shape parameter, and β is the scale parameter. The frequency curve of the gamma distribution is bell-shaped; that is, it has a discernible mode for $\rho > 1$ and thus for $\alpha_3 < 2$. It assumes a reverse J-shape for $\rho \leq 1$. As previously noted, it becomes the exponential distribution when $\rho = 1$.

When we make the standardizing transformation $Z = [X - E(X)]/\sqrt{V(X)}$, the pdf of the resulting standard gamma distribution $(0, 1, \alpha_3)$ with α_3 as the shape parameter becomes

$$g(z;0,1,\alpha_3) = \left(\frac{2}{\alpha_3}\right)^{4/\alpha_3^2}\left(z + \frac{2}{\alpha_3}\right)^{4/\alpha_3^2 - 1}\exp\left[-\frac{2}{\alpha_3}\left(z + \frac{2}{\alpha_3}\right)\right]$$

$$-\frac{2}{\alpha_3} < z < \infty$$

$$= 0 \quad \text{elsewhere} \tag{6.2.3}$$

We now turn our attention to ME, MLE, and MME, in turn.

6.3 MOMENT ESTIMATORS

The well-known ME are obtained by equating the first three sample moments to corresponding distribution moments. Thereby we have

$$\hat{\gamma} + \hat{\rho}\hat{\beta} = \bar{x}, \qquad \hat{\rho}\hat{\beta}^2 = s^2, \qquad \text{and} \qquad \frac{2}{\sqrt{\hat{\rho}}} = a_3 \tag{6.3.1}$$

It follows from (6.3.1) that

$$\hat{\rho} = \frac{4}{a_3^2}, \quad \hat{\beta} = \frac{sa_3}{2}, \quad \hat{\gamma} = \bar{x} - \frac{2s}{a_3} \tag{6.3.2}$$

where

$$\bar{x} = \frac{1}{n}\sum_1^n x_i$$

$$s^2 = \frac{1}{n-1}\sum_1^n (x_i - \bar{x})^2 \tag{6.3.3}$$

$$a_3 = \left[\sum_1^n \frac{(x_i - \bar{x})^3}{n}\right] \Big/ \left[\sum_1^n \frac{(x_i - \bar{x})^2}{n}\right]^{3/2}$$

Moment estimates fail to exist when a_3 is negative, and this might happen when α_3 is near zero and/or when sample sizes are small. This dilemma could be avoided by admitting a left-skewed reflection of the gamma distribution, but our concern here is with the right-skewed distribution with pdf (6.2.1). In most practical applications where skewness is minimal, the normal distribution would be preferred as a model and the question of existence is unlikely to be troublesome.

In practical applications, it sometimes happens that the moment estimate of γ is inadmissible since it is greater than the smallest sample observation x_1. This occurrence could be due to the fact that x_1 and perhaps some of the other small observations are "outliers." However, if the sample is quite large, and there is no reason to suspect the presence of outliers, we might wish to let $\gamma = x_1$ or $x_1 - \epsilon$ and then estimate ρ and β as

$$\hat{\beta} = \frac{s^2}{\bar{x} - x_1} \quad \text{and} \quad \hat{\rho} = \frac{\bar{x} - x_1}{\hat{\beta}} = \frac{(\bar{x} - x_1)^2}{s^2} \tag{6.3.4}$$

6.4 MAXIMUM LIKELIHOOD ESTIMATORS

When $\rho > 1$ and thus $\alpha_3 < 2$ (i.e., in the bell-shaped case), MLE equations based on random sample $\{x_i\}$, $i = 1, 2, \ldots, n$, are obtained by equating to zero partial derivatives of the loglikelihood function. We accordingly obtain

$$\frac{\partial \ln L}{\partial \gamma} = \frac{n}{\beta} - (\rho - 1)\sum_1^n (x_i - \gamma)^{-1} = 0 \tag{6.4.1}$$

$$\frac{\partial \ln L}{\partial \rho} = -n\psi(\rho) - n \ln \beta + \sum_{1}^{n} \ln(x_i - \gamma) = 0 \qquad (6.4.1)$$
(continued)

$$\frac{\partial \ln L}{\partial \beta} = \frac{-n\rho}{\beta} + \frac{1}{\beta^2} \sum_{1}^{n} (x_i - \gamma) = 0$$

where $\psi(\rho)$ is the digamma function, $\psi(\rho) = \partial \ln \Gamma(\rho)/\partial\rho$. Maximum likelihood estimates $\hat{\gamma}$, $\hat{\rho}$, and $\hat{\beta}$, are obtained as the simultaneous solution of the three equations of (6.4.1). Although the computation is more or less straightforward, as pointed out by Johnson and Kotz (1970), difficulties might be encountered when ρ is near to 1, even though it actually exceeds 1. Therefore Johnson and Kotz recommend that the MLE be employed only if $\rho > 2.5$ ($\alpha_3 < 1.265$). Note that the third equation of (6.4.1) reduces to $\gamma + \rho\beta = \bar{x}$.

A degree of simplification can be achieved by eliminating β from the first and third equations of (6.4.1) to obtain

$$\hat{\rho} = \left[1 - n \Big/ \left\{(\bar{x} - \hat{\gamma}) \sum_{1}^{n} (x_i - \hat{\gamma})^{-1}\right\}\right]^{-1}$$

$$\hat{\beta} = \frac{\bar{x} - \hat{\gamma}}{\hat{\rho}} \qquad (6.4.2)$$

6.4.1 The Two-Parameter Special Case with γ Known (= 0)

In this case, the second and third equations of (6.4.1) yield MLE equations

$$\ln \hat{\rho} - \psi(\hat{\rho}) = \ln \bar{x} - \frac{1}{n} \sum_{1}^{n} \ln x_i$$

$$\hat{\beta} = \frac{\bar{x}}{\hat{\rho}} \qquad (6.4.3)$$

The first equation of (6.4.3) can be solved iteratively for $\hat{\rho}$, and $\hat{\beta}$ then follows from the second of these equations. A first approximation to $\hat{\rho}$ for use in the iteration process might be read from the graph of Figure 6.1 or obtained by interpolation from Table 6.1. An alternative first approximation might be calculated from the first equation of (6.4.2) with $\hat{\gamma} = 0$.

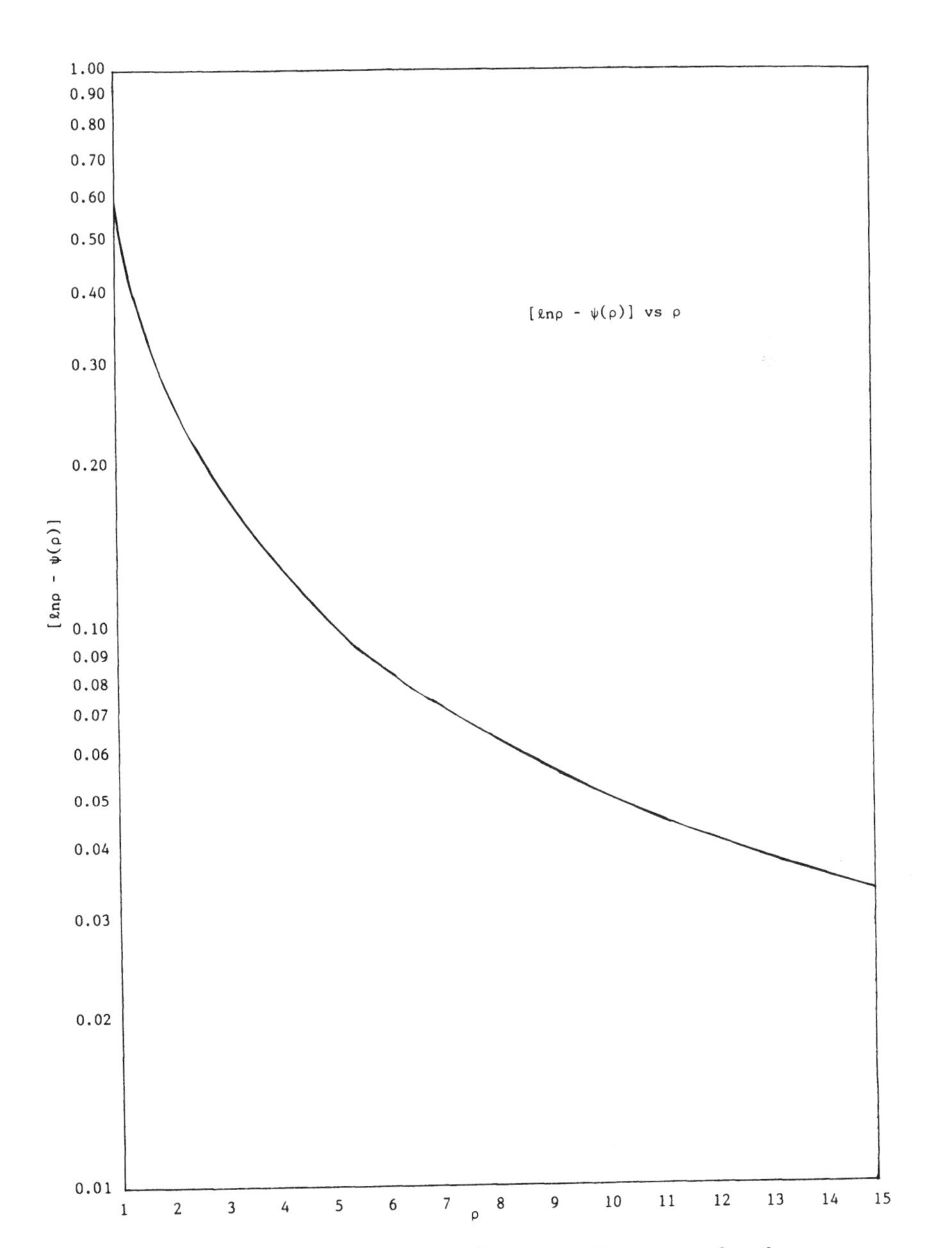

FIGURE 6.1 The estimating function $[\ln \rho - \psi(\rho)]$ versus ρ for the gamma distribution.

TABLE 6.1 The Estimating Function $[\ln \rho - \psi(\rho)]$

ρ	$[\ln \rho - \psi(\rho)]$	ρ	$[\ln \rho - \psi(\rho)]$	ρ	$[\ln \rho - \psi(\rho)]$
1	0.57721566	18	0.02803490	35	0.01435374
2	0.27036285	19	0.02654657	36	0.01395318
3	0.17582795	20	0.02520828	37	0.01357438
4	0.13017669	21	0.02399845	38	0.01321560
5	0.10332024	22	0.02289941	39	0.01287530
6	0.08564180	23	0.02189663	40	0.01255208
7	0.07312581	24	0.02097798	41	0.01224469
8	0.06380006	25	0.02013331	42	0.01195200
9	0.05658310	26	0.01935403	43	0.01167297
10	0.05083250	27	0.01863281	44	0.01140668
11	0.04614268	28	0.01796342	45	0.01115226
12	0.04224497	29	0.01734046	46	0.01090895
13	0.03895434	30	0.01675925	47	0.01067602
14	0.03613924	31	0.01621574	48	0.01045283
15	0.03370354	32	0.01570637	49	0.01023879
16	0.03157539	33	0.01522803	50	0.01003333
17	0.02970002	34	0.01477796	51	0.00983596

6.4.2 The Three-Parameter General Case with γ Unknown

In this case, we select a first approximation γ_1 and employ the first equation of (6.4.2) to calculate a corresponding value ρ_1. We subsequently calculate β_1 from the second equation of (6.4.2) by using γ_1 and ρ_1. Approximations γ_1, ρ_1, and β_1 are then substituted into the second equation of (6.4.1). If this result is zero, further calculations are unnecessary. Otherwise, we select a second approximation γ_2 and repeat the cycle of computations until we find two values γ_i and γ_j in a sufficiently narrow interval and such that $[\partial \ln L/\partial\rho]_i \gtrless 0 \gtrless [\partial \ln L/\partial\rho]_j$. Final estimates follow by linear interpolation. As a first approximation γ_1, we might choose the moment estimate γ^* as given in (6.3.2), provided $\gamma^* < x_1$, where x_1 is the smallest sample observation.

When $\rho \leq 1$ and thus $\alpha_3 \geq 2$ (i.e., in the reverse J-shape case), the likelihood function becomes infinite as $\gamma \to x_1$ (the smallest sample value), and MLE breaks down. Although we might set $\gamma = x_1$ and $E(X) = \bar{x}$, it follows that $\partial \ln L/\partial\rho \to -\infty$ as $\gamma \to x_1$ and thus $\hat{\rho}$ and $\hat{\beta}$ fail to exist. It is therefore necessary to use other estimators in this situation.

6.5 ASYMPTOTIC VARIANCES AND COVARIANCES

Asymptotic variances and covariances of the MLE $\hat{\gamma}$, $\hat{\beta}$, and $\hat{\rho}$ can be obtained by inverting the Fisher information matrix in which elements are negatives of expected values of second partials of the loglikelihood function. It thereby follows that

$$V(\hat{\gamma}, \hat{\beta}, \hat{\rho}) = \|a_{ij}\|^{-1}, \quad i, j = 1, 2, 3 \tag{6.5.1}$$

where

$$\begin{aligned}
a_{11} &= \frac{n}{\sigma^2}\frac{\rho}{\rho - 2} \\
a_{22} &= \frac{n}{\sigma^2}\rho^2 \\
a_{33} &= \frac{n}{\sigma^2}\rho\beta^2\psi'(\rho) \\
a_{12} &= a_{21} = \frac{n}{\sigma^2}\rho \\
a_{13} &= a_{31} = \frac{n}{\sigma^2}\left(\frac{\rho\beta}{\rho - 1}\right) \\
a_{23} &= a_{32} = \frac{n}{\sigma^2}\rho\beta
\end{aligned} \tag{6.5.2}$$

where $\psi'(\rho)$ is the trigamma function, i.e., the first derivative of the digamma function.

On carrying out the inversion of V, we have as asymptotic variances and covariances

$$V(\hat{\gamma}) = \frac{\sigma^2}{n}\phi_{11}, \quad V(\hat{\beta}) = \frac{\sigma^2}{n}\phi_{22}, \quad V(\hat{\rho}) = \frac{\sigma^2}{n\beta^2}\phi_{33} \tag{6.5.3}$$

$$\mathrm{Cov}(\hat{\gamma}, \hat{\beta}) = \frac{\sigma^2}{n}\phi_{12}, \quad \mathrm{Cov}(\hat{\gamma}, \hat{\rho}) = \frac{\sigma^2}{n\beta}\phi_{13}$$

$$\mathrm{Cov}(\hat{\beta}, \hat{\rho}) = \frac{\sigma^2}{n\beta}\phi_{23}$$

TABLE 6.2 Variance-Covariance Factors for Maximum Likelihood Estimates of Gamma Distribution Parameters[a]

α_3	ρ	ϕ_{11}	ϕ_{22}	ϕ_{33}	ϕ_{12}	ϕ_{13}	ϕ_{23}
0.32	39.0625	2026.5043	1.4746	8471.1637	53.7036	-4124.3022	-111.2797
0.34	34.6021	1564.8417	1.4714	6579.9761	47.0271	-3192.0778	-97.9103
0.36	30.8642	1223.9253	1.4679	5179.0026	41.4353	-2502.7921	-86.7092
0.38	27.7008	968.1494	1.4643	4124.3405	36.7061	-1984.9378	-77.2324
0.40	25.0000	773.5243	1.4605	3318.9110	32.6715	-1590.3115	-69.1440
0.42	22.6757	623.5494	1.4565	2695.8522	29.2026	-1285.7402	-62.1859
0.44	20.6612	506.6635	1.4523	2208.2496	26.1991	-1047.9671	-56.1577
0.46	18.9036	414.6290	1.4480	1822.6273	23.5819	-860.4125	-50.9012
0.48	17.3611	341.4876	1.4435	1514.7322	21.2883	-711.0755	-46.2906
0.50	16.0000	282.8694	1.4387	1266.7474	19.2675	-591.1488	-42.2249
0.52	14.7929	235.5289	1.4339	1065.4173	17.4785	-494.0867	-38.6219
0.54	13.7174	197.0275	1.4288	900.7639	15.8878	-414.9674	-35.4146
0.56	12.7551	165.5134	1.4236	765.1954	14.4677	-350.0503	-32.5474
0.58	11.8906	139.5666	1.4182	652.8780	13.1951	-296.4646	-29.9744
0.60	11.1111	118.0880	1.4127	559.2873	12.0509	-251.9866	-27.6571
0.62	10.4058	100.2198	1.4069	480.8844	11.0188	-214.8791	-25.5631
0.64	9.7656	85.2873	1.4011	414.8790	10.0851	-183.7744	-23.6651
0.66	9.1827	72.7559	1.3951	359.0545	9.2382	-157.5875	-21.9398
0.68	8.6505	62.1988	1.3889	311.6381	8.4681	-135.4519	-20.3671
0.70	8.1633	53.2735	1.3826	271.2024	7.7662	-116.6711	-18.9301
0.72	7.7160	45.7032	1.3761	236.5910	7.1252	-100.6816	-17.6138
0.74	7.3046	39.2632	1.3695	206.8618	6.5386	-87.0253	-16.4056
0.76	6.9252	33.7698	1.3628	181.2431	6.0009	-75.3273	--15.2942
0.78	6.5746	29.0721	1.3560	159.0995	5.5072	-65.2798	-14.2700
0.80	6.2500	25.0460	1.3490	139.9051	5.0532	-56.6283	-13.3243
0.82	5.9488	21.5883	1.3419	123.2229	4.6351	-49.1618	-12.4497
0.84	5.6689	18.6135	1.3347	108.6881	4.2497	-42.7046	-11.6395
0.86	5.4083	16.0499	1.3274	95.9948	3.8939	-37.1095	-10.8879
0.88	5.1653	13.8376	1.3200	84.8856	3.5653	-32.2532	-10.1896
0.90	4.9383	11.9260	1.3125	75.1431	3.2614	-28.0315	-9.5401
0.95	4.4321	8.1991	1.2933	55.6236	2.5968	-19.7084	-8.1034
1.00	4.0000	5.5950	1.2738	41.3553	2.0475	-13.7851	-6.8925
1.05	3.6281	3.7725	1.2539	30.8345	1.5927	-9.5510	-5.8663
1.10	3.3058	2.4987	1.2338	23.0221	1.2161	-6.5188	-4.9922
1.15	3.0246	1.6128	1.2137	17.1884	0.9047	-4.3493	-4.2449
1.20	2.7778	1.0021	1.1936	12.8143	0.6484	-2.8031	-3.6040
1.25	2.5600	0.5869	1.1738	9.5254	0.4386	-1.7097	-3.0530
1.30	2.3669	0.3104	1.1544	7.0487	0.2685	-.9459	-2.5784
1.35	2.1948	0.1318	1.1356	5.1835	0.1325	-.4225	-2.1692
1.40	2.0408	0.0219	1.1175	3.7806	0.0256	-.0741	-1.8162

[a]Valid only if $\rho > 2$.

where

$$\phi_{11} = \frac{\rho\psi'(\rho) - 1}{\rho M}, \qquad \phi_{22} = \frac{(\rho - 1)^2\psi'(\rho) - (\rho - 2)}{\rho(\rho - 1)^2(\rho - 2)M} \tag{6.5.4}$$

$$\phi_{33} = \frac{2}{\rho(\rho - 2)M}, \qquad \phi_{12} = \frac{1 - (\rho - 1)\psi'(\rho)}{\rho(\rho - 1)M}$$

$$\phi_{13} = \frac{-1}{\rho(\rho - 1)M}, \qquad \phi_{23} = \frac{-1}{\rho(\rho - 1)(\rho - 2)M}$$

where

$$M = \frac{2(\rho - 1)^2\psi'(\rho) - (2\rho - 3)}{(\rho - 1)^2(\rho - 2)} \tag{6.5.5}$$

The foregoing results agree with corresponding variances and covariances given by Johnson and Kotz (1970). Unfortunately, they are valid only when $\rho > 2$, but as already noted, the MLE are of doubtful utility unless $\rho > 2.5$. Consequently, when $\rho \leq 2.5$, it becomes important to consider other estimators, such as the MME. When they are applicable, the variances of (6.5.3) enable us to calculate approximate $1 - \alpha$ confidence intervals on the parameters as estimate $\pm z_{\alpha/2}\sqrt{V(\text{Est})}$, where $z_{\alpha/2}$ is the standard normal variate.

Note that the ϕ_{ij} are functions of ρ alone. In order to facilitate calculation of the variances and covariances, we give an abridged table (Table 6.2) of the ϕ_{ij} as functions of α_3, where $\alpha_3 = 2/\sqrt{\rho}$. We also note that for a two-parameter gamma distribution with $\gamma = 0$ known, then V is a 2×2 matrix and

$$V(\hat{\beta}) = \frac{\beta^2}{n}\left(\frac{\psi'(\rho)}{\rho\psi'(\rho) - 1}\right), \quad V(\hat{\rho}) = \frac{1}{n}\left(\frac{\rho}{\rho\psi'(\rho) - 1}\right) \tag{6.5.6}$$

$$\text{Cov}(\hat{\beta}, \hat{\rho}) = \frac{\sigma^2}{n\beta\rho}\left(\frac{-1}{\rho\psi'(\rho) - 1}\right) = \frac{-\beta}{n}\left(\frac{1}{\rho\psi'(\rho) - 1}\right)$$

In the case of the exponential distribution with $\rho = 1$ and $\gamma = 0$, then

$$V(\hat{\beta}) = \frac{\beta^2}{n} \tag{6.5.7}$$

6.6 MODIFIED MOMENT ESTIMATORS

The MME under consideration here were proposed in an effort to overcome the deficiencies previously noted with regard to ME and MLE for gamma distribution parameters. They are unbiased with respect to distribution mean and variance. They are applicable over the entire parameter space, and, as demonstrated in simulation studies by Cohen and Whitten (1982, 1986), their estimate variances are minimal or at least nearly minimal in comparison with corresponding variances of the ME and the MLE. With the aid of graphs and tables provided by Cohen and Whitten (1986), they are quite easy to calculate. The presentation which follows is an elaboration of this 1986 paper.

The estimating equations are $E(X) = \bar{x}$, $V(X) = s^2$, and $E[F(X_1)] = F(x_1)$, where $F(\cdot)$ is the cumulative distribution function, X_1 is the first-order statistic (a random variable), and x_1 is its observed (sample) value. $E(\cdot)$ denotes expected value. The modification thus consists of replacing the equation involving the third moment with one involving the first-order statistic. Since $E[F(X_1)] = 1/(n + 1)$, the MME equations become

$$\gamma + \rho\beta = \bar{x}, \qquad \rho\beta = s^2, \qquad F(x_1) = \frac{1}{n+1} \tag{6.6.1}$$

In order to make use of the third equation of (6.6.1), we employ the standardized pdf (6.2.3) and the standardized cdf (an incomplete gamma function), which we write as

$$G(z; 0, 1, \alpha_3) = \int_{-2/\alpha_3}^{z} g(t; 0, 1, \alpha_3)\, dt \tag{6.6.2}$$

With $\hat{E}(X) = \bar{x}$ and $\hat{\sigma}^2 = s^2$, we write $z_1 = (x_1 - \bar{x})/s$. Since $F(x_1) = G(z_1)$, it follows from (6.6.1) that

$$G(z_1; 0, 1, \hat{\alpha}_3) = \frac{1}{n+1} \qquad \hat{\beta} = \frac{s\hat{\alpha}_3}{2} \qquad \hat{\gamma} = \bar{x} - \frac{2s}{\hat{\alpha}_3} \qquad \hat{\rho} = \frac{4}{\hat{\alpha}_3^{\,2}} \tag{6.6.3}$$

The first equation of (6.6.3) must be solved for $\hat{\alpha}_3$. It could be solved directly for $\hat{\rho}$, but for comparisons with other skewed distributions, it seemed expedient to consider α_3 as the primary shape parameter and to consider ρ as a function of α_3. With $\hat{\alpha}_3$ thus determined, estimates $\hat{\beta}$, $\hat{\gamma}$, and $\hat{\rho}$ follow from the remaining equations of (6.6.3). As previously noted, $\hat{E}(X) = \bar{x}$ and $\hat{\sigma}^2 = s^2$.

Since $\gamma < x_1$, it follows from the third equation of (6.6.3) that $\hat{\alpha}_3 < 2s/(\bar{x} - x_1)$. This inequality provides an upper bound on $\hat{\alpha}_3$ which can facilitate solution of the first equation of (6.6.3) for this estimate. An equivalent lower bound on $\hat{\rho}$ follows since $\hat{\rho} > (\bar{x} - x_1)^2/s^2$, because $\alpha_3 = 2/\sqrt{\rho}$.

6.6.1 Computational Procedures

The first equation of (6.6.3) might be solved for $\hat{\alpha}_3$ by inverse interpolation in tables of the standardized gamma cdf, such as those included in the ap-

TABLE 6.3 Values of α_3 as a Function of z_1 and n for the Gamma Distribution

z_1 \ n	5	10	20	25	30	40	50	100	250	500	1000
-.50	3.9914	3.9993									
-.52	3.8327	3.8448									
-.54	3.6837	3.7013									
-.56	3.5428	3.5675									
-.58	3.4087	3.4423	3.4474								
-.60	3.2804	3.3245	3.3319	3.3325							
-.62	3.1567	3.2132	3.2235	3.2245	3.2250						
-.64	3.0367	3.1077	3.1216	3.1230	3.1237	3.1243					
-.66	2.9195	3.0072	3.0253	3.0273	3.0283	3.0293	3.0297				
-.68	2.8044	2.9111	2.9342	2.9368	2.9382	2.9396	2.9402				
-.70	2.6904	2.8188	2.8477	2.8511	2.8529	2.8548	2.8556				
-.72	2.5770	2.7299	2.7652	2.7696	2.7720	2.7744	2.7756	2.7772			
-.74	2.4632	2.6438	2.6864	2.6919	2.6949	2.6981	2.6996	2.7018			
-.76	2.3482	2.5602	2.6109	2.6176	2.6214	2.6254	2.6274	2.6303			
-.78	2.2309	2.4787	2.5383	2.5464	2.5510	2.5560	2.5585	2.5623			
-.80	2.1104	2.3990	2.4684	2.4779	2.4835	2.4896	2.4927	2.4975	2.4994		
-.82	1.9852	2.3208	2.4008	2.4120	2.4186	2.4259	2.4296	2.4357	2.4382		
-.84	1.8535	2.2438	2.3353	2.3483	2.3560	2.3646	2.3692	2.3766	2.3798		
-.86	1.7130	2.1677	2.2716	2.2866	2.2956	2.3056	2.3110	2.3200	2.3240	2.3250	
-.88	1.5602	2.0924	2.2096	2.2268	2.2371	2.2487	2.2549	2.2656	2.2706	2.2719	
-.90	1.3897	2.0175	2.1492	2.1685	2.1803	2.1936	2.2008	2.2134	2.2194	2.2210	2.2217
-.92	1.1919	1.9429	2.0900	2.1118	2.1250	2.1401	2.1484	2.1631	2.1703	2.1723	2.1732
-.94	.9453	1.8683	2.0320	2.0563	2.0712	2.0883	2.0977	2.1146	2.1231	2.1256	2.1267
-.96	.5713	1.7936	1.9751	2.0021	2.0187	2.0378	2.0484	2.0677	2.0777	2.0807	2.0821
-.98		1.7185	1.9191	1.9489	1.9673	1.9886	2.0005	2.0223	2.0339	2.0375	2.0392
-1.00		1.6429	1.8638	1.8967	1.9169	1.9405	1.9538	1.9783	1.9917	1.9959	1.9980
-1.02		1.5664	1.8093	1.8453	1.8675	1.8935	1.9082	1.9356	1.9508	1.9558	1.9582
-1.04		1.4890	1.7554	1.7947	1.8190	1.8475	1.8636	1.8940	1.9113	1.9170	1.9199
-1.06		1.4103	1.7019	1.7447	1.7712	1.8023	1.8200	1.8536	1.8730	1.8795	1.8829
-1.08		1.3302	1.6489	1.6954	1.7241	1.7580	1.7773	1.8142	1.8358	1.8432	1.8472
-1.10		1.2483	1.5962	1.6465	1.6777	1.7144	1.7354	1.7757	1.7996	1.8080	1.8126
-1.12		1.1645	1.5438	1.5981	1.6318	1.6714	1.6942	1.7381	1.7645	1.7739	1.7791
-1.14		1.0783	1.4916	1.5501	1.5863	1.6291	1.6537	1.7012	1.7302	1.7407	1.7466
-1.16		.9895	1.4396	1.5024	1.5413	1.5873	1.6137	1.6652	1.6968	1.7085	1.7150
-1.18		.8976	1.3876	1.4550	1.4967	1.5460	1.5744	1.6298	1.6643	1.6771	1.6844

(continued)

pendix as Table A.3.4. Earlier tables of the cdf were given by Salvosa (1936) and by Cohen et al. (1969). The Salvosa tables are limited to a maximum value of $\alpha_3 = 1.1$, but the Cohen-Helm-Sugg tables extend to a maximum of $\alpha_3 = 6.00$. Various computer programs are available for evaluating the incomplete gamma function of (6.6.2) when additional values of G( ) are required. Additional tables and a chart, which simplify this procedure, are reproduced from the Cohen-Whitten 1986 paper as Table 6.3 and Figure 6.2. Entries of $\hat{\alpha}_3$ are given directly as a function of z_1 and n in both table and chart. For any given random sample of size n from a gamma distribution, x_1, $\bar{x}$, s, and $z_1 = (x_1 - \bar{x})/s$ are required. We enter the table or the chart

TABLE 6.3 Continued

z_1 \ n	5	10	20	25	30	40	50	100	250	500	1000
-1.20		.8022	1.3356	1.4078	1.4524	1.5052	1.5355	1.5951	1.6324	1.6465	1.6546
-1.22		.7028	1.2836	1.3607	1.4084	1.4647	1.4972	1.5610	1.6013	1.6166	1.6256
-1.24		.5985	1.2315	1.3139	1.3647	1.4246	1.4592	1.5274	1.5708	1.5875	1.5973
-1.26		.4887	1.1793	1.2670	1.3211	1.3849	1.4217	1.4943	1.5409	1.5590	1.5697
-1.28		.3720	1.1268	1.2203	1.2777	1.3454	1.3845	1.4618	1.5116	1.5311	1.5428
-1.30		.2471	1.0742	1.1735	1.2344	1.3062	1.3476	1.4297	1.4828	1.5038	1.5165
-1.32		.1117	1.0212	1.1267	1.1913	1.2672	1.3111	1.3980	1.4545	1.4771	1.4908
-1.34			.9679	1.0798	1.1482	1.2285	1.2748	1.3667	1.4268	1.4508	1.4656
-1.36			.9142	1.0329	1.1051	1.1899	1.2387	1.3357	1.3994	1.4251	1.4409
-1.38			.8601	.9858	1.0621	1.1514	1.2029	1.3051	1.3725	1.3998	1.4168
-1.40			.8055	.9385	1.0190	1.1131	1.1673	1.2749	1.3460	1.3750	1.3931
-1.42			.7504	.8911	.9759	1.0749	1.1318	1.2449	1.3198	1.3505	1.3698
-1.44			.6947	.8434	.9327	1.0367	1.0965	1.2152	1.2940	1.3265	1.3469
-1.46			.6384	.7954	.8895	.9987	1.0613	1.1857	1.2685	1.3028	1.3245
-1.48			.5814	.7472	.8461	.9606	1.0262	1.1565	1.2434	1.2795	1.3024
-1.50			.5237	.6987	.8026	.9226	.9913	1.1275	1.2185	1.2565	1.2807
-1.52			.4652	.6498	.7589	.8846	.9564	1.0987	1.1940	1.2338	1.2593
-1.54			.4059	.6006	.7150	.8466	.9216	1.0701	1.1696	1.2114	1.2382
-1.56			.3456	.5509	.6710	.8086	.8868	1.0417	1.1456	1.1893	1.2174
-1.58			.2844	.5008	.6267	.7705	.8521	1.0135	1.1217	1.1674	1.1969
-1.60			.2221	.4503	.5822	.7324	.8174	.9854	1.0981	1.1458	1.1767
-1.62			.1587	.3992	.5374	.6942	.7827	.9575	1.0748	1.1245	1.1568
-1.64			.0941	.3476	.4923	.6559	.7481	.9297	1.0516	1.1033	1.1371
-1.66				.2955	.4470	.6175	.7134	.9020	1.0286	1.0824	1.1176
-1.68				.2427	.4013	.5790	.6787	.8744	1.0058	1.0617	1.0984
-1.70				.1893	.3552	.5404	.6440	.8469	.9831	1.0412	1.0793
-1.72				.1352	.3088	.5017	.6092	.8196	.9606	1.0209	1.0605
-1.74				.0804	.2620	.4628	.5744	.7923	.9383	1.0008	1.0419
-1.76					.2148	.4237	.5395	.7651	.9161	.9808	1.0235
-1.78					.1672	.3845	.5046	.7380	.8941	.9610	1.0052
-1.80					.1190	.3451	.4696	.7109	.8722	.9414	.9872
-1.82					.0704	.3055	.4345	.6839	.8504	.9219	.9693
-1.84						.2657	.3993	.6570	.8287	.9025	.9515
-1.86						.2257	.3640	.6301	.8072	.8833	.9339
-1.88						.1854	.3286	.6032	.7857	.8642	.9165

with n and z_1 and read $\hat{\alpha}_3$. In order to improve accuracy, we interpolate as necessary. Estimates $\hat{\beta}$, $\hat{\gamma}$, and $\hat{\rho}$ are then calculated from the last three equations of (6.6.3). In solving the first equation of (6.6.3) for $\hat{\alpha}_3$, remember that since $-2/\alpha_3 < z$, it follows that $\hat{\alpha}_3 < -2/z_1$, and no solution will be found outside this range.

In many applications, estimates obtained as described are sufficiently accurate to serve as final estimates. However, when greater accuracy is needed, a FORTRAN program prepared by the authors, and included here as Appendix A.7, might be employed, provided suitable computing facilities are available.

If occasions should arise when MLE might be desired, the MME provide excellent first approximations for use in iterative calculations of the

z_1 \ n	5	10	20	25	30	40	50	100	250	500	1000
-1.90						.1449	.2931	.5764	.7644	.8453	.8992
-1.92						.1042	.2574	.5496	.7431	.8264	.8820
-1.94						.0631	.2216	.5228	.7220	.8077	.8649
-1.96							.1857	.4961	.7009	.7891	.8480
-1.98							.1496	.4694	.6799	.7706	.8312
-2.00							.1134	.4427	.6590	.7522	.8145
-2.02							.0769	.4160	.6382	.7339	.7979
-2.04								.3893	.6174	.7157	.7814
-2.06								.3625	.5967	.6975	.7651
-2.08								.3358	.5761	.6795	.7488
-2.10								.3091	.5555	.6615	.7326
-2.12								.2824	.5350	.6436	.7165
-2.14								.2556	.5146	.6258	.7005
-2.16								.2289	.4942	.6081	.6846
-2.18								.2021	.4738	.5904	.6688
-2.20								.1753	.4535	.5728	.6530
-2.22								.1484	.4332	.5553	.6373
-2.24								.1216	.4130	.5378	.6217
-2.26								.0946	.3928	.5204	.6062
-2.28								.0677	.3726	.5030	.5907
-2.30									.3525	.4857	.5753
-2.32									.3324	.4684	.5599
-2.34									.3123	.4512	.5446
-2.36									.2923	.4341	.5294
-2.38									.2722	.4169	.5142
-2.40									.2522	.3999	.4991
-2.42									.2323	.3828	.4840
-2.44									.2123	.3658	.4690
-2.46									.1924	.3489	.4541
-2.48									.1724	.3319	.4392
-2.50									.1525	.3151	.4243
-2.55									.1028	.2730	.3873
-2.60									.0531	.2311	.3506
-2.70										.1478	.2778
-2.80										.0650	.2058

MLE. As pointed out in Section 6.4, however, use of the MLE is not recommended unless $\rho > 2.5$ ($\alpha_3 < 1.265$), whereas the MME are applicable for all possible values of ρ.

6.6.2 Exponential Distribution

As previously noted, the exponential distribution is a special case of the gamma distribution when $\rho = 1$. It is also a special case of the Weibull distribution when $\delta = 1$. Modified moment estimators that are easy to calculate and that are both best linear unbiased (BLUE) and minimum variance unbiased (MVUE) were given in (3.9.9). Estimate variances and covariance were given in (3.9.10).

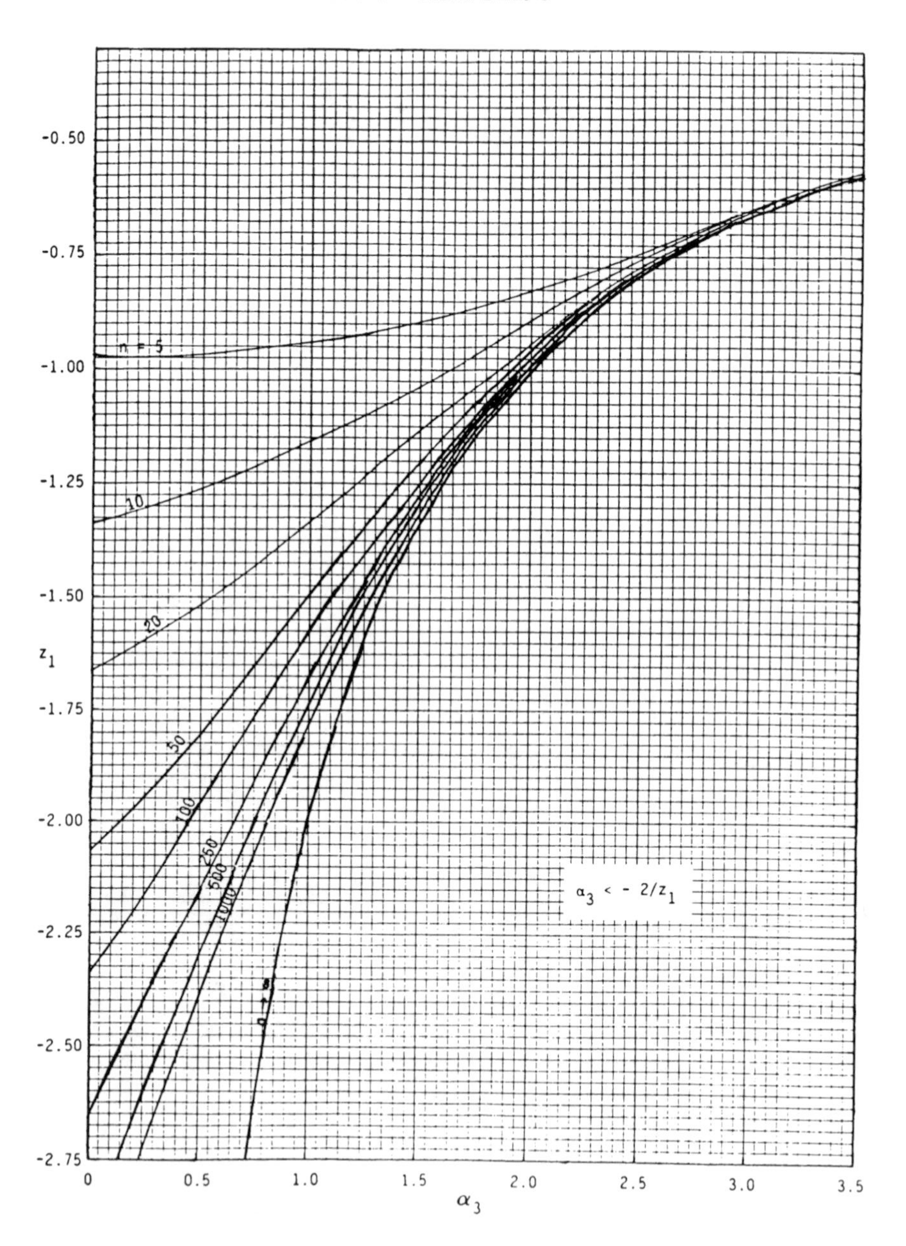

FIGURE 6.2 Graphs of α_3 as a function of z_1 and n for the gamma distribution. Reproduced from Cohen, A. C., and Whitten, B. J. (1986) with permission of the American Society for Quality Control.

6.7 ILLUSTRATIVE EXAMPLES

Practical applications of the MME are illustrated with the examples, which were also used as illustrations for parameter estimation in the Weibull, log-normal, and IG distributions. The data are recorded in Chapter 3 and except for summaries will not be repeated here.

Example 6.7.1. For this example, $n = 20$, $\bar{x} = 0.423125$, $s^2 = 0.0156948$, $s = 0.1252789$, $x_1 = 0.265$, $a_3 = 1.0673243$, and $z_1 = -1.262$. From Figure 6.2, with $n = 20$ and $z_1 = -1.262$, we read $\hat{\alpha}_3 = 1.174$. This value can be verified by linear interpolation in Table 6.3. With $\hat{\alpha}_3 = 1.174$, we subsequently employ (6.6.3) to calculate $\hat{\rho} = 2.902$, $\hat{\beta} = 0.0735$, $\hat{\gamma} = 0.210$, with $E(\hat{X}) = 0.4231$ and $\hat{\sigma} = 0.1253$. These values should be sufficiently accurate for most purposes. At least they might provide close first approximations for an iterative calculation of the MLE when these estimates are required. For use when greater computational accuracy is needed, the FORTRAN program of Cohen and Whitten (1982) is included as Appendix A.7. For this example, computer calculations using this program gave the exact values displayed in Table 6.4. Note that these values differ only slightly from the approximations.

For comparison, MLE calculated using the authors' FORTRAN program are presented along with the MME and ME in Table 6.4. Because of the small sample size and the fact that the ρ is small, some difficulty was encountered in retaining an adequate number of significant digits in the calculation of the MLE.

TABLE 6.4 Estimates for Example 6.7.1. Gamma Distribution

Estimator	$\hat{\gamma}$	$\hat{\rho}$	$\hat{\beta}$	$\hat{E}(X)$	$\sqrt{\hat{V}(X)}$	$\hat{\alpha}_3(X)$
ME	0.1882	3.5113	0.0669	0.4231	0.1253	1.0673
MLE	0.2627	1.1940	0.1343	0.4231	0.1468	1.8303
MME (approx)[a]	0.208	2.902	0.074	0.4231	0.1253	1.174
MME (exact)[b]	0.2096	2.9043	0.0735	0.4231	0.1253	1.1736

[a]Calculated by using graphs of Figure 6.2.
[b]Calculated by using authors' FORTRAN program. Asymptotic standard deviations calculated by using MME estimates in (6.5.3): $\sigma_{\hat{\gamma}} = 0.032$, $\sigma_{\hat{\rho}} = 1.47$, $\sigma_{\hat{\beta}} = 0.031$. From the central limit theorem: $\sigma_{\bar{x}} = 0.028$.

TABLE 6.5 Estimates for Example 6.7.2. Gamma Distribution

Estimator	$\hat{\gamma}$	$\hat{\rho}$	$\hat{\beta}$	$\hat{E}(X)$	$\sqrt{\hat{V}(X)}$	$\hat{\alpha}_3(X)$
ME	136.335	1.1517	73.060	220.4800	78.4056	1.8636
MLE (not calculated)						
MME (approx)[a]	147.6	0.863	84.40	220.48	78.4056	2.153
MME (exact)[b]	147.572	0.8647	84.317	220.4800	78.4056	2.1508
BLUE/MVUE[c]	145.1689	1	75.3111	220.4800	75.3111	2

[a]Calculated by using graphs of Figure 6.2.
[b]Calculated by using authors' FORTRAN program.
[c]Calculated by using (3.9.9) and assuming $\rho = 1$. Standard deviations of these estimates calculated from (3.9.10): $\sigma_{\hat{\gamma}} = 7.938$, $\sigma_{\hat{\beta}} = 25.104$, $\text{Cov}(\hat{\beta}, \hat{\gamma}) = -63.019$. Asymptotic variances of (6.5.3) not applicable since $\rho < 2$. From the central limit theorem: $\sigma_{\bar{X}} = 24.79$.

Example 6.7.2. For this example, $n = 10$, $\bar{x} = 220.48$, $s^2 = 6147.444$, $s = 78.405638$, $a_3 = 1.8635835$, $x_1 = 152.7$, and $z_1 = -0.864$. From Figure 6.2 or by interpolation in Table 6.3, with $n = 10$ and $z_1 = -0.864$, we read $\hat{\alpha}_3 = 2.153$. From (6.6.3), we calculate $\hat{\rho} = 0.863$, $\hat{\beta} = 84.40$, and $\hat{\gamma} = 147.6$, with $\hat{E}(X) = 220.48$ and $\hat{\sigma} = 78.41$. Computer computations with the FORTRAN program previously mentioned produced the exact values displayed in Table 6.5. Again these values differ only slightly from the approximations. Maximum likelihood estimates were not calculated for this example.

In Chapter 3 it was noted that this sample might have been selected from a two-parameter exponential distribution. Accordingly, estimates that were both BLUE and MVUE were calculated by employing estimating equations. Since the exponential distribution is a special case of both the Weibull and the gamma distributions in which $\delta = \rho = 1$ and $\alpha_3 = 2$, these estimates are also included in Table 6.5 for comparison with other estimates of gamma parameters obtained from the data of Example 6.7.2.

Example 6.7.3. Again, we employ the Shook data from Chapter 3 as an illustration. This time, we assume the sample to be from a gamma distribution, as did Miss Shook. In summary, the sample data are $n = 1000$, $\bar{x} = 118.74$, $s = 16.9175$, $a_3 = 0.976424$, $a_4 = 5.3206$, $x_1 = 78.70$, and $z_1 = -2.367$.

To calculate MME for parameters of the gamma distribution, we enter Table 6.3 with $n = 1000$ and $z_1 = -2.367$, and interpolate to obtain $\hat{\alpha}_3 = 0.5241$. It then follows that $\hat{\rho} = 4/\hat{\alpha}_3^2 = 14.562$, $\hat{\beta} = s/\sqrt{\hat{\rho}} = 4.433$, and

$\hat{\gamma} = \bar{x} - \hat{\rho}\hat{\beta} = 54.182$. A close approximation to the estimate given for α_3 might have been obtained from the graphs of Figure 6.2.

As mentioned previously, in the absence of compelling reasons for preferring a particular model on the basis of prior knowledge of the population which produced a given sample, it might be desirable to calculate several sets of estimates for different models, as we did here. The following tabulation presents a summary for comparing the various modified moment estimates obtained from the Shook example.

	Parameters estimated				
Distribution	Mean	Std. dev.	γ	α_3	α_4
Weibull	118.74	16.9175	74.993	0.23854	2.765
Lognormal	118.74	16.9175	30.455	0.5819	3.608
Inverse Gaussian	118.74	16.9175	28.224	0.5607	3.524
Gamma	118.74	16.9175	54.182	0.5241	3.412

On the basis of these comparisons, the Weibull estimate of γ seems most reasonable, but the estimates of α_3 and α_4 are at variance with the sample values a_3 and a_4. The gamma estimates of γ, α_3, and α_4 seem more in line with the sample values x_1, a_3, and a_4 and thus tend to justify Miss Shook's decision to fit a gamma distribution to these data. Although considerations such as these might be helpful in choosing a model for the description of sample data, the chi-square or the Kolmogorov-Smirnov goodness of fit test would be a more dependable test for determining whether the model chosen is acceptable.

6.8 REFLECTIONS

Modified moment estimators are likely to be more useful for estimating parameters of the gamma distribution than for estimating parameters of the IG distribution. Although ME are easy to calculate, use of the third moment introduces large estimates variances. Maximum likelihood estimators suffer from the same type of regularity problems that arise in the Weibull distribution. Maximum likelihood estimates do not exist unless the shape parameter ρ is greater than 1, and the asymptotic variances and covariances are not valid unless $\rho > 2$ $(\alpha_3 < \sqrt{2})$. The MME are applicable over the entire parameter space. Even when the MLE are applicable, they are not easy to calculate, and a good first approximation, such as might be provided by the MME, is required. The MME might not perform well for small values of α_3, but again we note that the normal distribution would usually be a better choice as a model when α_3 is very small.

7

Censored Sampling in the Exponential and Weibull Distributions

7.1 INTRODUCTION

This chapter is concerned with estimation of population parameters from censored samples, whereas the preceding chapters were concerned with complete samples. In life-span and reaction-time studies, it is common practice to cease observation before all specimens have failed. The resulting samples are accordingly said to be censored on the right. In a typical test, observation of survivors ceases with a single stage of censoring. Single-stage censoring has been fully considered by numerous writers, including Epstein and Sobel (1953), Epstein (1960), Gupta (1952), Hald (1949), Sarhan and Greenberg (1956, 1957, 1958), Cohen (1950, 1957, 1959, 1961), and Schneider (1986). Additional references to this and to related topics are listed in a bibliography compiled by Mendenhall (1958).

In many practical situations, the initial censoring results in withdrawal of only a portion of the survivors. Those which remain on test continue under observation until ultimate failure or until a subsequent stage of censoring is performed. For sufficiently large samples, censoring may be progressive through several stages. Such samples arise naturally when certain specimens must be withdrawn from a life test prior to failure for use as test objects in related experimentation. In other instances, progressively censored samples result from a compromise between the need for more rapid testing and the desire to include a few extreme life-spans in the sample data. When test facilities are limited and when prolonged tests are expensive, the early censoring of a substantial number of sample specimens is economically more realistic than continuing observation to obtain complete samples. In clinical trials, censoring is often the result of participant dropout. Singly censored samples are considered as special cases of progressively censored samples.

In progressive censoring, as in single-stage censoring, a distinction is made between censoring of Type I, in which times of censoring are predetermined, and Type II, in which censoring occurs when the number of survivors falls to predetermined levels. Estimating equations for the two types are identical, but there are minor differences in estimate variances. Progressively censored samples of a somewhat more complex type, in which both the times of censoring and the number of specimens withdrawn are the result of random causes, were considered by Sampford (1952) in connection with response-time studies involving animals.

Type II progressively censored samples have been considered by Herd (1956, 1957, 1960), who referred to them as "multiple censored samples," and by Roberts (1962a, b), who designated them as "hypercensored samples." Subsequent publications dealing with progressively censored samples have appeared, including Cohen (1963, 1965, 1966, 1975, 1976), Ringer and Sprinkle (1962), and Cohen and Norgaard (1977).

Modified moment estimators (MME) have been the estimators of primary interest for the complete samples considered in the preceding chapters, but modified maximum likelihood estimators (MMLE) are more appropriate when samples are censored. For the MMLE, we retain the equation involving the first-order statistic from the MME and substitute equations involving partials of the loglikelihood function for moment equations of the MME. Note that x_1, the smallest observation in a right-censored sample, is the observed value of the first-order statistic in a sample of size N. Where appropriate, maximum likelihood estimators (MLE) are also considered. However, some of the MLE are of limited value because of regularity problems.

7.2 PROGRESSIVELY CENSORED SAMPLES

Censored samples consist of both full-term (complete) and partial-term (censored) observations. Let N designate the total sample size, and let n designate the number of full-term observations. In a failure context, n is the number of failures. Suppose that censoring occurs progressively in k stages at times $T_1 < T_2 < \cdots < T_j < \cdots < T_k$, and that at the jth stage of censoring, c_j sample specimens selected randomly from the survivors at time T_j are removed (censored) from further observation. The sample data thus consist of the full-term observations $\{x_i\}$, $i = 1, \ldots, n$, plus $N - n$ partial-term observations $\{c_j, T_j\}$, $j = 1, \ldots, k$. It follows that

$$N = n + r, \quad \text{where } r = \sum_{1}^{k} c_j \tag{7.2.1}$$

The sum total (ST) of all observations, both full-term and partial-term, in a sample of the type described can be expressed as

$$ST = \sum_{1}^{n} x_i + \sum_{1}^{k} c_j T_j \tag{7.2.2}$$

Although it is standard procedure in life tests to place all N specimens "on test" at the same time, this is not an essential feature. It is only necessary that the recorded times x_i apply to specific sample specimens as indicated. In some applications, notably in the medical field, censoring might occur on individual items "one at a time." In other words, in such cases $c_j = 1$ for all j.

In Type I censoring, the T_j are fixed and the number of survivors at these times are observed values of random variables. In Type II censoring, the T_j coincide with times of failure and are random variables, whereas the number of survivors at these times are fixed.

The likelihood function, L(S), where S is a k-stage progressively censored sample, may be written as

$$L(S) = K \prod_{i=1}^{n} f(x_i) \prod_{j=1}^{k} [1 - F(T_j)]^{c_j} \tag{7.2.3}$$

where K is an ordering constant, f( ) is the density function, and F( ) is the distribution function.

In designating times (points) of censoring as $\{T_j\}$, we tend to obscure sample type distributions. However, these designations emphasize the fact that estimators for parameters of a given distribution are the same for both types of censored samples. In Type I samples, the T_j are sample constants; i.e., they are nonrandom. In Type II samples, the T_j are observed values of random variables, i.e., of order statistics. In a Type II singly censored sample, $T = x_n$, the observed value of the nth-order statistic in a sample of size N. Estimate variances and covariances depend upon expected values of sample data, and these differ between sample types.

In subsequent considerations, it is sometimes necessary to standardize observations. Standardized full-term observations are designated by z. In this notation, the standardized observed value of the first-order statistic becomes

$$z_1 = \frac{x_1 - E(X)}{\sqrt{V(X)}}$$

where E(X) is the distribution mean and V(X) is the distribution variance. Standardized, censoring points are designated by ξ. Thus,

$$\xi_j = \frac{T_j - E(X)}{\sqrt{V(X)}}, \quad j = 1, \ldots, k$$

In the special case of Type II progressive censoring, where the total sample is partitioned into equal subgroups of $r + 1$ units each, and where immediately following the first failure in a subgroup, the remaining r units of the subgroup are withdrawn, the density function of these "smallest observations" is well known [cf., for example, Hoel (1954, p. 304)]. Where $f(x)$ and $F(x)$ are the density and the distribution function of x, respectively, and u is the smallest observation in a sample of size $r + 1$, the density of u may be written as

$$h(u) = (r + 1)f(u)[1 - F(u)]^r \tag{7.2.4}$$

In practical applications, samples of this kind might arise from testing throwaway units, each consisting of $r + 1$ identical components, where the failure of any single component means failure of the unit.

7.3 CENSORED SAMPLES FROM THE EXPONENTIAL DISTRIBUTION

As noted in Chapters 3 and 6, the exponential distribution arises as a special case of both the Weibull and the gamma distributions in which the shape parameters are $\delta = 1$ for the Weibull, and $\rho = 1$ for the gamma distribution. Thereby $\alpha_3 = 2$. The resulting pdf and cdf follow as

$$f(x; \beta, \gamma) = \frac{1}{\beta} e^{-(x-\gamma)/\beta}, \quad \gamma < x < \infty \tag{7.3.1}$$

$$= 0 \quad \text{elsewhere}$$

and

$$F(x; \beta, \gamma) = 1 - e^{-(x-\gamma)/\beta} \tag{7.3.2}$$

The loglikelihood function for a progressively censored sample from this distribution is

$$\ln L = -n \ln \beta - \sum_1^n \frac{x_i - \gamma}{\beta} - \sum_1^k \frac{c_j(T_j - \gamma)}{\beta} + \ln K \tag{7.3.3}$$

7.3.1 Modified Maximum Likelihood Estimators

Estimating equations for the MMLE are $\partial \ln L/\partial\beta = 0$ and $E(X_1) = x_1$. From these equations it follows that

$$\hat{\beta} = \frac{ST - N\hat{\gamma}}{n}, \qquad \hat{\gamma} + \frac{\hat{\beta}}{N} = x_1 \tag{7.3.4}$$

where ST is given by (7.2.2).

On solving the two equations of (7.3.4), we obtain

$$\hat{\gamma} = \frac{nx_1 - ST/N}{n - 1}, \qquad \hat{\beta} = \frac{ST - Nx_1}{n - 1} \tag{7.3.5}$$

For a complete (uncensored) sample, $c_j = 0$ for all j, n = N, and ST = $n\bar{x}$. It then follows from (7.3.5) that

$$\hat{\gamma} = \frac{nx_1 - \bar{x}}{n - 1}, \qquad \hat{\beta} = \frac{n(x - x_1)}{n - 1}$$

These are recognized as the MME and also as the BLUE and the MVUE, which were given in (3.9.9). Thus, in this case, the MME and MMLE are identical.

Because of its simplicity and because of the ease with which estimates can be calculated, the exponential distribution is often a preferred model in reliability studies. Even when a more complex model might be appropriate, samples are often too small to permit detection of any differences that might exist between models. Furthermore, differences between the exponential, the Weibull, and gamma distributions diminish as $\alpha_3 \to 2$. In some situations, the exponential distribution is the preferred model simply because it provides a better fit to sample data than any other model that might be chosen. In other situations, theoretical considerations sometimes dictate that the exponential distribution is the true population distribution. For example, in some applications a constant hazard function is needed, and this calls for the exponential distribution.

In the special case for which $\gamma = 0$, the first equation of (7.3.4) is reduced to

$$\hat{\beta} = \frac{ST}{n} \tag{7.3.6}$$

which is recognized as the MLE for β based on a progressively censored sample from the one-parameter exponential distribution with origin at zero.

7.4 CENSORED SAMPLES FROM THE WEIBULL DISTRIBUTION

The pdf and the cdf of the three-parameter Weibull distribution are given in (3.2.1) and (3.2.2). The loglikelihood function for a progressively censored sample from this distribution can be written as

$$\ln L = -n \ln \theta + (\delta - 1) \sum_{1}^{n} \ln(x_i - \gamma) - \frac{1}{\theta} \sum{}^{*} (x_i - \gamma)^{\delta} + \ln K \tag{7.4.1}$$

where

$$\sum{}^{*} (x_i - \gamma)^{\delta} = \sum_{1}^{n} (x_i - \gamma)^{\delta} + \sum_{1}^{k} c_j (T_j - \gamma)^{\delta}$$

7.4.1 Three-Parameter Modified Maximum Likelihood Estimators

Here β has been replaced by θ, where $\theta = \beta^{\delta}$. Estimating equations for the MMLE are $\partial \ln L/\partial\delta = 0$, $\partial \ln L/\partial\theta = 0$, and $E(X_1) = x_1$. It follows that

$$\frac{\partial \ln L}{\partial \delta} = \frac{n}{\delta} + \sum_{1}^{n} \ln(x_i - \gamma) - \frac{1}{\theta} \sum{}^{*} (x_i - \gamma)^{\delta} \ln(x_i - \gamma) = 0$$

$$\frac{\partial \ln L}{\partial \theta} = -\frac{n}{\theta} + \frac{1}{\theta^2} \sum{}^{*} (x_i - \gamma)^{\delta} = 0 \tag{7.4.2}$$

$$E(X_1) = \gamma + \left(\frac{\theta}{N}\right)^{1/\delta} \Gamma_1 = x_1$$

where as in (7.4.1), Σ^* signifies summation over both full-term and partial-term observations; i.e.,

$$\sum{}^{*} (x_i - \gamma)^{\delta} = \sum_{1}^{n} (x_i - \gamma)^{\delta} + \sum_{1}^{k} c_j (T_j - \gamma)^{\delta} \tag{7.4.3}$$

$$\sum{}^{*} (x_i - \gamma)^{\delta} \ln(x_i - \gamma) = \sum_{1}^{n} (x_i - \gamma)^{\delta} \ln(x_i - \gamma) + \sum_{1}^{k} c_j (T_j - \gamma)^{\delta} \ln(T_j - \gamma)$$

Estimates $\hat{\theta}$, $\hat{\delta}$, and $\hat{\gamma}$ follow as the simultaneous solution of the three equations of (7.4.2). Various standard iterative procedures such as the Newton-Raphson method are available for solving these equations. However, calculations can be facilitated by first making a few algebraic simplifications.

Eliminate θ between the first and second equations of (7.4.2) to obtain

$$H_1(\gamma, \delta) = \sum^{*} (x_i - \gamma)^{\delta} \ln (x_i - \gamma) - \frac{1}{\delta} - \frac{1}{n} \sum_{1}^{n} \ln (x_i - \gamma) = 0 \qquad (7.4.4)$$

When $\gamma = 0$, this equation is identical with the corresponding maximum likelihood equation given by Cohen (1965).

Eliminate θ between the second and the third equations of (7.4.2) to obtain

$$H_2(\gamma, \delta) = \frac{\sum^{*} (x_i - \gamma)^{\delta} \ln (x_i - \gamma)}{nN[(x_1 - \gamma)/\Gamma_1]^{\delta}} - \frac{1}{\delta} - \frac{1}{n} \sum_{1}^{n} \ln (x_i - \gamma) = 0 \qquad (7.4.5)$$

Eliminate θ between the second and the third equations of (7.4.3) to obtain

$$H_3(\gamma, \delta) = nN \left[\frac{x_1 - \gamma}{\Gamma_1}\right]^{\delta} - \sum^{*} (x_i - \gamma)^{\delta} = 0 \qquad (7.4.6)$$

We now choose any pair from (7.4.4), (7.4.5), and (7.4.6) and solve for $\hat{\gamma}$ and $\hat{\delta}$. Suppose we elect to employ $H_2(\gamma, \delta) = 0$ and $H_3(\gamma, \delta) = 0$. Choose a first approximation $\gamma_1 < x_1$ and solve (7.4.6) for a corresponding first approximation δ_1. Substitute γ_1 and δ_1 into $H_2(\gamma, \delta) = 0$. If $H_2(\gamma_1, \delta_1) = 0$, our task is finished. Otherwise, we select a second approximation γ_2 and repeat the cycle of calculations. We continue until we find values γ_i, δ_i and γ_j, δ_j such that $|\gamma_i - \gamma_j|$ and $|\delta_i - \delta_j|$ are sufficiently small and such that

$$H_2(\gamma_i, \delta_i) \gtrless 0 \gtrless H_2(\gamma_j, \gamma_j)$$

Final estimates $\hat{\gamma}$ and $\hat{\delta}$ are then obtained by linear interpolation between γ_i and γ_j and between δ_i and δ_j. The estimate $\hat{\theta}$ can be obtained from the third equation of (7.4.2) as

$$\hat{\theta} = N \left[\frac{x_1 - \hat{\gamma}}{\hat{\Gamma}_1}\right]^{\hat{\delta}} \qquad (7.4.7)$$

and, if needed, $\hat{\beta} = \hat{\theta}^{1/\hat{\delta}}$.

In carrying out these calculations outlined, we note that x_1 is an upper bound on γ. In solving (7.4.6) for γ_1 and for subsequent approximations, note the range of permissible values for δ. If the distribution is bell-shaped and positively skewed, then $1 < \delta < 3.6023$. As a function of δ, we have $\alpha_3(1) = 2$ and $\alpha_3(3.6023) = 0$. When $\delta = 1$, our distribution is the exponential, and for all $\delta \leq 1$ the distribution is reverse J-shaped. For $\delta > 3.6023$, the skewness is negative; i.e., $\alpha_3 < 0$. These considerations might be helpful in choosing a first approximation to δ that will minimize the computational effort involved in arriving at final estimates.

7.4.2 Maximum Likelihood Estimators

When $\delta > 1$, and thus when the distribution is bell-shaped, the MLE can be found by simultaneously solving the equations $\partial \ln L/\partial\gamma = 0$, $\partial \ln L/\partial\delta = 0$, and $\partial \ln L/\partial\theta = 0$. The last two of these equations are the same as the first two equations of (7.4.2). The first of these equations is

$$\frac{\partial \ln L}{\partial\gamma} = \frac{\delta}{\theta}\sum{}^{*}(x_i - \gamma)^{\delta-1} - (\delta - 1)\sum_{1}^{n}(x_i - \gamma)^{-1} = 0 \tag{7.4.8}$$

where

$$\sum{}^{*}(x_i - \gamma)^{\delta-1} = \sum_{1}^{n}(x_i - \gamma)^{\delta-1} + \sum_{1}^{k} c_j(T_j - \gamma)^{\delta-1}$$

To calculate estimates, our procedure parallels that employed in calculating modified estimates. We choose a first approximation $\gamma_1 < x_1$ and solve (7.4.4) for a corresponding first approximation δ_1. We then substitute γ_1 and δ_1 into (7.4.8). If the result equals zero, these values are then the required estimates. Otherwise we select a new approximation and continue until we find pairs (γ_i, δ_i) and (γ_j, δ_j) in a sufficiently small region such that

$$\frac{\partial \ln L}{\partial\gamma}(\gamma_i, \delta_i) \gtrless 0 \lessgtr \frac{\partial \ln L}{\partial\gamma}(\gamma_j, \delta_j)$$

We subsequently calculate final estimates by linear interpolation. With $\hat{\gamma}$ and $\hat{\delta}$ thus determined, we employ the second equation of (7.4.2) to calculate

$$\hat{\theta} = \frac{1}{n}\sum{}^{*}(x_i - \hat{\gamma})^{\hat{\delta}} \quad \text{and} \quad \hat{\beta} = \hat{\theta}^{1/\hat{\delta}} \tag{7.4.9}$$

7.4.3 Two-Parameter Maximum Likelihood Estimators

Maximum likelihood estimators for Weibull parameters based on censored samples in the two-parameter special case where the origin is at zero were given by Cohen (1965). For a progressively censored sample, with $\gamma = 0$, applicable estimating equations follow from (7.4.4) and from (7.4.9) as

$$\frac{\sum^* x_i^{\hat{\delta}} \ln x_i}{\sum^* x_i^{\hat{\delta}}} - \frac{1}{\hat{\delta}} = \frac{1}{n}\sum_1^n \ln x_i \tag{7.4.10}$$

$$\hat{\theta} = \frac{1}{n}\sum^* x_i^{\hat{\delta}}$$

where Σ^* signifies summation over both full-term and partial-term observations. If we set $\gamma = 0$, the Σ^* are defined by (7.4.3). In this case, an iterative solution of the first equation of (7.4.10) for $\hat{\delta}$ is relatively simple. With $\hat{\delta}$ thus calculated, $\hat{\theta}$ follows from the second equation of (7.4.10) and $\hat{\beta} = \hat{\theta}^{1/\hat{\delta}}$.

Single-Stage Censoring. Single-stage censoring is merely a special case of progressive censoring in which $k = 1$, $r = c_1 = N - n$, and $T = T_1$.

7.5 THE HAZARD PLOT

The hazard plot developed by Nelson (1969, 1972, 1982) provides a simple graphical procedure for choosing an appropriate model and for approximating estimates of distribution parameters. For a given sample, full-term and censored observations are ordered with respect to magnitude and given a reverse rank number. Accordingly, the first observation is assigned the reverse rank N; the next is assigned reverse rank N - 1. The process is continued until the last observation is numbered 1. The reverse ranks thus represent the number of survivors immediately prior to the censoring of a corresponding observation. Estimates of the hazard or instantaneous failure rates as percentages are then calculated at times $\{x_i\}$, $i = 1, \ldots, n$, as

$$h(x_i) = 100\left(\frac{1}{m_i}\right) \tag{7.5.1}$$

where m_i is the reverse rank corresponding to x_i. The cumulative hazard estimate is calculated as

$$H(x_i) = \sum_{j=1}^{i} h(x_j) \tag{7.5.2}$$

A plot of the cumulative hazard function is then made on appropriate hazard cross-section paper. If the correct paper is chosen, the plotted points will lie on a straight line. Nelson (1969) has considered hazard plotting and the choice of hazard cross-section paper in detail. For the purpose of this discussion it is sufficient to state that ordinary rectangular coordinate paper is the appropriate hazard graph paper for the exponential distribution. Log-log graph paper is the hazard plotting paper for the Weibull distribution; and semilog paper, with life-spans plotted on the linear scale and the cumulative hazard function plotted on the logarithmic scale, is used for the extreme value distribution. Special plotting papers have been developed for other distributions, and these are described in detail by Nelson.

As a word of caution, remember that the underlying theory of hazard plots is based on the assumption that items selected for censoring are chosen at random from survivors at the time of censoring. Accordingly, the procedure presented here would not apply and might lead to erroneous results if items are selected for censoring because of indications that they are about to fail.

It is not essential that the cumulative hazard function be plotted on a scale that produces a straight line. The curved lines, which result from using ordinary rectangular scales, may also lead to correct interpretations. The important point to be remembered is that graphical techniques which involve fitting a line to plotted points by eye are more accurate if the fitted line is straight rather than curved.

As previously noted, the hazard function and the cumulative hazard function of the three-parameter Weibull distribution are

$$h(x) = \left(\frac{\delta}{\beta^{\delta}}\right)(x - \gamma)^{\delta-1}, \qquad H(x) = \left(\frac{x - \gamma}{\beta}\right)^{\delta} \tag{7.5.3}$$

If $\delta < 1$, the hazard function is a decreasing function of x and the cumulative hazard function is convex. If $\delta > 1$, the hazard function is an increasing function of x and the cumulative hazard function is concave. If $\delta = 1$, the hazard function is constant for all x and the cumulative hazard function is a straight line with positive slope. This situation is illustrated by Figures 7.1 and 7.2. Some points of special interest follow from (7.5.3) as

$$\begin{aligned} h(\gamma) &= 0, \quad \delta > 1 \\ h(\gamma) &= \infty, \quad \delta < 1 \\ H(\gamma) &= 0, \quad \delta \geq 1 \\ H(\beta + \gamma) &= 1 \end{aligned} \tag{7.5.4}$$

If we take logarithms of both sides of the second equation of (7.5.3), we obtain

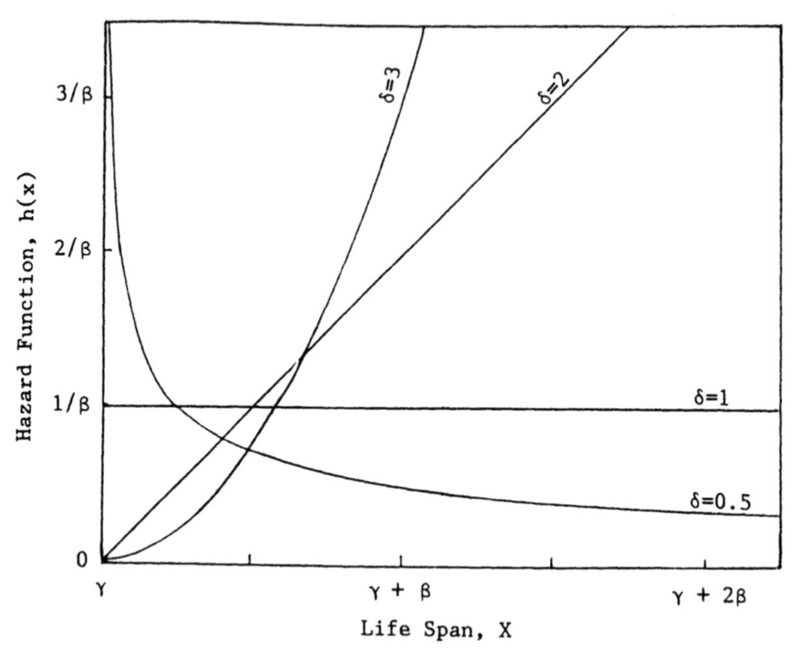

FIGURE 7.1 Weibull hazard functions.

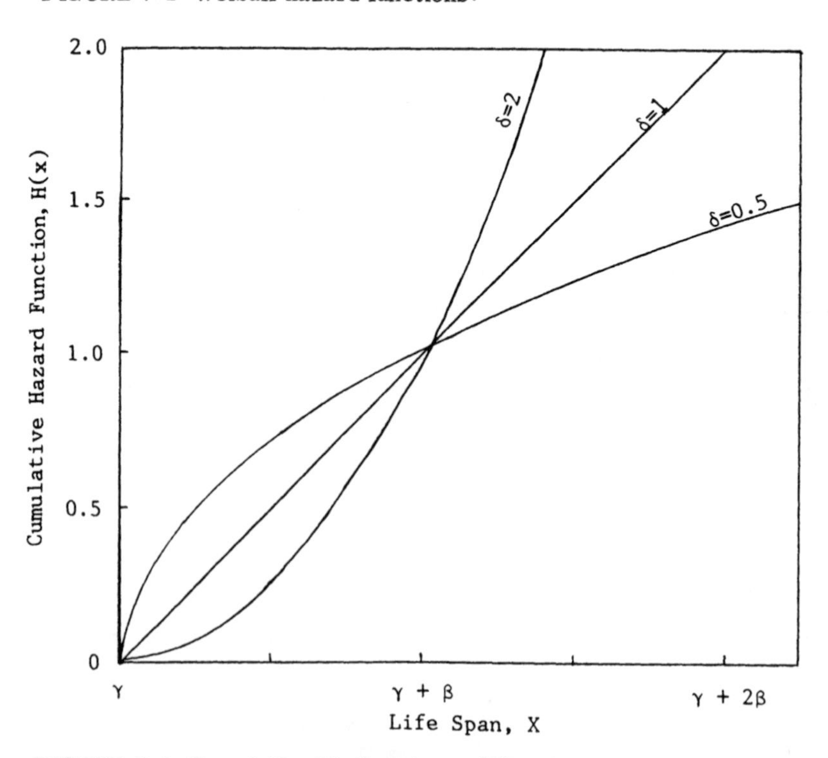

FIGURE 7.2 Cumulative Weibull hazard functions.

$$\ln(x - \gamma) = \ln \beta + \frac{1}{\delta} \ln H(x) \tag{7.5.5}$$

and

$$\ln H(x) = \delta \ln(x - \gamma) - \delta \ln \beta$$

Thus $\ln(x - \gamma)$ is a linear function of $\ln H(x)$, where $1/\delta$ is the slope. Inversely, $\ln H(x)$ is a linear function of $\ln(x - \gamma)$, where δ is the slope. Accordingly, as previously noted, Weibull hazard paper is log-log paper. Nelson incorporated a nomogram based on (7.5.5) with his version of log-log paper. This enabled users to determine graphically the shape parameter δ.

An alternative procedure which some might find advantageous, since it does not require use of special plotting paper, is as follows. From the cumulative hazard plot we read

$$\gamma^* = H^{-1}(0), \quad \beta^* + \gamma^* = H^{-1}(1), \quad \beta^* = H^{-1}(1) - H^{-1}(0) \tag{7.5.6}$$

Thus γ^* is the value of x corresponding to $H = 0$ and $\gamma^* + \beta^*$ is the value of x corresponding to $H = 1$ (or 100%). We now let $x = \beta^*/2 + \gamma^*$ and read $H[\beta^*/2 + \gamma^*]$ from the hazard plot. From (7.5.3) it follows that

$$H\left[\frac{\beta^*}{2} + \gamma^*\right] = 2^{-\delta^*} \tag{7.5.7}$$

If we take logarithms of both sides of (7.5.7) and simplify, we obtain

$$\delta^* = -\frac{\ln H[\beta^*/2 + \gamma^*]}{\ln 2}$$

and finally

$$\delta^* = -1.442695 \ln H\left[\frac{\beta^*}{2} + \gamma^*\right] \tag{7.5.8}$$

Thus, any time we have a cumulative hazard plot of a sample from a Weibull distribution, we can estimate γ and β from (7.5.6) and we can calculate δ^* from (7.5.8). This procedure can be carried out regardless of whether the hazard plot is on rectangular or log-log coordinate paper. It is necessary only that we be able to read $H = 0$ and $H = 1$ to calculate γ^* and β^* and then to read $H[\beta^*/2 + \gamma^*]$. However, if the plot is on log-log coordinate paper, the straight-line plot is more likely to yield accurate estimates.

Attention is invited to the pdf and the cdf of the three-parameter Weibull distribution when $x = \gamma + \beta$.

$$f(\gamma + \beta) = \left(\frac{\delta}{\beta^{\delta}}\right) e^{-1} = 0.3678794 \frac{\delta}{\beta^{\delta}}$$

$$F(\gamma + \beta) = 1 - e^{-1} = 0.6321206 \tag{7.5.9}$$

These results prevail for all values of β and δ. It is particularly noteworthy that the 63rd percentile (or more accurately the 63.21206th percentile) occurs at a distance β from the origin for all Weibull distributions regardless of the value of δ. Accordingly this result also applies to the exponential distribution.

7.6 ESTIMATE VARIANCES AND COVARIANCES

Except for a change in the scale parameter from β to θ, where $\theta = \beta^{\delta}$, the asymptotic variance-covariance matrix of the MLE $(\hat{\gamma}, \hat{\delta}, \hat{\theta})$ when samples are censored, is essentially the same as the matrix (3.8.1) for complete samples. For censored samples, however, it is expedient to replace expected values of the second partials with their sample evaluations calculated by substituting sample estimates $\hat{\gamma}$, $\hat{\delta}$, and $\hat{\theta}$ directly into these partials. Accordingly, in this case, the a_{ij} of (3.8.1), with $\hat{\beta}$ replaced by $\hat{\theta}$, are approximated as

$$\begin{aligned}
a_{11} &= -\frac{\partial^2 \ln L}{\partial \gamma^2}(\hat{\gamma}, \hat{\delta}, \hat{\theta}), \qquad a_{22} = -\frac{\partial^2 \ln L}{\partial \delta^2}(\hat{\gamma}, \hat{\delta}, \hat{\theta}) \\
a_{33} &= -\frac{\partial^2 \ln L}{\partial \theta^2}(\hat{\gamma}, \hat{\delta}, \hat{\theta}), \qquad a_{12} = a_{21} = -\frac{\partial^2 \ln L}{\partial \gamma \partial \delta}(\hat{\gamma}, \hat{\delta}, \hat{\theta}) \\
a_{13} &= a_{31} = -\frac{\partial^2 \ln L}{\partial \gamma \delta \theta}(\hat{\gamma}, \hat{\delta}, \hat{\theta}), \qquad a_{23} = a_{32} = -\frac{\partial^2 \ln L}{\partial \delta \partial \theta}(\hat{\gamma}, \hat{\delta}, \hat{\theta})
\end{aligned} \tag{7.6.1}$$

where

$$\begin{aligned}
\frac{\partial^2 \ln L}{\partial \gamma^2} &= -(\delta - 1) \sum_{1}^{n} (x_i - \gamma)^{-2} - \frac{\delta(\delta - 1)}{\theta} \sum^{*} (x_i - \gamma)^{\delta - 2} \\
\frac{\partial^2 \ln L}{\partial \delta^2} &= -\frac{n}{\delta^2} - \frac{1}{\theta} \sum^{*} (x_i - \gamma)^{\delta} [\ln (x_i - \gamma)]^2 \\
\frac{\partial^2 \ln L}{\partial \theta^2} &= \frac{n}{\theta^2} - \frac{2}{\theta^3} \sum^{*} (x_i - \gamma)^{\delta}
\end{aligned} \tag{7.6.2}$$

$$\frac{\partial^2 \ln L}{\partial\gamma\partial\delta} = \frac{\partial^2 \ln L}{\partial\delta\partial\gamma} = -\sum_1^n (x_i - \gamma)^{-1} + \frac{1}{\theta}\sum{}^*(x_i - \gamma)^{\delta-1}$$

$$+ \frac{\delta}{\theta}\sum{}^*(x_i - \gamma)^{\delta-1} \ln(x_i - \gamma)$$

$$\frac{\partial^2 \ln L}{\partial\gamma\partial\theta} = \frac{\partial^2 \ln L}{\partial\theta\partial\gamma} = -\frac{\delta}{\theta^2}\sum{}^*(x_i - \gamma)^{\delta-1}$$

$$\frac{\partial^2 \ln L}{\partial\delta\partial\theta} = \frac{\partial^2 \ln L}{\partial\theta\partial\delta} = \frac{1}{\theta^2}\sum{}^*(x_i - \gamma)^{\delta} \ln(x_i - \gamma)$$

and where, as previously defined, Σ^* signifies summation over both full-term and partial-term observations. Asymptotic variances and covariances calculated by substituting the a_{ij} approximations of (7.6.1) into (3.8.1) are not expected to differ very much from corresponding variances and covariances calculated from (3.8.4) as though the sample were complete of total size n, when in fact n is the number of full-term observations in a censored sample. For a censored sample from an exponential distribution, approximate variances and covariances can be calculated from (3.9.10), where n is the number of full-term observations in a censored sample.

Readers are again reminded that the asymptotic variances and covariances are not valid unless $\delta > 2$, and their accuracy might be subject to question unless δ is considerably greater than 2.

7.7 AN ILLUSTRATIVE EXAMPLE

In order to illustrate both the hazard plotting technique and parameter estimation from a progressively censored sample, we consider a sample that was originally given by Nelson (1969) of time to failure and/or time to censoring of 70 generator fans. Data for this sample together with calculated values of the hazard function h(x) and the cumulative hazard function H(x) are entered in Table 7.1. In summary, for this example, $N = 70$, $n = 12$, $x_1 = 4500$, $ST = 3{,}444{,}400$, $ST/n = 287{,}033.33$, $\Sigma_1^{12} x_i = 365{,}700$, and $\Sigma_1^k c_j T_j = 3{,}078{,}700$. A plot of the cumulative hazard function for this example on rectangular coordinate paper is given as Figure 7.3. A straight line is fitted by eye to the plotted points. The straight-line fit indicates that the exponential distribution is the appropriate model for these data. By reading from the fitted line, we have $H(267{,}500) = 100\%$ and $H(500) = 0$. Accordingly, $\gamma^* = 500$, $\gamma^* + \beta^* = 267{,}500$, and $\beta^* = 267{,}000$. We then

TABLE 7.1 A Progressively Censored Sample Consisting of Life Span Observations of 70 Generator Fans[a]

Rank	Reverse rank (m)	Time 1000 hr (x)	h(x) (%)	H(x) (%)	Rank	Reverse rank (m)	Time 1000 hr (x)	h(x) (%)	H(x) (%)
1	70	4.5	1.43	1.43	36	35	43.0+		
2	69	4.6+			37	34	46.0	2.94	18.78
3	68	11.5	1.47	2.90	38	33	48.5+		
4	67	11.5	1.49	4.39	39	32	48.5+		
5	66	15.6+			40	31	48.5+		
6	65	16.0	1.54	5.93	41	30	48.5+		
7	64	16.6+			42	29	50.0+		
8	63	18.5+			43	28	50.0+		
9	62	18.5+			44	27	50.0+		
10	61	18.5+			45	26	61.0+		
11	60	18.5+			46	25	61.0	4.00	22.78
12	59	18.5+			47	24	61.0+		
13	58	20.3+			48	23	61.0+		
14	57	20.3+			49	22	63.0+		
15	56	20.3+			50	21	64.5+		
16	55	20.7	1.82	7.75	51	20	64.5+		
17	54	20.7	1.85	9.60	52	19	67.0+		
18	53	20.8	1.89	11.49	53	18	74.5+		
19	52	22.0+			54	17	78.0+		
20	51	30.0+			55	16	78.0+		
21	50	30.0+			56	15	81.0+		
22	50	30.0+			57	14	81.0+		
23	48	30.0+			58	13	82.0+		
24	47	31.0	2.13	13.62	59	12	85.0+		
25	46	32.0+			60	11	85.0+		
26	45	34.5	2.22	15.84	61	10	85.0+		
27	44	37.5+			62	9	87.5+		
28	43	37.5+			63	8	87.5	12.50	35.28
29	42	41.5+			64	7	87.5+		
30	41	41.5+			65	6	94.0+		
31	40	41.5+			66	5	99.0+		
32	39	41.5+			67	4	101.0+		
33	38	43.0+			68	3	101.0+		
34	37	43.0+			69	2	101.0+		
35	36	43.0+			70	1	115.0+		

$h(x) = (100/m)$; $H(x) = \Sigma h(x)$.

[a]Given by Nelson (1969), and used here with permission of the author and the American Society for Quality Control.

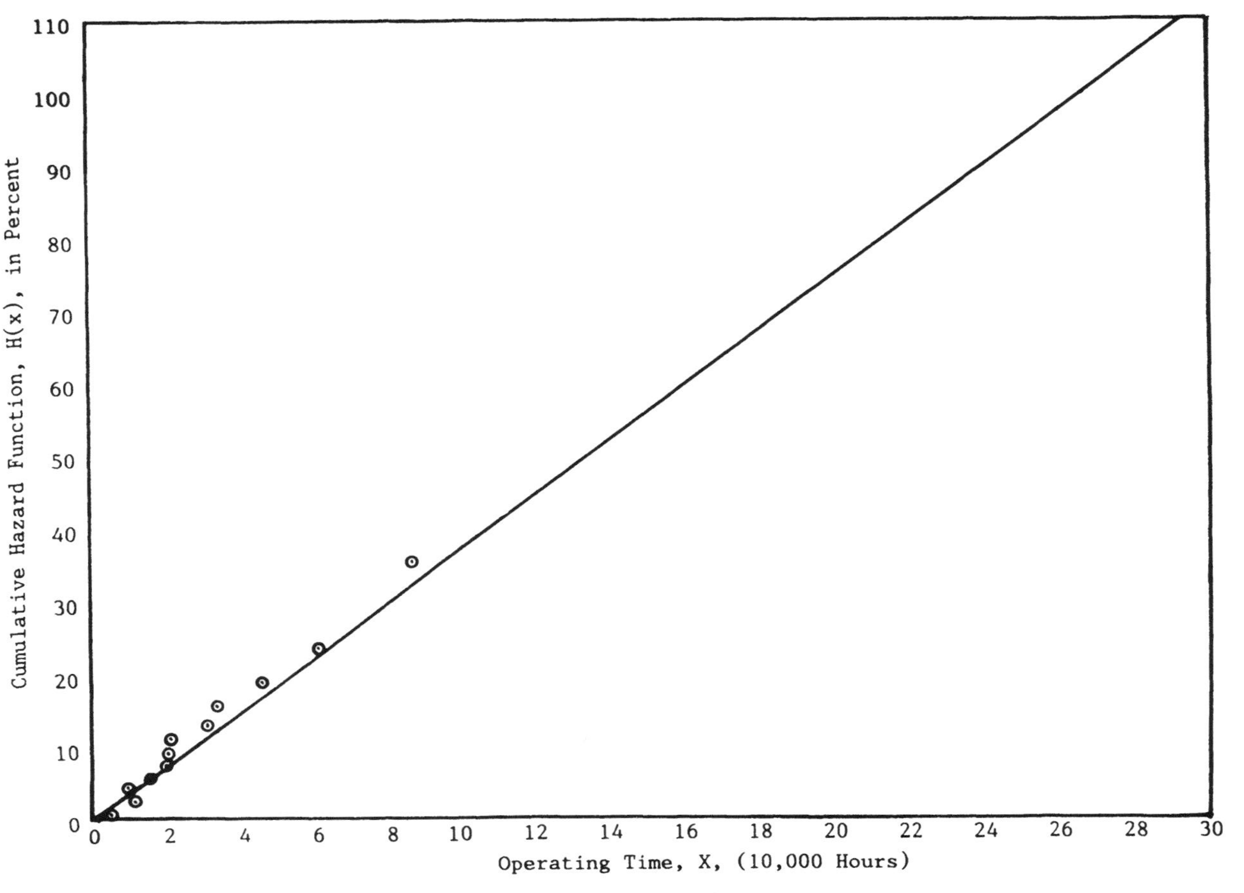

FIGURE 7.3 A cumulative hazard plot of the generator fan data.

calculate $\beta^*/2 + \gamma^* = 134{,}000$, and from the fitted line we read $H(134{,}000) = 0.50$. We now employ (7.5.8) to calculate

$$\delta^* = -1.442695 \ln 0.50 = 1.00$$

which confirms our earlier conclusion that this sample is from an exponential distribution. Having thus determined that the exponential distribution is an appropriate model for this example, we calculate MMLE by using the estimators of (7.3.5). We accordingly calculate

$$\hat{\gamma} = \frac{12(4500) - 3{,}444{,}400/70}{11} = 436$$

$$\hat{\beta} = \frac{3{,}444{,}400 - 70(4500)}{11} = 284{,}491$$

which are to be compared with the hazard plot estimates of $\gamma^* = 500$ and $\beta^* = 267{,}000$.

Estimates of the mean and standard deviation are $\hat{E}(X) = \hat{\gamma} + \hat{\beta} = 284{,}927$ and $\hat{\sigma}_X = \hat{\beta} = 284{,}491$. Estimate variances and covariance calculated from (3.9.10) with $n = 12$ are $V(\hat{\gamma}) = 6.1314 \times 10^8$, $V(\hat{\beta}) = 7.3577 \times 10^9$, and $Cov(\hat{\gamma}, \hat{\beta}) = -6.1314 \times 10^8$. Accordingly, $\sigma_{\hat{\gamma}} = 24{,}762$, $\sigma_{\hat{\beta}} = 85{,}777$, and $r_{\hat{\sigma},\hat{\beta}} = -0.29$.

If the estimator (7.5.8) had produced an estimate of $\delta > 1$, this would have indicated that the bell-shaped Weibull distribution was the appropriate model, and in that case, we might have calculated MLE as described in subsection 7.4.1. If the estimate of γ is zero, and (7.5.8) indicates that $\delta > 1$, the MLE $\hat{\delta}$ and $\hat{\theta}$ may be calculated in a simpler manner from (7.4.10).

In the case of an exponential distribution with $\gamma = 0$, the MLE of β is $\hat{\beta} = ST/n$, and for this example, this becomes $\hat{\beta} = 3{,}444{,}400/12 = 287{,}033$, a value that differs but slightly from the MMLE value.

The manufacturer had been considering the advisability of offering an 80,000 hour warranty and wished to know the extent of the resulting repair or replacement obligation. To make this determination, we substitute the MMLE $\hat{\gamma} = 436$ and $\hat{\beta} = 284{,}491$ into the cdf (7.3.2) and calculate $P\{X < 80{,}000\} = 1 - \exp\{-(80{,}000 - 436)/284{,}491\} = 0.2440$. We thereby conclude that an 80,000 hour warranty, without an improvement in quality to extend operating life, might require repair or replacement of as many as 24% of all fans sold.

Small discrepancies between graphical estimates obtained here and those given by Nelson (1969, 1982) are of a magnitude to be expected when hazard plots are fitted by eye.

8

Censored and Truncated Samples from the Normal and the Lognormal Distributions

8.1 INTRODUCTORY REMARKS

This chapter is a continuation of Chapter 4, in which the lognormal distribution was introduced and in which estimation of its parameters from complete samples was considered. In this chapter, we are concerned with estimation of lognormal and normal parameters from truncated and censored samples. Although our primary concern is with the lognormal distribution, because of the close relationship that exists between the normal and the lognormal distribution we are also led to consider estimation in the normal distribution when samples are truncated or censored.

8.2 MAXIMUM LIKELIHOOD ESTIMATION

The loglikelihood function of a progressively censored sample, as described in Chapter 7, from a three-parameter lognormal distribution with pdf (4.2.1) is

$$\ln L = -n \ln \sigma - \sum_{1}^{n} \ln (x_i - \gamma) - \frac{1}{2\sigma^2} \sum_{1}^{n} [\ln (x_i - \gamma) - \mu]^2 + \sum_{1}^{k} c_j \ln [1 - F(T_j)] + \ln K \tag{8.2.1}$$

where

$$F(T_j) = \int_{\gamma}^{T_j} f(x; \gamma, \mu, \sigma)\, dx = \Phi\left[\frac{\ln(T_j - \gamma) - \mu}{\sigma}\right] \tag{8.2.2}$$

where where $\Phi(\ )$ is the cdf of the standard normal distribution (0, 1). The total sample size is N, and the sample data consist of n full-term (complete) observations $\{x_i\}$, $i = 1, 2, \ldots, n$, plus $N - n$ partial-term (censored) observations. Censoring occurs progressively in k stages at times $T_1 < T_2 < \cdots < T_j < \cdots < T_k$. At time T_j, c_j sample specimens, selected randomly from the survivors at that time, are removed (i.e., censored) from further observation. Thus,

$$N - n = \sum_{1}^{k} c_j \tag{8.2.3}$$

Singly censored samples constitute a special case of progressively censored samples in which $k = 1$.

8.2.1 The Estimation Equations

Local maximum likelihood estimators (LMLE) based on progressively censored samples, are obtained as the simultaneous solution of the following equations:

$$\frac{\partial \ln L}{\partial \mu} = \frac{1}{\sigma^2}\sum_{1}^{n}[\ln(x_i - \gamma) - \mu] + \frac{1}{\sigma}\sum_{1}^{k} c_j Q_j = 0$$

$$\frac{\partial \ln L}{\partial \sigma} = -\frac{n}{\sigma} + \frac{1}{\sigma^3}\sum_{1}^{n}[\ln(x_i - \gamma) - \mu]^2 + \frac{1}{\sigma}\sum_{1}^{k} c_j \xi_j Q_j = 0 \tag{8.2.4}$$

$$\frac{\partial \ln L}{\partial \gamma} = \sum_{1}^{n}(x_i - \gamma)^{-1} + \frac{1}{\sigma^2}\sum_{1}^{n}\frac{\ln(x_i - \gamma) - \mu}{x_i - \gamma} + \frac{1}{\sigma}\sum_{1}^{k}\frac{c_j Q_j}{T_j - \gamma} = 0$$

where ln L is given by (8.2.1) and

$$Q_j = Q(\xi_j) = \frac{\phi(\xi_j)}{1 - \Phi(\xi)}, \qquad \xi_j = \frac{\ln(T_j - \gamma) - \mu}{\sigma} \tag{8.2.5}$$

In this notation, $\phi(\ )$ and $\Phi(\ )$ are the density and the cdf, respectively, of the standard normal distribution (0, 1), and $F(T_j) = \Phi(\xi_j)$.

A straightforward trial-and-error iterative procedure for simultaneously solving the three equations of (8.2.4) was outlined by Cohen (1976). However, these calculations can be reduced substantially by taking advantage of the simple relation that exists between the normal and the lognormal distribution. Thereby, estimation in the lognormal distribution is reduced to the more familiar problem of estimating normal distribution parameters.

When the random variable X has a lognormal distribution (γ, μ, σ^2), then $Y = \ln(X - \gamma)$ has a normal distribution (μ, σ^2). Consequently, in the analysis of lognormal sample data, we need only make the transformation $\{y_i\} = \{\ln(x_i - \gamma)\}$ with censoring points $\{D_j\} = \{\ln(T_j - \gamma)\}$, and the transformed sample becomes a normal sample to be analyzed by using applicable normal theory. When γ is known, the transformation and any subsequent estimation or other analysis is quite simple. The situation becomes more complicated when γ is unknown and must be estimated from the sample data. In the latter case, we begin with a first approximation $\gamma_1 < x_1$ and use applicable normal estimators to calculate conditional estimates $\mu(\gamma_1)$ and $\sigma(\gamma_1)$ from the transformed sample data. These conditional estimates can then be "tested" by imposing a third requirement or by substitution in a third estimating equation. This procedure was previously employed by Giesbrecht and Kempthorne (1976).

For LMLE from a progressively censored sample, the third equation of (8.2.4) would be employed as the "test" equation. If this equation is satisfied, our calculations are complete and $\hat{\gamma} = \gamma_1$, $\hat{\mu} = \mu(\gamma_1)$, $\hat{\sigma} = \sigma(\gamma_1)$. Otherwise we select a second approximation $\gamma_2 < \gamma_1$ and repeat the cycle of calculations. We continue until we find a pair of values γ_i and γ_j such that $|\gamma_i - \gamma_j|$ is sufficiently small and such that

$$\frac{\partial \ln L}{\partial \gamma}(\gamma_i) \lesseqgtr 0 \lesseqgtr \frac{\partial \ln L}{\partial \gamma}(\gamma_j)$$

Final estimates are then obtained by interpolation. As discussed in Chapter 4, because of regularity problems that are inherent in the lognormal distribution, estimates thus obtained are LMLE rather than global MLE. In many applications these estimates are satisfactory, but in some applications, modified maximum likelihood estimates (MMLE), considered in a later section of this chapter, might be preferred.

The test equation for the LMLE based on a progressively censored sample, as previously mentioned, is the third equation of (8.2.4). For other sample types, the test equations are given below.

Singly Right-Censored Sample

$$\frac{\sigma^2}{n}\frac{\partial \ln L}{\partial \gamma} = \frac{1}{n}\sum_{1}^{n}\frac{\ln(x_i - \gamma) - \mu}{x_i - \gamma} + \frac{\sigma^2}{n}\sum_{1}^{n}\frac{1}{x_i - \gamma} + \frac{\sigma c Q(\xi)}{n(T - \gamma)} = 0 \qquad (8.2.6)$$

Singly Right Truncated Sample

$$\frac{\sigma^2}{n}\frac{\partial \ln L}{\partial \gamma} = \frac{1}{n}\sum_1^n \frac{\ln(x_i - \gamma) - \mu}{x_i - \gamma} + \frac{\sigma^2}{n}\sum_1^n \frac{1}{x_i - \gamma} + \frac{\sigma Q(-\xi)}{T - \gamma} = 0 \qquad (8.2.7)$$

Doubly Censored Sample

$$\frac{\sigma^2}{n}\frac{\partial \ln L}{\partial \gamma} = \frac{1}{n}\sum_1^n \frac{\ln(x_i - \gamma) - \mu}{x_i - \gamma} + \frac{\sigma^2}{n}\sum_1^n \frac{1}{x_i - \gamma}$$

$$- \sigma\left[\frac{\Omega_1}{T_1 - \gamma} - \frac{\Omega_2}{T_2 - \gamma}\right] = 0 \qquad (8.2.8)$$

where Ω_1 and Ω_2 are defined in Section 8.4 by (8.4.19).

Doubly Truncated Sample

$$\frac{\sigma^2}{n}\frac{\partial \ln L}{\partial \gamma} = \frac{1}{n}\sum_1^n \frac{\ln(x_i - \gamma) - \mu}{x_i - \gamma} + \frac{\sigma^2}{n}\sum_1^n \frac{1}{x_i - \gamma}$$

$$- \sigma\left[\frac{\bar{Q}_1}{T_1 - \gamma} - \frac{\bar{Q}_2}{T_2 - \gamma}\right] = 0 \qquad (8.2.9)$$

where $\bar{Q}_1$ and $\bar{Q}_2$ are defined in Section 8.4 by (8.4.15). Conditional MLE $\mu(\gamma_j)$ and $\sigma(\gamma_j)$ can be calculated as described by Cohen (1957, 1959, 1963) for samples from a normal distribution that are truncated or censored. Estimators based on singly censored and singly truncated samples are derived in the 1959 paper. Additional tables of auxiliary estimating functions were provided by Cohen (1961). Estimates based on doubly truncated and doubly censored samples were presented in the 1957 paper, and estimators based on progressively censored samples were derived in the 1963 paper. When samples are complete, MLE for μ and σ are simply the mean and standard deviation of the transformed sample. Further details are included in Section 8.4.

8.3 MODIFIED MAXIMUM LIKELIHOOD ESTIMATORS FOR LOGNORMAL PARAMETERS BASED ON CENSORED SAMPLES

Modified maximum likelihood estimators which employ the first-order statistic are often to be preferred over the LMLE, which were considered in Section 8.2. Estimating equations in this case differ from corresponding

LMLE equations only in that the LMLE equation $\partial \ln L/\partial\gamma = 0$ is replaced by $E[\ln(X_1 - \gamma)] = \ln(x_1 - \gamma)$, which leads to

$$\gamma + e^{\mu+\sigma E(Z_{1;N})} = x_1 \tag{8.3.1}$$

where $E(Z_{1;N})$ is the expected value of the first-order statistic in a sample of size N from the standard normal distribution (0, 1). This equation is equivalent to the third complete sample estimating equation of (4.6.2). The two equations become identical when we make the substitutions $\beta = e^{\mu}$ and $\sigma = \sqrt{\ln \omega}$ in (4.6.2). An abridged table of $E(Z_{1;N})$ for selected values of N was extracted from tables of Harter (1961) and is included as Table 4.2.

Procedures for calculating the MMLE are essentially the same as those described in Section 8.2 for calculating the LMLE. We start with a first approximation γ_1, make the transformation $y_i = \ln(x_i - \gamma_1)$, and employ applicable normal distribution estimators to calculate conditional estimates $\mu(\gamma_1)$ and $\sigma(\gamma_1)$. We then substitute the values γ_1, $\mu(\gamma_1)$, and $\sigma(\gamma_1)$ into (8.3.1). If this test equation is satisfied, then no further calculation of estimates is needed. Otherwise, we select a second approximation γ_2 and repeat the cycle of calculations as described. We continue until we find two values γ_i and γ_j in a sufficiently narrow interval such that

$$\gamma_i + e^{\mu_i+\sigma_i E(Z_{1;N})} \lesseqgtr x_1 \lesseqgtr \gamma_j + e^{\mu_j+\sigma_j E(Z_{1;N})}$$

We then interpolate for final estimates.

The MMLE require that x_1 and the total sample size N be available from sample data. They are therefore applicable for complete samples and for samples that are censored only on the right. The LMLE are available for other types of samples.

An alternative MMLE was described by Cohen (1987), in which $E[F(X_1)] = F(x_1) = 1/(N + 1)$ is the replacement equation for $\partial \ln L/\partial\gamma = 0$. In its expanded form, this equation becomes

$$\gamma + e^{\mu+\sigma z_1} = x_1 \tag{8.3.2}$$

where z_1 is the standard normal variate corresponding to the cdf value of $1/(N + 1)$; i.e., $z_1 = \Phi^{-1}[1/(N + 1)]$. Equations (8.3.1) and (8.3.2) differ only in the substitution of z_1 in (8.3.2) for $E(Z_{1;N})$ in (8.3.1). The following abbreviated table provides comparisons between $\Phi^{-1}[1/(N + 1)]$ and $E[Z_{1;N}]$ for selected values of N.

N	$\Phi^{-1}(1/(N+1))$	$E(Z_{1,N})$
5	-0.9675	-1.16296
10	-1.34	-1.53875
20	-1.67	-1.86748
30	-1.85	-2.04276
40	-1.97	-2.16078
50	-2.06	-2.24907
75	-2.22	-2.40299
100	-2.33	-2.50759
200	-2.58	-2.74604
300	-2.72	-2.87777
400	-2.808	-2.96818

Note that $\Phi^{-1}(1/(N+1)) > E(Z_{1,N})$ for all N, and that the difference ranges from 0.196 for N = 5 to 0.160 for N = 400.

In most applications, the performance of (8.3.1) is slightly superior to that of (8.3.2), but in many instances the differences are insignificant. When sample sizes are small, (8.3.2) is sometimes preferred.

8.4 MAXIMUM LIKELIHOOD ESTIMATION IN THE NORMAL DISTRIBUTION

Since estimation in the lognormal distribution is intimately involved with estimation in the normal distribution, a brief account of MLE for normal distribution parameters based on truncated and censored samples of various types is included in this section. We let Y designate a random variable that is normal (μ, σ^2). In the context of this chapter, it represents a transformation from X, that is lognormal (γ, μ, σ^2). In the presentation which follows, we consider in turn samples that are singly truncate, singly censored, doubly truncated, doubly censored, and progressively censored. In all samples, N designates the total sample size, n designates the number of complete (full-term) observations, and N - n is the number of censored observations. In truncated samples and in complete samples N = n. We let D (or D_j in the case of multiple truncation or censoring) designate the point(s) of censoring or truncation. Samples to be considered are further described as follows.

Singly Truncated Samples. In singly truncated samples a terminus D is specified. Observation is possible only if $y \geq D$, in which truncation is said to be on the left, or if $y \leq D$, in which case truncation is said to be on the right. In this case, measurements are known for all sample specimens, and hence $N = n$. In certain applications it might be preferable to consider that the restriction (i.e., truncation) is imposed on the distribution rather than on the sample being observed. The adoption of this latter point of view involves no change in the estimators.

Singly Censored Samples. As in the singly truncated samples, a terminus D is identified, but in this case, sample specimens whose measurements fall in the restricted interval of the random variable may be identified and thus counted, though not otherwise measured. When the restricted (censored) interval consists of all values $y < D$, censoring is said to occur on the left. When the censored interval consists of all values $y > D$, censoring is said to occur on the right. The remaining specimens for which $y \geq D$ (or $y \leq D$) are fully measured without restriction. Samples of this type thus consist of N observations, of which n are fully measured and $N - n$ are censored. If D is fixed (i.e., predetermined), samples are said to be of Type I, in which case n is a random variable. If n is fixed, samples are said to be of Type II, and D is a random variable. In a singly right-censored sample of Type II, $D = y_n$. In a singly left-censored sample of Type II, $D = y_{(N-n+1)}$.

Doubly Truncated Samples. In doubly truncated samples both a lower terminus D_1 and an upper terminus D_2 are specified. Observation is possible only if $D_1 \leq y \leq D_2$, and $N = n$.

Doubly Censored Samples. With D_1 and D_2 as terminals, doubly censored samples bear the same relation to doubly truncated samples that singly censored samples bear to singly truncated samples. We let c_1 designate the number of censored observations such that $y < D_1$, and c_2 designates the number of censored observations such that $y > D_2$. Accordingly, $N - n = c_1 + c_2$. If D_1 and D_2 are fixed, then c_1 and c_2 are random variables and the sample is of Type I. If c_1 and c_2 are fixed, then D_1 and D_2 are random variables (i.e., order statistics) and the sample is of Type II.

Progressively Censored Samples. Progressively censored samples were described in Chapter 7. Censoring occurs of points D_j, $j = 1, \ldots, k$, with the censoring of c_j sample specimens at point D_j.

8.4.1 Estimators for Singly Truncated Samples

For both left and right truncation, the estimators are

$$\hat{\mu} = \bar{y} - \theta(\hat{\alpha})[\bar{y} - D], \qquad \hat{\sigma}^2 = s_y^2 + \theta(\hat{\alpha})[\bar{y} - D]^2 \tag{8.4.1}$$

where

TABLE 8.1 The Auxiliary Estimation Function $\theta(\alpha)$ for Singly Truncated Samples from the Normal Distribution (μ, σ^2)

α	.000	.001	.002	.003	.004	.005	.006	.007	.008	.009
.050	.000004	.000005	.000006	.000007	.000009	.000011	.000013	.000015	.000017	.000020
.060	.000024	.000027	.000031	.000036	.000041	.000047	.000053	.000060	.000067	.000075
.070	.000084	.000094	.000104	.000116	.000128	.000141	.000155	.000171	.000187	.000204
.080	.000223	.000242	.000263	.000285	.000309	.000334	.000360	.000388	.000417	.000448
.090	.000481	.000515	.000550	.000588	.000627	.000668	.000711	.000756	.000802	.000851
.100	.000902	.000954	.001009	.001066	.001125	.001187	.001250	.001316	.001384	.001455
.110	.001528	.001604	.001682	.001762	.001845	.001931	.002019	.002110	.002204	.002300
.120	.002400	.002502	.002607	.002715	.002826	.002939	.003056	.003176	.003299	.003425
.130	.003554	.003687	.003822	.003961	.004103	.004249	.004398	.004550	.004705	.004865
.140	.005027	.005193	.005363	.005536	.005713	.005893	.006078	.006265	.006457	.006652
.150	.006852	.007055	.007262	.007472	.007687	.007906	.008129	.008355	.008586	.008821
.160	.009060	.009303	.009551	.009802	.010058	.010318	.010583	.010852	.011125	.011402
.170	.011684	.011971	.012262	.012557	.012857	.013162	.013471	.013785	.014103	.014426
.180	.014754	.015087	.015425	.015767	.016114	.016467	.016824	.017186	.017553	.017925
.190	.018302	.018684	.019071	.019463	.019861	.020264	.020672	.021085	.021503	.021927
.200	.022356	.022791	.023231	.023677	.024128	.024584	.025046	.025514	.025987	.026466
.210	.026950	.027440	.027936	.028438	.028946	.029459	.029978	.030503	.031035	.031572
.220	.032115	.032664	.033219	.033780	.034347	.034921	.035501	.036087	.036679	.037278
.230	.037882	.038494	.039111	.039735	.040366	.041003	.041647	.042297	.042954	.043617
.240	.044287	.044964	.045648	.046338	.047035	.047739	.048450	.049168	.049893	.050625
.250	.051364	.052110	.052863	.053623	.054390	.055165	.055947	.056736	.057533	.058337
.260	.059148	.059967	.060794	.061627	.062469	.063318	.064175	.065039	.065911	.066791
.270	.067679	.068575	.069478	.070390	.071309	.072236	.073172	.074115	.075067	.076027
.280	.076995	.077972	.078956	.079950	.080951	.081961	.082979	.084006	.085042	.086086
.290	.087139	.088200	.089271	.090350	.091438	.092534	.093640	.094755	.095879	.097012
.300	.098153	.099305	.100465	.101634	.102813	.104002	.105199	.106406	.107623	.108849
.310	.110085	.111331	.112586	.113851	.115125	.116410	.117704	.119009	.120323	.121648
.320	.122983	.124327	.125682	.127048	.128423	.129809	.131206	.132613	.134030	.135459
.330	.136897	.138347	.139807	.141278	.142760	.144253	.145757	.147272	.148798	.150335
.340	.151884	.153444	.155015	.156597	.158191	.159797	.161414	.163043	.164683	.166336
.350	.168000	.169676	.171364	.173064	.174776	.176500	.178237	.179986	.181747	.183521
.360	.185307	.187106	.188917	.190741	.192578	.194427	.196290	.198165	.200054	.201955
.370	.203870	.205798	.207740	.209694	.211663	.213644	.215640	.217649	.219672	.221709
.380	.223759	.225824	.227903	.229996	.232103	.234224	.236360	.238510	.240675	.242854
.390	.245048	.247257	.249481	.251720	.253974	.256242	.258527	.260826	.263141	.265471

α	.000	.001	.002	.003	.004	.005	.006	.007	.008	.009
.400	.267817	.270178	.272555	.274948	.277357	.279782	.282222	.284679	.287153	.289642
.410	.292148	.294671	.297210	.299766	.302339	.304929	.307535	.310159	.312800	.315459
.420	.318134	.320828	.323539	.326267	.329014	.331778	.334560	.337361	.340179	.343016
.430	.345872	.348746	.351638	.354550	.357480	.360429	.363397	.366385	.369392	.372418
.440	.375464	.378530	.381615	.384720	.387845	.390990	.394156	.397342	.400548	.403776
.450	.407023	.410292	.413582	.416892	.420224	.423578	.426953	.430349	.433768	.437208
.460	.440670	.444154	.447661	.451190	.454742	.458316	.461913	.465533	.469177	.472843
.470	.476533	.480247	.483984	.487745	.491530	.495339	.499173	.503031	.506913	.510820
.480	.514753	.518710	.522692	.526700	.530733	.534793	.538878	.542988	.547126	.551289
.490	.555479	.559696	.563940	.568210	.572508	.576833	.581186	.585566	.589975	.594411
.500	.598876	.603369	.607891	.612442	.617022	.621631	.626269	.630937	.635635	.640362
.510	.645120	.649909	.654727	.659577	.664458	.669369	.674312	.679287	.684294	.689332
.520	.694403	.699507	.704643	.709811	.715013	.720249	.725518	.730820	.736157	.741528
.530	.746934	.752374	.757849	.763359	.768905	.774487	.780104	.785758	.791448	.797175
.540	.802938	.808739	.814578	.820454	.826368	.832320	.838311	.844340	.850409	.856517
.550	.862665	.868852	.875080	.881348	.887657	.894007	.900399	.906832	.913307	.919824
.560	.926384	.932986	.939632	.946321	.953054	.959831	.966653	.973519	.980431	.987388
.570	.994391	1.001439	1.008535	1.015677	1.022866	1.030103	1.037387	1.044720	1.052101	1.059531
.580	1.067011	1.074540	1.082119	1.089749	1.097429	1.105161	1.112944	1.120779	1.128667	1.136607
.590	1.144601	1.152648	1.160749	1.168905	1.177115	1.185381	1.193703	1.202080	1.210514	1.219006
.600	1.227554	1.236161	1.244826	1.253550	1.262333	1.271176	1.280080	1.289044	1.298069	1.307156
.610	1.316305	1.325517	1.334793	1.344132	1.353535	1.363003	1.372536	1.382136	1.391801	1.401534
.620	1.411334	1.421202	1.431139	1.441145	1.451221	1.461367	1.471585	1.481873	1.492234	1.502668
.630	1.513175	1.523756	1.534411	1.545142	1.555949	1.566832	1.577792	1.588831	1.599948	1.611144
.640	1.622420	1.633777	1.645214	1.656734	1.668337	1.680023	1.691794	1.703649	1.715590	1.727617
.650	1.739732	1.751935	1.764226	1.776607	1.789079	1.801641	1.814296	1.827044	1.839885	1.852821
.660	1.865852	1.878980	1.892205	1.905527	1.918949	1.932471	1.946094	1.959818	1.973646	1.987576
.670	2.001612	2.015753	2.030001	2.044357	2.058821	2.073395	2.088080	2.102877	2.117786	2.132810
.680	2.147949	2.163204	2.178576	2.194067	2.209678	2.225409	2.241263	2.257239	2.273341	2.289567
.690	2.305921	2.322404	2.339015	2.355758	2.372632	2.389641	2.406784	2.424063	2.441480	2.459036
.700	2.476732	2.494570	2.512552	2.530678	2.548951	2.567372	2.585942	2.604664	2.623537	2.642566
.710	2.661750	2.681091	2.700592	2.720254	2.740078	2.760067	2.780222	2.800545	2.821038	2.841703
.720	2.862541	2.883555	2.904746	2.926116	2.947668	2.969404	2.991325	3.013434	3.035732	3.058222
.730	3.080906	3.103787	3.126866	3.150146	3.173629	3.197317	3.221213	3.245320	3.269639	3.294173
.740	3.318926	3.343898	3.369094	3.394515	3.420165	3.446045	3.472160	3.498511	3.525102	3.551935

$$\hat{\alpha} = \frac{s_y^2}{(\bar{y} - D)^2} \tag{8.4.2}$$

and $\bar{y}$ and s_y^2 are the sample mean and variance, respectively. Table 8.1 contains entries of θ as a function of α. Accordingly, with $\hat{\alpha} = s^2/(\bar{y} - D)^2$ calculated from the sample data, we interpolate between entries in this table to obtain $\theta(\hat{\alpha})$, and the MLE follow from (8.4.1).

Detailed derivations of the estimators in (8.4.1) are given by Cohen (1959). Following are definitions of the pertinent functions involved in these derivations. Let ξ designate the standardized terminal

$$\xi = \frac{D - \mu}{\sigma} \tag{8.4.3}$$

Then $Q(\xi)$, the standardized hazard function evaluated at the terminal, is

$$Q(\xi) = \frac{\phi(\xi)}{1 - \Phi(\xi)} \tag{8.4.4}$$

where, consistent with the previously used notation, $\phi(\ )$ and $\Phi(\ )$ are the pdf and the cdf, respectively, of the standard normal distribution (0, 1). The auxiliary estimating function θ, with ξ as the argument, is

$$\theta(\xi) = \frac{Q(\xi)}{Q(\xi) - \xi} \tag{8.4.5}$$

The function $\alpha(\xi)$ is defined as

$$\alpha(\xi) = \frac{1 - Q(\xi)[Q(\xi) - \xi]}{[Q(\xi) - \xi]^2} \tag{8.4.6}$$

Maximum likelihood estimators of μ and σ^2 require that

$$\hat{\alpha} = \alpha(\hat{\xi}) = \frac{s_y^2}{(D - \bar{y})^2} \tag{8.4.7}$$

It has therefore been expedient to tabulate θ as a function of α rather than of ξ. Thus, with $s_y^2/(D - \bar{y})^2$ calculated from sample data, $\theta(\hat{\alpha})$ can be obtained directly from Table 8.1. Entries in this table extend from $\alpha = 0.050$ to $\alpha = 0.749$, with θ ranging from $\theta(0.050) = 0.000004$ to $\theta(0.749) = 3.551935$. The corresponding range of percent truncation is from less than 0.1% to 89%. Both $\alpha(\ )$ and $\theta(\ )$ increase as the degree of truncation increases, and it is most unlikely that values of α beyond the range of tabulated entries will be needed in any practical application involving parameter estimation from truncated normal distribution samples. For 50% truncation, $\xi = 0$, $Q(0) = \sqrt{2/\pi} = 0.79978846$, $\theta(0) = 1$, and $\alpha(0) = 0.5707963$. Most practical applications will involve a much less severe degree of truncation. The following table further exhibits α and θ in relation to percent of truncation. Note that calculations of θ become less accurate as the percent of truncation → 100.

Percent of Truncation, $\alpha(\ )$ and $\theta(\ )$ as Functions of $\eta = \pm\xi$[a]

$\eta = \pm\xi$	% Trun. $100\,\Phi(\eta)$	$\alpha(\eta)$	$\theta(\eta)$	$\eta = \pm\xi$	% Trun. $100\Phi(\eta)$	$\alpha(\eta)$	$\theta(\eta)$
-3.00	0.14	0.11	0.001	0	50.00	0.57	1.00
-2.75	0.30	0.13	0.003	0.25	59.87	0.61	1.32
-2.50	0.62	0.15	0.007	0.50	69.15	0.65	1.7
-2.25	1.22	0.18	0.014	0.75	77.34	0.69	2.3
-2.00	2.28	0.21	0.027	1.00	84.13	0.72	2.9
-1.75	4.01	0.25	0.05	1.25	89.44	0.75	3.6
-1.50	6.68	0.29	0.09	1.50	93.32	0.78	4.5
-1.25	10.57	0.33	0.14	1.75	95.99	0.80	5.3
-1.00	15.87	0.38	0.22	2.00	99.73	0.82	6.3
-0.75	22.66	0.43	0.35	2.25	98.78	0.83	7
-0.50	30.85	0.48	0.51	2.50	99.38	0.86	9
-0.25	40.13	0.53	0.75	3.00	99.86	0.88	12

[a]For left truncation, $\eta = \xi$ and for right truncation $\eta = -\xi$.

8.4.2 Estimators for Singly Censored Samples

For both left and right censoring, the estimators are

$$\hat{\mu} = \bar{y} - \lambda(h, \hat{\alpha})[\bar{y} - D], \quad \hat{\sigma}^2 = s_y^2 + \lambda(h, \hat{\alpha})[\bar{y} - D]^2 \tag{8.4.8}$$

where $\hat{\alpha}$ is given by (8.4.2) and h is the proportion of censored observations in the sample; i.e.,

$$h = \frac{N - n}{N} \tag{8.4.9}$$

The estimating function $\lambda(h, \xi)$ is analogous to $\theta(\xi)$, which arose in connection with estimation from truncated samples. It is defined as

$$\lambda(h, \xi) = \frac{\Omega(h, \xi)}{\Omega(h, \xi) - \xi} \tag{8.4.10}$$

where

$$\Omega(h, \xi) = \frac{h}{1 - h} Q(-\xi) \tag{8.4.11}$$

Although they were originally derived by Cohen (1959) for singly left-truncated and singly left-censored samples, the estimators of (8.4.1) and (8.4.8) are also applicable for singly right-truncated and singly right-censored samples. Since the normal distribution is symmetrical about the mean, it follows that truncation and censoring at $z = \xi$ on the left is equivalent to truncation and censoring on the right at $z = -\xi$. As a consequence of choosing to base derivations on left-restricted samples, the definition given for $\Omega(h, \xi)$ in (8.4.4) is the same as that subsequently given in (8.4.19) for $\Omega_1(a_1, \xi)$ when considering the left side of a doubly censored sample. The function $\alpha(h, \xi)$ is defined as

$$\alpha(h, \xi) = \frac{1 - \Omega(h, \xi)[\Omega(h, \xi) - \xi]}{[\Omega(h, \xi) - \xi]^2} \tag{8.4.12}$$

Maximum likelihood estimators of μ and σ^2 require that

$$\hat{\alpha} = \alpha(h, \hat{\xi}) = \frac{s_2^2}{(D - \bar{y})^2} \tag{8.4.13}$$

In order to facilitate calculation of estimators (8.4.8), it has therefore been expedient to tabulate λ in Table 8.2 as a function of h and α rather than as a function of h and ξ. With h and $\hat{\alpha}$ calculated from sample data, $\lambda(h, \hat{\alpha})$ can be obtained directly from Table 8.2, and estimates $\hat{\mu}$ and $\hat{\sigma}^2$ then follow from (8.4.8) for both left and right censoring and for both Type I and Type II censoring. It is necessary only that we remember that $D = y_n$ or y_{N-n+1} for Type II censoring.

TABLE 8.2 Auxiliary Estimation Function $\lambda(h, \alpha)$ for Singly Censored Samples from the Normal Distribution (μ, σ^2)

α \ h	.01	.02	.03	.04	.05	.06	.07	.08	.09	.10	.15
.00	.01010	.02040	.03090	.04161	.05251	.06363	.07495	.08649	.09824	.11020	.17342
.01	.01020	.02059	.03118	.04197	.05297	.06417	.07557	.08719	.09902	.11106	.17465
.02	.01029	.02077	.03145	.04233	.05341	.06469	.07618	.08787	.09978	.11190	.17586
.03	.01038	.02095	.03172	.04268	.05384	.06520	.07677	.08854	.10052	.11272	.17704
.04	.01047	.02113	.03197	.04302	.05426	.06570	.07734	.08919	.10125	.11352	.17821
.05	.01055	.02129	.03223	.04335	.05467	.06619	.07791	.08983	.10197	.11431	.17935
.06	.01064	.02146	.03247	.04367	.05507	.06667	.07846	.09046	.10267	.11508	.18047
.07	.01072	.02162	.03271	.04399	.05546	.06713	.07900	.09107	.10335	.11584	.18157
.08	.01080	.02178	.03294	.04430	.05585	.06759	.07953	.09168	.10403	.11659	.18266
.09	.01087	.02193	.03317	.04460	.05623	.06804	.08006	.09227	.10469	.11732	.18373
.10	.01095	.02208	.03340	.04490	.05660	.06848	.08057	.09285	.10534	.11804	.18479
.11	.01102	.02223	.03362	.04519	.05696	.06892	.08107	.09343	.10598	.11875	.18583
.12	.01110	.02238	.03384	.04548	.05732	.06934	.08157	.09399	.10661	.11944	.18685
.13	.01117	.02252	.03405	.04577	.05767	.06976	.08205	.09454	.10723	.12013	.18786
.14	.01124	.02266	.03426	.04604	.05802	.07018	.08254	.09509	.10785	.12081	.18886
.15	.01131	.02280	.03447	.04632	.05836	.07059	.08301	.09563	.10845	.12148	.18985
.16	.01138	.02293	.03467	.04659	.05869	.07099	.08348	.09616	.10905	.12214	.19082
.17	.01145	.02307	.03487	.04685	.05902	.07138	.08394	.09668	.10963	.12279	.19178
.18	.01151	.02320	.03507	.04712	.05935	.07177	.08439	.09720	.11021	.12343	.19273
.19	.01158	.02333	.03526	.04737	.05967	.07216	.08484	.09771	.11079	.12407	.19367
.20	.01164	.02346	.03545	.04763	.05999	.07254	.08528	.09822	.11135	.12469	.19460
.21	.01171	.02359	.03564	.04788	.06030	.07291	.08572	.09871	.11191	.12531	.19552
.22	.01177	.02371	.03583	.04813	.06061	.07329	.08615	.09921	.11246	.12592	.19643
.23	.01183	.02383	.03601	.04838	.06092	.07365	.08657	.09969	.11301	.12653	.19733
.24	.01189	.02396	.03620	.04862	.06122	.07401	.08700	.10017	.11355	.12713	.19822
.25	.01195	.02408	.03638	.04886	.06152	.07437	.08741	.10065	.11408	.12772	.19910
.26	.01201	.02420	.03656	.04909	.06182	.07473	.08783	.10112	.11461	.12831	.19997
.27	.01207	.02431	.03673	.04933	.06211	.07508	.08823	.10158	.11513	.12889	.20083
.28	.01213	.02443	.03691	.04956	.06240	.07542	.08864	.10205	.11565	.12946	.20169
.29	.01219	.02454	.03708	.04979	.06269	.07577	.08904	.10250	.11616	.13003	.20254
.30	.01224	.02466	.03725	.05002	.06297	.07611	.08943	.10295	.11667	.13059	.20338
.31	.01230	.02477	.03742	.05024	.06325	.07644	.08982	.10340	.11717	.13115	.20421
.32	.01236	.02488	.03758	.05047	.06353	.07678	.09021	.10384	.11767	.13170	.20503
.33	.01241	.02499	.03775	.05069	.06380	.07711	.09060	.10428	.11816	.13225	.20585
.34	.01247	.02510	.03791	.05090	.06408	.07743	.09098	.10472	.11865	.13279	.20666
.35	.01252	.02521	.03808	.05112	.06435	.07776	.09136	.10515	.11914	.13333	.20747
.36	.01257	.02532	.03824	.05133	.06461	.07808	.09173	.10557	.11962	.13386	.20826
.37	.01263	.02542	.03840	.05155	.06488	.07839	.09210	.10600	.12009	.13439	.20906
.38	.01268	.02553	.03855	.05176	.06514	.07871	.09247	.10642	.12057	.13491	.20984
.39	.01273	.02563	.03871	.05197	.06540	.07902	.09283	.10683	.12103	.13543	.21062

$\lambda(0, \alpha) = 0$ for all values of α. $h = (N - n)/N$.

TABLE 8.2 Continued

α \ h	.20	.25	.30	.35	.40	.45	.50	.60	.70	.80	.90
.00	.24268	.31862	.40210	.49414	.59607	.70957	.83684	1.14536	1.56148	2.17591	3.28261
.01	.24426	.32054	.40434	.49670	.59894	.71275	.84033	1.14947	1.56625	2.18139	3.28898
.02	.24581	.32243	.40655	.49923	.60178	.71590	.84378	1.15355	1.57098	2.18685	3.29532
.03	.24734	.32429	.40873	.50172	.60459	.71901	.84720	1.15759	1.57568	2.19227	3.30163
.04	.24885	.32612	.41089	.50419	.60736	.72210	.85060	1.16161	1.58035	2.19767	3.30792
.05	.25033	.32793	.41301	.50663	.61011	.72515	.85396	1.16559	1.58499	2.20304	3.31419
.06	.25179	.32972	.41511	.50904	.61283	.72817	.85729	1.16955	1.58960	2.20838	3.32043
.07	.25322	.33147	.41719	.51142	.61552	.73117	.86059	1.17347	1.59419	2.21369	3.32665
.08	.25464	.33321	.41924	.51378	.61818	.73414	.86386	1.17737	1.59874	2.21898	3.33284
.09	.25604	.33493	.42126	.51611	.62082	.73708	.86711	1.18124	1.60327	2.22424	3.33901
.10	.25741	.33662	.42326	.51842	.62343	.73999	.87033	1.18508	1.60777	2.22948	3.34516
.11	.25877	.33829	.42525	.52071	.62602	.74288	.87352	1.18890	1.61225	2.23469	3.35128
.12	.26012	.33995	.42720	.52297	.62858	.74575	.87669	1.19269	1.61669	2.23987	3.35739
.13	.26144	.34158	.42914	.52521	.63112	.74859	.87983	1.19645	1.62112	2.24503	3.36346
.14	.26275	.34320	.43106	.52743	.63364	.75140	.88295	1.20019	1.62552	2.25017	3.36952
.15	.26405	.34480	.43296	.52962	.63613	.75420	.88605	1.20390	1.62989	2.25528	3.37556
.16	.26533	.34638	.43484	.53180	.63860	.75697	.88912	1.20759	1.63424	2.26037	3.38157
.17	.26660	.34794	.43670	.53396	.64106	.75972	.89217	1.21126	1.63856	2.26543	3.38756
.18	.26785	.34949	.43855	.53610	.64349	.76245	.89519	1.21490	1.64287	2.27048	3.39353
.19	.26909	.35103	.44038	.53822	.64590	.76515	.89820	1.21852	1.64714	2.27550	3.39948
.20	.27031	.35255	.44219	.54032	.64829	.76784	.90118	1.22212	1.65140	2.28049	3.40541
.21	.27152	.35405	.44398	.54240	.65067	.77051	.90415	1.22570	1.65563	2.28547	3.41132
.22	.27273	.35554	.44576	.54447	.65302	.77315	.90709	1.22925	1.65985	2.29042	3.41721
.23	.27391	.35702	.44752	.54652	.65536	.77578	.91001	1.23279	1.66404	2.29536	3.42307
.24	.27509	.35848	.44927	.54855	.65768	.77839	.91292	1.23630	1.66821	2.30027	3.42892
.25	.27626	.35993	.45100	.55057	.65998	.78098	.91580	1.23979	1.67235	2.30516	3.43475
.26	.27741	.36137	.45272	.55257	.66227	.78356	.91867	1.24327	1.67648	2.31003	3.44056
.27	.27856	.36279	.45443	.55455	.66454	.78611	.92152	1.24672	1.68059	2.31488	3.44635
.28	.27969	.36421	.45612	.55653	.66679	.78865	.92434	1.25015	1.68467	2.31970	3.45212
.29	.28082	.36561	.45780	.55848	.66903	.79117	.92716	1.25357	1.68874	2.32451	3.45787
.30	.28193	.36700	.45946	.56042	.67125	.79368	.92995	1.25696	1.69279	2.32930	3.46360
.31	.28304	.36838	.46112	.56235	.67346	.79617	.93273	1.26034	1.69682	2.33407	3.46931
.32	.28414	.36975	.46276	.56427	.67565	.79864	.93549	1.26370	1.70082	2.33882	3.47501
.33	.28522	.37110	.46438	.56617	.67783	.80110	.93823	1.26704	1.70481	2.34355	3.48068
.34	.28630	.37245	.46600	.56806	.67999	.80354	.94096	1.27036	1.70879	2.34827	3.48634
.35	.28737	.37379	.46761	.56993	.68214	.80597	.94367	1.27367	1.71274	2.35296	3.49198
.36	.28844	.37511	.46920	.57179	.68427	.80838	.94637	1.27696	1.71668	2.35764	3.49761
.37	.28949	.37643	.47078	.57364	.68640	.81078	.94905	1.28023	1.72059	2.36230	3.50321
.38	.29053	.37774	.47235	.57548	.68851	.81316	.95172	1.28349	1.72449	2.36694	3.50880
.39	.29157	.37904	.47391	.57731	.69060	.81553	.95437	1.28673	1.72838	2.37156	3.51437

α \ h	.01	.02	.03	.04	.05	.06	.07	.08	.09	.10	.15
.40	.01278	.02574	.03887	.05217	.06566	.07933	.09319	.10725	.12150	.13595	.21139
.41	.01284	.02584	.03902	.05238	.06592	.07964	.09355	.10766	.12196	.13646	.21216
.42	.01289	.02594	.03917	.05258	.06617	.07994	.09391	.10806	.12242	.13697	.21292
.43	.01294	.02604	.03932	.05278	.06642	.08025	.09426	.10847	.12287	.13747	.21368
.44	.01299	.02614	.03947	.05298	.06667	.08055	.09461	.10887	.12332	.13797	.21443
.45	.01304	.02624	.03962	.05318	.06692	.08085	.09496	.10926	.12377	.13847	.21517
.46	.01309	.02634	.03977	.05338	.06717	.08114	.09530	.10966	.12421	.13896	.21591
.47	.01313	.02644	.03992	.05357	.06741	.08143	.09565	.11005	.12465	.13945	.21665
.48	.01318	.02654	.04006	.05377	.06765	.08173	.09598	.11044	.12509	.13994	.21738
.49	.01323	.02663	.04021	.05396	.06790	.08201	.09632	.11082	.12552	.14042	.21810
.50	.01328	.02673	.04035	.05415	.06813	.08230	.09666	.11121	.12595	.14090	.21882
.51	.01333	.02682	.04049	.05434	.06837	.08259	.09699	.11159	.12638	.14138	.21954
.52	.01337	.02692	.04064	.05453	.06861	.08287	.09732	.11196	.12681	.14185	.22025
.53	.01342	.02701	.04078	.05472	.06884	.08315	.09765	.11234	.12723	.14232	.22095
.54	.01347	.02710	.04092	.05490	.06907	.08343	.09797	.11271	.12765	.14278	.22166
.55	.01351	.02720	.04105	.05509	.06931	.08371	.09830	.11308	.12806	.14325	.22235
.56	.01356	.02729	.04119	.05527	.06954	.08398	.09862	.11345	.12848	.14371	.22305
.57	.01360	.02738	.04133	.05546	.06976	.08426	.09894	.11382	.12889	.14417	.22374
.58	.01365	.02747	.04146	.05564	.06999	.08453	.09926	.11418	.12930	.14462	.22442
.59	.01369	.02756	.04160	.05582	.07022	.08480	.09957	.11454	.12970	.14507	.22510
.60	.01374	.02765	.04173	.05600	.07044	.08507	.09989	.11490	.13011	.14552	.22578
.61	.01378	.02774	.04187	.05617	.07066	.08534	.10020	.11526	.13051	.14597	.22645
.62	.01383	.02783	.04200	.05635	.07088	.08560	.10051	.11561	.13091	.14641	.22712
.63	.01387	.02791	.04213	.05653	.07110	.08586	.10082	.11596	.13131	.14685	.22779
.64	.01391	.02800	.04226	.05670	.07132	.08613	.10112	.11631	.13170	.14729	.22845
.65	.01396	.02809	.04239	.05687	.07154	.08639	.10143	.11666	.13209	.14773	.22910
.66	.01400	.02817	.04252	.05705	.07175	.08665	.10173	.11701	.13248	.14816	.22976
.67	.01404	.02826	.04265	.05722	.07197	.08690	.10203	.11735	.13287	.14859	.23041
.68	.01409	.02834	.04278	.05739	.07218	.08716	.10233	.11769	.13326	.14902	.23106
.69	.01413	.02843	.04290	.05756	.07239	.08742	.10263	.11804	.13364	.14945	.23170
.70	.01417	.02851	.04303	.05773	.07260	.08767	.10292	.11837	.13402	.14987	.23234
.71	.01421	.02860	.04316	.05789	.07281	.08792	.10322	.11871	.13440	.15030	.23298
.72	.01425	.02868	.04328	.05806	.07302	.08817	.10351	.11905	.13478	.15072	.23361
.73	.01430	.02876	.04341	.05823	.07323	.08842	.10380	.11938	.13515	.15113	.23425
.74	.01434	.02885	.04353	.05839	.07344	.08867	.10409	.11971	.13553	.15155	.23487
.75	.01438	.02893	.04365	.05856	.07364	.08892	.10438	.12004	.13590	.15196	.23550
.76	.01442	.02901	.04377	.05872	.07385	.08916	.10467	.12037	.13627	.15237	.23612
.77	.01446	.02909	.04390	.05888	.07405	.08941	.10495	.12070	.13664	.15278	.23674
.78	.01450	.02917	.04402	.05904	.07425	.08965	.10524	.12102	.13700	.15319	.23735
.79	.01454	.02925	.04414	.05920	.07445	.08989	.10552	.12134	.13737	.15360	.23797

TABLE 8.2 Continued

α \ h	.20	.25	.30	.35	.40	.45	.50	.60	.70	.80	.90
.40	.29260	.38033	.47547	.57912	.69268	.81789	.95700	1.28995	1.73224	2.37616	3.51993
.41	.29363	.38161	.47701	.58093	.69475	.82023	.95963	1.29316	1.73609	2.38075	3.52546
.42	.29464	.38288	.47854	.58272	.69681	.82256	.96223	1.29636	1.73993	2.38532	3.53098
.43	.29565	.38414	.48006	.58450	.69886	.82488	.96483	1.29953	1.74374	2.38988	3.53649
.44	.29665	.38540	.48157	.58627	.70089	.82719	.96741	1.30270	1.74754	2.39441	3.54197
.45	.29765	.38665	.48307	.58803	.70292	.82948	.96998	1.30584	1.75133	2.39893	3.54744
.46	.29864	.38788	.48456	.58978	.70493	.83176	.97253	1.30898	1.75510	2.40344	3.55290
.47	.29962	.38912	.48605	.59152	.70693	.83402	.97507	1.31209	1.75885	2.40793	3.55834
.48	.30059	.39034	.48752	.59325	.70892	.83628	.97760	1.31520	1.76259	2.41240	3.56376
.49	.30156	.39156	.48899	.59497	.71090	.83852	.98012	1.31829	1.76631	2.41685	3.56917
.50	.30253	.39276	.49044	.59668	.71286	.84075	.98262	1.32136	1.77002	2.42129	3.57456
.51	.30348	.39396	.49189	.59838	.71482	.84297	.98511	1.32443	1.77371	2.42572	3.57993
.52	.30443	.39516	.49333	.60007	.71677	.84518	.98759	1.32748	1.77739	2.43013	3.58529
.53	.30538	.39635	.49476	.60175	.71870	.84738	.99006	1.33051	1.78106	2.43452	3.59064
.54	.30632	.39753	.49619	.60343	.72063	.84957	.99251	1.33353	1.78470	2.43890	3.59597
.55	.30725	.39870	.49760	.60509	.72255	.85174	.99495	1.33654	1.78834	2.44327	3.60128
.56	.30818	.39987	.49901	.60674	.72445	.85391	.99739	1.33954	1.79196	2.44762	3.60658
.57	.30910	.40103	.50041	.60839	.72635	.85606	.99981	1.34252	1.79557	2.45195	3.61187
.58	.31002	.40218	.50181	.61003	.72824	.85821	1.00222	1.34550	1.79916	2.45628	3.61714
.59	.31093	.40333	.50319	.61166	.73012	.86034	1.00462	1.34845	1.80274	2.46058	3.62240
.60	.31184	.40447	.50457	.61328	.73199	.86247	1.00700	1.35140	1.80631	2.46487	3.62764
.61	.31274	.40560	.50594	.61489	.73385	.86458	1.00938	1.35434	1.80987	2.46915	3.63287
.62	.31364	.40673	.50731	.61650	.73570	.86669	1.01175	1.35726	1.81341	2.47342	3.63808
.63	.31453	.40785	.50867	.61810	.73754	.86878	1.01411	1.36017	1.81694	2.47767	3.64328
.64	.31542	.40897	.51002	.61968	.73938	.87087	1.01645	1.36307	1.82045	2.48191	3.64847
.65	.31630	.41008	.51136	.62127	.74120	.87295	1.01879	1.36596	1.82395	2.48613	3.65364
.66	.31718	.41119	.51270	.62284	.74302	.87502	1.02112	1.36884	1.82744	2.49034	3.65880
.67	.31805	.41229	.51403	.62441	.74483	.87707	1.02343	1.37170	1.83092	2.49454	3.66394
.68	.31892	.41338	.51535	.62597	.74663	.87912	1.02574	1.37456	1.83439	2.49872	3.66907
.69	.31979	.41447	.51667	.62752	.74842	.88116	1.02804	1.37740	1.83784	2.50289	3.67419
.70	.32065	.41555	.51798	.62907	.75021	.88320	1.03032	1.38023	1.84128	2.50705	3.67930
.71	.32150	.41663	.51929	.63060	.75199	.88522	1.03260	1.38305	1.84471	2.51120	3.68439
.72	.32236	.41771	.52059	.63213	.75376	.88723	1.03487	1.38587	1.84813	2.51533	3.68947
.73	.32320	.41877	.52188	.63366	.75552	.88924	1.03713	1.38867	1.85154	2.51945	3.69453
.74	.32405	.41984	.52317	.63518	.75727	.89124	1.03938	1.39146	1.85493	2.52356	3.69958
.75	.32489	.42090	.52445	.63669	.75902	.89323	1.04162	1.39424	1.85832	2.52766	3.70462
.76	.32572	.42195	.52573	.63819	.76076	.89521	1.04386	1.39701	1.86169	2.53174	3.70965
.77	.32655	.42300	.52700	.63969	.76249	.89718	1.04608	1.39977	1.86505	2.53581	3.71467
.78	.32738	.42404	.52826	.64118	.76422	.89915	1.04830	1.40252	1.86840	2.53987	3.71967
.79	.32821	.42508	.52952	.64267	.76593	.90111	1.05051	1.40526	1.87174	2.54392	3.72466

α \ h	.01	.02	.03	.04	.05	.06	.07	.08	.09	.10	.15
.80	.01458	.02933	.04426	.05936	.07465	.09013	.10580	.12167	.13773	.15400	.23858
.81	.01462	.02941	.04438	.05952	.07485	.09037	.10608	.12199	.13809	.15440	.23918
.82	.01466	.02949	.04450	.05968	.07505	.09061	.10636	.12231	.13845	.15480	.23979
.83	.01470	.02957	.04461	.05984	.07525	.09085	.10664	.12262	.13881	.15520	.24039
.84	.01474	.02965	.04473	.06000	.07545	.09108	.10691	.12294	.13916	.15559	.24099
.85	.01478	.02972	.04485	.06015	.07564	.09132	.10719	.12325	.13952	.15599	.24158
.86	.01481	.02980	.04496	.06031	.07584	.09155	.10746	.12357	.13987	.15638	.24218
.87	.01485	.02988	.04508	.06046	.07603	.09179	.10773	.12388	.14022	.15677	.24277
.88	.01489	.02995	.04520	.06062	.07622	.09202	.10800	.12419	.14057	.15716	.24336
.89	.01493	.03003	.04531	.06077	.07641	.09225	.10827	.12450	.14092	.15755	.24394
.90	.01497	.03011	.04542	.06092	.07661	.09248	.10854	.12480	.14126	.15793	.24452
.91	.01500	.03018	.04554	.06107	.07680	.09271	.10881	.12511	.14161	.15832	.24511
.92	.01504	.03026	.04565	.06123	.07699	.09293	.10907	.12541	.14195	.15870	.24568
.93	.01508	.03033	.04576	.06138	.07717	.09316	.10934	.12572	.14229	.15908	.24626
.94	.01512	.03041	.04588	.06153	.07736	.09339	.10960	.12602	.14263	.15946	.24683
.95	.01515	.03048	.04599	.06168	.07755	.09361	.10987	.12632	.14297	.15983	.24740
.96	.01519	.03056	.04610	.06182	.07773	.09383	.11013	.12662	.14331	.16021	.24797
.97	.01523	.03063	.04621	.06197	.07792	.09406	.11039	.12692	.14365	.16058	.24854
.98	.01526	.03070	.04632	.06212	.07810	.09428	.11065	.12721	.14398	.16096	.24910
.99	.01530	.03078	.04643	.06227	.07829	.09450	.11090	.12751	.14431	.16133	.24966
1.00	.01534	.03085	.04654	.06241	.07847	.09472	.11116	.12780	.14465	.16170	.25022
1.50	.01699	.03417	.05153	.06908	.08682	.10476	.12290	.14125	.15981	.17858	.27585
2.00	.01842	.03703	.05583	.07483	.09403	.11343	.13304	.15287	.17291	.19318	.29806
2.50	.01969	.03958	.05967	.07996	.10046	.12117	.14210	.16325	.18463	.20624	.31794
3.00	.02085	.04191	.06317	.08464	.10633	.12823	.15037	.17273	.19532	.21816	.33611
3.50	.02192	.04406	.06641	.08897	.11176	.13477	.15802	.18150	.20522	.22919	.35294
4.00	.02293	.04607	.06943	.09302	.11684	.14088	.16517	.18970	.21448	.23951	.36870
4.50	.02387	.04797	.07229	.09684	.12162	.14665	.17192	.19744	.22321	.24925	.38356
5.00	.02477	.04977	.07499	.10046	.12616	.15211	.17832	.20478	.23150	.25849	.39766
5.50	.02562	.05148	.07757	.10391	.13049	.15733	.18442	.21177	.23940	.26730	.41112
6.00	.02644	.05312	.08004	.10721	.13464	.16232	.19026	.21847	.24696	.27573	.42400
6.50	.02723	.05470	.08242	.11039	.13862	.16711	.19587	.22491	.25423	.28383	.43639
7.00	.02798	.05622	.08470	.11345	.14245	.17173	.20128	.23111	.26123	.29165	.44832
7.50	.02871	.05768	.08691	.11640	.14616	.17619	.20651	.23711	.26800	.29919	.45986
8.00	.02942	.05910	.08905	.11926	.14975	.18052	.21157	.24291	.27455	.30650	.47103
8.50	.03011	.06048	.09112	.12204	.15323	.18471	.21648	.24854	.28091	.31359	.48186
9.00	.03078	.06182	.09314	.12473	.15661	.18878	.22125	.25402	.28709	.32048	.49240
9.50	.03143	.06313	.09510	.12736	.15991	.19275	.22589	.25934	.29311	.32719	.50265
10.00	.03206	.06440	.09701	.12992	.16312	.19662	.23042	.26454	.29897	.33373	.51265

TABLE 8.2 Continued

α \ h	.20	.25	.30	.35	.40	.45	.50	.60	.70	.80	.90
.80	.32903	.42612	.53078	.64415	.76764	.90306	1.05270	1.40799	1.87507	2.54796	3.72964
.81	.32984	.42715	.53203	.64562	.76935	.90500	1.05490	1.41071	1.87839	2.55198	3.73460
.82	.33065	.42817	.53327	.64709	.77105	.90694	1.05708	1.41342	1.88170	2.55599	3.73956
.83	.33146	.42919	.53451	.64855	.77274	.90887	1.05925	1.41613	1.88499	2.56000	3.74450
.84	.33227	.43021	.53574	.65000	.77442	.91079	1.06142	1.41882	1.88828	2.56399	3.74943
.85	.33307	.43122	.53697	.65145	.77610	.91270	1.06358	1.42150	1.89155	2.56797	3.75434
.86	.33387	.43223	.53819	.65290	.77777	.91460	1.06573	1.42418	1.89482	2.57193	3.75925
.87	.33466	.43323	.53941	.65433	.77943	.91650	1.06787	1.42685	1.89808	2.57589	3.76414
.88	.33546	.43423	.54062	.65577	.78109	.91840	1.07001	1.42950	1.90132	2.57984	3.76903
.89	.33624	.43523	.54183	.65719	.78274	.92028	1.07213	1.43215	1.90456	2.58377	3.77390
.90	.33703	.43622	.54303	.65861	.78439	.92216	1.07425	1.43479	1.90778	2.58770	3.77876
.91	.33781	.43721	.54423	.66003	.78603	.92403	1.07637	1.43742	1.91100	2.59161	3.78361
.92	.33859	.43819	.54542	.66144	.78766	.92589	1.07847	1.44004	1.91421	2.59551	3.78844
.93	.33936	.43917	.54661	.66284	.78929	.92775	1.08057	1.44266	1.91740	2.59940	3.79327
.94	.34014	.44015	.54780	.66424	.79091	.92960	1.08266	1.44526	1.92059	2.60329	3.79808
.95	.34091	.44112	.54898	.66564	.79252	.93145	1.08474	1.44786	1.92377	2.60716	3.80289
.96	.34167	.44209	.55015	.66703	.79413	.93329	1.08682	1.45045	1.92694	2.61102	3.80768
.97	.34243	.44305	.55132	.66841	.79574	.93512	1.08889	1.45303	1.93010	2.61487	3.81246
.98	.34319	.44401	.55249	.66979	.79734	.93694	1.09095	1.45560	1.93325	2.61871	3.81723
.99	.34395	.44497	.55365	.67116	.79893	.93876	1.09301	1.45817	1.93639	2.62254	3.82199
1.00	.34471	.44592	.55481	.67253	.80051	.94058	1.09506	1.46072	1.93952	2.62636	3.82674
1.50	.37929	.48973	.60812	.73566	.87383	1.02451	1.19009	1.57980	2.08610	2.80620	4.05201
2.00	.40934	.52788	.65466	.79092	.93818	1.09836	1.27394	1.68551	2.21714	2.96832	4.25750
2.50	.43629	.56213	.69651	.84068	.99621	1.16508	1.34983	1.78156	2.33672	3.11711	4.44765
3.00	.46092	.59348	.73485	.88631	1.04949	1.22641	1.41967	1.87018	2.44742	3.25541	4.62546
3.50	.48377	.62257	.77044	.92871	1.09903	1.28348	1.48472	1.95288	2.55096	3.38516	4.79305
4.00	.50516	.64982	.80381	.96848	1.14552	1.33707	1.54584	2.03071	2.64858	3.50779	4.95202
4.50	.52534	.67555	.83532	1.00605	1.18947	1.38775	1.60368	2.10443	2.74119	3.62434	5.10357
5.00	.54450	.69998	.86526	1.04176	1.23125	1.43595	1.65870	2.17465	2.82949	3.73564	5.24865
5.50	.56279	.72329	.89383	1.07585	1.27115	1.48200	1.71129	2.24181	2.91404	3.84235	5.38802
6.00	.58030	.74563	.92122	1.10853	1.30941	1.52616	1.76174	2.30628	2.99526	3.94498	5.52232
6.50	.59713	.76711	.94755	1.13996	1.34621	1.56866	1.81030	2.36836	3.07354	4.04398	5.65205
7.00	.61336	.78782	.97294	1.17027	1.38172	1.60966	1.85716	2.42831	3.14916	4.13970	5.77765
7.50	.62905	.80783	.99749	1.19958	1.41605	1.64931	1.90249	2.48632	3.22238	4.23245	5.89950
8.00	.64424	.82722	1.02127	1.22798	1.44932	1.68774	1.94643	2.54257	3.29342	4.32249	6.01791
8.50	.65898	.84603	1.04435	1.25555	1.48162	1.72506	1.98910	2.59722	3.36245	4.41005	6.13316
9.00	.67331	.86432	1.06679	1.28235	1.51302	1.76136	2.03060	2.65039	3.42965	4.49531	6.24549
9.50	.68726	.88213	1.08864	1.30845	1.54361	1.79671	2.07103	2.70220	3.49514	4.57845	6.35510
10.00	.70086	.89949	1.10995	1.33390	1.57344	1.83118	2.11047	2.75274	3.55906	4.65963	6.46220

8.4.3 Estimators for Doubly Truncated Samples

Maximum likelihood estimating equations for normal distribution parameters based on samples that are truncated on the left at D_1 and on the right at D_2, as derived by Cohen (1957), are

$$H_1(\xi_1, \xi_2) = \frac{\bar{Q}_1 - \bar{Q}_2 - \xi_1}{\xi_2 - \xi_1} = \frac{\nu_1}{w}$$
$$H_2(\xi_1, \xi_2) = \frac{1 + \xi_1\bar{Q}_1 - \xi_2\bar{Q}_2 - (\bar{Q}_1 - \bar{Q}_2)^2}{(\xi_2 - \xi_1)^2} = \frac{s^2}{w^2} \tag{8.4.14}$$

where

$$\bar{Q}_1 = \bar{Q}_1(\xi_1, \xi_2) = \frac{\phi(\xi_1)}{\Phi(\xi_2) - \Phi(\xi_1)}$$

and

$$\bar{Q}_2 = \bar{Q}_2(\xi_1, \xi_2) = \frac{\phi(\xi_2)}{\Phi(\xi_2) - \Phi(\xi_1)} \tag{8.4.15}$$

with

$$\xi_1 = \frac{D_1 - \mu}{\sigma} \quad \text{and} \quad \xi_2 = \frac{D_2 - \mu}{\sigma}$$
$$w = D_2 - D_1, \quad \nu_1 = \bar{y} - D_1 \tag{8.4.16}$$

and where $\bar{y}$ and s_y^2 are the mean and variance (unbiased), respectively, of the truncated sample.

The two equations of (8.3.14) can be solved simultaneously for estimates $\hat{\xi}_1$ and $\hat{\xi}_2$. It then follows that

$$\hat{\sigma} = \frac{w}{\hat{\xi}_2 - \hat{\xi}_1} \quad \text{and} \quad \hat{\mu} = D_1 - \hat{\sigma}\hat{\xi}_1 \tag{8.4.17}$$

Any one of various iterative procedures might be employed to solve (8.4.14). As an aid to facilitate their solution, Table 8.3, which contains entries of $H_1(\xi_1, \xi_2)$ and $H_2(\xi_1, \xi_2)$ for various values of the arguments ξ_1 and ξ_2, has been included. A less extensive tabulation of these functions was originally prepared by Thompson, Friedman, and Garelis (1954). A chart which enables one to read estimates $\hat{\xi}_1$ and $\hat{\xi}_2$ with one- or perhaps two-decimal accuracy was given by Cohen (1957). With permission from the <u>Biometrika</u> trustees, this chart is reproduced here as Figure 8.1. When

TABLE 8.3 Estimating Functions $H_1(\xi_1, \xi_2)$ and $H_2(\xi_1, \xi_2)$ for Doubly Truncated Samples from the Normal Distribution

ξ_1 \ ξ_2	.0	.1	.2	.3	.4	.5	.6	.7	.8	.9
.0	.000000	.499583	.498336	.496261	.493370	.489673	.485187	.479934	.473941	.467237
	.000000	.083305	.083221	.083075	.082863	.082577	.082207	.081742	.081170	.080477
-.1	.500417	.500000	.498754	.496684	.493802	.490122	.485661	.480441	.474492	.467845
	.083305	.083222	.083083	.082883	.082618	.082280	.081859	.081344	.080724	.079987
-.2	.501664	.501246	.500000	.497934	.495060	.491394	.486954	.481766	.475857	.469262
	.083221	.083083	.082890	.082638	.082323	.081936	.081469	.080911	.080250	.079475
-.3	.503739	.503316	.502066	.500000	.497131	.493475	.489054	.483892	.478020	.471471
	.083075	.082883	.082638	.082338	.081975	.081544	.081035	.080438	.079742	.078936
-.4	.506630	.506198	.504940	.502869	.500000	.496350	.491942	.486802	.480960	.474451
	.082863	.082618	.082323	.081975	.081569	.081098	.080551	.079921	.079195	.078364
-.5	.510327	.509878	.508606	.506525	.503650	.500000	.495599	.490474	.484655	.478180
	.082577	.082280	.081936	.081544	.081098	.080589	.080010	.079351	.078601	.077750
-.6	.514813	.514339	.513046	.510946	.508058	.504401	.500000	.494882	.489080	.482630
	.082207	.081859	.081469	.081035	.080551	.080010	.079403	.078721	.077952	.077087
-.7	.520066	.519559	.518234	.516108	.513198	.509526	.505118	.500000	.494206	.487773
	.081742	.081344	.080911	.080438	.079921	.079351	.078721	.078020	.077238	.076364
-.8	.526059	.525508	.524143	.521980	.519040	.515345	.510920	.505794	.500000	.493575
	.081170	.080724	.080250	.079742	.079195	.078601	.077952	.077238	.076447	.075571
-.9	.532763	.532155	.530738	.528529	.525549	.521820	.517370	.512227	.506425	.500000
	.080477	.079987	.079475	.078936	.078364	.077750	.077087	.076364	.075571	.074697
-1.0	.540138	.539462	.537981	.535714	.532684	.528913	.524428	.519259	.513440	.507006
	.079652	.079121	.078575	.078009	.077416	.076788	.076116	.075390	.074599	.073733
-1.1	.548142	.547383	.545826	.543491	.540399	.536576	.532047	.526844	.520999	.514549
	.078683	.078115	.077540	.076951	.076343	.075705	.075030	.074306	.073522	.072669
-1.2	.556724	.555870	.554224	.551808	.548645	.544759	.540178	.534932	.529053	.522578
	.077559	.076960	.076361	.075755	.075137	.074495	.073822	.073105	.072335	.071500
-1.3	.565831	.564866	.563119	.560611	.557366	.553409	.548766	.543468	.537548	.531042
	.076273	.075649	.075031	.074415	.073792	.073152	.072486	.071783	.071031	.070220
-1.4	.575400	.574311	.572450	.569839	.566502	.562464	.557752	.552395	.546427	.539882
	.074821	.074177	.073549	.072928	.072306	.071674	.071022	.070337	.069609	.068826

Top entry is $H_1(\xi_1, \xi_2)$ and bottom entry is $H_2(\xi_1, \xi_2)$. $H_1(\xi_1, \xi_2) = [\bar{Q}_1 - \bar{Q}_2 - \xi_1]/(\xi_2 - \xi_1)$; $H_2(\xi_1, \xi_2) = [1 + \xi_1\bar{Q}_1 - \xi_2\bar{Q}_2 - (\bar{Q}_1 - \bar{Q}_2)^2]/(\xi_2 - \xi_1)^2$.

ξ_1 \ ξ_2	1.0	1.1	1.2	1.3	1.4	1.5	1.6	1.7	1.8	1.9
.0	.459862	.451858	.443276	.434169	.424600	.414634	.404342	.393797	.383075	.372249
	.079652	.078683	.077559	.076273	.074821	.073201	.071415	.069471	.067381	.065161
−.1	.460538	.452617	.444130	.435134	.425689	.415861	.405718	.395334	.384782	.374134
	.079121	.078115	.076960	.075649	.074177	.072545	.070756	.068817	.066740	.064541
−.2	.462019	.454174	.445776	.436881	.427550	.417847	.407841	.397603	.387204	.376715
	.078575	.077540	.076361	.075031	.073549	.071912	.070126	.068198	.066140	.063967
−.3	.464286	.456509	.448192	.439389	.430161	.420571	.410688	.400581	.390319	.379972
	.078009	.076951	.075755	.074415	.072928	.071293	.069516	.067605	.065570	.063427
−.4	.467316	.459601	.451355	.442634	.433498	.424009	.414236	.404244	.394104	.383882
	.077416	.076343	.075137	.073792	.072306	.070680	.068918	.067027	.065020	.062911
−.5	.471087	.463424	.455241	.446591	.437536	.428136	.418459	.408569	.398535	.388423
	.076788	.075705	.074495	.073152	.071674	.070062	.068320	.066455	.064480	.062407
−.6	.475572	.467953	.459822	.451234	.442248	.432926	.423331	.413530	.403588	.393571
	.076116	.075030	.073822	.072486	.071022	.069428	.067711	.065877	.063937	.061905
−.7	.480741	.473156	.465068	.456532	.447605	.438347	.428824	.419098	.409235	.399298
	.075390	.074306	.073105	.071783	.070337	.068768	.067080	.065280	.063379	.061390
−.8	.486560	.479001	.470947	.462452	.453573	.444370	.434906	.425244	.415447	.405577
	.074599	.073522	.072335	.071031	.069609	.068069	.066414	.064653	.062795	.060852
−.9	.492994	.485451	.477422	.468958	.460118	.450959	.441543	.431933	.422190	.412376
	.073733	.072669	.071500	.070220	.068826	.067319	.065703	.063984	.062172	.060278
−1.0	.500000	.492466	.484452	.476011	.467200	.458075	.448698	.439129	.429431	.419661
	.072781	.071736	.070590	.069339	.067978	.066510	.064936	.063263	.061500	.059658
−1.1	.507534	.500000	.491994	.483568	.474776	.465677	.456330	.446794	.437129	.427395
	.071736	.070715	.069597	.068379	.067056	.065630	.064102	.062479	.060768	.058982
−1.2	.515548	.508006	.500000	.491581	.482802	.473720	.464394	.454882	.445244	.435536
	.070590	.069597	.068513	.067332	.066052	.064672	.063194	.061624	.059969	.058240
−1.3	.523989	.516432	.508419	.500000	.491227	.482156	.472845	.463350	.453730	.444041
	.069339	.068379	.067332	.066194	.064960	.063630	.062206	.060692	.059096	.057427
−1.4	.532800	.525224	.517198	.508773	.500000	.490934	.481631	.472147	.462540	.452864
	.067978	.067056	.066052	.064960	.063776	.062500	.061132	.059678	.058143	.056538

(continued)

TABLE 8.3 Continued

ξ_1 \ ξ_2	2.0	2.1	2.2	2.3	2.4	2.5	2.6	2.7	2.8	2.9
.0	.361395	.350582	.339876	.329338	.319020	.308968	.299220	.289803	.280740	.272043
	.062829	.060409	.057924	.055400	.052864	.050340	.047850	.045415	.043053	.040778
-.1	.363461	.352833	.342313	.331958	.321818	.311939	.302354	.293091	.284171	.275606
	.062239	.059856	.057414	.054939	.052456	.049987	.047554	.045177	.042872	.040651
-.2	.366206	.355743	.345386	.335193	.325211	.315482	.306039	.296910	.288113	.279660
	.061698	.059353	.056957	.054531	.052100	.049686	.047308	.044986	.042734	.040566
-.3	.369608	.359290	.349079	.339027	.329181	.319583	.310263	.301248	.292555	.284196
	.061194	.058891	.056541	.054165	.051786	.049425	.047101	.044832	.042632	.040513
-.4	.373646	.363456	.353370	.343441	.333714	.324227	.315012	.306092	.297487	.289206
	.060717	.058458	.056155	.053829	.051502	.049194	.046923	.044705	.042555	.040483
-.5	.378298	.368218	.358241	.348418	.338791	.329398	.320271	.311431	.302897	.294678
	.060255	.058042	.055787	.053512	.051237	.048981	.046762	.044595	.042493	.040466
-.6	.383540	.373556	.363671	.353937	.344394	.335080	.326024	.317249	.308772	.300602
	.059796	.057630	.055426	.053202	.050979	.048775	.046607	.044490	.042435	.040453
-.7	.389349	.379444	.369638	.359977	.350504	.341254	.332256	.323531	.315097	.306964
	.059328	.057212	.055058	.052887	.050717	.048566	.046449	.044380	.042372	.040435
-.8	.395695	.385857	.376114	.366514	.357097	.347898	.338945	.330259	.321856	.313747
	.058839	.056774	.054673	.052556	.050439	.048340	.046274	.044255	.042294	.040400
-.9	.402550	.392766	.383075	.373523	.364150	.354989	.346069	.337410	.329028	.320933
	.058317	.056305	.054259	.052197	.050134	.048089	.046075	.044105	.042190	.040340
-1.0	.409879	.400138	.390487	.380973	.371632	.362500	.353602	.344960	.336589	.328500
	.057751	.055795	.053805	.051799	.049792	.047801	.045839	.043920	.042052	.040247
-1.1	.417647	.407939	.398319	.388832	.379515	.370401	.361517	.352883	.344515	.336422
	.057132	.055233	.053302	.051353	.049404	.047468	.045559	.043690	.041871	.040111
-1.2	.425814	.416131	.406533	.397064	.387763	.378659	.369781	.361148	.352775	.344673
	.056450	.054611	.052740	.050851	.048960	.047080	.045226	.043409	.041640	.039925
-1.3	.434338	.424672	.415089	.405632	.396338	.387239	.378360	.369721	.361338	.353220
	.055698	.053922	.052112	.050285	.048453	.046632	.044834	.043070	.041350	.039683
-1.4	.443174	.433518	.423944	.414493	.405201	.396100	.387215	.378565	.370167	.362029
	.054873	.053160	.051415	.049650	.047880	.046118	.044377	.042667	.040999	.039379

ξ_1 \ ξ_2	.0	.1	.2	.3	.4	.5	.6	.7	.8	.9
-1.5	.585366	.584139	.582153	.579429	.575991	.571864	.567074	.561653	.555630	.549041
	.073201	.072545	.071912	.071293	.070680	.070062	.069428	.068768	.068069	.067319
-1.6	.595658	.594282	.592159	.589312	.585764	.581541	.576669	.571176	.565094	.558457
	.071415	.070756	.070126	.069516	.068918	.068320	.067711	.067080	.066414	.065703
-1.7	.606203	.604666	.602397	.599419	.595756	.591431	.586470	.580902	.574756	.568067
	.069471	.068817	.068198	.067605	.067027	.066455	.065877	.065280	.064653	.063984
-1.8	.616925	.615218	.612796	.609681	.605896	.601465	.596412	.590765	.584553	.577810
	.067381	.066740	.066140	.065570	.065020	.064480	.063937	.063379	.062795	.062172
-1.9	.627751	.625866	.623285	.620028	.616118	.611577	.606429	.600702	.594423	.587624
	.065161	.064541	.063967	.063427	.062911	.062407	.061905	.061390	.060852	.060278
-2.0	.638605	.636539	.633794	.630392	.626354	.621702	.616460	.610651	.604305	.597450
	.062829	.062239	.061698	.061194	.060717	.060255	.059796	.059328	.058839	.058317
-2.1	.649418	.647167	.644257	.640710	.636544	.631782	.626444	.620556	.614143	.607234
	.060409	.059856	.059353	.058891	.058458	.058042	.057630	.057212	.056774	.056305
-2.2	.660124	.657687	.654614	.650921	.646630	.641759	.636329	.630362	.623886	.616925
	.057924	.057414	.056957	.056541	.056155	.055787	.055426	.055058	.054673	.054259
-2.3	.670662	.668042	.664807	.660973	.656559	.651582	.646063	.640023	.633486	.626477
	.055400	.054939	.054531	.054165	.053829	.053512	.053202	.052887	.052556	.052197
-2.4	.680980	.678182	.674789	.670819	.666286	.661209	.655606	.649496	.642903	.635850
	.052864	.052456	.052100	.051786	.051502	.051237	.050979	.050717	.050439	.050134
-2.5	.691032	.688061	.684518	.680417	.675773	.670602	.664920	.658746	.652102	.645011
	.050340	.049987	.049686	.049425	.049194	.048981	.048775	.048566	.048340	.048089
-2.6	.700780	.697646	.693961	.689737	.684988	.679729	.673976	.667744	.661055	.653931
	.047850	.047554	.047308	.047101	.046923	.046762	.046607	.046449	.046274	.046075
-2.7	.710197	.706909	.703090	.698752	.693908	.688569	.682751	.676469	.669741	.662590
	.045415	.045177	.044986	.044832	.044705	.044595	.044490	.044380	.044255	.044105
-2.8	.719260	.715829	.711887	.707445	.702513	.697103	.691228	.684903	.678144	.670972
	.043053	.042872	.042734	.042632	.042555	.042493	.042435	.042372	.042294	.042190
-2.9	.727957	.724394	.720340	.715804	.710794	.705322	.699398	.693036	.686253	.679067
	.040778	.040651	.040566	.040513	.040483	.040466	.040453	.040435	.040400	.040340
-3.0	.736281	.732598	.728443	.723823	.718745	.713219	.707254	.700864	.694063	.686870
	.038601	.038526	.038489	.038482	.038497	.038523	.038552	.038574	.038580	.038561

(continued)

TABLE 8.3 Continued

ξ_1 \ ξ_2	1.0	1.1	1.2	1.3	1.4	1.5	1.6	1.7	1.8	1.9
-1.5	.541925	.534323	.526280	.517844	.509066	.500000	.490700	.481223	.471623	.461956
	.066510	.065630	.064672	.063630	.062500	.061280	.059973	.058580	.057109	.055569
-1.6	.551302	.543670	.535606	.527155	.518369	.509300	.500000	.490525	.480930	.471267
	.064936	.064102	.063194	.062206	.061132	.059973	.058727	.057399	.055994	.054521
-1.7	.560871	.553206	.545118	.536650	.527853	.518777	.509475	.500000	.490406	.480745
	.063263	.062479	.061624	.060692	.059678	.058580	.057399	.056137	.054800	.053395
-1.8	.570569	.562871	.554756	.546270	.537460	.528377	.519070	.509594	.500000	.490339
	.061500	.060768	.059969	.059096	.058143	.057109	.055994	.054800	.053532	.052197
-1.9	.580339	.572605	.564464	.555959	.547136	.538044	.528733	.519255	.509661	.500000
	.059658	.058982	.058240	.057427	.056538	.055569	.054521	.053395	.052197	.050933
-2.0	.590121	.582353	.574186	.565662	.556826	.547727	.538412	.528933	.519338	.509678
	.057751	.057132	.056450	.055698	.054873	.053970	.052989	.051933	.050804	.049610
-2.1	.599862	.592061	.583869	.575328	.566482	.557376	.548059	.538579	.528986	.519328
	.055795	.055233	.054611	.053922	.053160	.052323	.051410	.050423	.049364	.048240
-2.2	.609513	.601681	.593467	.584911	.576056	.566946	.557628	.548151	.538561	.528907
	.053805	.053302	.052740	.052112	.051415	.050643	.049797	.048878	.047888	.046833
-2.3	.619027	.611168	.602936	.594368	.585507	.576396	.567081	.557608	.548024	.538376
	.051799	.051353	.050851	.050285	.049650	.048943	.048162	.047310	.046388	.045401
-2.4	.628368	.620485	.612237	.603662	.594799	.585690	.576380	.566916	.557341	.547703
	.049792	.049404	.048960	.048453	.047880	.047236	.046519	.045732	.044875	.043955
-2.5	.637500	.629599	.621341	.612761	.603900	.594797	.585497	.576044	.566483	.556857
	.047801	.047468	.047080	.046632	.046118	.045535	.044880	.044155	.043363	.042507
-2.6	.646398	.638483	.630219	.621640	.612785	.603693	.594407	.584969	.575424	.565816
	.045839	.045559	.045226	.044834	.044377	.043852	.043257	.042592	.041861	.041066
-2.7	.655040	.647117	.638852	.630279	.621435	.612357	.603089	.593671	.584147	.574559
	.043920	.043690	.043409	.043070	.042667	.042198	.041659	.041052	.040379	.039644
-2.8	.663411	.655485	.647225	.638662	.629833	.620775	.611529	.602135	.592635	.583072
	.042052	.041871	.041640	.041350	.040999	.040582	.040097	.039545	.038927	.038247
-2.9	.671500	.663578	.655327	.646780	.637971	.628937	.619717	.610351	.600880	.591345
	.040247	.040111	.039925	.039683	.039379	.039011	.038577	.038076	.037511	.036884
-3.0	.679303	.671389	.663153	.654627	.645843	.636837	.627647	.618313	.608874	.599371
	.038509	.038415	.038271	.038073	.037815	.037493	.037106	.036653	.036136	.035559

ξ_1 \ ξ_2	2.0	2.1	2.2	2.3	2.4	2.5	2.6	2.7	2.8	2.9
-1.5	.452273	.442624	.433054	.423604	.414310	.405203	.396307	.387643	.379225	.371063
	.053970	.052323	.050643	.048943	.047236	.045535	.043852	.042198	.040582	.039011
-1.6	.461588	.451941	.442372	.432919	.423620	.414503	.405593	.396911	.388471	.380283
	.052989	.051410	.049797	.048162	.046519	.044880	.043257	.041659	.040097	.038577
-1.7	.471067	.461421	.451849	.442392	.433084	.423956	.415031	.406329	.397865	.389649
	.051933	.050423	.048878	.047310	.045732	.044155	.042592	.041052	.039545	.038076
-1.8	.480662	.471014	.461439	.451976	.442659	.433517	.424576	.415853	.407365	.399120
	.050804	.049364	.047888	.046388	.044875	.043363	.041861	.040379	.038927	.037511
-1.9	.490322	.480672	.471093	.461624	.452297	.443143	.434184	.425441	.416928	.408655
	.049610	.048240	.046833	.045401	.043955	.042507	.041066	.039644	.038247	.036884
-2.0	.500000	.490349	.480767	.471291	.461956	.452789	.443815	.435052	.426515	.418214
	.048359	.047059	.045721	.044357	.042977	.041593	.040214	.038851	.037510	.036200
-2.1	.509651	.500000	.490416	.480936	.471593	.462415	.453426	.444645	.436086	.427759
	.047059	.045829	.044560	.043264	.041950	.040629	.039312	.038008	.036723	.035466
-2.2	.519233	.509584	.500000	.490517	.481169	.471982	.462980	.454183	.445604	.437254
	.045721	.044560	.043360	.042130	.040881	.039624	.038368	.037122	.035893	.034689
-2.3	.528709	.519064	.509483	.500000	.490648	.481455	.472444	.463633	.455037	.446665
	.044357	.043264	.042130	.040966	.039781	.038586	.037390	.036202	.035028	.033877
-2.4	.538044	.528407	.518831	.509352	.500000	.490804	.481785	.472964	.464353	.455964
	.042977	.041950	.040881	.039781	.038659	.037525	.036388	.035256	.034136	.033036
-2.5	.547211	.537585	.528018	.518545	.509196	.500000	.490978	.482149	.473528	.465124
	.041593	.040629	.039624	.038586	.037525	.036450	.035370	.034293	.033227	.032177
-2.6	.556185	.546574	.537020	.527556	.518215	.509022	.500000	.491168	.482539	.474124
	.040214	.039312	.038368	.037390	.036388	.035370	.034346	.033322	.032307	.031306
-2.7	.564948	.555355	.545817	.536367	.527036	.517851	.508832	.500000	.491367	.482945
	.038851	.038008	.037122	.036202	.035256	.034293	.033322	.032350	.031385	.030432
-2.8	.573485	.563914	.554396	.544963	.535647	.526472	.517461	.508633	.500000	.491574
	.037510	.036723	.035893	.035028	.034136	.033227	.032307	.031385	.030467	.029560
-2.9	.581786	.572241	.562746	.553335	.544036	.534876	.525876	.517055	.508426	.500000
	.036200	.035466	.034689	.033877	.033036	.032177	.031306	.030432	.029560	.028696
-3.0	.589843	.580328	.570861	.561474	.552197	.543055	.534070	.525260	.516638	.508216
	.034926	.034242	.033515	.032752	.031961	.031150	.030326	.029496	.028668	.027847

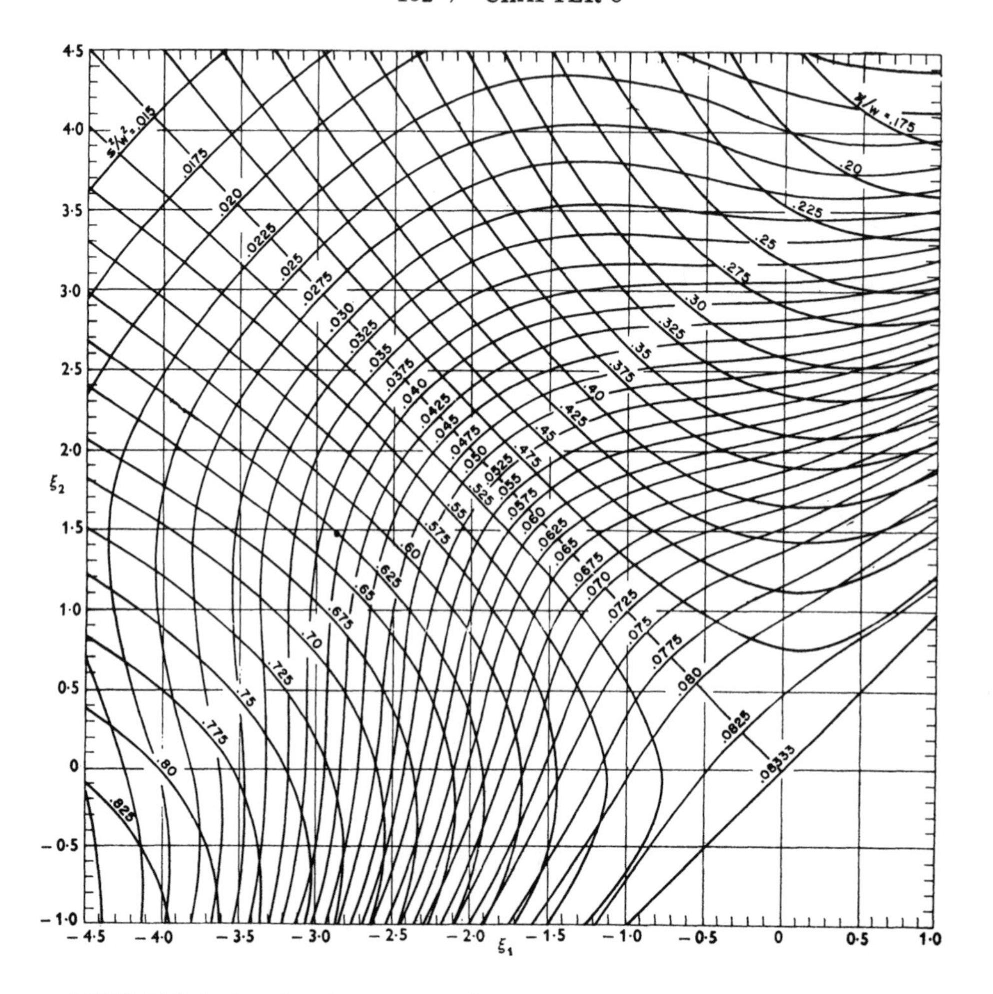

FIGURE 8.1 Graphs of estimating functions $H_1(\xi_1, \xi_2)$ and $H_2(\xi_1, \xi_2)$ for doubly truncated samples from the normal distribution. Reproduced from Cohen, A. C. (1957) with permission of the Biometrika Trustees.

we require greater accuracy than the chart can provide, values read from the chart can serve as first approximations to be improved through iteration.

8.4.4 Estimators for Doubly Censored Samples

The sample consists of a total of N observations, of which c_1 are censored on the left at D_1, c_2 are censored on the right at D_2, and there are n fully measured observations in the interval $D_1 \leq y \leq D_2$. Thus $N = n + c_1 + c_2$.

Maximum likelihood estimating equations for normal distribution parameters based on a sample of this type, as derived by Cohen (1957), are

$$\frac{\Omega_1 - \Omega_2 - \xi_1}{\xi_2 - \xi_1} = \frac{\nu_1}{w}$$
$$\frac{1 + \xi_1\Omega_1 - \xi_2\Omega_2 - (\Omega_1 - \Omega_2)^2}{(\xi_2 - \xi_1)^2} = \frac{s_y^2}{w^2} \tag{8.4.18}$$

where

$$\Omega_1 = \Omega_1(a_1, \xi_1) = \frac{a_1\,\phi(\xi_1)}{\Phi(\xi_1)} = a_1 Q(-\xi_1)$$
$$\Omega_2 = \Omega_2(a_2, \xi_2) = \frac{a_2\,\phi(\xi_2)}{1 - \Phi(\xi_2)} = a_2 Q(\xi_2) \tag{8.4.19}$$

where

$$a_1 = \frac{c_1}{n} \quad \text{and} \quad a_2 = \frac{c_2}{n} \tag{8.4.20}$$

where ξ_1, ξ_2, w, and ν_1 are given by (8.4.16) and $\bar{y}$ and s_y^2 are the mean and variance (unbiased), respectively, of the n fully measured observations. The two equations of (8.4.18) can be solved simultaneously for estimates $\hat{\xi}_1$ and $\hat{\xi}_2$. Estimates $\hat{\sigma}$ and $\hat{\mu}$ then follow from (8.4.17). First approximations to $\hat{\xi}_1$ and $\hat{\xi}_2$ for use in an iterative solution of (8.4.19) can be read from the chart of Figure 8.1 when we neglect information provided by the censored observations.

8.4.5 Maximum Likelihood Estimators for Progressively Censored Samples

Let N designate the total sample size and n the number of fully measured observations. In a life-span context, n is the number of failures. Suppose that censoring occurs progressively in k stages at points D_j, where j = 1, 2, . . ., k, and that c_j sample specimens selected randomly from the survivors at point (time) D_j are removed (censored) from further observation. Thus $N = n + \Sigma_1^k c_j$. In Type I censoring the D_j are fixed, and the number of survivors at these times are random variables. In Type II censoring, the D_j coincide with failure times and are random variables, whereas the number of survivors at these times are fixed. For both types, the c_j are fixed. Maximum likelihood estimating equations for normal distribution parameters based on a sample of this type, as obtained by Cohen (1963), are

$$\bar{y} = \mu - \sigma \sum_{1}^{k} a_j Q_j$$

$$s_y^{\ 2} = \sigma^2 \left[1 - \sum_{1}^{k} a_j \xi_j Q_j - \left(\sum_{1}^{k} a_j Q_j\right)^2\right] \tag{8.4.21}$$

where

$$a_j = \frac{c_j}{n}, \qquad \xi_j = \frac{D_j - \mu}{\sigma} \tag{8.4.22}$$

and

$$Q_j = Q(\xi_j) = \frac{\phi(\xi_j)}{1 - \Phi(\xi_j)}$$

and $\bar{y}$ and $s_y^{\ 2}$ are, respectively, the mean and variance (unbiased) of the n full-term observations in the sample. As previously noted, $\phi(\ )$ and $\Phi(\ )$ are the pdf and the cdf, respectively, of the standard normal distribution (0, 1).

Standard iterative procedures may be employed to solve the pair of equations in (8.4.21) simultaneously for the required estimates $\hat{\mu}$ and $\hat{\sigma}$. Newton's method is usually satisfactory. This iteration method is based on Taylor series expansions of the estimating equations in the vicinity of their simultaneous solution. With μ_0 and σ_0 designating approximate solutions, let

$$\hat{\mu} = \mu_0 + h \quad \text{and} \quad \hat{\sigma} = \sigma_0 + m \tag{8.4.23}$$

where h and m are corrections to be determined by the iteration process, and the symbol ^ is employed to distinguish estimates from the parameters being estimated. Using Taylor's theorem and neglecting powers of h and m above the first, we have

$$\begin{aligned} h \frac{\partial^2 L}{\partial \mu_0^{\ 2}} + m \frac{\partial^2 L}{\partial \mu_0\, \partial \sigma_0} &= -\frac{\partial L}{\partial \mu_0} \\ h \frac{\partial^2 L}{\partial \mu_0\, \partial \sigma_0} + m \frac{\partial^2 L}{\partial \sigma_0^{\ 2}} &= -\frac{\partial L}{\partial \sigma_0} \end{aligned} \tag{8.4.24}$$

Corrections h and m are then obtained by solving these two equations simultaneously. For the right-hand members of these equations, we have

$$\frac{\partial \ln L}{\partial \mu} = \frac{n}{\sigma}\left[\frac{\bar{y} - \mu}{\sigma} + \sum_{1}^{k} c_j Q_j\right]$$

$$\frac{\partial \ln L}{\partial \sigma} = \frac{n}{\sigma}\left[\frac{s_y^2 + (\bar{y} - \mu)^2}{\sigma^2} - 1 + \sum_1^k a_j \xi_j Q_j\right] \tag{8.4.25}$$

where ln L is the loglikelihood function for a progressively censored sample, as described, from a normal distribution (μ, σ^2).

For the coefficients on the left side of (8.4.24) we differentiate a second time to obtain

$$\begin{aligned}
\frac{\partial^2 \ln L}{\partial \mu^2} &= -\frac{n}{\sigma^2}\left[1 + \sum_1^k a_j A_j\right] \\
\frac{\partial^2 \ln L}{\partial \mu \partial \sigma} &= -\frac{n}{\sigma^2}\left[\frac{2(\bar{y} - \mu)}{\sigma} + \sum_1^k a_j B_j\right] \\
\frac{\partial^2 \ln L}{\partial \sigma^2} &= -\frac{n}{\sigma^2}\left[\frac{3\{s^2 + (\bar{y} - \mu)^2\}}{\sigma^2} - 1 + \sum_1^k a_j C_j\right]
\end{aligned} \tag{8.4.26}$$

where

$$\begin{aligned}
a_j &= \frac{c_j}{n} \\
A_j &= Q_j(Q_j - \xi_j) \\
B_j &= Q_j + \xi_j A_j \\
C_j &= \xi_j(Q_j + B_j)
\end{aligned} \tag{8.4.27}$$

The number of repetitions of the above iterative process required for a given degree of accuracy will depend largely on the initial approximations. In many instances, satisfactory first approximations μ_0 and σ_0 will be available from the sample percentiles. The pattern of censoring in specific examples will determine which percentiles are available, but, in general, a suitable first approximation to $\hat{\mu}$ may be calculated as

$$\mu_0 = \frac{P_j + P_{100-j}}{2}$$

where P_j is the jth percentile. It is desirable, though not necessary, to restrict j to the interval $25 \leq j \leq 50$.

In some circumstances, it might be desirable to employ the probit technique, which is discussed in a subsequent section, to obtain more

TABLE 8.4 The Functions Q, ξQ, A, B, and C

ξ	Q	ξQ	$A = Q(Q-\xi)$	$B = Q+\xi A$	$C = \xi(Q+B)$
−1.0	0.28760	−0.28760	0.37031	−0.083	−0.205
−0.5	0.50916	−0.25458	.51382	0.252	−0.381
0	0.79788	0	.63662	0.798	0
0.5	1.14108	0.57054	.73152	1.507	1.324
1.0	1.52514	1.52514	.80090	2.326	3.851
1.1	1.60580	1.76638	.81221	2.499	4.515
1.2	1.68755	2.02506	.82277	2.675	5.235
1.3	1.77033	2.30143	.83263	2.853	6.010
1.4	1.85406	2.59568	.84185	3.033	6.841
1.5	1.93868	2.90802	.85045	3.214	7.730
1.6	2.02413	3.23861	.85849	3.398	8.675
1.7	2.11036	3.58761	.86600	3.583	9.678
1.8	2.19731	3.95516	.87302	3.769	10.739
1.9	2.28495	4.34140	.87958	3.956	11.858
2.0	2.37322	4.74644	.88572	4.145	13.036
2.5	2.82274	7.05685	.91103	5.100	19.808
3.0	3.28310	9.84930	.92944	6.071	28.064
3.5	3.75137	13.12980	.94307	7.052	37.812

Source: Reproduced from Cohen, A. C. (1963).

accurate first approximations and thereby reduce the number of iterations necessary in calculating MLE.

As aids which might facilitate the iterative solution of (8.4.21), the functions Q, ξQ, A, B, and C were tabulated by Cohen (1963). This tabulation is reproduced here with permission of the American Statistical Association as Table 8.4.

Herd (1956) employed an iterative procedure for solving the estimating equations which, in some respects, is simpler than the traditional method of Newton, presented here. However, when close first approximations are available, Newton's method is usually satisfactory.

8.4.6 Probit Plot for Progressively Censored Samples

The probit plot, as employed by Cohen (1963), is a convenient graphical device for obtaining first approximations to MLE $\hat{\mu}$ and $\hat{\sigma}$ when samples from a normal distribution are progressively censored. In many applications, these estimates are sufficiently accurate for final estimates. The probit plot is a variation of the probability plot, which together with the cumulative hazard plot (considered in Chapter 7) has been used extensively by Nelson

(1982) and various quality control and reliability practitioners. Both the probability plot and the hazard plot are designed to achieve the same result. Although the probability plot is not limited to applications involving grouped data, it is well suited for use when data are grouped.

For convenience, we consider a sample in which data are grouped into k classes with boundaries D_0, D_1, . . ., D_k. Let f_j designate the number of failures observed in the jth class; i.e., during the time interval $D_{j-1} \le y \le D_j$. Let c_j designate the number of samples withdrawn (censored) at class boundary D_j. We assume that censoring can occur only at one of the boundaries, but does not necessarily occur at each boundary. Thus the total sample size is $N = n + \Sigma_1^k c_j$, with $n = \Sigma_1^k f_j$. Note that some of the c_j might be zeros.

Where f is the density of y, the conditional density subject to the restriction that $y > D_j$ may be written as

$$f(y \mid y > D_j) = \frac{f(y)}{1 - F_j} \tag{8.4.28}$$

It follows that expected frequencies in progressively censored samples of the types under consideration are

$$\begin{aligned} E(f_1) &= N \int_{D_0}^{D_1} f(y)\, dy = Np_1 \\ E(f_j) &= \left[N - \sum_{1}^{j-1} \frac{c_i}{1 - F_i} \right] p_j, \qquad j = 2, 3, \ldots, k \end{aligned} \tag{8.4.29}$$

where

$$p_j = \int_{D_{j-1}}^{D_j} f(y)\, dy$$

Accordingly, estimates of p_i may be obtained as

$$\begin{aligned} p_1^* &= \frac{f_1}{N} \\ p_j^* &= f_j \Big/ \left[N - \sum_{1}^{j-1} \frac{c_j}{1 - F_i^*} \right], \qquad j = 2, 3, \ldots, k \end{aligned} \tag{8.4.30}$$

with estimates of F_j given as

$$F_1^* = p_1^*, \quad F_j^* = F_{j-1}^* + p_j^*, \quad j = 2, 3, \ldots, k \tag{8.4.31}$$

The asterisks serve to distinguish estimates from parameters.

The F_i^* are transformed into probits by using a standard probit table or by using ordinary tables of normal curve areas, where the probit estimate z^* is

$$z_j^* = 5 + \xi_j^* \tag{8.4.32}$$

and where ξ_j^* is determined from the relation

$$F_j^* = \int_{-\infty}^{\xi_j^*} \phi(t)\, dt \tag{8.4.33}$$

The $k - 1$ pairs of values (D_j, z_j^*), $1 \le j \le k - 1$, when plotted on a rectangular grid should lie approximately on the probit regression line. This is a straight line, and its equations may be written in the form

$$z = 5 + \frac{1}{\sigma}(y - \mu) \tag{8.4.34}$$

Thus μ_0 is the value of y corresponding to the probit value of 5, and σ_0 is the reciprocal of the slope of the probit regression line.

When a high degree of accuracy is not required, estimates obtained from a probit regression line fitted by eye might be satisfactory without further improvement through iteration.

<u>An Illustrative Example</u>. As an illustration of the probit plot technique, we consider an example given by Cohen (1963) of a life test conducted on 316 biological specimens in a stress environment. For this example, a total of 316 specimens was observed under a specified stress. Life-spans, assumed to be normally distributed, were recorded in days. The resulting data were grouped as displayed in the accompanying table. Ten specimens, selected at random from the survivors after 36.5 days, were withdrawn. After 44.5 days, 10 additional specimens were similarly selected and withdrawn from further observation. The estimates p_j^* and F_j^* were calculated by (8.4.30) and (8.4.31). These values and the corresponding probit values are also included in the accompanying frequency table. Data for this example are summarized as $N = 316$, $n = \Sigma_1^{14} f_j = 296$, $\Sigma_1^{14} c_j = 20$, $\bar{y} = 39.2703$, $s_y^2 = 20.16344$.

The 13 points (D_j, Z_j), $j = 1, 2, \ldots, 13$, were plotted using rectangular coordinates, and the probit regression line was sketched by eye as

Life Distribution of Certain Biological Specimens in a Stress Environment

i	Boundaries D_i	Midpoints y_i	f_i	r_i	p_i^*	F_i^*	Probits z_i^*
0	24.5					0.000000	
1	26.5	25.5	1	0	0.003165	0.003165	2.270
2	28.5	27.5	1	0	0.003165	0.006330	2.507
3	30.5	29.5	4	0	0.012658	0.018988	2.925
4	32.5	31.5	18	0	0.056962	0.075950	3.567
5	34.5	33.5	18	0	0.056962	0.132912	3.887
6	36.5	35.5	37	10	0.117089	0.250001	4.326
7	38.5	37.5	45	0	0.148678	0.398679	4.743
8	40.5	39.5	57	0	0.188326	0.587005	5.220
9	42.5	41.5	39	0	0.128855	0.715860	5.571
10	44.5	43.5	43	10	0.142071	0.857931	6.071
11	46.5	45.5	20	0	0.086104	0.944035	6.590
12	48.5	47.5	9	0	0.038747	0.982782	7.116
13	50.5	49.5	1	0	0.004305	0.987087	7.229
14	52.5	51.5	3	0	0.012916	1.000000	
Totals			296	20			

shown in Figure 8.2. First approximations to $\hat{\mu}$ and $\hat{\sigma}$ follow from the sketched line as $\mu_0 = 39.8$ and $\sigma_0 = 4.55$. One cycle of iteration gave $\mu_1 = 39.6$ and $\sigma_1 = 4.60$ as new approximations, which were then used as the starting point for a second cycle of iteration. For this second cycle, (8.4.24) become

$$-14.591h + 0.801m = 0.261$$

$$0.801h - 28.067m = -0.331$$

On solving this pair of equations simultaneously, we obtain as new corrections $h = -0.017$ and $m = 0.011$. The final estimates of μ and σ thus become

$$\hat{\mu} = 39.6 - 0.017 = 39.583$$

$$\hat{\sigma} = 4.60 + 0.011 = 4.611$$

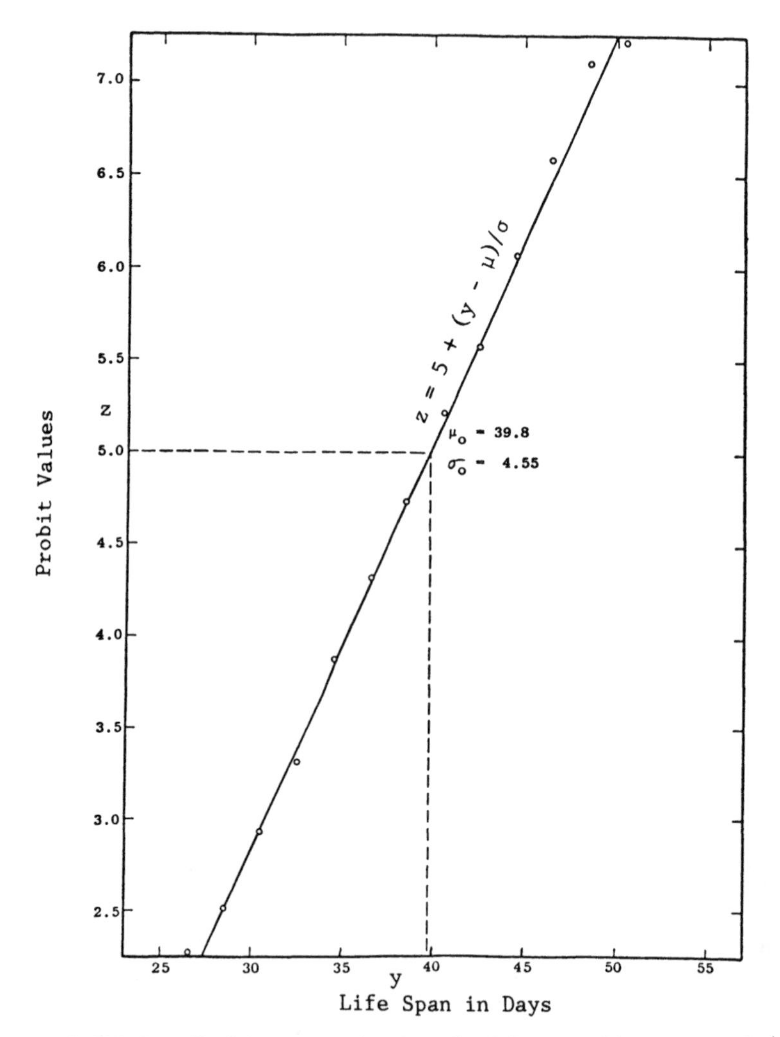

FIGURE 8.2 Probit regression line for life test of biological specimens in a stress environment. Adapted from Cohen, A. C. (1963).

The asymptotic variance-covariance matrix is now approximated as

$$\begin{bmatrix} 14.591 & -0.801 \\ -0.801 & 28.067 \end{bmatrix}^{-1} = \begin{bmatrix} 0.069 & 0.002 \\ 0.002 & 0.036 \end{bmatrix}$$

and hence $V(\hat{\mu}) = 0.069$, $V(\hat{\sigma}) = 0.036$, $\mathrm{Cov}(\hat{\mu}, \hat{\sigma}) = 0.002$. Note that elements of this information matrix are negatives of the coefficients of the Newton correction equations.

8.4.7 Asymptotic Variances and Covariances

Asymptotic variances and covariances of estimates of normal distribution parameters based on singly truncated and singly censored samples may be calculated as

$$
\begin{aligned}
V(\hat{\mu}) &= \frac{\sigma^2}{n}\mu_{11}, \quad \mathrm{Cov}(\hat{\mu}, \hat{\sigma}) = \frac{\sigma^2}{n}\mu_{12} \\
V(\hat{\sigma}) &= \frac{\sigma^2}{n}\mu_{22}, \quad \rho_{\hat{\mu},\hat{\sigma}} = \frac{\mu_{12}}{\sqrt{\mu_{11}\mu_{22}}}
\end{aligned}
\qquad (8.4.35)
$$

In order to facilitate these calculations in practical applications, Table 8.5, which contains entries of the μ_{ij} and ρ for both singly truncated and singly censored samples, has been reproduced here, with permission of the editors of Technometrics, from Cohen (1961). These factors were obtained by inverting information matrices in which elements are negatives of expected values of second partial derivatives of applicable loglikelihood functions.

For doubly truncated, doubly censored, and progressively censored samples, it does not seem feasible to obtain expected values of second partials of the loglikelihood functions. However, for large or even for moderately large samples, these expected values are approximately equal to the second partials evaluated for the calculated estimates $\hat{\mu}$, $\hat{\sigma}$, and $\hat{\xi}_j$. Therefore these approximations can be substituted into the appropriate information matrices, which can then be inverted to produce approximations to the estimate variances and covariances that are satisfactory when a high degree of precision is not required. Second partials of the loglikelihood equations for samples of these types are included here in order to facilitate these calculations.

For Doubly Truncated Samples

$$
\begin{aligned}
\frac{\partial^2 \ln L}{\partial \mu^2} &= -\frac{n}{\sigma^2}[1 + \xi_1\bar{Q}_1 - \xi_2\bar{Q}_2 - (\bar{Q}_1 - \bar{Q}_2)] \\
\frac{\partial^2 \ln L}{\partial \mu \partial \sigma} &= -\frac{n}{\sigma^2}\left[\frac{2(\bar{y} - \mu)}{\sigma} - (\bar{Q}_1 - \bar{Q}_2)(1 + \xi_1\bar{Q}_1 - \xi_2\bar{Q}_2) + \xi_1^2\bar{Q}_1 - \xi_2^2\bar{Q}_2\right] \\
\frac{\partial^2 \ln L}{\partial \sigma^2} &= -\frac{n}{\sigma^2}\left[\frac{3\{s_y^2 + (\bar{y} - \mu)^2\}}{\sigma^2} - (1 + \xi_1\bar{Q}_1 - \xi_2\bar{Q}_2)^2 + \xi_1^3\bar{Q}_1 - \xi_2^3\bar{Q}_2\right]
\end{aligned}
\qquad (8.4.36)
$$

For Doubly Censored Samples

$$
\frac{\partial^2 \ln L}{\partial \mu^2} = -\frac{1}{\sigma^2}[n + c_1 Q'(-\xi_1) + c_2 Q'(\xi_2)]
$$

TABLE 8.5 Variance and Covariance Factors for Estimates Based on Singly Truncated and Singly Censored Samples from the Normal Distribution

η	For Truncated Samples				For Censored Samples				Percent Rest.	η
	μ_{11}	μ_{12}	μ_{22}	ρ	μ_{11}	μ_{12}	μ_{22}	ρ		
-4.0	1.00054	-.001143	.502287	-.001613	1.00000	-.000006	.500030	-.000001	0.00	-4.0
-3.5	1.00313	-.005922	.510366	-.008277	1.00001	-.000052	.500208	-.000074	0.02	-3.5
-3.0	1.01460	-.024153	.536283	-.032744	1.00010	-.000335	.501180	-.000473	0.13	-3.0
-2.5	1.05738	-.081051	.602029	-.101586	1.00056	-.001712	.505280	-.002407	0.62	-2.5
-2.4	1.07437	-.101368	.622786	-.123924	1.00078	-.002312	.506935	-.003247	0.82	-2.4
-2.3	1.09604	-.126136	.646862	-.149803	1.00107	-.003099	.509030	-.004341	1.07	-2.3
-2.2	1.12365	-.156229	.674663	-.179434	1.00147	-.004121	.511658	-.005757	1.39	-2.2
-2.1	1.15880	-.192688	.706637	-.212937	1.00200	-.005438	.514926	-.007571	1.79	-2.1
-2.0	1.20350	-.236743	.743283	-.250310	1.00270	-.007123	.518960	-.009875	2.28	-2.0
-1.9	1.26030	-.289860	.785158	-.291388	1.00363	-.009266	.523899	-.012778	2.87	-1.9
-1.8	1.33246	-.353771	.832880	-.335818	1.00485	-.011971	.529899	-.016405	3.59	-1.8
-1.7	1.42405	-.430531	.887141	-.383041	1.00645	-.015368	.537141	-.020901	4.46	-1.7
-1.6	1.54024	-.522564	.948713	-.432293	1.00852	-.019610	.545827	-.026431	5.48	-1.6
-1.5	1.68750	-.632733	1.01846	-.482644	1.01120	-.024884	.556186	-.033181	6.68	-1.5
-1.4	1.87398	-.764405	1.09734	-.533054	1.01467	-.031410	.568471	-.041358	8.08	-1.4
-1.3	2.10982	-.921533	1.18642	-.582464	1.01914	-.039460	.582981	-.051193	9.68	-1.3
-1.2	2.40764	-1.10874	1.28690	-.629889	1.02488	-.049355	.600046	-.062937	11.51	-1.2
-1.1	2.78311	-1.33145	1.40009	-.674498	1.03224	-.061491	.620049	-.076861	13.57	-1.1
-1.0	3.25557	-1.59594	1.52746	-.715676	1.04168	-.076345	.643438	-.093252	15.87	-1.0
-0.9	3.84879	-1.90952	1.67064	-.753044	1.05376	-.094501	.670724	-.112407	18.41	-0.9
-0.8	4.59189	-2.28066	1.83140	-.786452	1.06923	-.116674	.702513	-.134620	21.19	-0.8
-0.7	5.52036	-2.71911	2.01172	-.815942	1.08904	-.143744	.739515	-.160175	24.20	-0.7
-0.6	6.67730	-3.23612	2.21376	-.841703	1.11442	-.176798	.782574	-.189317	27.43	-0.6
-0.5	8.11482	-3.84458	2.43990	-.864019	1.14696	-.217183	.832691	-.222233	30.85	-0.5
-0.4	9.89562	-4.55921	2.69271	-.883229	1.18876	-.266577	.891077	-.259011	34.46	-0.4
-0.3	12.0949	-5.39683	2.97504	-.899688	1.24252	-.327080	.959181	-.299607	38.21	-0.3
-0.2	14.8023	-6.37653	3.28997	-.913744	1.31180	-.401326	1.03877	-.343800	42.07	-0.2
-0.1	18.1244	-7.51996	3.64083	-.925727	1.40127	-.492641	1.13198	-.391156	46.02	-0.1
0.0	22.1875	-8.85155	4.03126	-.935932	1.51709	-.605233	1.24145	-.441013	50.00	0.0
0.1	27.1403	-10.3988	4.46517	-.944623	1.66743	-.744459	1.37042	-.492483	53.98	0.1
0.2	33.1573	-12.1927	4.94678	-.952028	1.86310	-.917165	1.52288	-.544498	57.93	0.2
0.3	40.4428	-14.2679	5.48068	-.958345	2.11857	-1.13214	1.70381	-.595891	61.79	0.3
0.4	49.2342	-16.6628	6.07169	-.963742	2.45318	-1.40071	1.91942	-.645504	65.54	0.4
0.5	59.8081	-19.4208	6.72512	-.968361	2.89293	-1.73757	2.17751	-.692299	69.15	0.5
0.6	72.4834	-22.5896	7.44658	-.972322	3.47293	-2.16185	2.48793	-.735459	72.57	0.6
0.7	87.6276	-26.2220	8.24204	-.975727	4.24075	-2.69858	2.86318	-.774443	75.80	0.7
0.8	105.66	-30.376	9.1178	-.97866	5.2612	-3.3807	3.3192	-.80899	78.81	0.8
0.9	127.07	-35.117	10.081	-.98119	6.6229	-4.2517	3.8765	-.83912	81.59	0.9
1.0	152.40	-40.515	11.138	-.98338	8.4477	-5.3696	4.5614	-.86502	84.13	1.0
1.1	182.29	-46.650	12.298	-.98529	10.903	-6.8116	5.4082	-.88703	86.43	1.1
1.2	217.42	-53.601	13.567	-.98694	14.224	-8.6818	6.4616	-.90557	88.49	1.2
1.3	258.61	-61.465	14.954	-.98838	18.735	-11.121	7.7804	-.92109	90.32	1.3
1.4	306.78	-70.347	16.471	-.98964	24.892	-14.319	9.4423	-.93401	91.92	1.4
1.5	362.91	-80.350	18.124	-.99074	33.339	-18.539	11.550	-.94473	93.32	1.5
1.6	428.11	-91.586	19.922	-.99171	44.986	-24.139	14.243	-.95361	94.52	1.6
1.7	503.57	-104.17	21.874	-.99256	61.132	-31.616	17.706	-.96097	95.54	1.7
1.8	591.03	-118.31	24.003	-.99332	83.638	-41.664	22.193	-.96706	96.41	1.8
1.9	691.78	-134.10	26.311	-.99398	115.19	-55.252	28.046	-.97211	97.13	1.9
2.0	807.71	-151.73	28.813	-.99457	159.66	-73.750	35.740	-.97630	97.72	2.0
2.1	940.38	-171.30	31.511	-.99509	222.74	-99.100	45.930	-.97979	98.21	2.1
2.2	1091.4	-192.92	34.405	-.99555	312.73	-134.08	59.526	-.98270	98.61	2.2
2.3	1265.4	-217.17	37.575	-.99596	441.92	-182.68	77.810	-.98514	98.93	2.3
2.4	1458.6	-243.23	40.858	-.99632	628.58	-250.68	102.59	-.98718	99.18	2.4
2.5	1677.8	-271.99	44.392	-.99665	899.99	-346.53	136.44	-.98890	99.38	2.5

When truncation or type I censoring occurs on the left, entries in this table corresponding to $\eta = \xi$ are applicable. For right truncated or type I right censored samples, read entries corresponding to $\eta = -\xi$, but delete negative signs from μ_{12} and ρ. For both type II left censored and type II right censored samples, read entries corresponding to Percent Restriction = 100h, but for right censoring delete negative signs from μ_{12} and ρ.

$$\frac{\partial^2 \ln L}{\partial\mu\partial\sigma} = -\frac{1}{\sigma^2}\left[\frac{2n(\bar{y}-\mu)}{\sigma} - c_1\lambda(-\xi_1) + c_2\lambda(\xi_2)\right]$$

$$\frac{\partial^2 \ln L}{\partial\sigma^2} = -\frac{1}{\sigma^2}\left[\frac{3n\{s_y^2 + (\bar{y}-\mu)^2\}}{\sigma^2} - n + c_1\zeta(-\xi_1) + c_2\zeta(\xi_2)\right] \qquad (8.4.37)$$

where

$$Q'(\xi) = Q(\xi)[Q(\xi) - \xi]$$
$$\lambda(\xi) = Q(\xi) + \xi Q'(\xi) \qquad (8.4.38)$$
$$\zeta(\xi) = \xi[Q(\xi) + \lambda(\xi)]$$

For Progressively Censored Samples. Applicable second partial derivatives were given earlier in (8.4.26) as coefficients of correction equations involved in the Newton iterative process for solving estimating equations (8.4.21).

For the special case of a progressively censored sample, when σ is known

$$V(\hat{\mu}) \sim \sigma^2 \Big/ n\left[1 + \sum_1^k a_j A_j\right] \qquad (8.4.39)$$

and for the special case in which μ is known,

$$V(\hat{\sigma}) \sim \frac{\sigma^2}{n}\left[\frac{3\{s^2 + (\bar{y}-\mu)^2\}}{\sigma^2} - 1 + \sum_1^k a_j C_j\right]^{-1} \qquad (8.4.40)$$

where C_j is defined in (8.4.27).

8.5 AN ILLUSTRATIVE EXAMPLE FROM A LOGNORMAL DISTRIBUTION

In order to illustrate computational procedures involved in calculating parameter estimates from a singly right-censored lognormal sample, an example originally given by Cohen (1951) has been selected. The complete sample consists of 20 randomly chosen observations from a lognormal distribution in which $\gamma = 100$, $\beta = 50$ ($\mu = \ln\beta = 3.912023$), and $\sigma = 0.4$ ($\alpha_3 = 1.3219144$). Individual observations, in order of magnitude, are tabulated below.

A Random Sample from a Lognormal Population

127.211	135.880	153.070	166.475
128.709	137.338	155.369	168.554
131.375	144.328	155.680	174.800
132.971	145.788	157.238	184.101
133.143	148.290	164.304	201.415

For the purpose of this illustration, the sample is considered to be Type II censored at $T = y_{18} = 174.800$. The censored sample thus consists of a total of $N = 20$ observations, of which $n = 18$ are fully measured and $N - n = 2$ are censored. The only information assumed to be known about the two censored observations is that both exceed 174.800. Only the alternative MMLE will be illustrated, since this is usually the preferred estimator when samples are small.

We select a first approximation $\gamma_1 = 110$ and proceed to calculate conditional estimates $\mu(\gamma_1)$ and $\sigma(\gamma_1)$. Hence, we make the transformation $\{y_i\} = \{\ln(x_i - 110)\}$, $i = 1, 2, \ldots, 18$. We then calculate $\bar{y}_{18}(110) = \Sigma_1^{18} y_i/18 = 3.5531876$, $s_y = \left(\Sigma_1^{18} [y_i - \bar{y}_{18}]^2/17\right)^{\frac{1}{2}} = 0.4191069$, and $D = \ln(174.800 - 110) = 4.1713056$. We subsequently calculate $h = (N - n)/N = 0.1$ and $\alpha(110) = s_y^2/(D - \bar{y})^2 = 0.4597342$. We enter Table 8.2 with these values and interpolate to obtain $\gamma(110) = 0.1389$. When the foregoing values are substituted into (8.4.8), as conditional estimates for μ and σ^2 with $\gamma = 110$, we calculate

$$\mu(110) = 3.5531876 - 0.1389(3.5531876 - 4.1713056) = 3.6390442$$

$$\sigma^2(110) = (0.4191069)^2 + 0.139(3.5531876 - 4.1713056)^2 = 0.2287201$$

and it follows that $\sigma(110) = 0.4782469$.

We substitute these conditional estimates into (8.3.2), with $z_1 = \Phi^{-1}[1/N + 1] = -1.67$ (from Table 8.2), and calculate

$$110 + e^{3.6390442+0.4782469(-1.67)} = 126.975$$

Since the value thus calculated is less than x_1 (= 127.211), we select a second approximation $\gamma_2 = 112$ and repeat the foregoing calculations to obtain

$$\mu(112) = 3.5793849 \quad \text{and} \quad \sigma(112) = 0.5092228$$

This time, substitution in (8.3.2) yields

$$112 + e^{3.5793849+0.5092228(-1.67)} = 127.317$$

a value which exceeds x_1. We accordingly conclude that $110 < \gamma < 112$, and we interpolate as summarized below for our final estimates.

γ	$\gamma + e^{\mu+\sigma(-1.67)}$	$\mu(\gamma)$	$\sigma(\gamma)$
112.00	127.317	3.5793849	0.5092228
111.38	$127.211 = x_1$	3.5978758	0.4996221
110.00	126.975	3.6390442	0.4782469

Final estimates, MMLE (alternative), then are $\hat{\gamma} = 111.4$, $\hat{\mu} = 3.598$, $\hat{\sigma} = 0.500$, $\hat{\omega} = 1.284$, $\hat{\alpha}_3 = 1.75$, $\hat{\beta} = 36.52$. These estimates compare favorably with those originally calculated by Cohen (1951) from the complete sample. Approximations to the estimate standard deviations calculated by using the MLE variances given in (4.5.2) with n equal to the number of complete observations (for this example n = 18) become $\sigma_{\hat{\gamma}} = 11.7$, $\sigma_{\hat{\beta}} = 14.0$, and $\sigma_{\hat{\mu}} = 0.38$. Even though these standard deviations are not strictly applicable, previous simulation studies indicate that they should be reasonably close to the correct values.

9

Censored Sampling in the Inverse Gaussian and Gamma Distributions

9.1 INTRODUCTION

This chapter is devoted to parameter estimation based on censored samples from the inverse Gaussian (IG) and gamma distributions. The IG distribution was introduced in Chapter 5, where an account of its properties and estimators for complete samples was given. The gamma distribution was introduced in Chapter 6, which contains similar results for this distribution. Some of the estimators for parameters of the gamma distribution presented in this chapter as well as those in Chapter 6 could be obtained as special cases of estimators given in Chapter 12 for parameters of the generalized gamma distribution.

9.2 CENSORED SAMPLING IN THE INVERSE GAUSSIAN DISTRIBUTION

Both maximum likelihood estimators (MLE) and modified estimators (MMLE) which employ the first-order statistic, are considered. In addition, we also present a procedure, due to Whitten et al. (1988), which involves estimation of censored observations in order to construct a pseudocomplete sample as a replacement for the original censored sample. Calculation of both MLE and MMLE is complicated as a result of problems encountered with differentiation under the integral sign involving the cumulative distribution function. The pseudocomplete sample technique avoids these complications, and it is likely to be preferred in the cases where it is applicable.

9.2.1 Maximum Likelihood Estimators

The loglikelihood function of a progressively censored sample from the three-parameter IG distribution with pdf (5.2.3) is

$$\ln L = \frac{3n}{2}\ln\mu - n\ln\sigma - \frac{3}{2}\sum_1^n \ln(x_i - \gamma) - \frac{\mu}{2\sigma^2}\sum_1^n \frac{(x_i - \gamma - \mu)^2}{x_i - \gamma}$$

$$+ \sum_1^k c_j \ln[1 - F(T_j)] + \text{constant} \tag{9.2.1}$$

Maximum likelihood estimates can be calculated by simultaneously solving the three estimating equations $\partial \ln L/\partial\gamma = 0$, $\partial \ln L/\partial\mu = 0$, and $\partial \ln L/\partial\sigma = 0$. On differentiating (9.2.1), we obtain

$$\frac{\partial \ln L}{\partial\gamma} = \frac{3}{2}\sum_1^n (x_i - \gamma)^{-1} + \frac{n\mu}{2\sigma^2} - \frac{\mu^3}{2\sigma^2}\sum_1^k \frac{c_j}{1 - F_j}\frac{\partial F_j}{\partial\gamma} = 0$$

$$\frac{\partial \ln L}{\partial\mu} = \frac{3n}{2\mu} - \frac{1}{2\sigma^2}\sum_1^n \frac{(x_i - \gamma - \mu)^2}{x_i - \gamma} + \frac{n\mu}{\sigma^2} - \frac{\mu}{\sigma^2}\sum_1^n (x_i - \gamma)^{-1}$$

$$- \sum_1^k \frac{c_j}{1 - F_j}\frac{\partial F_j}{\partial\mu} = 0 \tag{9.2.2}$$

$$\frac{\partial \ln L}{\partial\sigma} = -\frac{n}{\sigma} + \frac{\mu}{\sigma^3}\sum_1^n \frac{(x_i - \gamma - \mu)^2}{x_i - \gamma} - \sum_1^k \frac{c_j}{1 - F_j}\frac{\partial F_j}{\partial\sigma} = 0$$

In order to take partial derivatives of F_j, we introduce the standardizing transformation

$$Z = \frac{X - E(X)}{\sigma} = \frac{X - \gamma - \mu}{\sigma} \tag{9.2.3}$$

With X censored at T_j, it follows that Z is censored at ξ_j, where

$$\xi_j = \frac{T_j - \gamma - \mu}{\sigma} \tag{9.2.4}$$

The pdf $g(z; 0, 1, \alpha_3)$, with $\alpha_3 = 3\sigma/\mu$, was given in (5.2.5). We note that $F_j = F(T_j) = G(\xi_j) = G_j$, where $G(\ )$ is the cdf of Z,

$$G(\xi_j; 0, 1, \alpha_3) = \int_{-3/\alpha_3}^{\xi_j} g(z; 0, 1, \alpha_3)\, dz \tag{9.2.5}$$

We note further that

$$\frac{\partial \xi_j}{\partial \gamma} = \frac{-1}{\sigma}, \quad \frac{\partial \xi_j}{\partial \mu} = \frac{-1}{\sigma}, \quad \frac{\partial \xi_j}{\partial \sigma} = -\frac{\xi_j}{\sigma}, \quad \frac{\partial \alpha_3}{\partial \sigma} = \frac{3}{\mu}, \quad \frac{\partial \alpha_3}{\partial \mu} = -\frac{\alpha_3}{\mu} \tag{9.2.6}$$

The partial derivatives of F_j can now be expressed as

$$\frac{1}{1 - F_j}\frac{\partial F_j}{\partial \gamma} = \frac{1}{1 - G_j}\frac{\partial G_j}{\partial \gamma} = \frac{-g(\xi_j)}{\sigma(1 - G_j)}$$

$$\frac{1}{1 - F_j}\frac{\partial F_j}{\partial \mu} = \frac{1}{1 - G_j}\left[\int_{-3/\alpha_3}^{\xi_j} \frac{\partial g(z)}{\partial \mu}\, dz + g(\xi_j)\frac{\partial \xi_j}{\partial \mu}\right] \tag{9.2.7}$$

$$\frac{1}{1 - F_j}\frac{\partial F_j}{\partial \sigma} = \frac{1}{1 - G_j}\left[\int_{-3/\alpha_3}^{\xi_j} \frac{\partial g(z)}{\partial \sigma}\, dz + g(\xi_j)\frac{\partial \xi_j}{\partial \sigma}\right]$$

The occurrence of the integral terms in the last two equations of (9.2.7) is accounted for by the fact that g(z) is a function of α_3, which in turn is a function of μ and σ.

When a function $\phi(\alpha)$ is defined in terms of an integral such that

$$\phi(\alpha) = \int_{a(\alpha)}^{b(\alpha)} f(x; \alpha)\, dx \tag{9.2.8}$$

the derivative of ϕ with respect to α is

$$\frac{\partial \phi}{\partial \alpha} = \int_{a(\alpha)}^{b(\alpha)} \frac{\partial f}{\partial \alpha}\, dx + f(b, \alpha)\frac{\partial b}{\partial \alpha} - f(a, \alpha)\frac{\partial a}{\partial \alpha} \tag{9.2.9}$$

In most applications where $\phi(\,)$ is a cdf and $f(\,)$ is a pdf, $\partial f/\partial \alpha = 0$, $f(a, \alpha) = 0$, and only the middle term of (9.2.9) is retained. In the last two equations of (9.2.7), however, the first and second terms of (9.2.9) are both retained. Only the last term vanishes.

In order to simplify the integrals of (9.2.7), we let

$$W = \frac{3}{3 + z\alpha_3} \tag{9.2.10}$$

and thus

$$\frac{\partial W}{\partial \alpha_3} = \frac{-zW^2}{3}$$

The pdf of (5.2.5) can now be written as

$$g(z; 0, 1, \alpha_3) = \frac{1}{\sqrt{2\pi}} W^{3/2} \exp\left(\frac{-z^2}{2}\right) W, \qquad \frac{-3}{\alpha_3} < z \tag{9.2.11}$$

$$= 0, \qquad \text{otherwise}$$

We subsequently employ the foregoing results to simplify the expressions of (9.2.7), which are then substituted into (9.2.2). Thereby, the MLE equations become

$$\frac{3}{2}\sum_1^n (x_i - \gamma)^{-1} + \frac{n\mu}{2\sigma^2} - \frac{\mu^3}{2\sigma^2}\sum_1^n (x_i - \gamma)^{-2} + \frac{1}{\sigma}\sum_1^k c_j Q_j = 0$$

$$\frac{3n}{2\mu} - \frac{1}{2\sigma^2}\sum_1^n \frac{(x_i - \gamma - \mu)^2}{x_i - \gamma} + \frac{n\mu}{\sigma} - \frac{\mu^2}{\sigma} - \frac{\mu^2}{\sigma^2}\sum_1^n (x_i - \gamma)^{-1}$$

$$+ \frac{1}{\sigma}\sum_1^k c_j Q_j + \frac{9\sigma}{2\mu^2}\sum_1^k c_j D_j = 0 \tag{9.2.12}$$

$$-\frac{n}{\sigma} + \frac{\mu}{\sigma^3}\sum_1^n \frac{(x_i - \gamma - \mu)^2}{x_i - \gamma} + \sum_1^k c_j \xi_j Q_j - \frac{9}{2\mu}\sum_1^k c_j D_j = 0$$

where

$$Q_j = Q(\xi_j; \alpha_3) = \frac{g(\xi_j)}{1 - G(\xi_j)}$$

and (9.2.13)

$$D_j = D(\xi_j; \alpha_3) = \frac{1}{1 - G(\xi_j)} \int_{-3/\alpha_3}^{\xi_j} z\left\{\frac{z^2 - (3 + z\alpha_3)}{(3 + z\alpha_3)^2}\right\} g(z; 0, 1, \alpha_3)\, dz$$

Maximum likelihood estimates $\hat{\gamma}$, $\hat{\mu}$, and $\hat{\sigma}$ can be obtained as the simultaneous solution of the three estimating equations of (9.2.12). Unfortunately, explicit solutions are not available, and we must therefore resort to iterative procedures. Although other iterative methods might be employed, Newton's method will be satisfactory in most applications. The principal complication in these calculations concerns $D(\xi_j, \alpha_3)$. Evaluation of this function requires the use of numerical methods of integration, which might involve time-consuming calculations.

9.2.2 Estimation in the Two-Parameter Distribution

When the origin is known and can be set at zero, it is necessary only that the last two equations of (9.2.12), with $\gamma = 0$, be solved simultaneously for estimates $\hat{\mu}$ and $\hat{\sigma}$. In this case estimating equations are

$$\frac{3n}{2\mu} - \frac{1}{2\sigma^2}\sum_1^n \frac{(x_i - \mu)^2}{x_i} + \frac{n\mu}{\sigma} - \frac{\mu^2}{\sigma^2}\sum_1^n \frac{1}{x_i} + \frac{1}{\sigma}\sum_1^k c_j Q_j + \frac{9\sigma}{2\mu^2}\sum_1^k c_j D_j = 0$$

$$-\frac{n}{\sigma} + \frac{\mu}{\sigma^3}\sum_1^n \frac{(x_i - \mu)^2}{x_i} + \sum_1^k c_j \xi_j Q_j - \frac{9}{2\mu}\sum_1^k c_j D_j = 0 \tag{9.2.14}$$

With one less equation in the system, it is easier to solve (9.2.14) than to solve the three equations of (9.2.12). However, even here it is necessary to evaluate D_j with all of the attending complications.

9.2.3 Estimates from Singly Right-Censored Samples

For estimates from singly right-censored samples, $k = 1$ and $c_j = c$ for both three-parameter and two-parameter censored samples. Otherwise (9.2.12) and (9.2.14) are unchanged.

9.2.4 Modified Maximum Likelihood Estimators

For MMLE the estimating equation $\partial \ln L/\partial\gamma$ is replaced by $E[F(X_1)] = F(x_1)$, where $E[F(X_1)] = 1/(N + 1)$. After certain algebraic simplifications, this replacement becomes

$$\gamma + \mu + \sigma z_1 = x_1 \tag{9.2.15}$$

where

$$z_1 = G^{-1}\left[\frac{1}{N+1}; \alpha_3\right]; \text{ i.e., } \int_{-3/\alpha_3}^{z_1} g(z; 0, 1, \alpha_3)\, dz = \frac{1}{N+1} \qquad (9.2.16)$$

The complete set of estimating equations for progressively censored samples then consists of (9.2.15) plus the last two equations of (9.2.12). Thus, as in the case of MLE, evaluation of these estimates requires the simultaneous solution of a somewhat complex system of estimating equations. Iteration techniques are the same in both cases, and both involve the complications incidental to the evaluation of $D(\xi_j, \alpha_3)$

9.3 A PSEUDOCOMPLETE SAMPLE TECHNIQUE

The pseudocomplete sample technique of Whitten, Cohen, and Sundaraiyer (1988), whereby censored samples are "completed," offers an attractive alternative to MLE and MMLE when samples are singly censored. This alternative avoids the complications involved in solving systems of nonlinear equations and the complications involved in evaluating $D(\xi_j, \alpha_3)$. In order to complete a censored sample, we employ an iterative procedure to calculate estimates $\hat{x}_{n+i}$, $i = 1, \ldots, c$, of the c censored observations, which are added to the n complete observations. The resulting pseudocomplete sample of size N consists of the observed values $\{x_i\}$, $i = 1, 2, \ldots, n$, plus the estimated values $\{\hat{x}_{n+i}\}$, $i = 1, 2, \ldots, c$. Appropriate complete sample estimators can then be employed to calculate estimates of the distribution parameters from this sample. Thus ME, MLE, and MME which employ the first-order statistic, are available for the calculation of parameter estimates. Although this technique is applicable to various types of distributions, we are concerned here only with the IG distribution.

Estimating equations for completed values of the censored observations are

$$E[F(X_{n+i})] = F(\hat{x}_{n+i}) = \frac{n+i}{N+1} \qquad (9.3.1)$$

Equations (9.3.1) can be reduced to the equivalent form

$$\hat{x}_{n+i} = \hat{\mu} + \hat{\sigma}\hat{\xi}_{n+i} = T + \hat{\sigma}(\hat{\xi}_{n+i} - \hat{\xi}_n) \qquad (9.3.2)$$

where ξ_{n+i} are order statistics of the standardized distribution $(0, 1, \alpha_3)$, and where T is the point of censoring. Thus

$$G(\hat{\xi}_{n+i}; 0, 1, \alpha_3) = \frac{n+i}{N+1}, \qquad i = 0, 1, 2, \ldots, c \qquad (9.3.3)$$

where G( ) is the standardized cdf of the IG distribution given in (5.6.4). An alternative expression for $\hat{\xi}_{n+i}$ is

$$\hat{\xi}_{n+i} = G^{-1}\left[\frac{n+i}{N+1};\ \alpha_3\right]. \tag{9.3.4}$$

Let $x_{n+i}^{(j)}$, $i = 1, 2, \ldots, c$, designate the jth iteration of $\hat{x}_{n+i}$. For the first iteration, i.e., the first approximation, it is convenient to let $x_{n+1}^{(1)} = x_{n+2}^{(1)} = \cdots = x_{n+c}^{(1)} = T$. Of course, T is a lower bound on the x_{n+i}, and we could begin with first approximations that exceed this value. However, in most applications the choice of T for first approximations is convenient, and it usually results in reasonably rapid convergence to the final estimates.

First approximations to the mean and the standard deviation of the pseudocomplete sample of size N are

$$\bar{x}_N^{(1)} = \frac{1}{N}\sum_{i=1}^{n} x_i + cT$$

$$s_N^{(1)} = \sqrt{\frac{1}{N-1}\left[\sum_{i=1}^{n}\left(x_i - \bar{x}_N^{(1)}\right)^2 + c\left(T - \bar{x}_N^{(1)}\right)^2\right]} \tag{9.3.5}$$

For subsequent approximations, i.e., for $j = 2, 3, \ldots, k$,

$$\bar{x}_N^{(j)} = \sum_{i=1}^{n} x_i + \sum_{i=1}^{c} x_{n+i}^{(j)} \qquad N \tag{9.3.6}$$

$$s_N^{(j)} = \sqrt{\frac{1}{N-1}\left[\sum_{i=1}^{n}\left(x_i - \bar{x}_N^{(j)}\right)^2 + \sum_{i=1}^{c}\left(x_{n+i}^{(j)} - \bar{x}_N^{(j)}\right)^2\right]}$$

We must calculate $\{x_{n+i}^{(j)}\}$, $i = 1, 2, \ldots, c$, before $\bar{x}_N^{(j)}$ and $s_N^{(j)}$ can be evaluated, and these calculations depend on the type of distribution—in this case, the IG.

After calculating first approximations $\bar{x}_N^{(1)}$ and $s_N^{(1)}$, we employ complete sample estimators, presented in Chapter 5, to calculate corresponding approximations to α_3. For this purpose, we might use MLE, ME, or MME. We would need to calculate $a_3^{(j)}$ from approximations to the completed sample if we elect to employ ME. Since x_1 is available in the sample data, it is suggested that the MME be employed as estimators. Procedures for the calculation of α_3 are described in Chapter 5.

After calculating a first approximation $\alpha_3^{(1)}$, we employ (9.3.4) and calculate first approximations to ξ_{n+i} as

$$\xi_{n+i}^{(1)} = G^{-1}\left[\frac{n+i}{N+1};\ \alpha_3^{(1)}\right] \tag{9.3.7}$$

This can be accomplished by inverse interpolation in the cdf table of the standardized IG function given in Appendix A.3 as Table A.3.3.

Second approximations $x_{n+i}^{(2)}$, $i = 1, 2, \ldots, c$, follow as

$$x_{n+i}^{(2)} = T + s_N^{(1)}(\xi_{n+1}^{(1)} - \xi_n^{(1)}) \tag{9.3.8}$$

We repeat the cycle of calculations until, for all i, $|\hat{x}_{n+1}^{(j+1)} - \hat{x}_{n+i}^{(j)}|$ is less than a prescribed maximum error. At the conclusion of the jth cycle of iterations, $\bar{x}_N^{(j)}$ and $s_N^{(j)}$ are given by (9.3.6), and for use in calculating the MME we need

$$z_1^{(j)} = \frac{x_1 - \bar{x}_N^{(j)}}{s_N^{(j)}} \tag{9.3.9}$$

Final estimates of α_3, σ, μ, and γ are

$$\hat{\alpha}_3 = \alpha_3^{(k)}, \quad \hat{\sigma} = s_N^{(k)}, \quad \hat{\mu} = \frac{3s_N^{(k)}}{\hat{\alpha}_3}, \quad \hat{\gamma} = \bar{x}_N^{(k)} - \hat{\mu} \tag{9.3.10}$$

where (k) designates the final cycle of iterations.

9.3.1 Asymptotic Variances and Covariances

The asymptotic variances and covariances for complete sample estimates, given in (5.4.6), provide useful approximations to applicable variances and covariances of estimates obtained here. It is suggested, however, that the number of full-term observations (n) rather than the total number of observations (N) be substituted in the denominator of (5.4.6) when these calculations are made.

9.4 CENSORED SAMPLING IN THE GAMMA DISTRIBUTION

This section is concerned with parameter estimation in the gamma distribution when samples are censored. Both MLE and MMLE which employ the first-order statistic are considered. In certain respects, estimators for parameters of the gamma distribution are similar to corresponding estimators for IG parameters. For both distributions, calculation of estimates is somewhat complicated as a result of differentiation under the integral sign. Although estimators under consideration here might have been obtained as special cases of estimators for parameters of the generalized gamma distribution presented in Chapter 12, it seems expedient to proceed without further reference to the generalized distribution.

9.4.1 Maximum Likelihood Estimators

The loglikelihood function of a progressively censored sample from the three-parameter gamma distribution with pdf (6.2.1) is

$$\ln L = -n \ln \Gamma(\rho) - n\rho \ln \beta - \frac{1}{\beta}\sum_{1}^{n}(x_i - \gamma) + (\rho - 1)\sum_{1}^{n}\ln(x_i - \gamma)$$
$$+ \sum_{1}^{k} c_j \ln[1 - F_j] + \ln C \tag{9.4.1}$$

Because of regularity problems mentioned in Chapter 6, MLE are not appealing for parameters of this distribution. However, when $\rho > 1$, the distribution is bell-shaped and estimates can be calculated by simultaneously solving the equations $\partial \ln L/\partial\gamma = 0$, $\partial \ln L/\partial\beta = 0$, and $\partial \ln L/\partial\rho = 0$. These equations may be written as

$$\frac{\partial \ln L}{\partial\gamma} = \frac{n}{\beta} - (\rho - 1)\sum_{1}^{n}(x_i - \gamma)^{-1} - \sum_{1}^{k}\frac{c_j}{1 - F_j}\frac{\partial F_j}{\partial\gamma} = 0$$

$$\frac{\partial \ln L}{\partial\beta} = -\frac{n\rho}{\beta} + \frac{1}{\beta^2}\sum_{1}^{n}(x_i - \gamma) - \sum_{1}^{k}\frac{c_j}{1 - F_j}\frac{\partial F_j}{\partial\beta} = 0 \tag{9.4.2}$$

$$\frac{\partial \ln L}{\partial\rho} = -n\psi(\rho) - n\ln\beta + \sum_{1}^{n}\ln(x_i - \gamma) - \sum_{1}^{k}\frac{c_j}{1 - F_j}\frac{\partial F_j}{\partial\rho} = 0$$

where $\psi(\rho)$ is the digamma function, $\psi(\rho) = \partial \ln \Gamma(\rho)/\partial\rho = \Gamma'(\rho)/\Gamma(\rho)$. Except for terms involving partials of $F(x)$, these equations are the same as (6.4.1) for complete samples.

In order to evaluate the partials of $F(\xi_j)$, we need the standardized pdf $h(z; 0, 1, \rho)$. This equation was given as (6.2.3), with α_3 rather than ρ as the shape parameter. Since $\alpha_3 = 2/\sqrt{\rho}$ and thus $\rho = 4/\alpha_3^2$, the pdf can be expressed in a simpler algebraic form as

$$\begin{aligned} h(z; 0, 1, \rho) &= (\sqrt{\rho})^{\rho}(z + \sqrt{\rho})^{\rho-1}\exp\{-\sqrt{\rho}(z + \sqrt{\rho})\}, \quad -\sqrt{\rho} < z < \infty \\ &= 0, \quad \text{elsewhere} \end{aligned} \tag{9.4.3}$$

where

$$z = \frac{x - \gamma - \rho\beta}{\beta\sqrt{\rho}} \tag{9.4.4}$$

The standardized cdf becomes

$$H(z, \rho) = \int_{-\sqrt{\rho}}^{z} h(z; 0, 1, \rho)\, dz \tag{9.4.5}$$

When $X = T_j$, then $Z = \xi_j$ and $F(T_j) = H(\xi_j)$, where

$$\xi_j = \frac{T_j - \gamma - \rho\beta}{\beta\sqrt{\rho}} \tag{9.4.6}$$

The partials of F_j may now be expressed as

$$\begin{aligned}
\frac{\partial F_j}{\partial\gamma} &= \frac{\partial H_j}{\partial\xi_j}\frac{\partial\xi_j}{\partial\gamma} = -\frac{1}{\beta\sqrt{\rho}}\,h(\xi_j) \\
\frac{\partial F_j}{\partial\beta} &= \frac{\partial H_j}{\partial\xi_j}\frac{\partial\xi_j}{\partial\beta} = -\frac{T_j - \gamma}{\beta^2\sqrt{\rho}}\,h(\xi_j) \\
\frac{\partial F_j}{\partial\rho} &= \frac{\partial H_j}{\partial\rho} = \int_{-\sqrt{\rho}}^{\xi_j} \frac{\partial h(z; 0, 1, \rho)}{\partial\rho}\, dz + h(\xi_j; 0, 1, \rho)\,\frac{\partial\xi_j}{\partial\rho}
\end{aligned} \tag{9.4.7}$$

The expression given in (9.4.7) for $\partial F_j/\partial\rho$ can be simplified somewhat if we introduce the transformation

$$U = \frac{X - \gamma}{\beta} \tag{9.4.8}$$

With this transformation, the pdf of U follows from (6.2.1) as

$$\begin{aligned}
g(u; \rho) &= \frac{u^{\rho-1}e^{-u}}{\Gamma(\rho)}, \quad 0 < u < \infty \\
&= 0, \quad \text{elsewhere}
\end{aligned} \tag{9.4.9}$$

which we recognize as the pdf of a one-parameter gamma distribution with mean ρ and variance ρ. When $X = T_j$, then $U = \omega_j$, where

$$\omega_j = \frac{T_j - \gamma}{\beta} \tag{9.4.10}$$

The cdf of U is

$$I(u; \rho) = \int_0^u g(u, \rho)\, du \tag{9.4.11}$$

and

$$F(T_j) = H(\xi_j) = I(\omega_j) \tag{9.4.12}$$

Thus,

$$\frac{\partial F_j}{\partial \rho} = \frac{\partial I_j}{\partial \rho} = \int_0^{\omega_j} \frac{\partial g(u, \rho)}{\partial \rho}\, du$$

$$= -\psi(\rho) I(\omega_j) + \int_0^{\omega_j} (\ln u) g(u; \rho)\, du \tag{9.4.13}$$

We substitute the expressions for $\partial F_j / \partial \gamma$ and $\partial F_j / \partial \beta$ from (9.4.7) and the expression for $\partial F_j / \partial \rho$ from (9.4.13) into (9.4.2), and the resulting estimating equations become

$$\frac{n}{\beta} - (\rho - 1) \sum_1^n (x_i - \gamma)^{-1} + \frac{1}{\beta \sqrt{\rho}} \sum_1^k c_j Q_j = 0$$

$$-\frac{n\rho}{\beta} + \frac{1}{\beta^2} \sum_1^n (x_i - \gamma) + \frac{1}{\beta^2 \sqrt{\rho}} \sum_1^k c_j (T_j - \gamma) Q_j = 0 \tag{9.4.14}$$

$$-n\psi(\rho) - n \ln \beta + \sum_1^n \ln (x_i - \gamma) + \sum_1^k c_j [A_j \psi(\rho) - B_j] = 0$$

where

$$A_j = A(\omega_j, \rho) = \frac{I_j}{1 - I_j}$$

$$B_j = B(\omega_j, \rho) = \frac{1}{1 - I_j} \int_0^{\omega_j} (\ln u) g(u, \rho)\, du$$

(9.4.15)

$$Q_j = \frac{h(\xi_j)}{1 - H(\xi_j)}$$

$$\omega_j = \frac{T_j - \gamma}{\beta}$$

Note that

$$\xi_j = \frac{1}{\sqrt{\rho}}(\omega_j - \rho) \quad \text{and} \quad \omega_j = \sqrt{\rho}\,\xi_j + \rho \tag{9.4.16}$$

Maximum likelihood estimates can be obtained as the simultaneous solution of the three equations of (9.4.14). Standard iterative procedures, such as Newton's method, are available for this task, but the reader is cautioned that convergence problems might occur unless $\rho >> 1$. When $\rho = 1$, the exponential distribution emerges as a special case, and censored sample estimators for this distribution were considered in Chapter 7.

The functions $A(\omega, \rho)$ and $B(\omega, \rho)$ arise again in Chapter 12 in connection with the generalized gamma distribution, where they are tabulated as Tables 12.6 and 12.7. These tables are available as aids in the solution of (9.4.14).

9.4.2 Estimation in the Two-Parameter Gamma Distribution

When γ is known and may therefore be set to zero, we need only solve the last two equations of (9.4.14), with $\gamma = 0$, for the unknown estimates $\hat{\beta}$ and $\hat{\rho}$. With one less equation in the system, it is simpler to solve for two estimates than for three. However, even in this case, it is necessary to deal with B_j and its attending complications.

9.4.3 Estimation from Singly Right-Censored Samples

In singly right-censored samples $k = 1$ and $c_j = c$ for two- and three-parameter censored samples. Otherwise the estimating equations are the same as corresponding equations for progressively censored samples.

9.4.4 Modified Maximum Likelihood Estimates

For MMLE estimators, the estimating equation $\partial \ln L/\partial\gamma = 0$ is replaced by $E[F(X_1)] = F(x_1)$, where $E[F(X_1)] = 1/(N + 1)$. Estimating equations $\partial \ln L/\partial\beta = 0$ and $\partial \ln L/\partial\rho = 0$ are retained. Following a few simple algebraic steps, we obtain the replacement equation

$$\gamma + \rho\beta + \beta\sqrt{\rho}\, z_1 = x_1 \tag{9.4.17}$$

where

$$z_1 = H^{-1}\left[\frac{1}{N+1}; \alpha_3\right]; \text{ i.e., } \int_{-2/\alpha_3}^{z_1} h(z; 0, 1, \alpha_3)\, dz = \frac{1}{N+1} \qquad (9.4.18)$$

The complete system of estimating equations consists of (9.4.17) plus the last two equations of (9.4.14). As with MLE, calculation of these estimates requires the solution of a somewhat complex system of estimating equations. The same iteration techniques are applicable in both cases, and the complications involved in evaluating B_j are present in both cases.

9.4.5 Asymptotic Variance and Covariances

Useful approximations to variances and covariances of estimates based on censored samples from the gamma distribution can be calculated from corresponding complete sample formulas given in Chapter 6, where n, which appears in the denominators of (6.5.3), is the number of full-term observations in a given sample.

9.4.6 The Pseudocomplete Sample Technique

The pseudocomplete sample technique, which was introduced in Section 9.3 for singly censored samples from the IG distribution, can also be used to estimate parameters of the gamma distribution. The first cycle of estimates of censored observations is the same for all distributions. For the second and subsequent cycles in the iteration process, parameter estimates for the gamma distribution are calculated from appropriate complete sample estimators given in Chapter 6. Again, it is suggested that the MME be used, but the ME or the MLE could be employed. Estimates of ξ_{n+i} for the jth cycle of iteration are

$$\xi_{n+i}^{(j)} = H^{-1}\left[\frac{n+i}{N+1}; \alpha_3^{(j)}\right], \qquad j = 1, 2, 3, \ldots \qquad (9.4.19)$$

and

$$x_{n+i}^{(j)} = T + \sigma^{(j-1)}(\xi_{n+i}^{(j-1)} - \xi_n^{(j-1)}), \qquad j = 2, 3, 4, \ldots$$

Other details of the procedure are as described in Section 9.3 for parameters of the IG distribution.

9.5 AN ILLUSTRATIVE EXAMPLE

Practical application of the pseudocomplete sample technique in estimating parameters of skewed distribution from a singly right-censored sample is illustrated with a random sample from an IG distribution in which $\gamma = 10$, $\mu = 6$, $\sigma = 5$, and thus $\alpha_3 = 2.5$. The complete sample consists of N = 100 observations, tabulated here.

Random Sample from an IG Distribution
[$\gamma = 10$, $\mu = 6$, $\sigma = 5$, $\alpha_3 = 2.5$, $\bar{x} = 15.239$, $s = 5.146$, $a_3 = 2.314$]

10.8884	11.1417	11.1562	11.3311	11.3493
11.4726	11.5578	11.5839	11.6037	11.6113
11.6640	11.7501	11.8600	12.9766	12.1246
12.1868	12.2967	12.3286	12.3296	12.3327
12.3504	12.4423	12.5053	12.6288	12.6423
12.6492	12.6268	12.7462	12.8156	12.8532
13.1523	13.1642	13.1807	13.2550	13.3709
13.4310	13.5195	13.5336	13.5786	13.5987
13.6522	13.6539	13.6596	13.6825	13.7392
13.7657	13.7764	13.8136	13.8846	14.0068
14.0490	14.0568	14.3218	14.5941	14.6496
14.6811	14.7629	15.0891	15.2358	15.3520
15.4604	15.5547	15.6059	15.7032	15.7977
15.8906	15.8931	15.9440	16.1099	16.1922
16.2001	16.4352	16.4605	17.2224	17.2458
17.3007	17.4365	17.8224	18.1382	18.3316
18.4063	18.6581	18.8229	19.0166	19.0801
19.2521	19.3555	21.0255	22.1120	23.0403
24.3057	24.9244	25.2042	25.2134	25.8634
26.3125	27.5814	29.3053	32.5742	42.4409

For this sample, we calculate $\bar{x} = 15.239$, $s = 5.1458$, $a_3 = 2.3142$, $x_1 = 10.8884$, and $N = 100$.

For the purpose of this illustration, the sample was censored at $x_{95} = 25.8634$. Thus $n = 95$, $c = 5$, and $N = 100$. Parameter estimates calculated as described in Section 9.3 are $\hat{\gamma} = 9.905$, $\hat{\mu} = 5.954$, $\hat{\sigma} = 4.828$, and $\hat{\alpha}_3 = 2.433$. These estimates are to be compared with the complete sample MME $\hat{\gamma} = 10.009$, $\hat{\mu} = 5.914$, $\hat{\sigma} = 5.196$, and $\hat{\alpha}_3 = 2.610$, and with the population parameters $\gamma = 0$, $\mu = 6$, and $\alpha_3 = 2.5$.

Approximations to variances and covariances of the estimates obtained here can be calculated from (5.4.6). With $\hat{\alpha}_3 = 2.433$, we interpolate in Table 5.1 to obtain $\phi_{11}(2.433) = 0.42$, $\phi_{22}(2.433) = 1.42$, $\phi_{33}(2.433) = 2.25$, $\phi_{12}(2.433) = -0.42$, $\phi_{13}(2.433) = 0.34$, and $\phi_{23}(2.433) = -0.88$. We substitute these values together with $n = 95$ and $\hat{\sigma} = 4.828$ into (5.4.6) and calculate approximate variances and covariances as $V(\hat{\gamma}) = 0.10$, $V(\hat{\mu}) = 0.35$,

Successive Iterations of Pseudocomplete Sample from IG Distribution [$\gamma = 10$, $\mu = 6$, $\sigma = 5$, $E(X) = 16$, $\alpha_3 = 2.5$, $N = 100$, $h = 0.05$]

Item	Iteration number (j)								
	1	2	3	4	5	6	7	8	9
$\bar{x}_N^{(j)}$	15.6350	15.8134	15.8484	15.8567	15.8587	15.8592	15.8593	15.8593	15.8593
$\mu^{(j)}$	5.9760	5.9590	5.9553	5.9545	5.9544	5.9543	5.9543	5.9543	5.9543
$s_N^{(j)}$	5.1485	4.7899	4.7899	4.8186	4.8257	4.8274	4.8278	4.8279	4.8280
$\gamma^{(j)}$	9.6590	9.8931	9.8931	9.9021	9.9043	9.9049	9.9050	9.9050	9.9050
$\alpha_3^{(j)}$	2.0826	2.3515	2.4129	2.4277	2.4314	2.4322	2.4324	2.4325	2.4325
$z_1^{(j)}$	-1.1442	-1.0544	-1.0355	-1.0311	-1.0300	-1.0297	-1.0296	-1.0296	-1.0296
$\xi_{95}^{(j)}$	1.7737	1.7648	1.7619	1.7612	1.7610	1.7610	1.7609	1.7609	1.7609
$\xi_{96}^{(j)}$	1.9576	1.9580	1.9572	1.9570	1.9569	1.9569	1.9569	1.9569	1.9569
$\xi_{97}^{(j)}$	2.1847	2.1979	2.1998	2.2003	2.2004	2.2004	2.2004	2.2004	2.2004
$\xi_{98}^{(j)}$	2.4809	2.5123	2.5183	2.5197	2.5200	2.5201	2.5201	2.5201	2.5102
$\xi_{99}^{(j)}$	2.9043	2.9648	2.9771	2.9801	2.9808	2.9809	2.9810	2.9810	2.9810
$\xi_{100}^{(j)}$	3.6431	3.7609	3.7860	3.7920	3.7935	3.7938	3.7939	3.7940	3.7940
$x_{96}^{(j)}$	25.8634	26.6263	26.7661	26.7988	26.8067	26.8087	26.8092	26.8093	26.8093
$x_{97}^{(j)}$	25.8634	27.5687	27.8865	27.9610	27.9791	27.9836	27.9847	27.9849	27.9850
$x_{98}^{(j)}$	25.8634	28.7973	29.3552	29.4863	29.5182	29.5260	29.5280	29.5284	29.5285
$x_{99}^{(j)}$	25.8634	30.5537	31.4684	31.6841	31.7367	31.7495	31.7527	31.7535	31.7537
$x_{100}^{(j)}$	25.8634	33.6188	31.1869	35.5587	35.6493	35.6715	35.6770	35.6783	35.6787

$V(\hat{\sigma}) = 0.55$, $\mathrm{Cov}(\hat{\gamma}, \hat{\mu}) = -0.10$, $\mathrm{Cov}(\hat{\gamma}, \hat{\sigma}) = 0.08$, and $\mathrm{Cov}(\hat{\mu}, \hat{\sigma}) = -0.21$. The accuracy of these results could be improved slightly by substituting $\hat{\alpha}_3 = 2.433$ and $\hat{\sigma} = 4.828$ directly into (5.4.8), (5.4.7), and (5.4.6) without reference to Table 5.1.

10

The Rayleigh Distribution

10.1 INTRODUCTION

The Rayleigh distribution is frequently employed by engineers, physicists, and other scientists as a model for the analysis of data resulting from investigations involving wave propagation, radiation, and related inquiries. It is also applicable in the analysis of target error data. It was first derived by Lord Rayleigh (1919) in connection with a study of acoustical problems. More recently, this distribution has been the subject of accounts by Hirano (1986), Johnson and Kotz (1976), Kotz and Srinivasan (1969), Archer (1967), Siddiqui (1962), and many others. Although the Rayleigh designation was not mentioned, Cohen (1955) studied this distribution in connection with parameter estimation in the analysis of target data.

In its most general form, the Rayleigh distribution may be considered as the distribution of the distance X from the origin to a point $(Y_1, Y_2, \ldots, Y_p)$ in a p-dimensional Euclidean space where the components Y_j, $j = 1, 2, \ldots, p$, are independent random variables, each of which is normally distributed $(0, \sigma^2)$. The random variable X may thus be expressed as

$$X = \sqrt{\sum_{j=1}^{p} Y_j^2} \tag{10.1.1}$$

It is well known that $(X/\sigma)^2$ has a chi-square distribution with p degrees of freedom. The probability density function (pdf) of $(X/\sigma)^2$ may thus be written as

$$f\left(\frac{x}{\sigma}\right)^2 = \frac{2^{-p/2}}{\Gamma(p/2)}\left(\frac{x}{\sigma}\right)^{2(p/2-1)} \exp\left\{-\frac{1}{2}\left(\frac{x}{\sigma}\right)^2\right\}, \quad \frac{x}{\sigma} > 0$$
$$= 0, \quad \text{elsewhere} \tag{10.1.2}$$

It follows that X/σ has a chi distribution with p degrees of freedom, and the pdf of X (i.e., the pdf of the p-dimensional Rayleigh distribution) is

$$f(x; p, \sigma) = \frac{2^{-(p-2)/2}}{\sigma\Gamma(p/2)}\left(\frac{x}{\sigma}\right)^{p-1} \exp\left\{-\frac{1}{2}\left(\frac{x}{\sigma}\right)^2\right\}, \quad x > 0,\ \sigma > 0$$
$$= 0, \quad \text{elsewhere} \tag{10.1.3}$$

The cumulative distribution function (cdf) is $F[p/2,\ x^2/2\sigma^2]$, where

$$F(a, x) = \frac{1}{\Gamma(a)} \int_0^x t^{a-1} e^{-t}\, dt \tag{10.1.4}$$

The kth moment about the origin is

$$\mu'_k = 2^{k/2}\sigma^k\Gamma\left(\frac{k+p}{2}\right)\Big/\Gamma\left(\frac{p}{2}\right) \tag{10.1.5}$$

The expected value, variance, and mode of X are

$$E(X) = \sigma\sqrt{2}\,\Gamma\left(\frac{1+p}{2}\right)\Big/\Gamma\left(\frac{p}{2}\right)$$

$$V(X) = 2\sigma^2\left[\frac{\Gamma\,(2+p)/2}{\Gamma(p/2)} - \left(\frac{\Gamma((1+p)/2)}{\Gamma(p/2)}\right)^2\right] \tag{10.1.6}$$

$$Mo(X) = \sigma\sqrt{p-1}$$

Readers are reminded that $\Gamma(1/2) = \sqrt{\pi}$, $\Gamma(1+z) = z\Gamma(z)$, and thus $\Gamma(3/2) = \sqrt{\pi}/2$, $\Gamma(5/2) = 3\sqrt{\pi}/4$, $\Gamma(7/2) = 15\sqrt{\pi}/8$, etc. These results are needed in the evaluation of (10.1.6) for specific values of p.

10.2 MAXIMUM LIKELIHOOD AND MOMENT ESTIMATION IN THE p-DIMENSIONAL DISTRIBUTION

Let $\{x_i\}$, $i = 1, \ldots, n$, designate a random sample of size n from a distribution with pdf (10.1.3). The likelihood function of this sample is

$$L(x_1 \ldots x_n; \sigma) = 2^{-n(p-2)/2} \sigma^{-np} \left[\Gamma\left(\frac{p}{2}\right)\right]^{-n} \left[\prod_1^n x_i^{p-1}\right] \exp\left\{-\sum_1^n \frac{x_i^2}{2\sigma^2}\right\} \tag{10.2.1}$$

After taking logarithms, differentiating and equating the results to zero, we obtain

$$\frac{\partial \ln L}{\partial \sigma} = \frac{np}{\sigma} + \frac{1}{\sigma^3} \sum_1^n x_i^2 = 0 \tag{10.2.2}$$

The maximum likelihood estimator (MLE) of σ follows as

$$\hat{\sigma} = \sqrt{\sum_1^n \frac{x_i^2}{np}} \tag{10.2.3}$$

It has been pointed out by Hirano (1986) that for a fixed value of p, the distribution is a member of the one-parameter exponential family of distributions. He further pointed out that $\Sigma_{i=1}^n x_i^2$ is a sufficient statistic for σ and that the mean (expected value) and variance of this statistic are

$$E\left[\sum_1^n x_i^2\right] = np\sigma^2 \quad \text{and} \quad \text{Var}\left[\sum_1^n x_i^2\right] = 2np\sigma^4 \tag{10.2.4}$$

The moment estimator (ME) is obtained by equating E(X) from (10.1.6) to the sample mean $\bar{x}$. Thus we obtain

$$\sigma^* = \frac{\bar{x}\Gamma(p/2)}{\sqrt{2}\,\Gamma((p+1)/2)} \tag{10.2.5}$$

10.3 SPECIAL CASES

Although the general form of the Rayleigh distribution with $p > 3$ might have limited applications, the special cases in which $p = 1$, 2, and 3, respectively, are of sufficient interest to warrant further attention.

10.3.1 The Folded Normal Distribution

The Rayleigh distribution with pdf (10.1.3), in which $p = 1$, is sometimes called the folded Gaussian, the folded normal, or the half-normal distribution. In this case the pdf becomes

$$f(x; 1, \sigma) = \frac{1}{\sigma}\sqrt{\frac{2}{\pi}}\exp\left(-\frac{x^2}{2\sigma^2}\right), \qquad 0 < x < \infty \tag{10.3.1}$$

The expected value, variance, coefficient of variation, α_3, α_4, and mode of this distribution are

$$\begin{aligned} E(X) &= \frac{\sqrt{2/\pi}}{\sigma} \doteq 0.797885\sigma \\ V(X) &= \frac{\sigma^2(\pi - 2)}{\pi} \doteq 0.363380\sigma^2 \\ CV(X) &= \sqrt{\frac{\pi - 2}{2}} \doteq 0.755511 \\ \alpha_3(X) &= \frac{\sqrt{2}(4 - \pi)}{(\pi - 2)^{3/2}} \doteq 0.995272 \\ \alpha_4(X) &= \frac{3\pi^2 - 4\pi - 12}{(\pi - 2)^2} \doteq 3.869177 \\ Mo(X) &= 0 \end{aligned} \tag{10.3.2}$$

The cdf can be expressed in terms of the standard normal distribution as

$$F(x; 1, \sigma) = 2\Phi\left(\frac{x}{\sigma}\right) - 1 \tag{10.3.3}$$

where $\Phi(\ )$ is the cdf of the standard normal distribution (0, 1).

10.3.2 The Two-Dimensional Rayleigh Distribution

Although this distribution is only a single special case of the p-dimensional distribution, it is often labeled with the Rayleigh designation as though it were the only one to be considered. With $p = 2$, the pdf of (10.1.3) is reduced to

$$\begin{aligned} f(x; 2, \sigma) &= \frac{x}{\sigma^2}\exp\left\{-\frac{1}{2}\left(\frac{x}{\sigma}\right)^2\right\}, \qquad 0 < x < \infty \\ &= 0, \qquad \text{elsewhere} \end{aligned} \tag{10.3.4}$$

This distribution is further recognized as a special case of the two-parameter Weibull distribution (β, δ), which was considered in Chapter 3.

Here, $\delta = 2$ and $\beta^2 = 2\sigma^2$. Moments and related properties of the two-dimensional Rayleigh distribution can be obtained either from the Weibull properties or from the p-dimensional Rayleigh distribution. The Weibull parameterization $(\beta, 2)$ enables the pdf to be written as

$$\begin{aligned} f(x; 2, \beta) &= \frac{2}{\beta^2} x \exp\left\{-\left(\frac{x}{\beta}\right)^2\right\}, \quad 0 < x < \infty \\ &= 0, \quad \text{elsewhere} \end{aligned} \tag{10.3.5}$$

or with $\theta = \beta^2$ as

$$\begin{aligned} f(x; 2, \theta) &= \frac{2}{\theta} x \exp\left(-\frac{x^2}{\theta}\right), \quad 0 < x < \infty \\ &= 0, \quad \text{elsewhere} \end{aligned} \tag{10.3.6}$$

We thus have three equivalent parametrizations for this distribution. With σ as the parameter, the cdf is

$$F(x; 2, \sigma) = 1 - \exp\left(-\frac{x^2}{2\sigma^2}\right) \tag{10.3.7}$$

The kth moment about the origin is

$$\mu_k' = 2^{k/2}\sigma^k\Gamma_k = (\sigma\sqrt{2})^k\left(\frac{k}{2}\right)\Gamma\left(\frac{k}{2}\right) \tag{10.3.8}$$

where, as in Chapter 3, $\Gamma_k = \Gamma(1 + k/2)$. Other characteristics of interest are

$$\begin{aligned} E(X) &= \sigma\sqrt{\frac{\pi}{2}} \doteq 1.253314\sigma \\ V(X) &= \frac{\sigma^2(4 - \pi)}{2} \doteq 0.429204\sigma^2 \\ CV(X) &= \frac{\sqrt{V(X)}}{E(X)} = \sqrt{\frac{4 - \pi}{\pi}} \doteq 0.522723 \\ \alpha_3(X) &= \frac{2(\pi - 3)\sqrt{\pi}}{(4 - \pi)^{3/2}} \doteq 0.631110 \\ \alpha_4(X) &= \frac{32 - 3\pi^2}{(4 - \pi)^2} \doteq 3.245089 \\ Mo(X) &= \sigma \\ Me(X) &= \sigma\sqrt{\ln 4} \doteq 1.17741\sigma \end{aligned} \tag{10.3.9}$$

The pth percentile is

$$x_p = [-2\sigma^2 \ln(1 - p)]^{1/2}, \qquad 0 < p < 1 \tag{10.3.10}$$

The hazard function, $h(x) = f(x)/[1 - F(x)]$, which is of interest in reliability analysis, is the increasing function

$$h(x) = \frac{x}{\sigma^2} \tag{10.3.11}$$

10.3.3 The Maxwell-Boltzmann Distribution (p = 3)

This distribution is the special case of the Rayleigh distribution in which p = 3, and the pdf follows from (10.1.3) as

$$\begin{aligned} f(x; 3, \sigma) &= \frac{1}{\sigma}\sqrt{\frac{2}{\pi}}\left(\frac{x}{\sigma}\right)^2 \exp\left\{-\frac{1}{2}\left(\frac{x}{\sigma}\right)^2\right\}, \qquad 0 < x < \infty \\ &= 0, \qquad \text{elsewhere} \end{aligned} \tag{10.3.12}$$

Moments and other characteristics of interest follow from (10.1.5) and from (10.1.6) as

$$\begin{aligned} E(X) &= 2\sigma\sqrt{\frac{2}{\pi}} \doteq 1.595769\sigma \\ V(X) &= \frac{\sigma^2(3\pi - 8)}{\pi} \doteq 0.453521\sigma^2 \\ CV(X) &= \sqrt{\frac{3\pi - 8}{8}} \doteq 0.422016 \\ \alpha_3(X) &= \frac{\sqrt{2}(32 - 10\pi)}{(3\pi - 8)^{3/2}} \doteq 0.485693 \\ \alpha_4(X) &= \frac{15\pi^2 + 16\pi - 192}{(3\pi - 8)^2} \doteq 3.108164 \\ Mo(X) &= \sigma\sqrt{2} \end{aligned} \tag{10.3.13}$$

This distribution is of special interest to physicists and engineers.

10.4 COMPLETE SAMPLE ESTIMATORS WHEN p = 1, 2, AND 3

Complete sample estimators for the generalized p-dimensional distribution were given in Section 10.2. The MLE and ME based on complete samples from the special case distributions in which p = 1, 2, and 3, respectively, are listed next.

Folded Normal Distribution, p = 1

$$\text{MLE: } \hat{\sigma} = \sum_1^n \sqrt{\frac{x_i^2}{n}}$$

$$\text{ME: } \sigma^* = \bar{x}\sqrt{\frac{\pi}{2}} = 1.253314\bar{x}$$

Rayleigh Distribution, p = 2

$$\text{MLE: } \hat{\sigma} = \sum_1^n \sqrt{\frac{x_i^2}{2n}}$$

$$\text{ME: } \sigma^* = \bar{x}\sqrt{\frac{2}{\pi}} = 0.797885\bar{x}$$

Maxwell-Boltzmann Distribution, p = 3

$$\text{MLE: } \hat{\sigma} = \sqrt{\sum_1^n \frac{x_i^2}{3n}}$$

$$\text{ME: } \sigma^* = \bar{x}\sqrt{\frac{\pi}{8}} = 0.626657\bar{x}$$

10.5 TWO-PARAMETER RAYLEIGH DISTRIBUTION

With the addition of a threshold parameter, γ, the two-dimensional Rayleigh distribution with pdf (10.3.4) becomes

$$f(x;\gamma,\sigma) = \frac{x-\gamma}{\sigma^2}\exp\left\{-\frac{1}{2}\left(\frac{x-\gamma}{\sigma}\right)^2\right\}, \quad \gamma < x < \infty$$
$$= 0, \quad \text{elsewhere} \qquad (10.5.1)$$

In this form, it might compete with other positively skewed distributions as a model for use in life-span and related applications where the origin or threshold is different from zero. In estimating γ, we are led to consider the first-order statistic x_1 in a random ordered sample $\{x_i\}$, $i = 1, 2, \ldots, n$. As noted in Chapter 3, when $X \sim W(\gamma, \beta, \delta)$, the first-order statistic $X_1 \sim W(\gamma, \beta', \delta)$, where $\beta' = \beta/n^{1/\delta}$. Since in the special case under consideration here, $\beta^2 = 2\sigma^2$ and $\delta = 2$, it follows that $\beta' = \sigma\sqrt{2/n}$. Accordingly, $X_1 \sim W(\gamma, \beta', 2)$ and

$$
\begin{aligned}
f(x_{1:n}; \gamma, \sigma) &= \frac{n}{\sigma^2}(x_1 - \gamma) \exp\left\{-\frac{n}{2}\left(\frac{x_1 - \gamma}{\sigma}\right)^2\right\}, \quad \gamma < x_1 < \infty \\
&= 0, \quad \text{elsewhere}
\end{aligned}
\tag{10.5.2}
$$

Note that (10.5.2) is also the pdf of a two-parameter, two-dimensional Rayleigh distribution with parameter $\sigma/\sqrt{n}$. It follows that

$$
E(X_1) = \gamma + \sigma\sqrt{\frac{\pi}{2n}} \tag{10.5.3}
$$

10.6 ESTIMATION IN THE TWO-PARAMETER RAYLEIGH DISTRIBUTION

10.6.1 Maximum Likelihood Estimators

The likelihood function of a random sample $\{x_i\}$, $i = 1, 2, \ldots, n$ from a population with pdf (10.5.1) is

$$
L(x_1, x_2, \cdot, x_n; \gamma, \sigma) = \sigma^{-2n} \prod_{i=1}^{n} (x_i - \gamma) \exp\left\{-\frac{1}{2}\sum_{1}^{n}\left(\frac{x_i - \gamma}{\sigma}\right)^2\right\} \tag{10.6.1}
$$

On taking logarithms, we obtain the loglikelihood function

$$
\ln L = -2n \ln \sigma + \sum_{1}^{n} \ln(x_i - \gamma) - \frac{1}{2\sigma^2}\sum_{1}^{n}(x_i - \gamma)^2 \tag{10.6.2}
$$

On differentiating (10.6.2) and equating to zero, we obtain the estimating equations

$$\frac{\partial \ln L}{\partial \sigma} = -\frac{2n}{\sigma} + \frac{1}{\sigma^3}\sum_1^n (x_i - \gamma)^2 = 0$$

$$\frac{\partial \ln L}{\partial \gamma} = -\sum_1^n (x_i - \gamma)^{-1} + \frac{1}{\sigma^2}\sum_1^n (x_i - \gamma) = 0 \tag{10.6.3}$$

These equations are subsequently simplified to yield

$$\frac{n(\bar{x} - \hat{\gamma})}{\sum_1^n (x_i - \hat{\gamma})^{-1}} - \frac{1}{2n}\sum_1^n (x_i - \hat{\gamma})^2 = 0$$

$$\hat{\sigma}^2 = \frac{1}{2n}\sum_1^n (x_i - \hat{\gamma})^2 \tag{10.6.4}$$

The first equation of (10.6.4) can be solved for $\hat{\gamma}$ by trial and error or by various iterative procedures. As a first approximation, we need a value $\gamma_1 < x_1$, where x_1 is the smallest sample observation. If the first approximation is too large, we select a smaller second approximation and continue through additional cycles of calculations until we find a pair of values γ_i and γ_j, one too large and one too small and such that $|\gamma_i - \gamma_j|$ is sufficiently small. Linear interpolation between γ_i and γ_j will then produce a final estimate $\hat{\gamma}$. With $\hat{\gamma}$ thus calculated, $\hat{\sigma}$ follows from the second equation of (10.6.4).

10.6.2 Moment Estimators

Moment estimators $E(X) = \bar{x}$ and $V(X) = s^2$, where $s^2 = \Sigma_1^n (x_i - \bar{x})^2/(n - 1)$ and $\bar{x} = \Sigma_1^n x_i/n$, become

$$\sigma^* = s\sqrt{\frac{2}{4 - \pi}}, \qquad \gamma^* = \bar{x} - \sigma^*\sqrt{\frac{\pi}{2}} \tag{10.6.5}$$

These explicit estimators can be calculated quite easily from the sample data. The estimate γ^* is subject to the restriction $\gamma^* < x_1$.

10.6.3 Modified Moment Estimators

As estimating equations, we employ $E(X_1) = x_1$ and $E(X) = \bar{x}$. We thus have

$$\hat{\sigma}\sqrt{\frac{\pi}{2}} + \hat{\gamma} = \bar{x}, \qquad \hat{\sigma}\sqrt{\frac{\pi}{2n}} + \hat{\gamma} = x_1 \tag{10.6.6}$$

The simultaneous solution of these two equations yields

$$\hat{\sigma} = \frac{\bar{x} - x_1}{\sqrt{\pi/2} - \sqrt{\pi/2n}}$$
$$\hat{\gamma} = \bar{x} - \hat{\sigma}\sqrt{\frac{\pi}{2}} \tag{10.6.7}$$

These explicit estimators are easily calculated from sample data.

10.6.4 Modified Maximum Likelihood Estimators

For MMLE we employ $\partial \ln L/\partial\sigma = 0$ and $E(X_1) = x_1$. From (10.6.3) and (10.5.3) the estimating equations become

$$\frac{-2n}{\sigma} + \frac{1}{\sigma^3}\sum_1^n (x_i - \gamma)^2 = 0$$
$$\gamma + \sigma\sqrt{\frac{\pi}{2n}} = x_1 \tag{10.6.8}$$

On eliminating σ between these two equations, we obtain

$$\frac{2n(x_1 - \gamma)^2}{\pi} = \frac{1}{2n}\sum_1^n (x_i - \gamma)^2 \tag{10.6.9}$$

in which γ is the only unknown quantity. It is a relatively simple matter to solve (10.6.9) for $\hat{\gamma}$. The trial-and-error procedure previously described or any standard iterative method can be used. With $\hat{\gamma}$ thus determined, $\hat{\sigma}$ follows from the first equation of (10.6.8) as

$$\hat{\sigma}^2 = \frac{1}{2n}\sum_1^n (x_i - \hat{\gamma})^2 \tag{10.6.10}$$

Any one of the above estimators should be satisfactory in most instances. However, in view of the ease with which it can be calculated, a slight preference might be expressed for the MME.

10.7 TRUNCATED SAMPLES

The likelihood function of a truncated sample consisting of n randomly selected observations from a population with pdf (10.1.3), when each observation is subject to the restriction $0 < x \leq T$, is

$$L(x_1, \ldots, x_n; p, \sigma) = 2^{-n(p-2)/2}\sigma^{-np}\left[\Gamma\left(\frac{p}{2}\right)\right]^{-n} \cdot \left[\prod_{i=1}^{n} x_i^{p-1}\right] \exp\left\{-\frac{1}{2\sigma^2}\sum_{1}^{n} x_i^2\right\}[F_p(T)]^{-n} \tag{10.7.1}$$

where

$$F_p(T;\sigma) = \frac{2^{-(1-2)/2}}{\Gamma(p/2)}\int_0^T \left\{\left(\frac{x}{\sigma}\right)^{p-1} \exp\left[-\frac{1}{2}\left(\frac{x}{\sigma}\right)^2\right]\right\}\frac{dx}{\sigma} \tag{10.7.2}$$

If we let $z = x/\sigma$ and thus $z_0 = T/\sigma$, then (10.7.2) reduces to

$$F_p(T;\sigma) = G_p(z_0) = \int_0^T g_p(z)\, dx \tag{10.7.3}$$

where

$$g_p(z) = \frac{2^{-(p-2)/2}}{\Gamma(p/2)} z^{p-1} \exp\left\{-\frac{z^2}{2}\right\}, \qquad 0 < z < \infty$$
$$= 0, \qquad \text{elsewhere} \tag{10.7.4}$$

On taking logarithms of (10.7.1), differentiating with respect to σ, and equating to zero, we obtain

$$\frac{\partial \ln L}{\partial\sigma} = \frac{nz_0}{\sigma}\frac{g_p(z_0)}{G_p(z_0)} - \frac{np}{\sigma} + \frac{1}{\sigma^3}\sum_{1}^{n} x_i^2 = 0 \tag{10.7.5}$$

We substitute $\sigma = T/z_0$ into (10.7.5) and simplify to obtain

$$\frac{1}{nT^2}\sum_{1}^{n} x_i^2 = J_p(\hat{z}_0) \tag{10.7.6}$$

where

$$J_p(z_0) = \frac{1}{z_0}\left[\frac{p}{z_0} - \frac{g_p(z_0)}{G_p(z_0)}\right] \qquad (10.7.7)$$

Therefore, σ has been replaced as the unknown parameter by z_0, and we must solve (10.7.6) for the MLE $\hat{z}_0$. With $\hat{z}_0$ thus determined, the estimate of σ follows as

$$\hat{\sigma} = \frac{T}{\hat{z}_0} \qquad (10.7.8)$$

The solution of (10.7.6) can be accomplished by standard iterative methods or by interpolation in applicable tables of the estimating function $J_p(z_0)$.

10.7.1 Truncation in the Two-Dimensional Rayleigh Distribution

When $p = 2$, it follows that

$$g_2(z_0) = z_0 \exp\left(-\frac{z_0^2}{2}\right) = z_0\sqrt{2\pi}\,\phi(z_0) \qquad (10.7.9)$$

and

$$G_2(z_0) = 1 - \sqrt{2\pi}\,\phi(z_0) \qquad (10.7.10)$$

where $\phi(z)$ is the pdf of the standard normal distribution (0, 1). In this case, the estimating equation (10.7.6) becomes

$$\frac{1}{nT^2}\sum_1^n x_i^2 = \frac{2}{\hat{z}_0^2} - \frac{\phi(\hat{z}_0)}{\phi(0) - \phi(\hat{z}_0)} = J_2(\hat{z}_0) \qquad (10.7.11)$$

For any given value of z, it is a relatively simple task to evaluate $J_2(z)$ from an ordinary table of standard normal curve ordinates. Thus, trial-and-error procedures coupled with linear interpolation can be employed to solve (10.7.11) for $\hat{z}_0$. In view of the special interest attached to this case, the estimating function $J_2(z)$ has been tabulated and is included here as Table 10.1. A graph of this function is included in Figure 10.1. With $\Sigma_1^n x_i^2/nT^2$ calculated from sample data, $\hat{z}_0$ can be read from the graph or determined more precisely by interpolating in Table 10.1. The values thus obtained should be sufficiently accurate for most practical purposes. When greater accuracy is needed, additional values of $J_2(z)$ can easily be calculated to permit interpolation over a smaller interval. With $\hat{z}_0$ determined, $\hat{\sigma}$ follows from (10.7.8) as $\hat{\sigma} = T/\hat{z}_0$.

TABLE 10.1 Truncated Sample Estimating Function $J_2(z)$, for Rayleigh Distribution, $J_2(z) = 2/z^2 - \phi(z)/[\phi(0) - \phi(z)]$

z	.00	.01	.02	.03	.04	.05	.06	.07	.08	.09
.0	.50000	.50000	.49998	.49996	.49993	.49990	.49985	.49980	.49973	.49966
.1	.49958	.49950	.49940	.49930	.49918	.49906	.49893	.49880	.49865	.49850
.2	.49833	.49816	.49798	.49780	.49760	.49740	.49718	.49696	.49673	.49650
.3	.49625	.49600	.49573	.49546	.49518	.49490	.49460	.49430	.49398	.49366
.4	.49333	.49300	.49265	.49230	.49193	.49156	.49118	.49080	.49040	.49000
.5	.48959	.48917	.48874	.48830	.48785	.48740	.48694	.48647	.48599	.48550
.6	.48501	.48450	.48399	.48347	.48295	.48241	.48186	.48131	.48075	.48018
.7	.47960	.47902	.47842	.47782	.47721	.47659	.47597	.47533	.47469	.47404
.8	.47338	.47271	.47204	.47135	.47066	.46996	.46925	.46854	.46781	.46708
.9	.46634	.46559	.46484	.46407	.46330	.46252	.46174	.46094	.46014	.45933
1.0	.45851	.45768	.45684	.45600	.45515	.45429	.45343	.45255	.45167	.45078
1.1	.44989	.44898	.44807	.44715	.44623	.44529	.44435	.44340	.44245	.44148
1.2	.44051	.43953	.43855	.43756	.43656	.43555	.43453	.43351	.43248	.43145
1.3	.43041	.42936	.42830	.42724	.42617	.42509	.42401	.42292	.42182	.42072
1.4	.41961	.41850	.41737	.41624	.41511	.41397	.41282	.41167	.41051	.40934
1.5	.40817	.40699	.40581	.40462	.40342	.40222	.40102	.39981	.39859	.39737
1.6	.39614	.39490	.39366	.39242	.39117	.38992	.38866	.38740	.38613	.38485
1.7	.38358	.38229	.38101	.37972	.37842	.37712	.37582	.37451	.37319	.37188
1.8	.37056	.36923	.36791	.36658	.36524	.36390	.36256	.36122	.35987	.35852
1.9	.35717	.35581	.35445	.35309	.35172	.35035	.34898	.34761	.34624	.34486
2.0	.34348	.34210	.34072	.33933	.33795	.33656	.33517	.33378	.33239	.33100
2.1	.32960	.32821	.32681	.32541	.32402	.32262	.32122	.31982	.31842	.31702
2.2	.31562	.31422	.31282	.31142	.31002	.30863	.30723	.30583	.30443	.30304
2.3	.30164	.30024	.29885	.29746	.29607	.29468	.29329	.29190	.29052	.28913
2.4	.28775	.28637	.28499	.28361	.28224	.28087	.27950	.27813	.27677	.27540
2.5	.27404	.27269	.27133	.26998	.26863	.26729	.26595	.26461	.26327	.26194
2.6	.26061	.25929	.25796	.25665	.25533	.25402	.25271	.25141	.25011	.24882
2.7	.24753	.24624	.24496	.24368	.24241	.24114	.23987	.23861	.23736	.23611
2.8	.23486	.23362	.23238	.23115	.22992	.22870	.22748	.22627	.22506	.22386
2.9	.22267	.22147	.22029	.21911	.21793	.21676	.21559	.21444	.21328	.21213
3.0	.21099	.20985	.20872	.20759	.20647	.20535	.20424	.20314	.20204	.20095
3.1	.19986	.19878	.19770	.19663	.19557	.19451	.19346	.19241	.19137	.19033
3.2	.18930	.18828	.18726	.18625	.18524	.18424	.18324	.18225	.18127	.18029
3.3	.17932	.17835	.17739	.17644	.17549	.17454	.17361	.17267	.17175	.17083
3.4	.16991	.16900	.16810	.16720	.16631	.16542	.16454	.16367	.16280	.16193
3.5	.16107	.16022	.15937	.15853	.15769	.15686	.15604	.15521	.15440	.15359
3.6	.15278	.15199	.15119	.15040	.14962	.14884	.14807	.14730	.14654	.14578
3.7	.14503	.14428	.14354	.14280	.14207	.14134	.14061	.13990	.13918	.13848
3.8	.13777	.13707	.13638	.13569	.13501	.13433	.13365	.13298	.13231	.13165
3.9	.13099	.13034	.12969	.12905	.12841	.12778	.12714	.12652	.12590	.12528
4.0	.12466	.12405	.12345	.12285	.12225	.12166	.12107	.12048	.11990	.11933
4.1	.11875	.11818	.11762	.11706	.11650	.11595	.11539	.11485	.11431	.11377
4.2	.11323	.11270	.11217	.11165	.11112	.11061	.11009	.10958	.10907	.10857
4.3	.10807	.10757	.10708	.10659	.10610	.10562	.10514	.10466	.10418	.10371
5.0	.08000	.07968	.07936	.07905	.07873	.07842	.07811	.07780	.07750	.07719

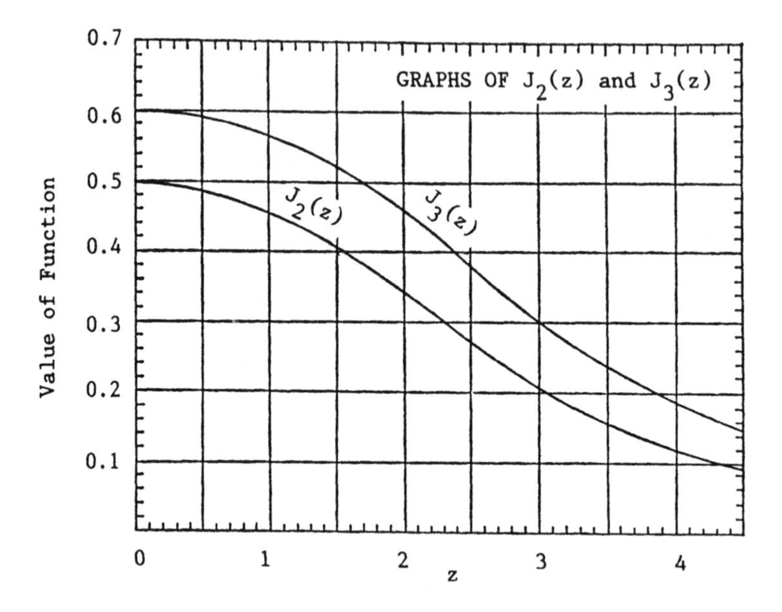

FIGURE 10.1 Truncated sample estimating functions for Rayleigh distribution. Adapted from Cohen, A. C. (1955).

10.7.2 Truncation in the Three-Dimensional Rayleigh Distribution

When $p = 3$, $g_3(z) = (2z^2/\sqrt{2\pi})\exp(-z^2/2) = 2z^2\phi(z)$, and $G_3(z) = 2[\Phi(z) - z\phi(z)] - 1$, where, as previously stated, $\phi(z)$ and $\Phi(z)$ are the pdf and the cdf, respectively, of the standard normal distribution (0, 1). The estimating equation for this case subsequently follows from (10.7.6) and (10.7.7) as

$$\frac{1}{nT^2}\sum_{1}^{n} x_i^2 = \frac{3}{\hat{z}_0^{\,2}} - \frac{\hat{z}_0\,\phi(\hat{z}_0)}{\Phi(\hat{z}_0) - \hat{z}_0\,\phi(\hat{z}_0) - 0.5} = J_3(\hat{z}_0) \tag{10.7.12}$$

Equation (10.7.12) can be solved for the MLE $\hat{z}_0$ by following the same procedure as that for solving (10.7.11) in the two-dimensional case. To facilitate these calculations, the estimating function $J_3(z)$ is tabulated in Table 10.2, and a graph is included in Figure 10.1 along with the graph of $J_2(z)$.

TABLE 10.2 Truncated Sample Estimating Function $J_3(z)$ for Rayleigh Distribution, $J_3(z) = 3/z^2 - z\phi(z)/[\Phi(z) - z\phi(z) - 0.5]$

z	.00	.01	.02	.03	.04	.05	.06	.07	.08	.09
.0	.60000	.60043	.60001	.59997	.59995	.59991	.59988	.59983	.59978	.59972
.1	.59966	.59959	.59951	.59942	.59933	.59923	.59912	.59901	.59889	.59876
.2	.59863	.59849	.59834	.59818	.59802	.59785	.59768	.59750	.59731	.59711
.3	.59691	.59670	.59648	.59626	.59603	.59579	.59554	.59529	.59503	.59477
.4	.59450	.59422	.59393	.59364	.59333	.59303	.59271	.59239	.59206	.59173
.5	.59138	.59103	.59068	.59031	.58994	.58956	.58918	.58878	.58838	.58798
.6	.58756	.58714	.58671	.58628	.58584	.58539	.58493	.58446	.58399	.58351
.7	.58303	.58254	.58203	.58153	.58101	.58049	.57996	.57942	.57888	.57833
.8	.57777	.57720	.57663	.57605	.57546	.57487	.57426	.57365	.57304	.57241
.9	.57178	.57114	.57049	.56984	.56918	.56851	.56783	.56715	.56646	.56576
1.0	.56505	.56434	.56362	.56289	.56215	.56141	.56065	.55990	.55913	.55836
1.1	.55757	.55679	.55599	.55519	.55437	.55355	.55273	.55189	.55105	.55020
1.2	.54935	.54848	.54761	.54673	.54585	.54495	.54405	.54314	.54223	.54130
1.3	.54037	.53943	.53848	.53753	.53657	.53560	.53462	.53364	.53265	.53165
1.4	.53064	.52963	.52861	.52758	.52654	.52550	.52445	.52339	.52233	.52126
1.5	.52018	.51909	.51800	.51690	.51579	.51467	.51355	.51242	.51128	.51014
1.6	.50899	.50783	.50666	.50549	.50431	.50313	.50193	.50073	.49953	.49832
1.7	.49710	.49587	.49464	.49340	.49215	.49090	.48964	.48837	.48710	.48582
1.8	.48453	.48324	.48194	.48064	.47933	.47801	.47669	.47536	.47403	.47269
1.9	.47134	.46999	.46864	.46727	.46590	.46453	.46315	.46177	.46037	.45898
2.0	.45758	.45617	.45476	.45335	.45192	.45050	.44907	.44763	.44619	.44475
2.1	.44330	.44184	.44039	.43892	.43746	.43599	.43451	.43303	.43155	.43006
2.2	.42857	.42708	.42558	.42408	.42258	.42107	.41956	.41805	.41653	.41501
2.3	.41349	.41197	.41044	.40891	.40738	.40584	.40431	.40277	.40123	.39968
2.4	.39814	.39659	.39504	.39350	.39194	.39039	.38884	.38728	.38573	.38417
2.5	.38261	.38106	.37950	.37794	.37638	.37482	.37326	.37170	.37014	.36858
2.6	.36702	.36546	.36390	.36234	.36078	.35922	.35767	.35611	.35456	.35300
2.7	.35145	.34990	.34835	.34680	.34526	.34371	.34217	.34063	.33909	.33756
2.8	.33602	.33449	.33296	.33143	.32991	.32839	.32687	.32535	.32384	.32233
2.9	.32082	.31932	.31782	.31632	.31482	.31333	.31185	.31037	.30889	.30741
3.0	.30594	.30447	.30301	.30155	.30010	.29865	.29720	.29576	.29432	.29289
3.1	.29146	.29004	.28862	.28721	.28580	.28439	.28300	.28160	.28021	.27883
3.2	.27745	.27608	.27471	.27335	.27199	.27064	.26930	.26796	.26662	.26529
3.3	.26397	.26265	.26134	.26004	.25874	.25744	.25615	.25487	.25359	.25232
3.4	.25106	.24980	.24855	.24730	.24606	.24483	.24360	.24238	.24116	.23995
3.5	.23875	.23755	.23636	.23518	.23400	.23282	.23166	.23050	.22934	.22820
3.6	.22705	.22592	.22479	.22367	.22255	.22144	.22034	.21924	.21815	.21706
3.7	.21598	.21491	.21384	.21278	.21173	.21068	.20964	.20860	.20757	.20655
3.8	.20553	.20452	.20352	.20252	.20152	.20053	.19955	.19858	.19761	.19664
3.9	.19569	.19473	.19379	.19285	.19191	.19099	.19006	.18915	.18823	.18733
4.0	.18643	.18553	.18464	.18376	.18288	.18201	.18115	.18028	.17943	.17858
4.1	.17773	.17689	.17606	.17523	.17441	.17359	.17277	.17197	.17116	.17037
4.2	.16957	.16878	.16800	.16722	.16645	.16568	.16492	.16416	.16341	.16266
4.3	.16192	.16118	.16045	.15972	.15899	.15827	.15756	.15684	.15614	.15544
5.0	.11999	.11951	.11903	.11856	.11809	.11762	.11716	.11670	.11624	.11578

10.8 CENSORED SAMPLES

We consider a random sample consisting of a total of N observations, of which n are equal to or less than T and are precisely measured; n_0 censored observations are greater than T, but the extent of their excess is unknown. Our sample thus consists of the ordered observations $\{x_i\}$, $i = 1, \ldots, n$, such that $x_i \leq T$ plus n_0 censored observations, each of which exceeds T. Of course, $N = n + n_0$. The two types of censoring described in Chapter 7 are also recognized here. In Type I censoring, T is a fixed constant, and n and $n_0 = N - n$ are observed values of random variables. In Type II censoring, n and n_0 are fixed constants, and $T = x_n$ (the nth-order statistic) is the observed value of a random variable. For both types we let T designate the point of censoring, and the likelihood function for a censored sample from a population with pdf (10.1.3) is

$$L = k2^{-n(p-2)/2}\sigma^{-np}\left[\Gamma\left(\frac{p}{2}\right)\right]^{-n} \exp\left\{-\sum_{1}^{n}\frac{x_i^2}{2\sigma^2}\right\}\prod_{i=1}^{n} x_i^{p-1}[1 - F_p(T)]^{n_0} \tag{10.8.1}$$

where k is an ordering constant which does not involve the unknown parameter σ, and $F_p(\cdot)$ is the cdf of X. On taking logarithms of (10.8.1), differentiating with respect to σ, and equating to zero, we obtain the MLE equation.

$$\frac{\partial \ln L}{\partial\sigma} = \frac{-n_0}{1 - F_p(T)}\,\frac{\partial F_p(T)}{\partial\sigma} - \frac{np}{\sigma} + \frac{1}{\sigma^3}\sum_{1}^{n} x_i^2 = 0 \tag{10.8.2}$$

After substituting $\sigma = T/z_0$ into (10.8.2) and simplifying, we get

$$\frac{1}{nT^2}\sum_{1}^{n} x_i^2 = \frac{1}{z_0}\left[\frac{p}{z_0} - \left(\frac{n_0}{n}\right)\frac{g_p(z_0)}{1 - G_p(z_0)}\right] = H_p(z_0) \tag{10.8.3}$$

where z_0 is the new unknown parameter. We now turn our attention to the special cases with p = 2 and 3, respectively.

10.8.1 Censored Samples from Population of Dimension 2

When p = 2, it follows from (10.7.9) that

$$\frac{g_2(z_0)}{1 - G_2(z_0)} = z_0 \tag{10.8.4}$$

and the MLE equation (10.8.3) becomes

$$\frac{1}{nT^2}\sum_{1}^{n} x_i^2 = \frac{2}{\hat{z}_0^{\,2}} - \frac{n_0}{n} \tag{10.8.5}$$

We substitute $\hat{z}_0 = T/\hat{\sigma}$ into (10.8.5) and simplify to obtain the explicit MLE

$$\hat{\sigma} = \sqrt{\frac{1}{2n}\left[\sum_{1}^{n} x_i^2 + n_0 T^2\right]} \tag{10.8.6}$$

10.8.2 Censored Samples from Population of Dimension 3

When $p = 3$, it follows from (10.7.4) that

$$\frac{g_3(z_0)}{1 - G_3(z_0)} = \frac{z_0^{\,2}\,\phi(z_0)}{z_0\,\phi(z_0) + 1 - \Phi(z_0)} \tag{10.8.7}$$

and estimating equation (10.8.3) becomes

$$\frac{1}{nT^2}\sum_{1}^{n} x_i^2 = \frac{3}{\hat{z}_0^{\,2}} - h\left[\frac{\hat{z}_0\,\phi(\hat{z}_0)}{\hat{z}_0\,\phi(\hat{z}_0) + 1 - \Phi(\hat{z}_0)}\right] = H_3(h, \hat{z}_0) \tag{10.8.8}$$

where $h = n_0/n$, the ratio of censored to uncensored observations. It is thus necessary to solve (10.8.8) for $\hat{z}_0$, and the MLE of σ follows as

$$\hat{\sigma} = \frac{T}{\hat{z}_0} \tag{10.8.9}$$

Note that h as used here is slightly different from the usage in previous chapters, where $h = n_0/N$.

To facilitate the solution of (10.8.8) in practical applications, a table of $H_3(h, z)$ for appropriate values of h and z is included here as Table 10.3. With $\Sigma_1^n x_i^2/nT^2$ and h available from sample data, an estimate of z_0 can be read directly from this table with sufficient accuracy for most practical applications. When greater precision is required, additional values of $H_3(h, z)$ as needed can be calculated from the right side of (10.8.8) with the aid of ordinary tables of the pdf and the cdf of the standard normal distribution (0, 1).

TABLE 10.3 Censored Sample Estimating Function $H_3(h, z)$ for Rayleigh Distribution, $H_3(h, z) = 3/z^2 - hz\phi(z)/[z\phi(z) + 1 - \Phi(z)]$; $h = n_0/n$

z \ h	.01	.02	.03	.04	.05	.06	.07	.08	.09
.6	8.32912	8.32490	8.32068	8.31647	8.31225	8.30804	8.30382	8.29960	8.29539
.7	6.11770	6.11296	6.10821	6.10346	6.09872	6.09397	6.08923	6.08448	6.07973
.8	4.68228	4.67705	4.67183	4.66660	4.66138	4.65615	4.65093	4.64571	4.64048
.9	3.69805	3.69240	3.68674	3.68109	3.67543	3.66978	3.66412	3.65847	3.65282
1.0	2.99396	2.98792	2.98188	2.97584	2.96980	2.96376	2.95772	2.95168	2.94564
1.1	2.47295	2.46657	2.46018	2.45380	2.44741	2.44103	2.43464	2.42826	2.42187
1.2	2.07664	2.06994	2.06325	2.05656	2.04986	2.04317	2.03647	2.02978	2.02308
1.3	1.76818	1.76121	1.75423	1.74726	1.74029	1.73332	1.72635	1.71938	1.71241
1.4	1.52339	1.51617	1.50896	1.50174	1.49452	1.48730	1.48008	1.47286	1.46564
1.5	1.32589	1.31845	1.31101	1.30357	1.29613	1.28869	1.28125	1.27380	1.26636
1.6	1.16423	1.15659	1.14895	1.14131	1.13367	1.12603	1.11839	1.11075	1.10311
1.7	1.03024	1.02242	1.01460	1.00678	.99896	.99114	.98332	.97550	.96768
1.8	.91794	.90996	.90198	.89400	.88602	.87803	.87005	.86207	.85409
1.9	.82290	.81477	.80664	.79851	.79039	.78226	.77413	.76600	.75787
2.0	.74174	.73348	.72522	.71696	.70870	.70044	.69218	.68392	.67566
2.1	.67189	.66351	.65513	.64675	.63838	.63000	.62162	.61324	.60486
2.2	.61135	.60286	.59437	.58588	.57740	.56891	.56042	.55193	.54344
2.3	.55852	.54993	.54135	.53276	.52417	.51559	.50700	.49841	.48983
2.4	.51216	.50348	.49480	.48613	.47745	.46877	.46010	.45142	.44274
2.5	.47124	.46248	.45372	.44496	.43621	.42745	.41869	.40993	.40117
2.6	.43495	.42612	.41728	.40845	.39962	.39078	.38195	.37311	.36428
2.7	.40262	.39372	.38481	.37591	.36701	.35810	.34920	.34030	.33140
2.8	.37369	.36472	.35575	.34679	.33782	.32886	.31989	.31092	.30196
2.9	.34769	.33867	.32964	.32062	.31160	.30257	.29355	.28452	.27550
3.0	.32426	.31518	.30610	.29702	.28794	.27886	.26979	.26071	.25163
3.1	.30305	.29392	.28479	.27566	.26654	.25741	.24828	.23915	.23002
3.2	.28380	.27462	.26545	.25627	.24710	.23793	.22875	.21958	.21041
3.3	.26627	.25705	.24783	.23862	.22940	.22018	.21097	.20175	.19254
3.4	.25026	.24100	.23175	.22249	.21324	.20398	.19473	.18547	.17621
3.5	.23561	.22631	.21702	.20773	.19844	.18914	.17985	.17056	.16127
3.6	.22216	.21283	.20350	.19418	.18485	.17552	.16620	.15687	.14754
3.7	.20978	.20042	.19106	.18171	.17235	.16299	.15363	.14427	.13491
3.8	.19837	.18898	.17959	.17021	.16082	.15143	.14204	.13265	.12327
3.9	.18782	.17841	.16899	.15958	.15016	.14075	.13133	.12191	.11250
4.0	.17806	.16862	.15918	.14973	.14029	.13085	.12141	.11197	.10253
4.1	.16900	.15953	.15007	.14060	.13114	.12167	.11221	.10274	.09327
4.2	.16058	.15109	.14160	.13211	.12263	.11314	.10365	.09416	.08467
4.3	.15274	.14323	.13372	.12421	.11470	.10519	.09568	.08617	.07666
4.4	.14543	.13590	.12637	.11684	.10731	.09778	.08825	.07872	.06919
4.5	.13860	.12905	.11950	.10995	.10040	.09085	.08131	.07176	.06221
4.6	.13221	.12264	.11308	.10351	.09394	.08438	.07481	.06524	.05568
4.7	.12622	.11664	.10706	.09747	.08789	.07831	.06872	.05914	.04956
4.8	.12061	.11101	.10141	.09181	.08221	.07261	.06301	.05341	.04381
4.9	.11533	.10572	.09610	.08649	.07688	.06726	.05765	.04803	.03842
5.0	.11037	.10074	.09111	.08149	.07186	.06223	.05260	.04297	.03334

(continued)

z \ h	.10	.15	.20	.25	.30	.35	.40	.45	.50
.6	8.29117	8.27009	8.24901	8.22792	8.20684	8.18576	8.16468	8.14360	8.12252
.7	6.07499	6.05126	6.02753	6.00380	5.98007	5.95634	5.93260	5.90887	5.88514
.8	4.63526	4.60914	4.58301	4.55689	4.53077	4.50465	4.47853	4.45241	4.42629
.9	3.64716	3.61889	3.59062	3.56235	3.53408	3.50581	3.47754	3.44926	3.42099
1.0	2.93960	2.90940	2.87920	2.84900	2.81881	2.78861	2.75841	2.72821	2.69801
1.1	2.41549	2.38356	2.35164	2.31971	2.28778	2.25586	2.22393	2.19201	2.16008
1.2	2.01639	1.98292	1.94945	1.91598	1.88250	1.84903	1.81556	1.78209	1.74862
1.3	1.70544	1.67058	1.63573	1.60087	1.56602	1.53116	1.49631	1.46145	1.42660
1.4	1.45842	1.42233	1.38623	1.35014	1.31405	1.27795	1.24186	1.20576	1.16967
1.5	1.25892	1.22172	1.18451	1.14730	1.11010	1.07289	1.03569	.99848	.96128
1.6	1.09547	1.05726	1.01906	.98086	.94265	.90445	.86625	.82804	.78984
1.7	.95986	.92076	.88166	.84256	.80346	.76435	.72525	.68615	.64705
1.8	.84611	.80620	.76629	.72638	.68647	.64656	.60665	.56674	.52683
1.9	.74975	.70911	.66847	.62783	.58719	.54655	.50591	.46527	.42463
2.0	.66740	.62610	.58480	.54351	.50221	.46091	.41961	.37831	.33701
2.1	.59648	.55458	.51269	.47079	.42889	.38699	.34510	.30320	.26130
2.2	.53496	.49252	.45008	.40764	.36520	.32276	.28032	.23788	.19544
2.3	.48124	.43831	.39538	.35244	.30951	.26658	.22364	.18071	.13778
2.4	.43407	.39068	.34730	.30392	.26053	.21715	.17377	.13038	.08700
2.5	.39241	.34862	.30482	.26103	.21724	.17344	.12965	.08585	.04206
2.6	.35545	.31128	.26711	.22294	.17877	.13460	.09043	.04626	.00209
2.7	.32249	.27798	.23346	.18895	.14443	.09992	.05540	.01089	-.03363
2.8	.29299	.24816	.20333	.15850	.11366	.06883	.02400	-.02083	-.06566
2.9	.26647	.22135	.17623	.13110	.08598	.04086	-.00426	-.04939	-.09451
3.0	.24255	.19716	.15177	.10638	.06098	.01559	-.02980	-.07519	-.12058
3.1	.22090	.17526	.12962	.08398	.03834	-.00730	-.05294	-.09858	-.14422
3.2	.20123	.15536	.10949	.06363	.01776	-.02811	-.07398	-.11985	-.16572
3.3	.18332	.13724	.09116	.04508	-.00100	-.04709	-.09317	-.13925	-.18533
3.4	.16696	.12068	.07440	.02812	-.01815	-.06443	-.11071	-.15699	-.20327
3.5	.15198	.10551	.05905	.01259	-.03387	-.08033	-.12679	-.17325	-.21972
3.6	.13822	.09159	.04495	-.00168	-.04831	-.09494	-.14157	-.18821	-.23484
3.7	.12556	.07877	.03198	-.01482	-.06161	-.10840	-.15519	-.20198	-.24877
3.8	.11388	.06694	.02000	-.02694	-.07388	-.12082	-.16776	-.21469	-.26163
3.9	.10308	.05601	.00893	-.03815	-.08523	-.13230	-.17938	-.22646	-.27354
4.0	.09309	.04588	-.00133	-.04854	-.09574	-.14295	-.19016	-.23736	-.28457
4.1	.08381	.03648	-.01085	-.05818	-.10551	-.15283	-.20016	-.24749	-.29482
4.2	.07518	.02774	-.01970	-.06714	-.11459	-.16203	-.20947	-.25691	-.30436
4.3	.06715	.01960	-.02795	-.07550	-.12305	-.17060	-.21815	-.26569	-.31324
4.4	.05966	.01201	-.03564	-.08329	-.13094	-.17859	-.22624	-.27389	-.32154
4.5	.05266	.00491	-.04283	-.09058	-.13832	-.18606	-.23381	-.28155	-.32930
4.6	.04611	-.00172	-.04956	-.09739	-.14523	-.19306	-.24089	-.28873	-.33656
4.7	.03997	-.00795	-.05586	-.10378	-.15170	-.19962	-.24754	-.29545	-.34337
4.8	.03421	-.01378	-.06178	-.10978	-.15778	-.20577	-.25377	-.30177	-.34977
4.9	.02880	-.01927	-.06734	-.11541	-.16349	-.21156	-.25963	-.30770	-.35578
5.0	.02371	-.02443	-.07257	-.12072	-.16886	-.21700	-.26515	-.31329	-.36144

(continued)

TABLE 10.3 Continued

z \ h	.55	.60	.65	.70	.75	.80	.85	.90	.95
.6	8.10143	8.08035	8.05927	8.03819	8.01711	7.99602	7.97494	7.95386	7.93278
.7	5.86141	5.83768	5.81395	5.79022	5.76649	5.74276	5.71903	5.69530	5.67157
.8	4.40017	4.37404	4.34792	4.32180	4.29568	4.26956	4.24344	4.21732	4.19119
.9	3.39272	3.36445	3.33618	3.30791	3.27964	3.25137	3.22310	3.19482	3.16655
1.0	2.66781	2.63761	2.60741	2.57721	2.54701	2.51681	2.48662	2.45642	2.42622
1.1	2.12815	2.09623	2.06430	2.03238	2.00045	1.96853	1.93660	1.90467	1.87275
1.2	1.71515	1.68168	1.64820	1.61473	1.58126	1.54779	1.51432	1.48085	1.44738
1.3	1.39174	1.35689	1.32203	1.28718	1.25232	1.21747	1.18261	1.14776	1.11290
1.4	1.13357	1.09748	1.06138	1.02529	.98920	.95310	.91701	.88091	.84482
1.5	.92407	.88686	.84966	.81245	.77525	.73804	.70084	.66363	.62642
1.6	.75163	.71343	.67523	.63702	.59882	.56062	.52241	.48421	.44601
1.7	.60795	.56885	.52975	.49065	.45155	.41245	.37334	.33424	.29514
1.8	.48692	.44701	.40710	.36719	.32728	.28737	.24746	.20755	.16765
1.9	.38399	.34336	.30272	.26208	.22144	.18080	.14016	.09952	.05888
2.0	.29571	.25441	.21311	.17181	.13052	.08922	:04792	.00662	-.03468
2.1	.21941	.17751	.13561	.09372	.05182	.00992	-.03197	-.07387	-.11577
2.2	.15300	.11056	.06812	.02568	-.01676	-.05920	-.10164	-.14408	-.18652
2.3	.09484	.05191	.00898	-.03396	-.07689	-.11982	-.16276	-.20569	-.24862
2.4	.04362	.00024	-.04315	-.08653	-.12991	-.17330	-.21668	-.26006	-.30345
2.5	-.00174	-.04553	-.08932	-.13312	-.17691	-.22071	-.26450	-.30829	-.35209
2.6	-.04208	-.08625	-.13042	-.17460	-.21877	-.26294	-.30711	-.35128	-.39545
2.7	-.07814	-.12266	-.16717	-.21169	-.25620	-.30072	-.34523	-.38975	-.43426
2.8	-.11049	-.15533	-.20016	-.24499	-.28982	-.33465	-.37948	-.42431	-.46915
2.9	-.13963	-.18476	-.22988	-.27500	-.32012	-.36525	-.41037	-.45549	-.50062
3.0	-.16597	-.21136	-.25675	-.30215	-.34754	-.39293	-.43832	-.48371	-.52910
3.1	-.18986	-.23550	-.28114	-.32678	-.37242	-.41805	-.46369	-.50933	-.55497
3.2	-.21159	-.25746	-.30332	-.34919	-.39506	-.44093	-.48680	-.53267	-.57854
3.3	-.23141	-.27749	-.32357	-.36965	-.41573	-.46182	-.50790	-.55398	-.60006
3.4	-.24955	-.29582	-.34210	-.38838	-.43466	-.48094	-.52721	-.57349	-.61977
3.5	-.26618	-.31264	-.35910	-.40556	-.45202	-.49848	-.54495	-.59141	-.63787
3.6	-.28147	-.32810	-.37473	-.42136	-.46800	-.51463	-.56126	-.60789	-.65452
3.7	-.29556	-.34235	-.38914	-.43593	-.48272	-.52951	-.57630	-.62309	-.66989
3.8	-.30857	-.35551	-.40245	-.44939	-.49633	-.54327	-.59021	-.63714	-.68408
3.9	-.32061	-.36769	-.41477	-.46185	-.50892	-.55600	-.60308	-.65016	-.69723
4.0	-.33178	-.37898	-.42619	-.47340	-.52061	-.56781	-.61502	-.66223	-.70943
4.1	-.34215	-.38948	-.43681	-.48413	-.53146	-.57879	-.62612	-.67345	-.72078
4.2	-.35180	-.39924	-.44668	-.49413	-.54157	-.58901	-.63645	-.68390	-.73134
4.3	-.36079	-.40834	-.45589	-.50344	-.55099	-.59854	-.64609	-.69364	-.74119
4.4	-.36919	-.41684	-.46449	-.51214	-.55979	-.60744	-.65509	-.70274	-.75039
4.5	-.37704	-.42479	-.47253	-.52028	-.56802	-.61577	-.66351	-.71126	-.75900
4.6	-.38440	-.43223	-.48006	-.52790	-.57573	-.62356	-.67140	-.71923	-.76707
4.7	-.39129	-.43921	-.48713	-.53504	-.58296	-.63088	-.67880	-.72672	-.77463
4.8	-.39776	-.44576	-.49376	-.54176	-.58975	-.63775	-.68575	-.73375	-.78174
4.9	-.40385	-.45192	-.49999	-.54807	-.59614	-.64421	-.69228	-.74036	-.78843
5.0	-.40958	-.45772	-.50587	-.55401	-.60215	-.65030	-.69844	-.74658	-.79473

10.9 RELIABILITY OF ESTIMATES

10.9.1 Complete Samples

For complete samples, $np\hat{\sigma}^2/\sigma^2$ has a chi-square distribution with np degrees of freedom (d.f.). To prove this, we note that each component Y_i of X is normal $(0, \sigma)$ and thus Y_j/σ is normal (0, 1). Let y_{ji} designate the ith observation of y_j, and $\Sigma_{i=1}^{n} (y_{ji}/\sigma)^2$, which is $(y_j/\sigma)^2$ summed over all sample observations, has a chi-square distribution with d.f. = n. Since the sum of independent chi-square distributions is itself a chi-square distribution with degrees of freedom equal to the sum of the respective degrees of freedom of the component distributions, and since

$$\sum_{j=1}^{p} \sum_{i=1}^{n} \left(\frac{y_{ji}}{\sigma}\right)^2 = \frac{np\hat{\sigma}^2}{\sigma^2}$$

the stated result follows.

Confidence intervals for σ can be determined as

$$P\left\{\sqrt{\frac{np\hat{\sigma}^2}{\chi_2^{\ 2}}} < \sigma < \sqrt{\frac{np\hat{\sigma}^2}{\chi_1^{\ 2}}}\right\} = 1 - \alpha \tag{10.9.1}$$

where $\hat{\sigma}$ is given by (10.2.3) and $\chi_1^{\ 2}$ and $\chi_2^{\ 2}$ are read from standard chi-square tables with d.f. = np such that $P[\chi^2 > \chi_1^{\ 2}] = 1 - \alpha/2$, and $P[\chi^2 > \chi_2^{\ 2}] = \alpha/2$.

Since $np\hat{\sigma}^2/\sigma^2$ has a chi-square distribution with d.f. = np, then $\hat{\sigma}\sqrt{np}/\sigma$ has a chi distribution with the same degrees of freedom. Kendall (1948, p. 294) gives the moments of the chi distribution, and by using his results, the exact variance of $\hat{\sigma}$ becomes

$$V(\hat{\sigma}) = \frac{\sigma^2}{2np}\left[2np - \left(\frac{2\Gamma((np + 1)/2)}{\Gamma(np/2)}\right)^2\right] \tag{10.9.2}$$

An expansion based on an extended form of Stirling's formula, also given by Kendall, permits (10.9.2) to be written as

$$V(\hat{\sigma}) = \frac{\sigma^2}{2np}\left[1 - \frac{1}{4np} + \cdots\right] \tag{10.9.3}$$

For the mean, Kendall's results give

$$E(\hat{\sigma}) = \sigma\left[1 - \frac{1}{4np} + \frac{1}{32(np)^2} + \cdots\right] \tag{10.9.4}$$

For large values of n, confidence intervals for σ can be approximated from the normal distribution with mean and variance as given by (10.9.3) and (10.9.4).

Similar exact sampling results are not available for restricted samples, but the asymptotic variance for each of the MLE, including that based on a complete sample, is given by

$$\text{Asy. Var. } (\hat{\sigma}) = -\left[E\left(\frac{\partial^2 \ln L}{\partial\sigma^2}\right)\right]^{-1} \tag{10.9.5}$$

For restricted samples, this leads to specific variances as given below.

10.9.2 Truncated Samples

Although the result in this case is not claimed to be obvious, straightforward algebraic manipulation gives

$$E\left(\frac{\partial^2 \ln L}{\partial\sigma^2}\right) = -\frac{2pE(n)}{\sigma^2}\left\{1 - \frac{z_0 g_p(z_0)}{2pG_p(z_0)}\left[z_0^2 - (p-2) + \frac{z_0 g_p(z_0)}{G_p(z_0)}\right]\right\} \tag{10.9.6}$$

By using this result, the asymptotic variance in the two-dimensional case is given by (10.9.5) as

$$\text{Asy. Var. } (\hat{\sigma}) = \frac{\sigma^2}{4E(n)}\left\{1 - \frac{z_0^4}{4}\,\frac{\phi(0)\phi(z_0)}{[\phi(0) - \phi(z_0)]^2}\right\}^{-1} \tag{10.9.7}$$

10.9.3 Censored Samples

Again in this case, after some straightforward, though involved, algebraic manipulation, we find

$$E\left(\frac{\partial^2 \ln L}{\partial\sigma^2}\right) = -\frac{2pE(n)}{\sigma^2}\left\{1 - \frac{z_0 \phi(z_0)}{2pG_p(z_0)}\left[z_0^2 - (p-2) - \frac{z_0 g_p(z_0)}{1 - G_p(z_0)}\right]\right\} \tag{10.9.8}$$

With this result, the corresponding asymptotic variance for a two-dimensional population is given by (10.9.5) as

$$\text{Asy. Var. } (\hat{\sigma}) = \frac{\sigma^2}{4E(n)} \tag{10.9.9}$$

Throughout this chapter, n designates the number of measured observations in each sample considered, and E(n), which appears in (10.9.6)-(10.9.9), is the expected value of n. With E(n) appropriately evaluated, the same variance formulas are applicable regardless of the sampling scheme employed. Values of E(n) for three sampling schemes that seem most likely to occur are given below.

(i) The experiment is continued until n measured observations are obtained in the range $0 \leq x \leq T$. Since n is a fixed number, $E(n) = n$, and T is fixed. The sample is thus truncated, and the number of rejected observations is unknown.

(ii) A fixed total of, say, $N = n + n_0$ observations is selected, but only those observations for which $0 \leq x \leq T$, where T is fixed, are measured. The number of measured observations, n, in this case is a random variable, and $E(n) = NG_p(T)$. The sample in this case is Type I censored.

(iii) The n smallest observations out of a total N are measured. In this case both n and N are fixed numbers, but T is a random variable. As already mentioned, the estimators in this case are the same as those given for Type I censored samples. Hence $T = x_n$, where x_n is the largest of the n measured observations in a given sample. With n fixed, $E(n) = n$ as in case (i).

10.10 ILLUSTRATIVE EXAMPLES

To illustrate the practical application of results obtained in this paper, we employ a sample selected from a population distributed according to (10.1.3) with $p = 2$ and $\sigma = 10$. The actual selection was made by using a table of random numbers, and although certain approximations were involved in obtaining observations, the sample appears to be adequate for the intended purpose. By appropriately adding information, the same basic data of the truncated sample also serves to illustrate estimation from censored and complete samples.

10.10.1 Truncated Sample

In this case, we let $T = 26$, and the sample consists of $n = 25$ observations, each of which is less than or equal to T. $\Sigma_1^{25} x_i^2 = 3561$, and $\Sigma_1^{25} x_i^2/nT^2 = 3561/(25)(26)^2 = 0.21071$. Interpolating from $J_2(z)$ in Table 10.1, we have $\hat{z}_0 = 3.003$, a result which might be read with only slightly less accuracy from the corresponding graph of Figure 10.1. From (10.7.8), we have $\hat{\sigma} = 26/3.003 = 8.66$. Using this value for σ, which for illustrative purposes is assumed to be unknown, we obtain $V(\hat{\sigma})$ from (10.9.7) as 1.27481 and $\hat{\sigma}_{\hat{\sigma}} = \sqrt{V(\sigma)} = 1.129$.

10.10.2 Censored Sample

To the truncated sample, we add information of the occurrence of $n_0 = 2$ observations for which $x > T$. Otherwise, the sample remains unchanged. The required estimate is computed from (10.8.5) as $\hat{\sigma} = \sqrt{[3561 + 2(26)]/2(25)} = 9.91$. By using this value for σ, which is still assumed to be unknown (for purpose of this illustration), we approximate $V(\hat{\sigma})$ from (7.9.9) as $(9.91)^2/4(25) = 0.9821$, and $\hat{\sigma}_{\hat{\sigma}} = 0.991$.

10.10.3 Complete Sample

To the censored sample actual measurements of the two censored observations were added to produce a complete sample with $n = 27$, and for the sample thus formed, $\Sigma_1^{27} x_i^2 = 5019$. From (10.1.7) with $p = 2$, we compute $\hat{\sigma} = \sqrt{5019/2(27)} = 9.64$, a value which differs by only a small amount from the censored sample estimate of the previous paragraph. The variance of this estimate is computed from (10.9.3) as $V(\hat{\sigma}) = 0.8520$, and $\hat{\sigma}_{\hat{\sigma}} = 0.923$.

The above results are summarized in the following table along with 0.95 confidence limits based on the variances as computed and the normal curve approximation for the distribution of $\hat{\sigma}$ in each of the cases considered.

Type of sample	$\hat{\sigma}$	0.95 Confidence interval
Truncated	8.66	$6.45 < \sigma < 10.87$
Censored	9.91	$7.97 < \sigma < 11.85$
Complete	9.64	$7.79 < \sigma < 11.41$

10.11 PARAMETER ESTIMATION IN THE TWO-PARAMETER RAYLEIGH DISTRIBUTION OF DIMENSION 2 WHEN SAMPLES ARE CENSORED

The pdf of the two-parameter Rayleigh distribution was given in (10.5.1). The likelihood function of a random sample of size N from this distribution that is censored at $X = T$ with n_0 censored observations and n measured observations that are not greater than T and where $N = n + n_0$ is

$$L = k\sigma^{-2n}\left[\prod_{i=1}^{n} (x_i - \gamma) \exp\right]\left\{-\frac{1}{2\sigma^2}\sum_{1}^{n}(x_i - \gamma)^2\right\}[1 - F(T)]^{n_0} \tag{10.11.1}$$

where k is an ordering constant. We substitute $F(T) = 1 - \exp\{-(T-\gamma)^2/2\sigma^2\}$ into (10.11.1) and take logarithms to obtain

$$\ln L = -2n \ln \sigma + \sum_1^n \ln(x_i - \gamma) - \frac{1}{2\sigma^2}\sum_1^n (x_i - \gamma)^2 - \frac{n_0(T-\gamma)^2}{2\sigma^2} + \ln k \tag{10.11.2}$$

10.11.1 Maximum Likelihood Estimators

The MLE equations $\partial \ln L/\partial\gamma = 0$ and $\partial \ln L/\partial\sigma = 0$ follow from (10.11.2) as

$$\frac{\partial \ln L}{\partial \gamma} = -\sum_1^n (x_i - \gamma)^{-1} + \frac{1}{\sigma^2}\sum_1^n (x_i - \gamma) + \frac{n_0}{\sigma^2}(T-\gamma) = 0$$

$$\frac{\partial \ln L}{\partial \sigma} = -\frac{2n}{\sigma} + \frac{1}{\sigma^3}\sum_1^n (x_i - \gamma)^2 + \frac{n_0}{\sigma^3}(T-\gamma)^2 = 0 \tag{10.11.3}$$

We eliminate σ between the two equations of (10.11.3) and simplify to obtain the following equation in which $\hat{\gamma}$ is the only unknown.

$$\frac{\sum_1^n x_i + n_0 T - N\hat{\gamma}}{\sum_1^n (x_i - \hat{\gamma})^{-1}} - \frac{1}{2n}\left[\sum_1^n (x_i - \hat{\gamma})^2 + n_0(T-\hat{\gamma})^2\right] = 0 \tag{10.11.4}$$

From the second equation of (10.11.3) we obtain

$$\hat{\sigma}^2 = \frac{1}{2n}\left[\sum_1^n (x_i - \hat{\gamma})^2 + n_0(T-\hat{\gamma})^2\right] \tag{10.11.5}$$

It is a relatively simple task to solve (10.11.4) for $\hat{\gamma}$. The trial-and-error technique which has been previously described can be employed, but various other iterative procedures are also applicable. With $\hat{\gamma}$ thus determined, $\hat{\sigma}^2$ follows directly from (10.11.5).

10.11.2 Modified Maximum Likelihood Estimators

Estimating equations in this case are $\partial \ln L/\partial\sigma = 0$ and $E(X_1) = x_1$. The first of these equations is included in (10.11.3), where it appears as the second equation. The expected value of X_1 in a sample of size n was given in (10.5.3). Here the total sample size is N, and the second estimating equation becomes

$$E(X_1) = \gamma + \sigma\sqrt{\frac{\pi}{N}} = x_1 \tag{10.11.6}$$

We eliminate σ between (10.11.6) and the second equation of (10.11.3). Thereby we obtain the following equation in which $\hat{\gamma}$ is the only unknown quantity:

$$\frac{N(x_1 - \hat{\gamma})^2}{\pi} - \frac{1}{2n}\left[\sum_1^n (x_i - \hat{\gamma})^2 + n_0(T - \hat{\gamma})^2\right] = 0 \tag{10.11.7}$$

This equation can be solved by the same procedures as those employed in solving (10.11.4) and other corresponding equations. With $\hat{\gamma}$ thus determined, $\hat{\sigma}^2$ is calculated as

$$\hat{\sigma}^2 = \frac{1}{2n}\left[\sum_1^n (x_i - \hat{\gamma})^2 + n_0(T - \hat{\gamma})^2\right] \tag{10.11.8}$$

This result is identical with (10.11.5).

10.12 SOME CONCLUDING REMARKS

Estimators presented in this chapter for parameters of the two-dimensional Rayleigh distribution can also be obtained as special cases of the Weibull estimators presented in Chapter 3 for complete samples and in Chapter 7 for censored samples. With $\delta = 2$ and $\theta = 2\sigma^2$, maximum likelihood estimating equations (10.6.4) follow as the simultaneous solution of the first and third equations of (3.3.2). Likewise, estimating equations (10.11.3) can be obtained as a special case from (7.4.8) and the second equation of (7.4.2). The MMLE of (10.11.7) and (10.11.8) can be obtained as a corresponding special case from the second and third equations of (7.4.2).

11

The Pareto Distribution

11.1 INTRODUCTION

The Pareto distribution is a reverse J-shaped positively skewed distribution that is of special interest to economists. It is named after an Italian-born Swiss professor of economics (1848-1923), who formulated it as a model for the distribution of incomes. In recent years it has been the subject of research papers by numerous writers. Among these are Hagstroem (1960), Harris (1960), Malik (1966, 1967, 1970), Mandelbrodt (1960), Harter and Moore (1967), and Srivastava (1965). Volume 1 by Johnson and Kotz (1970) contains an excellent expository account of this distribution together with a discussion of some of its applications. A lengthy list of references is also included.

11.2 SOME FUNDAMENTALS

In the parameterization of interest here, the pdf of this distribution is written as

$$f(x;\gamma,\alpha) = \alpha\gamma^{\alpha}x^{-(\alpha+1)}, \qquad 0 \le \gamma \le x, \quad \alpha > 0 \tag{11.2.1}$$

$$= 0, \qquad \text{elsewhere}$$

and the cdf is

$$F(x;\gamma,\alpha) = 1 - \left(\frac{\gamma}{x}\right)^{\alpha} \tag{11.2.2}$$

where γ is a threshold parameter and α is essentially a shape parameter. Johnson and Kotz (1970) point out that this distribution is a special case of Pearson's Type VI distribution.

The rth moment about the origin is

$$\mu'_r = \frac{\alpha\gamma^r}{\alpha - r}, \qquad r < \alpha \tag{11.2.3}$$

The moments thus fail to exist unless $\alpha > r$. The expected value (mean) variance, α_3, α_4, mode, median, and mean deviation are

$$\begin{aligned}
E(X) &= \frac{\alpha\gamma}{\alpha - 1}, \qquad \alpha > 1 \\
V(X) &= \frac{\alpha\gamma^2}{(\alpha - 1)^2(\alpha - 2)}, \qquad \alpha > 2 \\
\alpha_3(X) &= 2\frac{\alpha + 1}{\alpha - 3}\sqrt{\frac{\alpha - 2}{\alpha}}, \qquad \alpha > 3 \\
\alpha_4(X) &= \frac{3(\alpha - 2)(3\alpha^2 + \alpha + 2)}{\alpha(\alpha - 3)(\alpha - 4)}, \qquad \alpha > 4 \\
Mo(X) &= \gamma, \quad Me(X) = 2^{1/\alpha}\gamma \\
MD(X) &= 2\gamma(\alpha - 1)^{-1}(1 - \alpha^{-1})^{\alpha-1}, \qquad \alpha > 1
\end{aligned} \tag{11.2.4}$$

The hazard or failure rate function for this distribution is

$$h(x) = \frac{\alpha}{x} \tag{11.2.5}$$

which is a decreasing function. Thus the Pareto distribution might be a suitable model in situations where product or systems development results in improved performance as development proceeds. Note that the Pareto distribution is reverse J-shaped, and thus the maximum or modal value occurs at the origin with $f(\gamma) = \alpha/\gamma$. To better display important distinguishing characteristics of this distribution, we tabulate α_3 and α_4 as functions of α in Table 11.1, and the mean and variance as functions of α and γ in Table 11.2.

11.2.1 The First-Order Statistic

The cdf of the first order statistic $X_{1:n}$ in a random sample of size n is

TABLE 11.1 Pareto Distribution Skewness and Kurtosis, α_3 and α_4, as Functions of α

α	α_3	α_4	α	α_3	α_4
4.1	6.63629	789.66519	12.0	2.63718	15.48611
4.2	6.27247	387.09524	12.2	2.62383	15.31782
4.4	5.69738	188.41558	12.4	2.61104	15.15796
4.6	5.26267	123.78261	12.6	2.59875	15.00591
4.8	4.92203	92.26389	12.8	2.58696	14.86111
5.0	4.64758	73.80000	13.0	2.57563	14.72308
5.2	4.42153	61.76224	13.2	2.56472	14.59134
5.4	4.23196	53.33862	13.4	2.55423	14.46547
5.6	4.07059	47.13874	13.6	2.54411	14.34510
5.8	3.93150	42.39901	13.8	2.53437	14.22988
6.0	3.81032	38.66667	14.0	2.52496	14.11948
6.2	3.70375	35.65689	14.2	2.51589	14.01361
6.4	3.60927	33.18199	14.4	2.50712	13.91200
6.6	3.52491	31.11344	14.6	2.49865	13.81440
6.8	3.44911	29.36046	14.8	2.49046	13.72057
7.0	3.38062	27.85714	15.0	2.48253	13.63030
7.2	3.31841	26.55456	15.2	2.47486	13.54340
7.4	3.26165	25.41567	15.4	2.46743	13.45968
7.6	3.20965	24.41190	15.6	2.46022	13.37897
7.8	3.16183	23.52092	15.8	2.45324	13.30112
8.0	3.11769	22.72500	16.0	2.44647	13.22596
8.2	3.07683	22.00992	16.2	2.43990	13.15337
8.4	3.03889	21.36412	16.4	2.43351	13.08322
8.6	3.00356	20.77813	16.6	2.42731	13.01539
8.8	2.97058	20.24412	16.8	2.42129	12.94976
9.0	2.93972	19.75556	17.0	2.41544	12.88623
9.2	2.91079	19.30694	17.2	2.40974	12.82470
9.4	2.88360	18.89362	17.4	2.40420	12.76508
9.6	2.85801	18.51163	17.6	2.39881	12.70728
9.8	2.83386	18.15759	17.8	2.39356	12.65122
10.0	2.81106	17.82857	18.0	2.38845	12.59683
10.2	2.78947	17.52203	18.2	2.38347	12.54402
10.4	2.76902	17.23577	18.4	2.37861	12.49273
10.6	2.74961	16.96786	18.6	2.37388	12.44290
10.8	2.73116	16.71661	18.8	2.36927	12.39447
11.0	2.71360	16.48052	19.0	2.36476	12.34737
11.2	2.69688	16.25828	20.0	2.34381	12.13015
11.4	2.68092	16.04870	30.0	2.21843	12.13015
11.6	2.66568	15.85076	50.0	2.12637	10.06002
11.8	2.65112	15.66351	100.0	2.06154	9.50385

TABLE 11.2 Pareto Distribution Mean and Variance as Functions of α and γ. Top entry is the mean, and bottom entry is the variance.

α \ γ	1	2	3	4	5	6	7	8
2.2	1.8333	3.6667	5.5000	7.3333	9.1667	11.0000	12.8333	14.6667
	7.6389	30.5556	68.7500	122.2222	190.9722	275.0000	374.3056	488.8889
2.4	1.7143	3.4286	5.1429	6.8571	8.5714	10.2857	12.0000	13.7143
	3.0612	12.2449	27.5510	48.9796	76.5306	110.2041	150.0000	195.9184
2.6	1.6250	3.2500	4.8750	6.5000	8.1250	9.7500	11.3750	13.0000
	1.6927	6.7708	15.2344	27.0833	42.3177	60.9375	82.9427	108.3333
2.8	1.5556	3.1111	4.6667	6.2222	7.7778	9.3333	10.8889	12.4444
	1.0802	4.3210	9.7222	17.2840	27.0062	38.8889	52.9321	69.1358
3.0	1.5000	3.0000	4.5000	6.0000	7.5000	9.0000	10.5000	12.0000
	.7500	3.0000	6.7500	12.0000	18.7500	27.0000	36.7500	48.0000
3.2	1.4545	2.9091	4.3636	5.8182	7.2727	8.7273	10.1818	11.6364
	.5510	2.2039	4.9587	8.8154	13.7741	19.8347	26.9972	35.2617
3.4	1.4167	2.8333	4.2500	5.6667	7.0833	8.5000	9.9167	11.3333
	.4216	1.6865	3.7946	6.7460	10.5407	15.1786	20.6597	26.9841
3.6	1.3846	2.7692	4.1538	5.5385	6.9231	8.3077	9.6923	11.0769
	.3328	1.3314	2.9956	5.3254	8.3210	11.9822	16.3092	21.3018
3.8	1.3571	2.7143	4.0714	5.4286	6.7857	8.1429	9.5000	10.8571
	.2693	1.0771	2.4235	4.3084	6.7319	9.6939	13.1944	17.2336
4.0	1.3333	2.6667	4.0000	5.3333	6.6667	8.0000	9.3333	10.6667
	.2222	.8889	2.0000	3.5556	5.5556	8.0000	10.8889	14.2222
4.2	1.3125	2.6250	3.9375	5.2500	6.5625	7.8750	9.1875	10.5000
	.1864	.7457	1.6779	2.9830	4.6609	6.7116	9.1353	11.9318
4.4	1.2941	2.5882	3.8824	5.1765	6.4706	7.7647	9.0588	10.3529
	.1586	.6344	1.4273	2.5375	3.9648	5.7093	7.7710	10.1499
4.6	1.2778	2.5556	3.8333	5.1111	6.3889	7.6667	8.9444	10.2222
	.1365	.5461	1.2286	2.1842	3.4129	4.9145	6.6892	8.7369
4.8	1.2632	2.5263	3.7895	5.0526	6.3158	7.5789	8.8421	10.1053
	.1187	.4749	1.0685	1.8995	2.9679	4.2738	5.8172	7.5979
5.0	1.2500	2.5000	3.7500	5.0000	6.2500	7.5000	8.7500	10.0000
	.1042	.4167	.9375	1.6667	2.6042	3.7500	5.1042	6.6667
5.2	1.2381	2.4762	3.7143	4.9524	6.1905	7.4286	8.6667	9.9048
	.0921	.3685	.8291	1.4739	2.3030	3.3163	4.5139	5.8957
5.4	1.2273	2.4545	3.6818	4.9091	6.1364	7.3636	8.5909	9.8182
	.0820	.3281	.7383	1.3126	2.0509	2.9533	4.0198	5.2504
5.6	1.2174	2.4348	3.6522	4.8696	6.0870	7.3043	8.5217	9.7391
	.0735	.2941	.6616	1.1762	1.8378	2.6465	3.6022	4.7049

α \ γ	1	2	3	4	5	6	7	8
5.8	1.2083 .0662	2.4167 .2650	3.6250 .5962	4.8333 1.0599	6.0417 1.6562	7.2500 2.3849	8.4583 3.2461	9.6667 4.2398
6.0	1.2000 .0600	2.4000 .2400	3.6000 .5400	4.8000 .9600	6.0000 1.5000	7.2000 2.1600	8.4000 2.9400	9.6000 3.8400
6.2	1.1923 .0546	2.3846 .2184	3.5769 .4913	4.7692 .8735	5.9615 1.3648	7.1538 1.9653	8.3462 2.6750	9.5385 3.4939
6.4	1.1852 .0499	2.3704 .1995	3.5556 .4489	4.7407 .7981	5.9259 1.2470	7.1111 1.7957	8.2963 2.4442	9.4815 3.1924
6.6	1.1786 .0458	2.3571 .1830	3.5357 .4118	4.7143 .7320	5.8929 1.1438	7.0714 1.6471	8.2500 2.2418	9.4286 2.9281
6.8	1.1724 .0421	2.3448 .1685	3.5172 .3790	4.6897 .6738	5.8621 1.0528	7.0345 1.5161	8.2069 2.0635	9.3793 2.6952
7.0	1.1667 .0389	2.3333 .1556	3.5000 .3500	4.6667 .6222	5.8333 .9722	7.0000 1.4000	8.1667 1.9056	9.3333 2.4889
7.5	1.1538 .0323	2.3077 .1291	3.4615 .2905	4.6154 .5164	5.7692 .8069	6.9231 1.1619	8.0769 1.5815	9.2308 2.0656
8.0	1.1429 .0272	2.2857 .1088	3.4286 .2449	4.5714 .4354	5.7143 .6803	6.8571 .9796	8.0000 1.3333	9.1429 1.7415
8.5	1.1333 .0232	2.2667 .0930	3.4000 .2092	4.5333 .3720	5.6667 .5812	6.8000 .8369	7.9333 1.1391	9.0667 1.4879
9.0	1.1250 .0201	2.2500 .0804	3.3750 .1808	4.5000 .3214	5.6250 .5022	6.7500 .7232	7.8750 .9844	9.0000 1.2857
9.5	1.1176 .0175	2.2353 .0701	3.3529 .1578	4.4706 .2805	5.5882 .4383	6.7059 .6311	7.8235 .8591	8.9412 1.1220
10.0	1.1111 .0154	2.2222 .0617	3.3333 .1389	4.4444 .2469	5.5556 .3858	6.6667 .5556	7.7778 .7562	8.8889 .9877
15.0	1.0714 .0059	2.1429 .0235	3.2143 .0530	4.2857 .0942	5.3571 .1472	6.4286 .2119	7.5000 .2885	8.5714 .3768
20.0	1.0526 .0031	2.1053 .0123	3.1579 .0277	4.2105 .0492	5.2632 .0769	6.3158 .1108	7.3684 .1508	8.4211 .1970
25.0	1.0417 .0019	2.0833 .0075	3.1250 .0170	4.1667 .0302	5.2083 .0472	6.2500 .0679	7.2917 .0925	8.3333 .1208
30.0	1.0345 .0013	2.0690 .0051	3.1034 .0115	4.1379 .0204	5.1724 .0318	6.2069 .0459	7.2414 .0624	8.2759 .0815
35.0	1.0294 .0009	2.0588 .0037	3.0882 .0083	4.1176 .0147	5.1471 .0229	6.1765 .0330	7.2059 .0450	8.2353 .0587

$$F(X_{1:n}) = 1 - \left(\frac{\gamma}{x_1}\right)^{n\alpha}, \quad x_1 \geq \gamma \tag{11.2.6}$$

The corresponding pdf is

$$f(x_{1:n}) = n\alpha\gamma^{n\alpha}x_1^{-(n\alpha+1)} \tag{11.2.7}$$

Thus if X is distributed according to a Pareto distribution (γ, α), it follows that $X_{1:n}$ is distributed according to a Pareto distribution $(\gamma, n\alpha)$, and the expected value of $X_{1:n}$ is

$$E(X_{1:n}) = \frac{n\alpha\gamma}{n\alpha - 1} \tag{11.2.8}$$

Its variance is

$$V(X_{1:n}) = \frac{n\alpha\gamma^2}{(n\alpha - 1)^2(n\alpha - 2)} \tag{11.2.9}$$

Other characteristics of interest follow from (11.2.4) when α is replaced by $n\alpha$.

11.3 PARAMETER ESTIMATION FROM COMPLETE SAMPLES

If γ is known, a maximum likelihood estimator (MLE) of α can be found by differentiating the loglikelihood function and equating the result to zero. If, however, γ is unknown, regularity problems are encountered when we attempt to estimate this parameter by the method of maximum likelihood. In this case we can use moment estimators (ME) or modified estimators (MME) which employ the first-order statistic. Moment estimators are easy to calculate, but in most instances modified estimators are more efficient and are preferred for that reason.

<u>Moment Estimators</u>. The ME equations are $E(X) = \bar{x}$ and $V(X) = s^2$, where $\bar{x}$ and s^2 are the sample mean and variance (unbiased), respectively. From (11.2.4) these equations become

$$\frac{\alpha\gamma}{\alpha - 1} = \bar{x}, \qquad \frac{\alpha\gamma^2}{(\alpha - 1)^2(\alpha - 2)} = s^2 \tag{11.3.1}$$

We eliminate γ between these two equations and simplify to obtain the following quadratic equation in α:

$$\alpha^2 - 2\alpha - \left(\frac{\bar{x}}{s}\right)^2 = 0 \tag{11.3.2}$$

Since it is necessary that $\alpha > 2$, the larger root of this equation, which we write as

$$\alpha^* = 1 + \sqrt{1 + \left(\frac{\bar{x}}{s}\right)^2} \tag{11.3.3}$$

is the required estimator of α. The corresponding estimator of γ follows from the first equations of (11.3.1) as

$$\gamma^* = \frac{\bar{x}(\alpha^* - 1)}{\alpha^*} \tag{11.3.4}$$

<u>Maximum Likelihood Estimators</u>. The likelihood function of a random sample $\{x_i\}$, $i = 1, 2, \ldots, n$, from a population with pdf (11.2.1) is

$$L = \prod_{i=1}^{n} \frac{\alpha\gamma^{\alpha}}{x_i^{\alpha+1}} \tag{11.3.5}$$

On taking logarithms of (11.3.5), we obtain the loglikelihood function

$$\ln L = n \ln \alpha + n\alpha \ln \gamma - \sum_{1}^{n} (\alpha + 1) \ln x_i \tag{11.3.6}$$

When γ is known, we can differentiate this equation with respect to α, equate the result to zero, and thereby obtain the following estimating equation for α:

$$\frac{\partial \ln L}{\partial \alpha} = \frac{n}{\alpha} + n \ln \gamma - \sum_{1}^{n} \ln x_i = 0 \tag{11.3.7}$$

From this equation it follows that

$$\hat{\alpha} = n \Bigg/ \left[\sum_{1}^{n} \ln\left(\frac{x_i}{\gamma}\right)\right] \tag{11.3.8}$$

Unfortunately, when γ is unknown, maximum likelihood estimation breaks down and we must resort to ME or to some type of modified estimators.

Modified Maximum Likelihood Estimators. As estimating equations, we employ $\partial \ln L/\partial\alpha = 0$ and $E(X_1) = x_1$. From (11.3.8) and (11.2.7) we have

$$\alpha = n \Big/ \left[\sum_1^n \ln \left(\frac{x_i}{\gamma} \right) \right] \tag{11.3.9}$$

$$\frac{n\alpha\gamma}{n\alpha - 1} = x_1$$

We eliminate α between these two equations and simplify to obtain the following equation in which $\hat{\gamma}$ is the only unknown quantity:

$$\frac{n(x_1 - \hat{\gamma})}{x_1} - \frac{1}{n} \sum_1^n \ln \left(\frac{x_i}{\hat{\gamma}} \right) = 0 \tag{11.3.10}$$

Any of the previously mentioned iterative procedures can be used to solve this equation for $\hat{\gamma}$, and with $\hat{\gamma}$ thus determined $\hat{\alpha}$ follows from either equation of (11.3.9). It is perhaps simpler to employ the second equation, which is readily reduced to

$$\hat{\alpha} = \frac{x_1}{n(x_1 - \hat{\gamma})} \tag{11.3.11}$$

As a first approximation to $\hat{\gamma}$ in (11.3.10), we might choose a value γ_1 that is slightly less than x_1.

11.3.4 Modified Moment Estimators

For MME we employ $E(X) = \bar{x}$ and $E(X_1) = x_1$. Accordingly, from (11.2.4) and (11.2.7) we have

$$\alpha\gamma = \bar{x}(\alpha - 1), \qquad n\alpha\gamma = x_1(n\alpha - 1) \tag{11.3.12}$$

On solving these two equations simultaneously for $\hat{\alpha}$ and $\hat{\gamma}$, we obtain

$$\hat{\alpha} = \frac{n\bar{x} - x_1}{n(\bar{x} - x_1)}$$

$$\hat{\gamma} = \bar{x}\,\frac{\hat{\alpha} - 1}{\hat{\alpha}} = \frac{(n - 1)x_1\bar{x}}{n\bar{x} - x_1} \tag{11.3.13}$$

11.4 PARAMETER ESTIMATION FROM TRUNCATED CENSORED SAMPLES

We will consider only truncation or censoring on the right at a point T.

11.4.1 Truncated Samples

The likelihood function of a random sample consisting of n observations $\{x_i\}$, $i = 1, \ldots, n$, from a distribution with pdf (11.2.1) truncated at T such that $\gamma \leq x \leq T$ becomes

$$L = \prod_{1}^{n} \alpha\gamma^{\alpha} x_i^{-(\alpha+1)} [F(T)]^{-1} \tag{11.4.1}$$

We substitute $F(T) = 1 - (\gamma/T)^{\alpha}$ into (11.4.1) and take logarithms to obtain the loglikelihood function

$$\ln L = n \ln \alpha + n\alpha \ln \gamma - (\alpha + 1) \sum^{n} \ln x_i - n \ln\left[1 - \left(\frac{\gamma}{T}\right)^{\alpha}\right] \tag{11.4.2}$$

When γ is known, we differentiate with respect to α and equate the result to zero. Thereby we obtain the estimating equation

$$\frac{\partial \ln L}{\partial \alpha} = \frac{n}{\alpha} + n \ln \gamma - \sum_{1}^{n} \ln x_i + \frac{n(\gamma/T)^{\alpha} \ln (\gamma/T)}{1 - (\gamma/T)^{\alpha}} = 0 \tag{11.4.3}$$

A slight simplification yields

$$\frac{1}{\hat{\alpha}} + \frac{(\gamma/T)^{\hat{\alpha}} \ln (\gamma/T)}{1 - (\gamma/T)^{\hat{\alpha}}} = \frac{1}{n} \sum^{n} \ln x_i - \ln \gamma \tag{11.4.4}$$

With γ known, (11.4.4) can be solved for $\hat{\alpha}$ by employing standard iterative procedures. A first approximation for use in the iterative calculations can be obtained from (11.3.8), which was derived for complete (untruncated) samples. If the truncation is not too severe, this approximation should be reasonably close to the required solution of (11.4.4).

When γ is unknown, we might choose as an estimate a value that is slightly less than x_1. The problem of estimating γ is much improved if the sample is censored rather than truncated. In the censored case with the total sample size N known, we are able to make use of the expected value of the first-order statistic.

11.4.2 Censored Samples

Let N designate the total sample size. Let c designate the number of censored observations greater than T, and let n designate the number of complete observations less than or equal to T. The likelihood function of a random sample of this type from a population with pdf (11.2.1) is

$$L = k[1 - F(T)]^c \prod_1^n \alpha\gamma^\alpha x_i^{-(\alpha+1)} \tag{11.4.5}$$

where k is an ordering constant that does not depend on α or γ, and $N = n + c$.

Substitute $F(T) = 1 - (\gamma/T)^\alpha$ in (11.4.5) and take logarithms to obtain

$$\ln L = \ln k + n \ln \alpha + n\alpha \ln \gamma - (\alpha + 1) \sum^n \ln x_i + c\alpha \ln\left(\frac{\gamma}{T}\right) \tag{11.4.6}$$

We differentiate this equation with respect to α and equate the result to zero. We thus obtain the estimating equation

$$\frac{\partial \ln L}{\partial \alpha} = \frac{n}{\alpha} + n \ln \gamma - \sum_1^n \ln x_i + c \ln\left(\frac{\gamma}{T}\right) = 0 \tag{11.4.7}$$

When γ is known, we therefore have the MLE

$$\hat{\alpha} = n\left[\sum_1^n \ln x_i - N \ln \gamma + c \ln T\right]^{-1} \tag{11.4.8}$$

When γ is unknown we are faced with the same estimation problem as that encountered with truncated samples. However, in this instance we are able to fall back on the expected value of X_1 in a sample of size N.

Modified Maximum Likelihood Estimators (MMLE) in Censored Samples. As estimating equations, we employ $\partial \ln L/\partial\alpha = 0$ and $E(X_{1:n}) = x_1$. Thus from (11.4.8) and (11.2.8), we write

$$\alpha = n\left[\sum_1^n \ln x_i - N \ln \gamma + c \ln T\right]^{-1} \tag{11.4.9}$$

$$\frac{N\alpha\gamma}{N\alpha - 1} = x_1$$

We eliminate α between the two equations of (11.4.9) and thereby obtain the following equation in which $\hat{\gamma}$ is the only unknown:

$$\frac{x_1}{N(x_1 - \hat{\gamma})} = n \Bigg/ \left[\sum_{1}^{n}{}' \ln x_i - N \ln \hat{\gamma} + c \ln T\right] \tag{11.4.10}$$

Any of the various iterative methods previously described can be used to solve (11.4.10) for $\hat{\gamma}$. As a first approximation it is only necessary that we choose a value less than x_1. With $\hat{\gamma}$ thus calculated, the second equation of (11.4.9) enables us to calculate $\hat{\alpha}$ as

$$\hat{\alpha} = \frac{x_1}{N(x_1 - \hat{\gamma})} \tag{11.4.11}$$

11.5 RELIABILITY OF ESTIMATES

As noted in Section 11.2, maximum likelihood estimation breaks down when γ is unknown. A second estimating equation corresponding to $\partial \ln L/\partial\gamma = 0$ cannot be obtained in the usual way since ln L is unbounded with respect to γ. Since γ is a lower bound on X, ln L must be maximized subject to the constraint

$$\hat{\gamma} \leq x_1 \tag{11.5.1}$$

Accordingly the value of $\hat{\gamma}$ which maximizes (11.3.5) subject to (11.5.1) is

$$\hat{\gamma} = x_1 \tag{11.5.2}$$

It has been shown by Quant (1966) that $\hat{\gamma}$ and $\hat{\alpha}$, as given by (11.5.2) and (11.3.8), are consistent estimators. However, $\hat{\gamma}$ is biased. In small samples, the bias might be quite large. For this reason, the MMLE of Section 11.3 are preferred.

It is sometimes convenient to express the estimator $\hat{\alpha}$ in terms of the geometric mean G, where

$$G = \left(\prod_{1}^{n} x_i\right)^{1/n} \tag{11.5.3}$$

When G is substituted into (11.3.8), we have the alternative expression

$$\hat{\alpha} = \ln\left(\frac{G}{\gamma}\right) \tag{11.5.4}$$

Johnson and Kotz (1970) point out that G is a sufficient statistic for α when γ is known, and $\hat{\gamma}$ as given by (11.5.2) is sufficient for γ when α is known. Malik (1970) has shown that when both α and γ are unknown $(\hat{\alpha}, \hat{\gamma})$ is a joint pair of sufficient statistics for (α, γ) and $\hat{\alpha}$ is stochastically independent of $\hat{\gamma}$. Again, however, it must be emphasized that $\hat{\gamma}$ as given by (11.5.2) is not unbiased.

The pdf of $\hat{\alpha}$ as given by Johnson and Kotz (1970) is

$$f(\hat{\alpha}) = \frac{\alpha^{n-1} n^{n-1}}{\Gamma(n-1)\hat{\alpha}^{n}} \exp\left(-\frac{n\alpha}{\hat{\alpha}}\right), \qquad 0 \le \hat{\alpha} \tag{11.5.5}$$

It thus follows that $2n\alpha/\hat{\alpha}$ is distributed as χ^2 with 2n degrees of freedom. The expected value and the variance of $\hat{\alpha}$ are, respectively,

$$E(\hat{\alpha}) = \frac{n\alpha}{n-2} \quad \text{and} \quad V(\hat{\alpha}) = \frac{n^2\alpha^2}{(n-2)^2(n-3)} \tag{11.5.6}$$

Furthermore, the distribution of $\hat{\alpha}$ tends to normality as n tends to infinity. A $100(1-\alpha)\%$ confidence interval on α is given as

$$\frac{\hat{\alpha}\chi^2_{2n,\alpha/2}}{2n} < \alpha < \frac{\hat{\alpha}\chi^2_{2n,1-\alpha/2}}{2n} \tag{11.5.7}$$

The variance of the MLE $\hat{\gamma} = x_1$ from (11.2.8) is

$$V(\hat{\gamma}) = \frac{n\alpha\gamma^2}{(n\alpha-1)^2(n\alpha-2)} \tag{11.5.8}$$

Accordingly for large n, we might take advantage of the approach to normality to approximate a 95% confidence interval on γ as

$$\hat{\gamma} - 1.96\frac{\hat{\gamma}}{n\hat{\alpha}-1}\sqrt{\frac{n\hat{\alpha}}{n\hat{\alpha}-2}} \le \gamma \le \hat{\gamma} + 1.96\frac{\hat{\gamma}}{n\hat{\alpha}-1}\sqrt{\frac{n\hat{\alpha}}{n\hat{\alpha}-2}} \tag{11.5.9}$$

Although the confidence interval of (11.5.9) applies to complete (uncensored) samples, we need only replace n with N to obtain a corresponding confidence interval that is applicable to censored samples. Furthermore, these same confidence intervals should provide reasonable approximations to confidence intervals based on modified estimates of γ.

11.6 AN ILLUSTRATIVE EXAMPLE

As an illustrative example, we have chosen the results of a salary survey in a large corporation. The sample consists of monthly take-home compensation paid to 1000 professional employees as recorded to the nearest dollar. This example has no significance other than as a vehicle for illustrating computational procedures.

Monthly compensation in dollars	1,000 1,999	2,000 2,999	3,000 3,999	4,000 4,999	5,000 5,999	6,000 6,999	7,000 7,999	8,000 8,999	9,000 9,999
Frequencies	869	85	28	10	4	2	1	1	0

In summary, $n = 1000$, $\bar{x} = 1709.50$, $s = 665.09$, $x_1 = 1000.45$. Moment estimates for this sample calculated from (11.3.3) and (11.3.4) are $\alpha^* = 3.758$ and $\gamma^* = 1254.60$. However, $\gamma^* > x_1$, and therefore this estimate is inadmissible. Modified moment estimates calculated from (11.3.13) are $\hat{\alpha} = 2.4096$ and $\hat{\gamma} = 1000.03$. These estimates appear to be more reasonable. When they are substituted into the formula for V(X) given in (11.2.4), the MME for the variance is calculated to be $\hat{V}(X) = 2,960,869.6$, and $\hat{\sigma}_X = \sqrt{\hat{V}(X)} = 1720.72$, a value that is approximately equal to the sample mean.

12

The Generalized Gamma Distribution

12.1 INTRODUCTION

The generalized gamma distribution was first proposed by Stacy (1962). It was proposed independently by Cohen (1969) as a generalized Weibull distribution. This distribution has also been considered by Stacy and Mirham (1965), Harter (1967), Arora (1973), Hsieh (1977), by Whitten and Cohen (1981), and perhaps others. Stacy approached the problem as one of adding a Weibull shape parameter to the gamma distribution, whereas Cohen considered the addition of a gamma shape parameter to the Weibull distribution. Both approaches yield the same end result. With two shape parameters, the generalized distribution is more flexible than either the gamma or the Weibull distribution alone. However, with an additional parameter, estimation in the generalized distribution is somewhat more complex. In its most general form, the resultant distribution is a four-parameter distribution with a threshold parameter, two shape parameters, and a scale parameter. If the threshold or origin is set at zero, only three parameters remain, and the estimation problem becomes correspondingly simpler.

In a parameterization that relates to the Weibull and the gamma distributions as they appear in this volume, the pdf of the three-parameter generalized distribution with origin at zero is

$$f(x;\beta,\rho,\delta) = \frac{\delta}{\beta^{\rho\delta}\Gamma(\rho)} x^{\rho\delta-1}\exp\left\{-\left(\frac{x}{\beta}\right)^{\delta}\right\}, \qquad 0 < x < \infty \tag{12.1.1}$$

$$= 0 \quad \text{elsewhere}$$

where δ and ρ are shape parameters and β is a scale parameter. For some purposes it is expedient to replace the scale parameter β with an alternative scale parameter $\theta = \beta^{\delta}$. The cdf of this distribution as given by Arora (1973) is

$$F(x; \beta, \rho, \delta) = I\left[\left(\frac{x}{\beta}\right)^{\delta}; \rho\right] \tag{12.1.2}$$

where I(,) is the incomplete gamma function

$$I(z; a) = \int_0^z \frac{e^{-t} t^{a-1}}{\Gamma(a)} dt, \qquad a > 1 \tag{12.1.3}$$

The kth moment of this distribution about the origin is

$$\mu'_k = \beta^k \left[\frac{\Gamma(\rho + k/\delta)}{\Gamma(\rho)}\right] \tag{12.1.4}$$

If we adopt the notation

$$G_k(\rho, \delta) = \frac{\Gamma(\rho + k/\delta)}{\Gamma(\rho)} \tag{12.1.5}$$

previously employed by Whitten and Cohen (1981), the expected value (mean), mode, variance, α_3, α_4, and the coefficient of variation (CV) can be written, with $G_k(\rho, \delta)$ abbreviated to G_k, as

$$E(X) = \beta G_1$$

$$Mo(X) = \beta\left(\frac{\rho\delta - 1}{\delta}\right)^{1/\delta}, \qquad \rho\delta > 1$$

$$V(X) = \beta^2(G_2 - G_1^2)$$

$$\alpha_3(X) = \frac{G_3 - 3G_2G_1 + 2G_1^3}{(G_2 - G_1^2)^{3/2}} \tag{12.1.6}$$

$$\alpha_4(X) = \frac{G_4 - 4G_3G_1 + 6G_2G_1^2 - 3G_1^4}{(G_2 - G_1^2)^2}$$

$$CV(X) = \frac{\sqrt{V(X)}}{E(X)} = \frac{1}{G_1}\sqrt{G_2 - G_1^2}$$

TABLE 12.1 Values of ρ for Various Combinations of α_3 and δ

α_3 \ δ	.2	.3	.4	.5	.6	.7	.8	.9	1.0
.10	19619.42	8105.71	4227.47	2501.22	1600.59	1079.86	756.33	544.46	400.00
.20	4919.30	2031.74	1059.10	626.30	400.59	270.15	189.15	136.13	100.00
.30	2197.12	906.72	472.28	279.07	178.37	120.20	84.12	60.51	44.44
.40	1244.27	512.94	266.89	157.52	100.58	67.72	47.35	34.05	25.00
.50	803.19	330.66	171.81	101.28	64.58	43.43	30.34	21.79	16.00
.60	563.53	231.62	120.15	70.71	45.02	30.23	21.09	15.14	11.11
.70	418.97	171.88	88.99	52.27	33.22	22.27	15.52	11.13	8.16
.80	325.08	133.08	68.76	40.30	25.56	17.10	11.90	8.52	6.25
.90	260.67	106.46	54.87	32.09	20.31	13.56	9.42	6.74	4.94
1.00	214.54	87.40	44.93	26.21	16.55	11.02	7.64	5.46	4.00
1.10	180.35	73.28	37.57	21.85	13.76	9.15	6.33	4.51	3.31
1.20	154.31	62.52	31.96	18.54	11.64	7.72	5.33	3.79	2.78
1.30	134.00	54.12	27.58	15.95	9.99	6.60	4.55	3.24	2.37
1.40	117.84	47.45	24.10	13.89	8.67	5.72	3.93	2.79	2.04
1.50	104.79	42.05	21.29	12.23	7.61	5.00	3.43	2.43	1.78
1.60	94.02	37.61	18.98	10.87	6.74	4.42	3.02	2.14	1.56
2.00	65.67	24.91	12.89	7.27	4.45	2.88	1.95	1.37	1.00
2.25	54.90	21.47	10.58	5.91	3.58	2.30	1.55	1.08	.79
2.50	47.07	18.24	8.90	4.93	2.96	1.88	1.26	.88	.64
3.00	36.59	13.92	6.66	3.62	2.13	1.34	.88	.61	.44

α_3 \ δ	1.1	1.2	1.3	1.4	1.5	1.6	1.7	1.8	1.9
.10	298.36	255.05	171.09	130.75	100.20	76.83	58.82	44.87	34.03
.20	74.60	56.29	42.84	32.79	25.20	19.41	14.96	11.52	8.87
.30	33.16	25.04	19.09	14.65	11.31	8.76	6.82	5.33	4.19
.40	18.66	14.11	10.77	8.30	6.44	5.03	3.96	3.15	2.52
.50	11.95	9.04	6.92	5.36	4.18	3.30	2.63	2.12	1.73
.60	8.30	6.29	4.83	3.75	2.95	2.35	1.90	1.55	1.28
.70	6.10	4.63	3.57	2.79	2.21	1.77	1.44	1.19	1.00
.80	4.67	3.55	2.75	2.16	1.72	1.39	1.15	.96	.81
.90	3.69	2.81	2.18	1.72	1.38	1.13	.93	.79	.67
1.00	2.99	2.29	1.78	1.41	1.14	.93	.78	.66	.57
1.10	2.48	1.89	1.48	1.18	.95	.79	.66	.56	.49
1.20	2.08	1.60	1.25	1.00	.81	.67	.57	.49	.43
1.30	1.77	1.36	1.07	.86	.70	.59	.50	.43	.37
1.40	1.53	1.18	.93	.75	.61	.51	.44	.38	.33
1.50	1.33	1.03	.81	.65	.54	.45	.39	.34	.29
1.60	1.17	.91	.72	.58	.48	.70	.35	.30	.26
2.00	.75	.58	.74	.38	.32	.27	.23	.20	.18
2.25	.60	.46	.37	.30	.25	.22	.19	.17	.15
2.50	.48	.38	.30	.25	.21	.19	.16	.14	.12
3.00	.34	.26	.21	.17	.15	.13	.11	.10	.09

TABLE 12.1 Continued

α_3 \ δ	2.0	2.4	2.7	3.0	3.5	4.0	5.0	5.5
.10	25.61	7.42	2.88	1.43	.72	.48	.29	.24
.20	6.83	2.51	1.37	.88	.54	.38	.24	.21
.30	3.31	1.46	.92	.65	.43	.32	.21	.18
.40	2.05	1.02	.69	.51	.35	.27	.18	.16
.50	1.43	.77	.55	.42	.30	.23	.16	.14
.60	1.08	.62	.45	.35	.25	.20	.14	.12
.70	.85	.51	.38	.30	.22	.17	.12	.11
.80	.70	.43	.33	.26	.19	.15	.11	.10
.90	.58	.37	.28	.23	.17	.14	.10	.09
1.00	.50	.32	.25	.20	.15	.12	.09	.08
1.10	.43	.28	.22	.18	.14	.11	.08	.07
1.20	.37	.25	.19	.16	.12	.10	.07	.06
1.30	.33	.22	.17	.14	.11	.09	.07	.06
1.40	.29	.20	.16	.13	.10	.08	.06	.05
1.50	.26	.18	.14	.12	.09	.07	.06	.05
1.60	.24	.16	.13	.11	.08	.06	.05	.04
2.00	.16	.11	.09	.08	.06	.05	.04	.03
2.25	.13	.09	.08	.06	.05	.04	.03	.03
2.50	.11	.08	.06	.05	.04	.03	.03	.02
3.00	.08	.06	.05	.04	.03	.02	.02	.02

Source: Reproduced from Hsieh, P. I. (1977).

Note that α_3, α_4, and CV are functions of the two shape parameters only. The distribution is bell-shaped when $\rho\delta > 1$. Otherwise it is reverse J-shaped. Table 12.1, reproduced here with permission from Hsieh (1977), contains entries of ρ for various combinations of α_3 and δ for which $\alpha_3 > 0$.

12.2 PARAMETER ESTIMATION IN THE THREE-PARAMETER DISTRIBUTION

Moment and maximum likelihood estimators are the standard estimators here as they are for parameters of the distributions considered in previous chapters. Moment estimators are simple to derive, but their sampling errors are sometimes unacceptably large. Maximum likelihood estimators are more efficient, but are somewhat more complex to derive and to calculate from sample data. Moreover, they sometimes lead to regularity problems. We are therefore led to consider modified estimators, which combine ME and MLE equations with functions of order statistics.

12.2.1 Moment Estimators

The ME equations are $E(X) = \bar{x}$, $V(X) = s^2$, and $\alpha_3 = a_3$, where

$$\bar{x} = \sum_{1}^{n} \frac{x_i}{n}, \qquad s^2 = \sum_{1}^{n} \frac{(x_i - \bar{x})^2}{n - 1}$$

and

$$a_3 = \frac{1}{n} \sum_{1}^{n} (x_i - \bar{x})^3 \Big/ \left[\frac{1}{n} \sum_{1}^{n} (x_i - \bar{x})^2 \right]^{3/2}$$

From (12.1.6) these equations become

$$\begin{aligned} \bar{x} &= \beta G_1 \\ s^2 &= \beta^2 (G_2 - G_1^2) \\ a_3 &= \frac{G_3 - 3G_2 G_1 + 2G_1^3}{(G_2 - G_1^2)^{3/2}} \end{aligned} \tag{12.2.1}$$

We eliminate β between the first two equations of (12.2.1) to obtain

$$1 + \left(\frac{s}{\bar{x}}\right)^2 = \frac{G_2}{G_1^2} \tag{12.2.2}$$

A simultaneous solution of (12.2.2) and the third equation of (12.2.1) will yield the moment estimates δ^* and ρ^*.

Table 12.1, which contains entries of ρ as a function of α_3 and δ, and Table 12.3, which contains entries of CV as a function of ρ and δ, are available to aid in the calculation of estimates. Table 12.2, which contains entries of α_3 as a function of ρ and δ, might also be helpful in certain calculations.

The following routine for the calculation of estimates is relatively simple to execute. With $\bar{x}$, s, and a_3 available from the sample data, we calculate the sample CV = $s/\bar{x}$. We select a first approximation δ_1 and enter Table 12.1 with a_3 and δ_1. We interpolate for a corresponding approximation ρ_1. We then enter Table 12.3 with δ_1 and ρ_1 and interpolate for CV. If the value thus obtained is equal to the sample value $s/\bar{x}$, no further calculations are necessary. Otherwise, we select a second approximation δ_2 and repeat this cycle of calculations. We continue until we find two values δ_i and δ_j in a sufficiently narrow interval and such that

$$CV(\delta_i, \rho_i) \gtrless \frac{s}{\bar{x}} \lessgtr CV(\delta_j, \rho_j)$$

TABLE 12.2 Mean, Standard Deviation, α_3, and α_4 of the Generalized Gamma Distribution

δ \ ρ	.0	.1	.5	1.0	2.0	3.0	4.0
.25	.1570	.3283	6.5625	24.0000	120.0000	360.0000	840.0000
	11.5910	16.9188	88.7412	199.3590	590.3219	1297.9985	2438.6882
	690.6998	478.3141	110.2550	60.0917	29.6518	19.0788	13.9530
	**	**	*	*	2904.5957	1106.1203	553.2657
.50	.0256	.0525	.7500	2.0000	6.0000	12.0000	20.0000
	.3954	.5705	2.4495	4.4721	9.1652	14.6969	20.9762
	51.1772	35.8752	10.1041	6.6188	4.3020	3.3567	2.8257
	5486.3206	2689.4695	207.0000	87.7200	37.4082	23.4815	17.3058
.75	.0226	.0456	.5307	1.1906	2.7782	4.6303	6.6882
	.1959	.2799	1.0205	1.6108	2.6445	3.5899	4.4842
	20.5207	14.4828	4.4742	3.1212	2.1777	1.7663	1.5236
	708.5041	354.0614	36.0918	18.9870	10.7250	8.0633	6.7592
1.00	.0250	.0500	.5000	1.0000	2.0000	3.0000	4.0000
	.1581	.2236	.7071	1.0000	1.4142	1.7321	2.0000
	12.6491	8.9443	2.8284	2.0000	1.4142	1.1547	1.0000
	243.0000	123.0000	15.0000	9.0000	6.0000	5.0000	4.5000
1.25	.0288	.0571	.5063	.9314	1.6765	2.3471	2.9730
	.1480	.2073	.5780	.7498	.9521	1.0871	1.1918
	9.2877	6.5580	2.0396	1.4295	1.0028	.8159	.7051
	123.9409	63.2318	8.7470	5.8022	4.3696	3.9036	3.6738
1.50	.0332	.0654	.5234	.9027	1.5046	2.0061	2.4519
	.1465	.2034	.5067	.6129	.7172	.7783	.8224
	7.4557	5.2458	1.5638	1.0720	.7375	.5949	.5118
	77.3474	39.6433	6.0683	4.3904	3.6363	3.4078	3.2992
1.75	.0380	.0741	.5436	.8906	1.3995	1.7994	2.1422
	.1486	.2044	.4601	.5252	.5763	.6017	.6185
	6.3053	4.4144	1.2376	.8207	.5487	.4372	.3736
	54.2864	27.8947	4.6772	3.6585	3.2620	3.1575	3.1115
2.00	.0428	.0830	.5642	.8862	1.3293	1.6617	1.9386
	.1522	.2076	.4263	.4633	.4825	.4887	.4917
	5.5132	3.8371	.9953	.6311	.4057	.3179	.2692
	41.0658	21.1284	3.8692	3.2451	3.0593	3.0251	3.0136
2.25	.0478	.0920	.5841	.8857	1.2794	1.5637	1.7954
	.1566	.2119	.3997	.4165	.4156	.4107	.4062
	4.9318	3.4100	.8053	.4812	.2928	.2240	.1873
	32.6924	16.8292	3.3665	3.0015	2.9484	2.9555	2.9634
2.50	.0529	.1011	.6029	.8873	1.2422	1.4906	1.6894
	.1614	.2166	.3779	.3797	.3654	.3538	.3451
	4.4846	3.0789	.6505	.3586	.2010	.1480	.1210
	26.9959	13.8984	3.0406	2.8568	2.8902	2.9215	2.9399

Top line of each entry is E(X), second line is $\sqrt{V(X)}$, third line is α_3, and bottom line is α_4.

TABLE 12.2 Continued

δ \ ρ	.0	.1	.5	1.0	2.0	3.0	4.0
2.75	.0579	.1101	.6205	.8899	1.2134	1.4341	1.6079
	.1662	.2214	.3594	.3496	.3263	.3106	.2994
	4.1282	2.8131	.5209	.2559	.1246	.0849	.0663
	22.9077	11.7930	2.8248	2.7733	2.8638	2.9086	2.9322
3.00	.0630	.1190	.6368	.8930	1.1906	1.3891	1.5434
	.1711	.2262	.3433	.3246	.2949	.2767	.2641
	3.8360	2.5937	.4100	.1681	.0598	.0318	.0202
	19.8500	10.2180	2.6813	2.7295	2.8572	2.9086	2.9339
3.25	.0680	.1279	.6520	.8963	1.1721	1.3525	1.4912
	.1759	.2309	.3289	.3032	.2691	.2494	.2361
	3.5911	2.4086	.3135	.0920	.0041	-.0138	-.0191
	17.4868	9.0016	2.5873	2.7121	2.8633	2.9166	2.9413
3.50	.0731	.1366	.6661	.8997	1.1568	1.3221	1.4480
	.1806	.2354	.3160	.2847	.2476	.2270	.2133
	3.3820	2.2496	.2283	.0251	-.0444	-.0532	-.0531
	15.6113	8.0376	2.5282	2.7127	2.8775	2.9295	2.9522
3.75	.0781	.1452	.6792	.9031	1.1440	1.2965	1.4117
	.1852	.2397	.3043	.2686	.2294	.2082	.1944
	3.2008	2.1111	.1524	-.0342	-.0871	-.0877	-.0827
	14.0901	7.2572	2.4945	2.7259	2.8969	2.9453	2.9651
4.00	.0831	.1536	.6914	.9064	1.1330	1.2746	1.3808
	.1896	.2437	.2936	.2543	.2136	.1923	.1786
	3.0418	1.9888	.0841	-.0872	-.1249	-.1182	-.1089
	12.8335	6.6142	2.4795	2.7478	2.9195	2.9629	2.9791
4.25	.0881	.1620	.7027	.9096	1.1236	1.2558	1.3542
	.1938	.2475	.2837	.2415	.2000	.1787	.1651
	2.9006	1.8796	.0221	-.1350	-.1587	-.1454	-.1321
	11.7793	6.0765	2.4786	2.7759	2.9441	2.9814	2.9936
4.50	.0930	.1702	.7133	.9126	1.1154	1.2393	1.3311
	.1979	.2511	.2745	.2301	.1880	.1669	.1534
	2.7742	1.7814	-.0345	-.1784	-.1890	-.1697	-.1529
	10.8830	5.6212	2.4883	2.8081	2.9698	3.0003	3.0082
4.75	.0979	.1783	.7231	.9154	1.1082	1.2248	1.3108
	.2019	.2545	.2660	.2197	.1774	.1565	.1433
	2.6601	1.6923	-.0864	-.2179	-.2165	-.1916	-.1716
	10.1124	5.2313	2.5062	2.8432	2.9959	3.0192	3.0228
5.00	.1027	.1862	.7323	.9182	1.1018	1.2120	1.2928
	.2057	.2577	.2580	.2103	.1679	.1473	.1344
	2.5563	1.6109	-.1344	-.2541	-.2415	-.2115	-.1885
	9.4432	4.8943	2.5304	2.8803	3.0221	3.0378	3.0370

TABLE 12.2 Continued

δ \ ρ	.0	.1	.5	1.0	2.0	3.0	4.0
5.25	.1076	.1940	.7410	.9208	1.0961	1.2005	1.2768
	.2093	.2607	.2505	.2017	.1594	.1392	.1265
	2.4614	1.5360	-.1788	-.2874	-.2643	-.2296	-.2038
	8.8568	4.6006	2.5594	2.9185	3.0481	3.0560	3.0509
5.50	.1123	.2017	.7491	.9232	1.0911	1.1902	1.2624
	.2128	.2635	.2435	.1938	.1518	.1319	.1195
	2.3740	1.4669	-.2201	-.3182	-.2853	-.2461	-.2179
	8.3391	4.3427	2.5922	2.9574	3.0736	3.0738	3.0643
5.75	.1171	.2092	.7568	.9255	1.0865	1.1810	1.2494
	.2162	.2661	.2369	.1865	.1448	.1253	.1133
	2.2932	1.4027	-.2587	-.3467	-.3045	-.2613	-.2307
	7.8789	4.1150	2.6278	2.9964	3.0985	3.0910	3.0773
6.00	.1217	.2166	.7640	.9277	1.0823	1.1725	1.2377
	.2195	.2685	.2306	.1798	.1385	.1193	.1076
	2.2182	1.3428	-.2948	-.3733	-.3223	-.2753	-.2426
	7.4672	3.9126	2.6656	3.0355	3.1228	3.1076	3.0898

* Values are greater than 10,000
** Values are greater than 100,000

TABLE 12.2 Continued

δ \ ρ	5.0	6.0	10.0	25.0	50.0	100.0
.25	1680.0000	3024.0000	17160.0000	*	**	**
	4139.5652	6538.1392	26186.1337	*	**	**
	11.0054	9.1189	5.5936	2.7122	1.7325	1.1617
	326.3377	214.7949	73.4839	17.2729	8.5074	5.4076
.50	30.0000	42.0000	110.0000	650.0000	2550.0000	10100.0000
	27.9285	35.4965	71.1337	262.4881	724.7758	2024.9938
	2.4789	2.2312	1.6747	1.0250	.7161	.5032
	13.8994	11.7657	7.8575	4.7884	3.8673	3.4269
.75	8.9175	11.2956	22.0152	73.7459	185.0175	465.1886
	5.3425	6.1734	9.3062	19.6867	34.9063	62.0422
	1.3592	1.2384	.9554	.6018	.4249	.3002
	5.9871	5.4773	4.4709	3.5822	3.2900	3.1447
1.00	5.0000	6.0000	10.0000	25.0000	50.0000	100.0000
	2.2361	2.4495	3.1623	5.0000	7.0711	10.0000
	.8944	.8165	.6325	.4000	.2828	.2000
	4.2000	4.0000	3.6000	3.2400	3.1200	3.0600
1.25	3.5676	4.1384	6.2598	13.0909	22.8288	39.7789
	1.2787	1.3537	1.5851	2.0954	2.5833	3.1826
	.6298	.5744	.4441	.2804	.1981	.1400
	3.5370	3.4463	3.2664	3.1060	3.0529	3.0264
1.50	2.8606	3.2420	4.5906	8.5121	13.5420	21.5204
	.8572	.8862	.9704	1.1362	1.2775	1.4351
	.4558	.4148	.3194	.2008	.1417	.1001
	3.2361	3.1949	3.1146	3.0450	3.0224	3.0111
1.75	2.4482	2.7280	3.6824	6.2618	9.3278	13.8780
	.6310	.6410	.6684	.7169	.7545	.7934
	.3313	.3007	.2302	.1440	.1014	.0716
	3.0860	3.0699	3.0398	3.0152	3.0075	3.0037
2.00	2.1809	2.3990	3.1230	4.9751	7.0534	9.9875
	.4934	.4946	.4968	.4987	.4994	.4997
	.2374	.2147	.1630	.1012	.0712	.0502
	3.0085	3.0058	3.0020	3.0003	3.0001	3.0000
2.25	1.9948	2.1722	2.7484	4.1606	5.6758	7.7331
	.4023	.3990	.3892	.3710	.3573	.3440
	.1638	.1473	.1106	.0680	.0476	.0335
	2.9693	2.9738	2.9835	2.9932	2.9966	2.9983
2.50	1.8583	2.0069	2.4818	3.6065	4.7703	6.3020
	.3382	.3325	.3167	.2896	.2703	.2523
	.1044	.0930	.0685	.0413	.0287	.0202
	2.9516	2.9596	2.9757	2.9903	2.9952	2.9976

TABLE 12.2 Continued

δ \ ρ	5.0	6.0	10.0	25.0	50.0	100.0
2.75	1.7541	1.8816	2.2834	3.2087	4.1381	5.3305
	.2908	.2839	.2652	.2343	.2132	.1940
	.0554	.0483	.0338	.0194	.0133	.0092
	2.9464	2.9557	2.9740	2.9899	2.9950	2.9975
3.00	1.6720	1.7835	2.1305	2.9110	3.6758	4.6364
	.2546	.2471	.2270	.1949	.1737	.1547
	.0143	.0107	.0049	.0012	.0004	.0001
	2.9485	2.9579	2.9759	2.9908	2.9955	2.9978
3.25	1.6059	1.7047	2.0093	2.6809	3.3253	4.1202
	.2261	.2183	.1978	.1658	.1450	.1269
	-.0208	-.0212	-.0197	-.0143	-.0105	-.0076
	2.9551	2.9638	2.9798	2.9925	2.9964	2.9982
3.50	1.5514	1.6401	1.9110	2.4982	3.0516	3.7238
	.2032	.1953	.1748	.1435	.1236	.1065
	-.0511	-.0487	-.0409	-.0275	-.0198	-.0142
	2.9644	2.9718	2.9849	2.9947	2.9974	2.9988
3.75	1.5058	1.5862	1.8297	2.3501	2.8327	3.4112
	.1843	.1765	.1564	.1260	.1071	.0911
	-.0774	-.0727	-.0593	-.0390	-.0279	-.0199
	2.9751	2.9810	2.9907	2.9970	2.9986	2.9993
4.00	1.4672	1.5405	1.7616	2.2277	2.6542	3.1593
	.1686	.1609	.1412	.1120	.0941	.0791
	-.1007	-.0938	-.0754	-.0491	-.0351	-.0249
	2.9866	2.9908	2.9967	2.9995	2.9999	3.0000
4.25	1.4339	1.5014	1.7035	2.1250	2.5060	2.9525
	.1552	.1477	.1286	.1006	.0836	.0696
	-.1213	-.1125	-.0897	-.0580	-.0413	-.0293
	2.9985	3.0008	3.0029	3.0020	3.0011	3.0006
4.50	1.4050	1.4675	1.6536	2.0377	2.3812	2.7802
	.1438	.1365	.1180	.0911	.0751	.0619
	-.1397	-.1292	-.1024	-.0660	-.0469	-.0332
	3.0104	3.0108	3.0089	3.0044	3.0023	3.0012
4.75	1.3798	1.4379	1.6102	1.9627	2.2749	2.6345
	.1339	.1268	.1089	.0832	.0679	.0555
	-.1562	-.1441	-.1138	-.0731	-.0519	-.0368
	3.0222	3.0206	3.0149	3.0068	3.0035	3.0018
5.00	1.3574	1.4117	1.5721	1.8975	2.1832	2.5099
	.1253	.1183	.1010	.0764	.0619	.0503
	-.1711	-.1577	-.1241	-.0795	-.0564	-.0399
	3.0337	3.0302	3.0206	3.0091	3.0047	3.0024

TABLE 12.2 Continued

δ \ ρ	5.0	6.0	10.0	25.0	50.0	100.0
5.25	1.3375	1.3885	1.5385	1.8405	2.1035	2.4022
	.1177	.1109	.0942	.0706	.0568	.0458
	-.1847	-.1699	-.1334	-.0852	-.0604	-.0428
	3.0448	3.0395	3.0262	3.0113	3.0058	3.0029
5.50	1.3198	1.3678	1.5085	1.7901	2.0336	2.3084
	.1109	.1044	.0882	.0655	.0525	.0420
	-.1971	-.1811	-.1419	-.0905	-.0641	-.0454
	3.0556	3.0485	3.0316	3.0134	3.0068	3.0036
5.75	1.3037	1.3491	1.4817	1.7453	1.9717	2.2259
	.1049	.0985	.0829	.0611	.0487	.0388
	-.2084	-.1914	-.1497	-.0953	-.0675	-.0478
	3.0659	3.0571	3.0367	3.0154	3.0078	3.0038
6.00	1.2892	1.3322	1.4575	1.7052	1.9167	2.1529
	.0994	.0933	.0782	.0572	.0453	.0359
	-.2188	-.2008	-.1568	-.0997	-.0706	-.0500
	3.0759	3.0654	3.0415	3.0174	3.0088	3.0041

* Values are greater than 10,000
** Values ard greater than 100,000

We then interpolate for the final estimates δ^* and ρ^*. The estimate β^* can be calculated from the first equation of (12.2.1) as $\beta^* = \bar{x}/G_1(\rho^*, \delta^*)$. We can evaluate $G_1(\ )$ with the aid of ordinary tables of the gamma function, such as those given by Abramowitz and Stegun (1964). Satisfactory first approximations to δ can usually be found from a cursory examination of Tables 12.1-12.3. When a_4 is available from the sample data, a first approximation can be read from Figure 12.1.

12.2.2 Maximum Likelihood Estimators

The likelihood function of a random sample $\{x_i\}$, $i = 1, 2, \ldots, n$, from a population with pdf (12.1.1) is

$$L(x_1, \ldots, x_n) = \left(\frac{\delta}{\beta^{\rho\delta}\Gamma(\rho)}\right)^n \exp\left\{-\sum_1^n \left(\frac{x_i}{\beta}\right)^\delta\right\} \prod_1^n x_i^{\rho\delta-1} \qquad (12.2.3)$$

and the loglikelihood function is

$$\ln L = n \ln \delta - n\rho\delta \ln \beta - n \ln \Gamma(\rho) - \sum_1^n \left(\frac{x_i}{\beta}\right)^\delta + (\rho\delta - 1) \sum_1^n \ln x_i \qquad (12.2.4)$$

TABLE 12.3 Coefficient of Variation of the Generalized Gamma Distribution: $CV = (G_2 - G_1^2)^{\frac{1}{2}}/G_1$

δ \ ρ	0.025	0.05	0.5	1.0	2.0	3.0	4.0
.25	73.8424	51.5415	13.5225	8.3066	4.9193	3.6056	2.9032
.50	15.4288	10.8672	3.2660	2.2361	1.5275	1.2247	1.0488
.75	8.6810	6.1340	1.9229	1.3529	.9519	.7753	.6705
1.00	6.3246	4.4721	1.4142	1.0000	.7071	.5774	.5000
1.25	5.1326	3.6277	1.1414	.8050	.5679	.4632	.4009
1.50	4.4069	3.1116	.9681	.6790	.4767	.3880	.3354
1.75	3.9133	2.7596	.8465	.5897	.4118	.3344	.2887
2.00	3.5524	2.5014	.7555	.5227	.3630	.2941	.2536
2.25	3.2746	2.3023	.6844	.4703	.3249	.2626	.2262
2.50	3.0526	2.1428	.6269	.4279	.2942	.2374	.2043
2.75	2.8701	2.0115	.5792	.3929	.2689	.2166	.1862
3.00	2.7166	1.9009	.5390	.3634	.2477	.1992	.1711
3.25	2.5851	1.8061	.5045	.3383	.2296	.1844	.1583
3.50	2.4708	1.7236	.4745	.3165	.2141	.1717	.1473
3.75	2.3703	1.6509	.4481	.2974	.2005	.1606	.1377
4.00	2.2809	1.5862	.4247	.2805	.1886	.1509	.1293
4.25	2.2007	1.5281	.4037	.2656	.1780	.1423	.1219
4.50	2.1282	1.4756	.3849	.2521	.1686	.1346	.1153
4.75	2.0622	1.4277	.3678	.2400	.1601	.1278	.1093
5.00	2.0018	1.3839	.3523	.2291	.1524	.1215	.1040
5.25	1.9463	1.3435	.3381	.2191	.1455	.1159	.0991
5.50	1.8949	1.3062	.3251	.2099	.1391	.1108	.0947
5.75	1.8472	1.2716	.3130	.2015	.1333	.1061	.0907
6.00	1.8028	1.2393	.3019	.1938	.1280	.1018	.0870

δ \ ρ	5.0	6.0	10.0	25.0	50.0	100.0
.25	2.4640	2.1621	1.5260	.8696	.5907	.4089
.50	.9309	.8452	.6467	.4038	.2842	.2005
.75	.5991	.5465	.4227	.2670	.1887	.1334
1.00	.4472	.4082	.3162	.2000	.1414	.1000
1.25	.3584	.3271	.2532	.1601	.1132	.0800
1.50	.2997	.2733	.2114	.1335	.0943	.0667
1.75	.2577	.2350	.1815	.1145	.0809	.0572
2.00	.2262	.2061	.1591	.1002	.0708	.0500
2.25	.2017	.1837	.1416	.0892	.0630	.0445
2.50	.1820	.1657	.1276	.0803	.0567	.0400
2.75	.1658	.1509	.1161	.0730	.0515	.0364
3.00	.1523	.1385	.1066	.0670	.0472	.0334
3.25	.1408	.1281	.0985	.0618	.0436	.0308
3.50	.1310	.1191	.0915	.0574	.0405	.0286
3.75	.1224	.1113	.0855	.0536	.0378	.0267
4.00	.1149	.1044	.0802	.0503	.0355	.0250
4.25	.1083	.0984	.0755	.0473	.0334	.0236
4.50	.1024	.0930	.0713	.0447	.0315	.0223
4.75	.0971	.0882	.0676	.0424	.0299	.0211
5.00	.0923	.0838	.0643	.0403	.0284	.0200
5.25	.0880	.0799	.0612	.0383	.0270	.0191
5.50	.0840	.0763	.0585	.0366	.0258	.0182
5.75	.0804	.0730	.0559	.0350	.0247	.0174
6.00	.0771	.0700	.0536	.0336	.0237	.0167

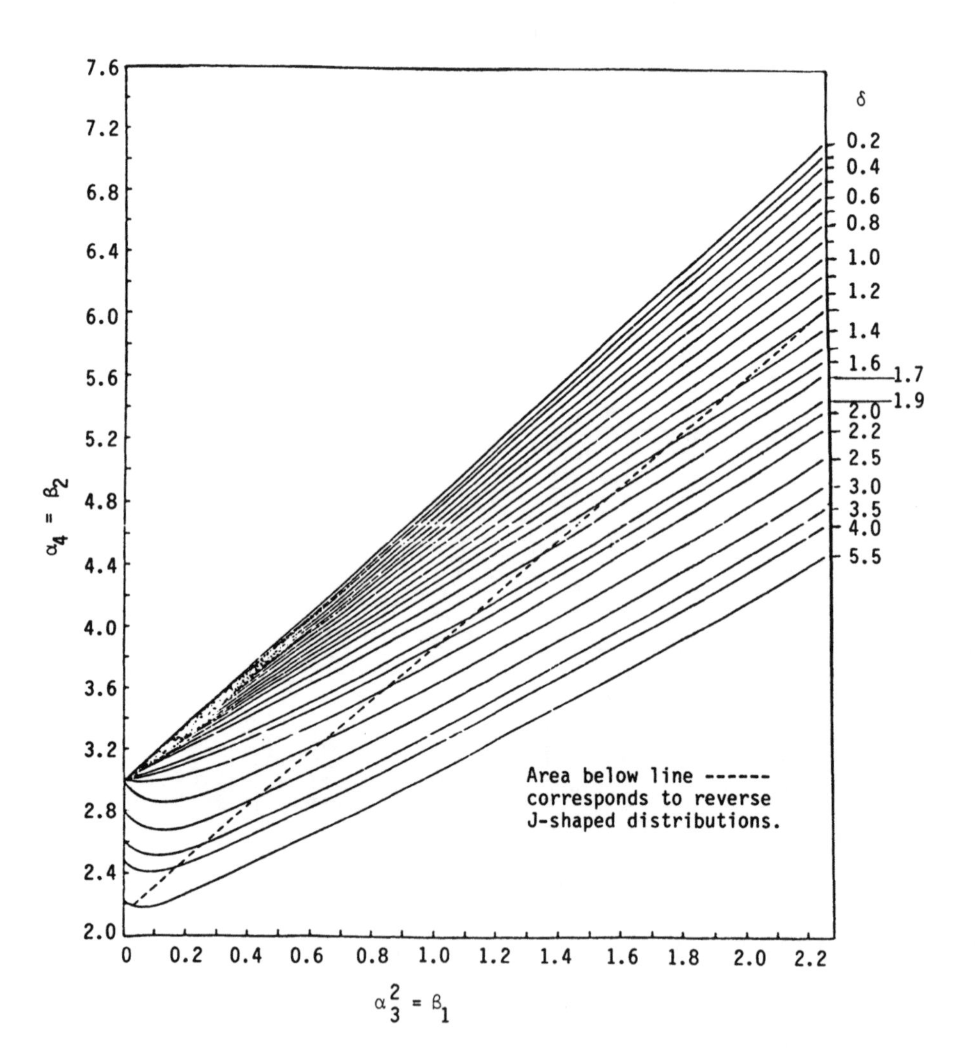

FIGURE 12.1 β_1 versus β_2 curves for the generalized gamma distribution. Adapted with permission from Arora, M. S. (1973).

When $\rho\delta > 1$, and the distribution is thus bell-shaped, MLE equations can be obtained by differentiating (12.2.4) with respect to the parameters β, ρ, and δ and then equating the derivatives to zero. However, we can obtain equivalent estimating equations that are much easier to solve if we replace the scale parameter β with the alternative $\theta = \beta^{\delta}$. With this change, the loglikelihood function of (12.2.4) is

$$\ln L = n \ln \delta - n\rho \ln \theta - n \ln \Gamma(\rho) - \frac{1}{\theta}\sum_{1}^{n} x_i^{\delta} + (\rho\delta - 1)\sum_{1}^{n} \ln x_i \tag{12.2.5}$$

We differentiate (12.2.5) with respect to θ, ρ, and δ in turn and equate to zero, thereby obtaining

$$\begin{aligned} \frac{\partial \ln L}{\partial \theta} &= -\frac{n\rho}{\theta} + \frac{1}{\theta^2}\sum_{1}^{n} x_i^{\delta} = 0 \\ \frac{\partial \ln L}{\partial \rho} &= -n \ln \theta - n\psi(\rho) + \delta \sum_{1}^{n} \ln x_i = 0 \\ \frac{\partial \ln L}{\partial \delta} &= \frac{n}{\delta} - \frac{1}{\theta}\sum_{1}^{n} x_i^{\delta} \ln x_i + \rho \sum_{1}^{n} \ln x_i = 0 \end{aligned} \tag{12.2.6}$$

where $\psi(\rho)$ is the digamma function, $\psi(\rho) = \partial \ln \Gamma(\rho)/\partial\rho = \Gamma'(\rho)/\Gamma(\rho)$.

It is necessary that we solve the three equations of (12.2.6) simultaneously for estimates $\hat{\theta}$, $\hat{\rho}$, and $\hat{\delta}$. It then follows that $\hat{\beta} = \hat{\theta}^{1/\hat{\delta}}$. Since a solution does not exist in closed form, we might resort to an iterative procedure such as that described by Arora (1973). However, a trial-and-error procedure that is relatively easy to apply might sometimes be preferred. To facilitate this solution, we make some minor simplifications of the estimating equations. We solve the first equation of (12.2.6) for $\hat{\theta}$ and then eliminate θ between the first and third equations. The second equation is retained in its original form. We thus have

$$\begin{aligned} \hat{\theta} &= \sum_{1}^{n} \frac{x_i^{\hat{\delta}}}{n\hat{\rho}} \\ \sum_{1}^{n} x_i^{\hat{\delta}} - n\hat{\rho} \sum_{1}^{n} x_i^{\hat{\delta}} \ln x_i \Big/ \left[\frac{n}{\hat{\delta}} + \hat{\rho}\sum_{1}^{n} \ln x_i\right] &= 0 \\ D(\hat{\theta}, \hat{\rho}, \hat{\delta}) = \hat{\delta}\sum_{1}^{n} \ln x_i - n \ln \hat{\theta} - n\psi(\hat{\rho}) &= 0 \end{aligned} \tag{12.2.7}$$

To solve these equations, we begin with a first approximation δ_1 and substitute it into the second equation. We then solve this equation for an approximation ρ_1. We subsequently substitute δ_1 and ρ_1 into the first equation and solve for a corresponding approximation θ_1. We then substitute the three approximations δ_1, ρ_1, and θ_1 into the third equation. If this equation is satisfied, then $\hat{\delta} = \delta_1$, $\hat{\rho} = \rho_1$, $\hat{\theta} = \theta_1$, and our calculations are complete. Otherwise we select a second approximation δ_2 and repeat the cycle of calculations. We continue until we find a pair of approximations δ_i and δ_j in a sufficiently narrow interval such that

$$D(\theta_i, \rho_i, \delta_i) \gtrless 0 \gtrless D(\theta_j, \rho_j, \delta_j)$$

We subsequently interpolate for final estimates $\hat{\theta}$, $\hat{\rho}$, and $\hat{\delta}$, and calculate $\hat{\beta} = \hat{\theta}^{1/\hat{\delta}}$

12.2.3 Modified Estimators

Numerous modifications of the ME and the MLE are possible, although not all of these have desirable properties. The number of estimating equations must equal the number of parameters to be estimated. Hence, we can choose from various combinations of moment, maximum likelihood, and order statistic estimating equations. When a threshold parameter is to be estimated, an appropriate function of the first-order statistic is desirable. However, when the origin (threshold) is known, the first-order statistic loses some of its advantage in the estimating equations, so we resort to other modifications. The modified estimators presented here have been chosen in an effort to provide estimates that are easy to calculate, unbiased (or at least almost unbiased), and exhibit minimum or near minimum variance.

12.2.4 Modified Maximum Likelihood Estimators

A modification of the MLE that employs the estimating equations $\partial \ln L/\partial\rho = 0$, $\partial \ln L/\partial\delta = 0$, and $E(X) = \bar{x}$ is relatively easy to calculate and is less likely to be adversely affected by regularity problems than is the MLE. With $\theta = \beta^{\delta}$, the estimating equations can be simplified to

$$-n\delta \ln\left(\frac{\bar{x}}{G_1}\right) - n\psi(\rho) + \delta \sum_1^n \ln x_i = 0$$

$$\frac{n}{\delta} - \left(\frac{G_1}{\bar{x}}\right)^{\delta} \sum_1^n x_i^{\delta} \ln x_i + \rho \sum_1^n \ln x_i = 0 \tag{12.2.8}$$

$$\beta = \frac{\bar{x}}{G_1}$$

The first two equations of (12.2.8) can be solved simultaneously for $\hat{\delta}$ and $\hat{\rho}$. The third equation can then be employed to calculate $\hat{\beta}$.

12.2.5 Modified Moment Estimators

The modification considered here employs $E(X) = \bar{x}$, $V(X) = s^2$, and $\partial \ln L/\partial\delta = 0$ as estimating equations. Although other modifications, such as one in which $\partial \ln L/\partial\delta = 0$ is replaced with $\partial \ln L/\partial\rho = 0$, might have merit, we limit our attention here to the modification as stated. The estimating equations can be simplified to

$$\beta = \frac{\bar{x}}{G_1}$$

$$\frac{G_2 - G_1^2}{G_1^2} = \frac{s^2}{\bar{x}^2} \qquad (12.2.9)$$

$$\frac{n}{\delta} - \left(\frac{G_1}{\bar{x}}\right)^{\delta} \sum_1^n x_i^{\delta} \ln x_i + \rho \sum_1^n \ln x_i = 0$$

The second and third equations of (12.2.9) can be solved simultaneously for estimates $\hat{\rho}$ and $\hat{\delta}$. The first equation can then be employed to calculate $\hat{\beta}$.

12.3 THE FOUR-PARAMETER DISTRIBUTION

The addition of a threshold parameter transforms the pdf of (12.1.1) into that of the more general four-parameter distribution with pdf

$$f(x;\gamma,\beta,\rho,\delta) = \frac{\delta}{\beta^{\rho\delta}\Gamma(\rho)}(x-\gamma)^{\rho\delta-1}\exp\left\{-\left[\frac{x-\gamma}{\beta}\right]^{\delta}\right\}, \qquad \gamma < x < \infty$$
$$= 0, \qquad \text{elsewhere} \qquad (12.3.1)$$

The expected value in the four-parameter distribution becomes

$$E(X) = \gamma + \beta G_1 \qquad (12.3.2)$$

but $V(X)$, $\alpha_3(X)$, and $\alpha_4(X)$ are invariant under this change of origin and remain as given in (12.1.6). The coefficient of variation, however, becomes $CV(X) = \sqrt{V(X)}/E(X-\gamma)$, so

$$CV(X) = \frac{1}{G_1}\sqrt{G_2 - G_1^2} \tag{12.3.3}$$

which is identical with (12.1.6).

12.3.1 Estimation in the Four-Parameter Distribution

Parameter estimation in the four-parameter distribution becomes considerably more complex than in the three-parameter distribution. Now, in addition to estimating β, ρ, and δ as described for the three-parameter distribution, we must also estimate the threshold parameter γ.

Moment Estimators for the Four-Parameter Distribution. The ME equations are $E(X) = \bar{x}$, $V(X) = s^2$, $\alpha_3(X) = a_3$, and $\alpha_4(X) = a_4$. These may be expressed as

$$\frac{G_3 - 3G_2G_1 + 2G_1^3}{(G_2 - G_1^2)^{3/2}} = a_3$$

$$\frac{G_4 - 4G_2G_1^2 + 6G_3G_1 - 3G_1^4}{(G_2 - G_1^2)^2} = a_4 \tag{12.3.4}$$

$$\beta^* = \frac{s}{\sqrt{G_2 - G_1^2}}$$

$$\gamma^* = \bar{x} - \frac{sG_1}{\sqrt{G_2 - G_1^2}}$$

where a_3 and a_4 are the third and fourth standard moments of the sample. The first two equations of (12.3.4) can be solved simultaneously for estimates ρ^* and δ^*. First approximations to these values for use in an iterative solution might be read from entries in Table 12.2. With ρ^* and δ^* thus calculated, they can be substituted into the third equation for the calculation of β^* and into the fourth equation for the calculation of γ^*.

As aids which greatly simplify these calculations, Table 12.4 and Figure 12.1, with α_4 as a function of α_3 and δ, are available. With a_3 and a_4 calculated from the sample data, an estimate of δ can be obtained from these sources. A reasonably accurate value can be read from Figure 12.1 and a more precise value can be obtained by interpolation from Table 12.4. With an estimate of δ thus available, an estimate of ρ can be obtained by interpolation from Table 12.1, which contains entries of ρ as a function of α_3 and δ. Estimates of β and γ then follow from the last two equations of (12.3.4). The functions $G_1(\rho, \delta)$ and $G_2(\rho, \delta)$, which are involved in these calculations, are defined in (12.1.5), and they can be evalu-

TABLE 12.4 Skewness versus Kurtosis (α_3 versus α_4) for the Generalized Gamma Distribution

α_3 \ δ	.5000	.6000	.7000	.8000	.9000	1.0000	1.1000	1.2000	1.3000	1.4000	1.5000	1.6000	1.7000
.10	3.0169	3.0167	3.0162	3.0158	3.0155	3.0150	3.0145	3.0138	3.0131	3.0122	3.0111	3.0098	3.0082
.20	3.0672	3.0661	3.0649	3.0635	3.0619	3.0600	3.0578	3.0553	3.0524	3.0489	3.0447	3.0396	3.0334
.30	3.1514	3.1489	3.1461	3.1429	3.1392	3.1350	3.1302	3.1245	3.1180	3.1102	3.1010	3.0901	3.0770
.40	3.2694	3.2649	3.2598	3.2540	3.2475	3.2400	3.2314	3.2215	3.2100	3.1966	3.1810	3.1627	3.1414
.50	3.4214	3.4142	3.4061	3.3970	3.3867	3.3750	3.3616	3.3463	3.3287	3.3085	3.2853	3.2588	3.2288
.60	3.6077	3.5971	3.5852	3.5719	3.5569	3.5400	3.5208	3.4991	3.4744	3.4465	3.4150	3.3800	3.3415
.70	3.8285	3.8137	3.7972	3.7788	3.7581	3.7350	3.7090	3.6799	3.6472	3.6109	3.5709	3.5274	3.4810
.80	4.0841	4.0642	4.0421	4.0176	3.9903	3.9600	3.9262	3.8888	3.8475	3.8023	3.7536	3.7020	3.6483
.90	4.3749	4.3489	4.3202	4.2885	4.2536	4.2150	4.1725	4.1260	4.0754	4.0211	3.9638	3.9045	3.8441
1.00	4.7011	4.6679	4.6315	4.5916	4.5479	4.5000	4.4478	4.3914	4.3311	4.2676	4.2018	4.1350	4.0684
1.10	5.0632	5.0216	4.9763	4.9269	4.8732	4.8150	4.7523	4.6853	4.6148	4.5419	4.4677	4.3938	4.3213
1.20	5.4615	5.4102	5.3546	5.2945	5.2297	5.1600	5.0858	5.0077	4.9267	4.8442	4.7618	4.6809	4.6027
1.30	5.8964	5.8339	5.7667	5.6945	5.6172	5.5350	5.4484	5.3585	5.2666	5.1745	5.0839	4.9961	4.9124
1.40	6.3682	6.2930	6.2126	6.1268	6.0358	5.9400	5.8402	5.7379	5.6348	5.5330	5.4341	5.3395	5.2501
1.50	6.8773	6.7876	6.6924	6.5917	6.4856	6.3750	6.2611	6.1458	6.0313	5.9195	5.8122	5.7108	5.6157
1.60	7.4241	7.3182	7.2064	7.0890	6.9665	6.8400	6.7112	6.5824	6.4559	6.3340	6.2183	6.1099	6.0091
1.70	8.0089	7.8847	7.7546	7.6190	7.4786	7.3350	7.1904	7.0475	6.9088	6.7766	6.6523	6.5366	6.4299
1.80	8.6320	8.4876	8.3372	8.1815	8.0218	7.8600	7.6988	7.5412	7.3899	7.2471	7.1140	6.9910	6.8781
1.90	9.2938	9.1269	8.9542	8.7768	8.5962	8.4150	8.2363	8.0635	7.8992	7.7455	7.6033	7.4728	7.3536
2.00	9.9946	9.8028	9.6057	9.4047	9.2018	9.0000	8.8030	8.6143	8.4367	8.2718	8.1202	7.9819	7.8562
2.10	10.7347	10.5155	10.2919	10.0653	9.8385	9.6150	9.3988	9.1937	9.0023	8.8258	8.6647	8.5183	8.3857
2.20	11.5143	11.2653	11.0127	10.7587	10.5064	10.2600	10.0238	9.8017	9.5960	9.4077	9.2366	9.0819	8.9422
2.30	12.3337	12.0521	11.7683	11.4849	11.2055	10.9350	10.6780	10.4382	10.2178	10.0172	9.8359	9.6725	9.5256
2.40	13.1932	12.8762	12.5587	12.2438	11.9358	11.6400	11.3613	11.1033	10.8676	10.6544	10.4625	10.2902	10.1356
2.50	14.0930	13.7376	13.3840	13.0355	12.6973	12.3750	12.0737	11.7969	11.5455	11.3192	11.1164	10.9349	10.7724
2.60	15.0333	14.6366	14.2442	13.8601	13.4900	13.1400	12.8153	12.5189	12.2514	12.0117	11.7976	11.6065	11.4358
2.70	16.0144	15.5732	15.1393	14.7174	14.3138	13.9350	13.5861	13.2695	12.9854	12.7317	12.5059	12.3050	12.1258
2.80	17.0363	16.5474	16.0694	15.6076	15.1688	14.7600	14.3860	14.0486	13.7472	13.4793	13.2415	13.0303	12.8424
2.90	18.0994	17.5594	17.0345	16.5306	16.0550	15.6150	15.2150	14.8562	14.5371	14.2544	14.0042	13.7824	13.5854
3.00	19.2037	18.6093	18.0347	17.4864	16.9724	16.5000	16.0732	15.6923	15.3549	15.0570	14.7940	14.5613	14.3550

(continued)

TABLE 12.4 Continued

α_3 \ δ	1.8000	1.9000	2.0000	2.1000	2.2000	2.3000	2.4000	2.5000	2.6000	2.7000	2.8000	2.9000	3.0000
.10	3.0062	3.0036	3.0003	2.9959	2.9899	2.9818	2.9704	2.9547	2.9335	2.9063	2.8738	2.8375	2.7991
.20	3.0258	3.0163	3.0044	2.9896	2.9712	2.9486	2.9217	2.8907	2.8564	2.8199	2.7822	2.7443	2.7069
.30	3.0614	3.0426	3.0203	2.9943	2.9645	2.9314	2.8955	2.8579	2.8194	2.7809	2.7429	2.7059	2.6703
.40	3.1166	3.0883	3.0565	3.0215	2.9839	2.9445	2.9043	2.8640	2.8242	2.7855	2.7482	2.7124	2.6783
.50	3.1952	3.1584	3.1188	3.0773	3.0347	2.9918	2.9494	2.9079	2.8678	2.8294	2.7928	2.7581	2.7252
.60	3.2999	3.2559	3.2104	3.1643	3.1183	3.0732	3.0295	2.9875	2.9474	2.9093	2.8733	2.8394	2.8074
.70	3.4324	3.3825	3.3324	3.2828	3.2344	3.1878	3.1431	3.1007	3.0606	3.0228	2.9872	2.9537	2.9223
.80	3.5936	3.5388	3.4849	3.4325	3.3822	3.3342	3.2888	3.2460	3.2057	3.1679	3.1325	3.0993	3.0682
.90	3.7839	3.7248	3.6675	3.6127	3.5606	3.5115	3.4652	3.4219	3.3813	3.3434	3.3079	3.2748	3.2438
1.00	4.0032	3.9401	3.8799	3.8228	3.7690	3.7186	3.6714	3.6274	3.5864	3.5481	3.5124	3.4791	3.4480
1.10	4.2514	4.1846	4.1214	4.0621	4.0066	3.9548	3.9066	3.8617	3.8200	3.7812	3.7451	3.7114	3.6800
1.20	4.5282	4.4577	4.3917	4.3300	4.2726	4.2194	4.1699	4.1241	4.0815	4.0420	4.0053	3.9711	3.9393
1.30	4.8333	4.7592	4.6902	4.6261	4.5667	4.5118	4.4610	4.4140	4.3704	4.3300	4.2925	4.2577	4.2252
1.40	5.1664	5.0886	5.0165	4.9499	4.8884	4.8317	4.7793	4.7310	4.6862	4.6448	4.6064	4.5707	4.5375
1.50	5.5274	5.4457	5.3704	5.3011	5.2373	5.1786	5.1245	5.0746	5.0285	4.9859	4.9464	4.9098	4.8757
1.60	5.9159	5.8303	5.7516	5.6794	5.6131	5.5522	5.4962	5.4447	5.3971	5.3532	5.3125	5.2747	5.2396
1.70	6.3318	6.2420	6.1597	6.0845	6.0155	5.9523	5.8943	5.8409	5.7917	5.7462	5.7042	5.6652	5.6290
1.80	6.7749	6.6807	6.5947	6.5162	6.4444	6.3787	6.3184	6.2631	6.2120	6.1650	6.1215	6.0812	6.0437
1.90	7.2450	7.1462	7.0563	6.9743	6.8995	6.8311	6.7685	6.7110	6.6580	6.6092	6.5641	6.5223	6.4835
2.00	7.7420	7.6384	7.5443	7.4587	7.3807	7.3095	7.2443	7.1845	7.1295	7.0788	7.0320	6.9886	6.9484
2.10	8.2658	8.1572	8.0587	7.9693	7.8879	7.8137	7.7458	7.6835	7.6263	7.5736	7.5249	7.4799	7.4381
2.20	8.8162	8.7024	8.5993	8.5059	8.4209	8.3435	8.2727	8.2079	8.1484	8.0935	8.0429	7.9961	7.9526
2.30	9.3932	9.2739	9.1661	9.0684	8.9797	8.8990	8.8252	8.7576	8.6956	8.6385	8.5858	8.5371	8.4919
2.40	9.9967	9.8717	9.7589	9.6568	9.5642	9.4799	9.4030	9.3325	9.2679	9.2084	9.1536	9.1028	9.0558
2.50	10.6267	10.4957	10.3777	10.2710	10.1743	10.0863	10.0060	9.9326	9.8652	9.8033	9.7461	9.6933	9.6443
2.60	11.2830	11.1459	11.0224	10.9110	10.8099	10.7181	10.6344	10.5578	10.4876	10.4230	10.3634	10.3084	10.2574
2.70	11.9657	11.8221	11.6931	11.5766	11.4711	11.3752	11.2879	11.2080	11.1348	11.0675	11.0054	10.9481	10.8950
2.80	12.6746	12.5244	12.3895	12.2678	12.1577	12.0577	11.9665	11.8833	11.8069	11.7368	11.6721	11.6124	11.5570
2.90	13.4098	13.2528	13.1117	12.9846	12.8697	12.7653	12.6703	12.5835	12.5039	12.4308	12.3634	12.3012	12.2435
3.00	14.1712	14.0070	13.8597	13.7270	13.6071	13.4982	13.3991	13.3086	13.2257	13.1495	13.0793	13.0145	12.9544

ated with the aid of standard tables of the gamma function, such as those given by Abramowitz and Stegun (1964). Estimates calculated as thus described will be sufficiently accurate for many practical purposes. When greater accuracy is required, they can be improved by iteration.

The principal objections to ME involve large sampling errors resulting from use of higher moments and the tendency of these estimators to produce inadmissible estimates of the threshold parameter γ. Sampling errors can be reduced by increasing sample sizes, but that is not always feasible. Even in samples as large as 500 to 1000, these errors are often of appreciable magnitude.

Inadmissible estimates of γ in which the estimated value is greater than the smallest sample observation occur frequently. When this happens, we might simply reduce such estimates to equal x_1 and then estimate the remaining parameters from the first three sample moments. Otherwise, we might prefer to employ the modified moment estimators (MME).

Maximum Likelihood Estimation. For the four-parameter distribution, the loglikelihood function of (12.2.4) becomes

$$\ln L = n \ln \delta - n\rho\delta \ln \beta - n \ln \Gamma(\rho) - \sum_1^n \left[\frac{x_i - \gamma}{\beta}\right]^\delta + (\rho\delta - 1) \sum_1^n \ln (x_i - \gamma) \tag{12.3.5}$$

or, with θ as the scale parameter,

$$\ln L = n \ln \delta - n\rho \ln \theta - n \ln \Gamma(\rho) - \frac{1}{\theta} \sum_1^n (x_i - \gamma)^\delta + (\rho\delta - 1) \sum_1^n \ln (x_i - \gamma)$$

When $\rho\delta > 1$—that is, when the distribution is bell-shaped—the MLE equations are $\partial \ln L/\partial\gamma = 0$, $\partial \ln L/\partial\theta = 0$, $\partial \ln L/\partial\rho = 0$, and $\partial \ln L/\partial\delta = 0$. The first equation becomes

$$\frac{\partial \ln L}{\partial\gamma} = \frac{\delta}{\theta} \sum_1^n (x_i - \gamma)^{\delta-1} - (\rho\delta - 1) \sum_1^n (x_i - \gamma)^{-1} = 0 \tag{12.3.6}$$

The last three equations are as given in (12.2.6) with x_i replaced by $x_i - \gamma$. As previously mentioned, Arora (1973) outlined an iterative procedure for solving these equations. However, his procedure involves laborious calculations.

In large samples ($n \geq 100$), we might set $\hat{\gamma} = x_1$, the smallest sample observation, or perhaps $\hat{\gamma} = x_1 - \epsilon$, where ϵ is an arbitrarily small quantity, and then estimate the three remaining parameters as described for the three-parameter distribution. In small samples, however, γ might be considerably smaller than x_1. Improved results can usually be obtained by employing modified estimators based on the first-order statistic, as was done with other distributions in preceding chapters.

Modified Maximum Likelihood Estimators. Although numerous modifications of the MLE are possible, our attention is limited to the one in which the estimating equations are $\partial \ln L/\partial \rho = 0$, $\partial \ln L/\partial \delta = 0$, $E(X) = \bar{x}$, and $E[F(X_1)] = F(x_1)$. In their expanded forms, these equations become

$$-n \ln \theta - n\psi(\rho) + \delta \sum_1^n \ln (x_i - \gamma) = 0$$

$$\frac{n}{\delta} - \frac{1}{\theta} \sum_1^n (x_i - \gamma)^\delta \ln (x_i - \gamma) + \rho \sum_1^n \ln (x_i - \gamma) = 0 \qquad (12.3.7)$$

$$\gamma + \theta^{1/\delta} G_1 = \bar{x}, \qquad F(x_1; \gamma, \theta, \rho, \delta) = \frac{1}{n+1}$$

Eliminate θ between the first and third equations and between the second and third equations to get

$$-n\delta \ln \left[\frac{\bar{x} - \gamma}{G_1}\right] - n\psi(\rho) + \delta \sum_1^n \ln (x_i - \gamma) = 0$$

$$\frac{n}{\delta} - \left[\frac{G_1}{\bar{x} - \gamma}\right]^\delta \sum_1^n (x_i - \gamma)^\delta \ln (x_i - \gamma) + \rho \sum_1^n \ln (x_i - \gamma) = 0 \qquad (12.3.8)$$

For fixed values of γ, these two equations can be solved simultaneously for conditional solutions $\hat{\rho}(\gamma)$ and $\hat{\delta}(\gamma)$. The third equation of (12.3.7) can then be solved for $\hat{\theta}(\gamma)$. These values can be "tested" by substitution in the fourth equation of (12.3.7). For final estimates, we interpolate between γ_i, θ_i, ρ_i, δ_i and γ_j, θ_j, ρ_j, δ_j where $\gamma_i - \gamma_j$ is sufficiently small and where

$$F(x_1; \gamma_i, \theta_i, \rho_i, \delta_i) \gtrless \frac{1}{n+1} \gtrless F(x_1; \gamma_j, \theta_j, \rho_j, \delta_j)$$

We subsequently calculate $\hat{\beta} = \hat{\theta}^{1/\hat{\delta}}$.

Modified Moment Estimators. Estimating equations for MME differ from those for the ME only in that the ME equation involving the fourth sample moment is replaced with an equation involving the first-order statistic. Estimating equations thus are

$$\gamma + \beta G_1 = \bar{x}, \qquad \beta^2 (G_2 - G_1^2) = s^2$$

$$\frac{G_3 - 3G_2G_1 + 2G_1^3}{(G_2 - G_1^2)^{3/2}} = a_3, \qquad F(x_1; \gamma, \beta, \rho, \delta) = \frac{1}{n+1} \tag{12.3.9}$$

These are the preferred estimators when each of the four parameters γ, β (or θ), ρ, and δ must be estimated. Computational procedures for these estimators are discussed in detail in Section 12.5, which follows consideration of asymptotic sampling errors of the MLE.

12.4 ESTIMATE VARIANCES AND COVARIANCES

The asymptotic variance-covariance matrix of MLE is obtained in the usual manner by inverting the Fisher information matrix in which the elements a^{ij} are negatives of expected values of second-order partial derivatives of the loglikelihood function with respect to the parameters. In the most general case, when each of the four parameters (γ, θ, ρ, δ) is estimated, the variance-covariance matrix is

$$V(\hat{\gamma}, \hat{\theta}, \hat{\rho}, \hat{\delta}) = \|a_{ij}\| = \|a^{ij}\|^{-1}, \qquad i, j = 1, 2, 3, 4 \tag{12.4.1}$$

With the loglikelihood function as given by the second equation of (12.3.5), the second partial derivatives are

$$\begin{aligned}
\frac{\partial^2 \ln L}{\partial\gamma^2} &= \frac{-\delta(\delta - 1)}{\theta}\sum_1^n (x_i - \gamma)^{\delta-2} - (\rho\delta - 1)\sum_1^n (x_i - \gamma)^{-2} \\
\frac{\partial^2 \ln L}{\partial\theta^2} &= \frac{n\rho}{\theta^2} - \frac{2}{\theta^3}\sum_1^n (x_i - \gamma)^{\delta} \\
\frac{\partial^2 \ln L}{\partial\rho^2} &= -n\psi'(\rho) \\
\frac{\partial^2 \ln L}{\partial\delta^2} &= \frac{-n}{\delta^2} - \frac{1}{\theta}\sum_1^n (x_i - \gamma)^{\delta}\ln^2(x_i - \gamma) \\
\frac{\partial^2 \ln L}{\partial\gamma\partial\theta} &= \frac{\partial^2 \ln L}{\partial\theta\partial\gamma} = \frac{-\delta}{\theta^2}\sum_1^n (x_i - \gamma)^{\delta-1} \\
\frac{\partial^2 \ln L}{\partial\gamma\partial\rho} &= \frac{\partial^2 \ln L}{\partial\rho\partial\gamma} = -\delta\sum_1^n (x_i - \gamma)^{-1}
\end{aligned} \tag{12.4.2}$$

$$\frac{\partial^2 \ln L}{\partial\gamma\partial\delta} = \frac{\partial^2 \ln L}{\partial\delta\,\partial\gamma}$$

$$= \frac{1}{\theta}\left[\delta \sum_1^n (x_i - \gamma)^{\delta-1} \ln (x_i - \gamma) + \sum_1^n (x_i - \gamma)^{\delta-1}\right] - \rho \sum_1^n (x_i - \gamma)^{-1}$$

$$\frac{\partial^2 \ln L}{\partial\theta\partial\rho} = \frac{\partial^2 \ln L}{\partial\rho\,\partial\theta} = -\frac{n}{\theta}$$

$$\frac{\partial^2 \ln L}{\partial\theta\partial\delta} = \frac{\partial^2 \ln L}{\partial\delta\,\partial\theta} = \frac{1}{\theta^2}\sum_1^n (x_i - \gamma)^\delta \ln (x_i - \gamma)$$

$$\frac{\partial^2 \ln L}{\partial\rho\partial\delta} = \frac{\partial^2 \ln L}{\partial\delta\,\partial\rho} = \sum_1^n \ln (x_i - \gamma)$$

where $\psi(\cdot)$ is the digamma function, $\psi(z) = d \ln \Gamma(z)/dz = \Gamma'(z)/\Gamma(z)$, and $\psi'(\cdot)$ is the trigamma function, $\psi'(z) = d\psi(z)/dz$. From these definitions, it follows that $\Gamma'(z) = \Gamma(z)\psi(z)$ and $\Gamma''(z) = \Gamma(z)[\psi'(z) + \psi^2(z)]$.

Elements a^{ij} of the information matrix are obtained by taking expected values of the second partials of (12.4.2) and affixing negative signs. Thereby, it follows that

$$a^{11} = \frac{n}{\theta^{2/\delta}} G_{-2}[\delta(\rho\delta - 2) + 1], \quad a^{22} = \frac{n\rho}{\theta^2}, \quad a^{33} = n\psi'(\rho)$$

$$a^{44} = \frac{n}{\delta^2}\{1 + \rho\psi'(\rho + 1) + \rho[\ln \theta + \psi(\rho + 1)]^2\}$$

$$a^{12} = a^{21} = \frac{n\delta}{\theta^{1+1/\delta}} G_{\delta-1}, \quad a^{13} = a^{31} = \frac{n\delta}{\theta^{1/\delta}} G_{-1}$$

$$a^{14} = a^{41} = \frac{n\rho}{\theta^{1/\delta}} G_{-1}\left\{\rho - \left(\rho - \frac{1}{\delta}\right)\left[1 + \ln \theta + \psi\left(\rho + 1 - \frac{1}{\delta}\right)\right]\right\} \tag{12.4.3}$$

$$a^{23} = a^{32} = \frac{n}{\theta}, \quad a^{24} = a^{42} = \frac{-n\rho}{\delta\theta}[\ln \theta + \psi(\rho + 1)]$$

$$a^{34} = a^{43} = \frac{-n}{\delta}[\ln \theta + \psi(\rho)]$$

where the G_k, as defined by (12.1.5), are combinations of gamma functions. The $\psi(\cdot)$ and the $\psi'(\cdot)$ are the digamma and the trigamma functions as previously defined. The variances and covariances a_{ij} are obtained by inverting the information matrix $\|a^{ij}\|$ as indicated by (12.4.1). Unfortunately, these results are valid only if $\rho\delta > 2$, and, due to computational difficulties involved in evaluating near singular matrices, it is not recommended that they be employed unless $\rho\delta > 4$. In the special cases of the Weibull distribution, which follows from (12.1.1) when $\rho = 1$, and of the gamma distribution, which follows when $\delta = 1$, the corresponding reduced variance-covariance matrices are valid only if $\delta > 2$ and $\rho > 2$, respectively. Again, computational difficulties are likely to be encountered unless $\delta > 2.5$ and $\rho > 2.5$ in these two special cases.

The following well-known recursive formulas have been employed in deriving some of the preceding results, and they are included as a convenience for readers.

$$\psi(z + 1) = \psi(z) + \frac{1}{z}, \qquad \Gamma(z + 1) = z\Gamma(z)$$

$$G_k = \left(\rho + \frac{k - \delta}{\delta}\right) G_{k-\delta}$$

The Variance-Covariance Matrix $V(\hat{\gamma}, \hat{\beta}, \hat{\rho}, \hat{\delta})$. With $\beta = \theta^{1/\delta}$ as the scale parameter, the variance-covariance matrix becomes

$$V(\hat{\gamma}, \hat{\beta}, \hat{\rho}, \hat{\delta}) = \|b_{ij}\| = \|b^{ij}\|^{-1}, \qquad i, j = 1, 2, 3, 4 \tag{12.4.4}$$

In this case, the applicable loglikelihood function is given by the first equation of (12.3.5), and the second partial derivatives are

$$\frac{\partial^2 \ln L}{\partial \gamma^2} = \frac{\delta(\delta - 1)}{\beta^\delta} \sum_1^n (x_i - \gamma)^{\delta-2} - (\rho\delta - 1) \sum_1^n (x_i - \gamma)^{-2}$$

$$\frac{\partial^2 \ln L}{\partial \beta^2} = \frac{n\rho\delta}{\beta^2} - \frac{\delta(\delta + 1)}{\beta^{\delta+2}} \sum_1^n (x_i - \gamma)^{\delta}$$

$$\frac{\partial^2 \ln L}{\partial \rho^2} = -n\psi'(\rho)$$

$$\frac{\partial^2 \ln L}{\partial \delta^2} = \frac{-n}{\delta^2} - \sum_1^n \left(\frac{x_i - \gamma}{\beta}\right)^\delta \ln^2\left(\frac{x_i - \gamma}{\beta}\right)$$

$$\frac{\partial^2 \ln L}{\partial \gamma \partial \beta} = \frac{\partial^2 \ln L}{\partial \beta \partial \gamma} = \frac{-\delta^2}{\beta^{\delta+1}} \sum_1^n (x_i - \gamma)^{\delta-1}$$

$$\frac{\partial^2 \ln L}{\partial\gamma\partial\rho} = \frac{\partial^2 \ln L}{\partial\rho\partial\gamma} = -\delta\sum_1^n (x_i - \gamma)^{-1}$$

$$\frac{\partial^2 \ln L}{\partial\gamma\partial\delta} = \frac{\partial^2 \ln L}{\partial\delta\partial\gamma} \tag{12.4.5}$$

$$= \frac{1}{\beta^\delta}\left[\delta\sum_1^n (x_i - \gamma)^{\delta-1} \ln(x_i - \gamma) + (1 - \delta \ln \beta)\sum_1^n (x_i - \gamma)^{\delta-1}\right]$$

$$-\rho\sum_1^n (x_i - \gamma)^{-1}$$

$$\frac{\partial^2 \ln L}{\partial\beta\partial\rho} = \frac{\partial^2 \ln L}{\partial\rho\partial\beta} = \frac{-n\delta}{\beta}$$

$$\frac{\partial^2 \ln L}{\partial\beta\partial\delta} = \frac{\partial^2 \ln L}{\partial\delta\partial\beta}$$

$$= \frac{1}{\beta^{\delta+1}}\left[\delta\sum_1^n (x_i - \gamma)^\delta \ln(x_i - \gamma) + (1 - \delta \ln \beta)\sum_1^n (x_i - \gamma)^\delta\right] - \frac{n\rho}{\beta}$$

$$\frac{\partial^2 \ln L}{\partial\rho\partial\delta} = \frac{\partial^2 \ln L}{\partial\delta\partial\rho} = -n \ln \beta + \sum_1^n \ln(x_i - \gamma)$$

Elements b^{ij} of the information matrix are obtained by taking expected values of the second partials of (12.4.5) and affixing negative signs. Thereby, with β as the scale parameter, it follows that

$$b^{11} = \frac{n}{\beta^2} G_{-2}[\delta(\rho\delta - 2) + 1], \quad b^{22} = \frac{n\rho\delta^2}{\beta^2}, \quad b^{33} = n\psi'(\rho)$$

$$b^{44} = \frac{n}{\delta^2}\{1 + \rho[\psi'(\rho + 1) + \psi^2(\rho + 1)]\}$$

$$b^{12} = b^{21} = \frac{n\delta^2}{\beta^2} G_{\delta-1}, \quad b^{13} = b^{31} = \frac{n\delta}{\beta} G_{-1} \tag{12.4.6}$$

$$b^{14} = b^{41} = \frac{n}{\beta} G_{-1}\left\{\rho - \left(\rho - \frac{1}{\delta}\right)\left[1 + \psi\left(\rho + 1 - \frac{1}{\delta}\right)\right]\right\}$$

$$b^{23} = b^{32} = \frac{n\delta}{\beta}, \quad b^{24} = b^{42} = \frac{-n\rho}{\beta}\psi(\rho + 1), \quad b^{34} = b^{43} = \frac{-n}{\delta}\psi(\rho)$$

Variance-Covariance Matrix When γ Is Known. When γ is known and the origin thus can be set to zero, the first row and the first column of the variance-covariance matrices (12.4.1) and (12.4.4) are eliminated, and the resulting 3×3 matrices can be expressed as

$$V(\hat{\theta}, \hat{\rho}, \hat{\delta}) = \begin{vmatrix} \frac{n\rho}{\theta^2} & \frac{n}{\theta} & \frac{-n\rho}{\delta\theta}[\ln\theta + \psi(\rho+1)] \\ \frac{n}{\theta} & n\psi'(\rho) & \frac{-n}{\delta}[\ln\theta + \psi(\rho)] \\ \frac{-n\rho}{\delta\theta}[\ln\theta + \psi(\rho+1) & \frac{-n}{\delta}[\ln\theta + \psi(\rho)] & a^{44} \end{vmatrix}^{-1} \quad (12.4.7)$$

where $a^{44} = (n/\delta^2)\{1 + \rho\psi'(\rho+1) + \rho[\ln\theta + \psi(\rho+1)]^2\}$.

$$V(\hat{\beta}, \hat{\rho}, \hat{\delta}) = \begin{vmatrix} \frac{n\rho\delta^2}{\beta^2} & \frac{n\delta}{\beta} & \frac{-n\rho}{\beta}\psi(\rho+1) \\ \frac{n\delta}{\beta} & n\psi'(\rho) & \frac{-n}{\delta}\psi(\rho) \\ \frac{-n\rho}{\beta}\psi(\rho+1) & \frac{-n}{\delta}\psi(\rho) & b^{44} \end{vmatrix}^{-1} \quad (12.4.8)$$

where $b^{44} = (n/\delta^2)\{1 + \rho[\psi'(\rho+1) + \psi^2(\rho+1)]^2\}$.

If we specialize our distribution further, i.e., let $\rho = 1$ to obtain the Weibull distribution or let $\delta = 1$ to obtain the gamma distribution, the matrices of (12.4.7) and (12.4.8) are reduced to the expressions given in Chapter 3 for the Weibull distribution and to those given in Chapter 6 for the gamma distribution.

For the Weibull distribution with origin zero, the variance-covariance matrices are reduced to

$$V(\hat{\theta}, \hat{\delta}) = \begin{vmatrix} \frac{n}{\theta^2} & \frac{-n}{\delta\theta}[\ln\theta + \psi(2)] \\ \frac{-n}{\delta\theta}[\ln\theta + \psi(2)] & \frac{n}{\delta^2}\{1 + \psi'(2) + [\ln\theta + \psi(2)]^2\} \end{vmatrix}^{-1} \quad (12.4.9)$$

and

$$V(\hat{\beta}, \hat{\delta}) = \begin{vmatrix} \frac{n\delta^2}{\beta^2} & \frac{-n}{\beta}\psi(2) \\ \frac{-n}{\beta}\psi(2) & \frac{n}{\delta^2}\{1 + \psi'(2) + \psi^2(2)\} \end{vmatrix}^{-1} \tag{12.4.10}$$

On completing the inversion, we obtain variances and covariances that are identical to those given in Chapter 3 for the Weibull distribution.

For the gamma distribution with origin zero, the variance-covariance matrix is reduced to

$$V(\hat{\beta}, \hat{\rho}) = \begin{vmatrix} \frac{n}{\beta^2} & \frac{n\rho}{\beta} \\ \frac{n}{\beta} & n\psi'(\rho) \end{vmatrix}^{-1} \tag{12.4.11}$$

in agreement with corresponding results given in Chapter 6. For this distribution, with $\delta = 1$ it follows that $\beta = \theta$.

Although the asymptotic variances and covariances derived in this section are strictly applicable only for MLE, numerous simulation studies have shown that they closely approximate corresponding variances and covariances of the various modified estimators presented in this chapter.

12.5 SIMPLIFIED COMPUTATIONAL PROCEDURES FOR THE MODIFIED MOMENT ESTIMATORS

The MME are particularly attractive since they are unbiased with respect to mean and variance, and their variances are near minimal. They are relatively easy to calculate with the aid of tables and charts presented here. These computational aids have been patterned after those presented in Chapter 5 for the IG distribution and in Chapter 6 for the gamma distribution. The estimating equations are $E(X) = \bar{x}$, $V(X) = s^2$, $\alpha_3(X) = a_3$, and $E[F(X_1)] = F(x_1) = 1/(n+1)$, where $\bar{x}$, s^2, and a_3 are defined as in earlier chapters. In order to facilitate construction of the tables and/or charts, it is necessary that our distribution be transformed into standard units with mean zero and unit variance. Accordingly, with

$$Z = \frac{X - E(X)}{\sqrt{V(X)}} \tag{12.5.1}$$

the pdf of (12.3.1) becomes

$$h(z; 0, 1, \rho, \delta) = \begin{cases} \dfrac{\delta}{\Gamma(\rho)} J(Jz + G_1)^{\rho\delta - 1} \exp[-(Jz + G_1)^{\delta}], & -E < z \\ 0, & \text{elsewhere} \end{cases} \tag{12.5.2}$$

where

$$J = \sqrt{G_2 - G_1^2} \quad \text{and} \quad E = \frac{G_1}{J} \tag{12.5.3}$$

The corresponding cdf can be written as

$$H(z; 0, 1, \rho, \delta) = \int_{-E}^{z} h(t; 0, 1, \rho, \delta)\, dt \tag{12.5.4}$$

If we make the transformation $w = (Jt + G_1)^{\delta}$, the cdf of (12.5.4) becomes

$$H(z; 0, 1, \rho, \delta) = \int_0^{(Jz+G_1)^{\delta}} w^{\rho-1} \exp(-w)\, dw$$

$$= I[(Jz + G_1)^{\delta}; \rho] \tag{12.5.5}$$

which is recognized as an incomplete gamma function. As previously written, ρ and δ are the primary shape parameters. In the discussion that follows, it is expedient to consider α_3 and δ as the shape parameters of interest, where $\alpha_3(X)$ is defined in (12.1.6) as a function of ρ and δ. Estimating equations of the MME become

$$\begin{aligned} &\gamma + \beta G_1 = \bar{x}, \quad \beta^2(G_2 - G_1^2) = s^2, \quad \alpha_3(X) = a_3 \\ &z_1 = \frac{x_1 - \bar{x}}{s}, \quad H(z_1; 0, 1, \alpha_3, \delta) = \frac{1}{n+1} \end{aligned} \tag{12.5.6}$$

We simplify these equations to

$$\begin{aligned} &H(z_1; 0, 1, \hat{\alpha}_3, \hat{\delta}) = \frac{1}{n+1}, \quad \hat{\beta} = \frac{s}{\sqrt{\hat{G}_2 - \hat{G}_1^2}} \\ &\hat{\gamma} = \bar{x} - \hat{\beta}\hat{G}_1, \quad \hat{\alpha}_3 = a_3 \end{aligned} \tag{12.5.7}$$

With δ assumed known ($= \delta_1$) we solve the first equation of (12.5.7) for $\alpha_3^{(1)}$. If $\alpha_3^{(1)} = a_3$, then $\hat{\delta} = \delta_1$ and we proceed to calculate $\hat{\rho}$, $\hat{\beta}$, and $\hat{\gamma}$. Otherwise, we select a second assumed value for δ ($= \delta_2$) and repeat the

TABLE 12.5 Skewness, α_3, as a Function of z_1, n, and δ: $z_1 = (x_1 - \bar{x})/s$

$\delta = 0.2$

z_1 \ n	5	10	20	25	30	40	50	100	250	500	1000
-.42	10.4504	11.0809	11.3532	11.4046	11.4384	11.4802	11.5053	11.5561	11.5886	11.6008	11.6078
-.44	9.5792	10.2340	10.5212	10.5761	10.6124	10.6576	10.6849	10.7409	10.7775	10.7915	10.7997
-.46	8.8021	9.4795	9.7808	9.8390	9.8777	9.9262	9.9557	10.0167	10.0574	10.0734	10.0828
-.48	8.1056	8.8039	9.1185	9.1799	9.2209	9.2726	9.3041	9.3701	9.4150	9.4329	9.4437
-.50	7.4782	8.1963	8.5233	8.5877	8.6309	8.6856	8.7192	8.7900	8.8390	8.8589	8.8711
-.52	6.9105	7.6472	7.9860	8.0533	8.0986	8.1562	8.1917	8.2673	8.3204	8.3423	8.3559
-.54	6.3947	7.1491	7.4990	7.5691	7.6163	7.6767	7.7141	7.7943	7.8514	7.8754	7.8904
-.56	5.9240	6.6955	7.0559	7.1285	7.1777	7.2408	7.2800	7.3647	7.4257	7.4517	7.4682
-.58	5.4928	6.2809	6.6512	6.7264	6.7774	6.8430	6.8840	6.9730	7.0379	7.0660	7.0839
-.60	5.0962	5.9006	6.2804	6.3579	6.4107	6.4788	6.5214	6.6146	6.6834	6.7134	6.7329
-.62	4.7300	5.5506	5.9395	6.0193	6.0738	6.1442	6.1884	6.2857	6.3582	6.3903	6.4112
-.64	4.3906	5.2275	5.6252	5.7072	5.7632	5.8359	5.8817	5.9829	6.0591	6.0931	6.1155
-.66	4.0748	4.9284	5.3345	5.4185	5.4761	5.5510	5.5983	5.7033	5.7831	5.8190	5.8429
-.68	3.7798	4.6506	5.0649	5.1510	5.2101	5.2870	5.3357	5.4445	5.5277	5.5656	5.5909
-.70	3.5031	4.3920	4.8143	4.9023	4.9628	5.0418	5.0919	5.2042	5.2908	5.3306	5.3574
-.72	3.2424	4.1504	4.5807	4.6705	4.7324	4.8133	4.8648	4.9806	5.0706	5.1121	5.1404
-.74	2.9956	3.9244	4.3623	4.4540	4.5173	4.6001	4.6529	4.7720	4.8653	4.9086	4.9383
-.76	2.7609	3.7121	4.1579	4.2513	4.3159	4.4006	4.4546	4.5770	4.6734	4.7186	4.7497
-.78	2.5363	3.5125	3.9659	4.0612	4.1270	4.2135	4.2688	4.3943	4.4938	4.5408	4.5732
-.80	2.3200	3.3241	3.7854	3.8823	3.9494	4.0376	4.0941	4.2228	4.3254	4.3740	4.4078
-.82	2.1103	3.1460	3.6152	3.7139	3.7822	3.8721	3.9298	4.0615	4.1670	4.2173	4.2525
-.84	1.9049	2.9772	3.4545	3.5548	3.6243	3.7159	3.7748	3.9095	4.0178	4.0698	4.1063
-.86	1.7015	2.8169	3.3023	3.4043	3.4751	3.5684	3.6283	3.7659	3.8771	3.9307	3.9685
-.88	1.4968	2.6641	3.1581	3.2618	3.3337	3.4286	3.4897	3.6301	3.7441	3.7993	3.8384
-.90	1.2864	2.5183	3.0211	3.1265	3.1996	3.2961	3.3583	3.5015	3.6182	3.6750	3.7153
-.92	1.0623	2.3787	2.8907	2.9978	3.0720	3.1702	3.2335	3.3795	3.4989	3.5572	3.5987
-.94	.8074	2.2448	2.7664	2.8752	2.9507	3.0504	3.1148	3.2635	3.3855	3.4453	3.4881
-.96	.4572	2.1160	2.6477	2.7582	2.8349	2.9363	3.0018	3.1531	3.2777	3.3390	3.3830
-.98		1.9918	2.5342	2.6465	2.7244	2.8274	2.8939	3.0479	3.1750	3.2378	3.2830
-1.00		1.8717	2.4254	2.5395	2.6186	2.7233	2.7908	2.9475	3.0771	3.1414	3.1877
-1.02		1.7553	2.3210	2.4370	2.5174	2.6236	2.6923	2.8515	2.9836	3.0493	3.0968
-1.04		1.6420	2.2207	2.3386	2.4202	2.5281	2.5978	2.7597	2.8943	2.9613	3.0100
-1.06		1.5316	2.1242	2.2440	2.3269	2.4365	2.5073	2.6717	2.8087	2.8772	2.9270
-1.08		1.4235	2.0311	2.1529	2.2372	2.3484	2.4203	2.5873	2.7267	2.7965	2.8475
-1.10		1.3174	1.9412	2.0651	2.1507	2.2637	2.3366	2.5063	2.6481	2.7192	2.7713
-1.12		1.2128	1.8542	1.9804	2.0673	2.1821	2.2561	2.4283	2.5725	2.6450	2.6981
-1.14		1.1093	1.7700	1.8984	1.9868	2.1033	2.1785	2.3533	2.4998	2.5737	2.6279
-1.16		1.0064	1.6884	1.8191	1.9090	2.0273	2.1036	2.2811	2.4299	2.5051	2.5603
-1.18		.9036	1.6090	1.7422	1.8336	1.9538	2.0313	2.2114	2.3626	2.4390	2.4953
-1.20		.8002	1.5319	1.6676	1.7606	1.8827	1.9613	2.1441	2.2976	2.3754	2.4327
-1.22		.6955	1.4567	1.5951	1.6897	1.8138	1.8936	2.0790	2.2349	2.3139	2.3723
-1.24		.5888	1.3833	1.5246	1.6209	1.7470	1.8280	2.0161	2.1743	2.2546	2.3141
-1.26		.4788	1.3116	1.4559	1.5539	1.6821	1.7644	1.9552	2.1157	2.1973	2.2578
-1.28		.3641	1.2414	1.3888	1.4887	1.6190	1.7026	1.8962	2.0590	2.1419	2.2034
-1.30		.2424	1.1726	1.3233	1.4251	1.5577	1.6425	1.8389	2.0041	2.0882	2.1507
-1.32		.1105	1.1051	1.2593	1.3631	1.4979	1.5841	1.7834	1.9509	2.0363	2.0997
-1.34			1.0387	1.1966	1.3025	1.4397	1.5272	1.7294	1.8992	1.9859	2.0504
-1.36			.9733	1.1351	1.2432	1.3828	1.4718	1.6769	1.8491	1.9370	2.0025
-1.38			.9088	1.0747	1.1851	1.3273	1.4177	1.6258	1.8004	1.8895	1.9560
-1.40			.8450	1.0154	1.1282	1.2730	1.3649	1.5760	1.7530	1.8434	1.9109
-1.42			.7819	.9570	1.0723	1.2199	1.3133	1.5276	1.7070	1.7986	1.8671
-1.44			.7193	.8994	1.0174	1.1679	1.2629	1.4803	1.6621	1.7550	1.8244
-1.46			.6572	.8426	.9634	1.1169	1.2135	1.4342	1.6184	1.7126	1.7830
-1.48			.5953	.7865	.9103	1.0668	1.1651	1.3891	1.5758	1.6713	1.7426
-1.50			.5335	.7310	.8579	1.0177	1.1177	1.3451	1.5343	1.6310	1.7033
-1.52			.4719	.6760	.8062	.9694	1.0712	1.3020	1.4937	1.5917	1.6650
-1.54			.4101	.6214	.7551	.9218	1.0255	1.2599	1.4541	1.5534	1.6277
-1.56			.3480	.5671	.7045	.8750	.9806	1.2186	1.4155	1.5160	1.5912
-1.58			.2856	.5132	.6545	.8289	.9365	1.1782	1.3776	1.4794	1.5556
-1.60			.2226	.4594	.6049	.7833	.8930	1.1385	1.3406	1.4437	1.5209

TABLE 12.5 Continued

z_1 \ n	5	10	20	25	30	40	50	100	250	500	1000
-1.62			.1588	.4057	.5557	.7384	.8502	1.0996	1.3044	1.4088	1.4870
-1.64			.0940	.3520	.5068	.6940	.8080	1.0615	1.2690	1.3746	1.4538
-1.66				.2982	.4581	.6501	.7664	1.0240	1.2342	1.3412	1.4213
-1.68				.2443	.4097	.6066	.7253	.9872	1.2002	1.3084	1.3895
-1.70				.1901	.3613	.5635	.6848	.9509	1.1668	1.2764	1.3585
-1.72				.1355	.3131	.5208	.6446	.9153	1.1340	1.2449	1.3280
-1.74				.0806	.2648	.4784	.6050	.8802	1.1018	1.2141	1.2982
-1.76					.2165	.4362	.5657	.8457	1.0703	1.1839	1.2689
-1.78					.1681	.3943	.5267	.8117	1.0392	1.1542	1.2402
-1.80					.1195	.3527	.4881	.7781	1.0087	1.1251	1.2121
-1.82					.0705	.3111	.4499	.7450	.9787	1.0964	1.1845
-1.84						.2698	.4118	.7124	.9492	1.0683	1.1574
-1.86						.2284	.3741	.6801	.9202	1.0407	1.1308
-1.88						.1872	.3365	.6483	.8916	1.0135	1.1046
-1.90						.1459	.2991	.6168	.8634	.9868	1.0789
-1.92						.1046	.2619	.5857	.8357	.9605	1.0537
-1.94						.0633	.2248	.5549	.8084	.9346	1.0288
-1.96							.1878	.5244	.7814	.9091	1.0044
-1.98							.1509	.4942	.7548	.8840	.9803
-2.00							.1141	.4644	.7285	.8593	.9566
-2.02							.0769	.4348	.7026	.8349	.9333
-2.04								.4054	.6771	.8109	.9103
-2.06								.3763	.6518	.7872	.8877
-2.08								.3474	.6268	.7638	.8654
-2.10								.3187	.6022	.7407	.8434
-2.12								.2902	.5778	.7179	.8217
-2.14								.2620	.5537	.6954	.8003
-2.16								.2338	.5298	.6732	.7792
-2.18								.2059	.5062	.6513	.7584
-2.20								.1781	.4828	.6296	.7378
-2.22								.1504	.4597	.6081	.7175
-2.24								.1228	.4367	.5869	.6974
-2.26								.0955	.4140	.5660	.6776
-2.28								.0680	.3915	.5452	.6581
-2.30									.3692	.5247	.6387
-2.32									.3471	.5044	.6196
-2.34									.3251	.4843	.6007
-2.36									.3033	.4644	.5820
-2.38									.2817	.4447	.5635
-2.40									.2603	.4251	.5452
-2.42									.2390	.4058	.5271
-2.44									.2179	.3866	.5091
-2.46									.1969	.3676	.4914
-2.48									.1760	.3488	.4738
-2.50									.1553	.3301	.4564
-2.52									.1347	.3115	.4391
-2.54									.1142	.2931	.4221
-2.56									.0938	.2749	.4051
-2.58									.0735	.2568	.3884
-2.60									.0540	.2388	.3717
-2.62										.2210	.3552
-2.64										.2032	.3389
-2.66										.1856	.3227
-2.68										.1682	.3066
-2.70										.1508	.2906
-2.72										.1335	.2748
-2.74										.1164	.2591
-2.76										.0994	.2435
-2.78										.0823	.2280
-2.80										.0653	.2126

TABLE 12.5 Continued

δ = 0.5

z_1 \ n	5	10	20	25	30	40	50	100	250	500	1000
-.30	10.3281	10.3335									
-.32	9.6118	9.6223	9.6234	9.6235							
-.34	8.9771	8.9952	8.9976	8.9977	8.9978	8.9979					
-.36	8.4095	8.4379	8.4424	8.4427	8.4429	8.4430	8.4431				
-.38	7.8977	7.9391	7.9466	7.9473	7.9476	7.9479	7.9480				
-.40	7.4325	7.4895	7.5012	7.5024	7.5030	7.5036	7.5038	7.5040			
-.42	7.0067	7.0818	7.0989	7.1008	7.1018	7.1027	7.1031	7.1036			
-.44	6.6146	6.7099	6.7335	6.7363	6.7378	6.7393	6.7400	6.7409	6.7412		
-.46	6.2514	6.3688	6.4000	6.4040	6.4062	6.4084	6.4095	6.4110	6.4115	6.4115	
-.48	5.9131	6.0543	6.0942	6.0995	6.1026	6.1058	6.1074	6.1097	6.1104	6.1106	6.1106
-.50	5.5966	5.7631	5.8125	5.8195	5.8235	5.8279	5.8301	5.8334	5.8347	5.8349	5.8350
-.52	5.2991	5.4923	5.5520	5.5608	5.5660	5.5717	5.5746	5.5793	5.5812	5.5816	5.5817
-.54	5.0184	5.2394	5.3102	5.3210	5.3274	5.3347	5.3385	5.3448	5.3474	5.3481	5.3483
-.56	4.7526	5.0023	5.0849	5.0978	5.1057	5.1146	5.1194	5.1276	5.1312	5.1321	5.1325
-.58	4.4998	4.7794	4.8743	4.8895	4.8989	4.9096	4.9155	4.9258	4.9307	4.9320	4.9325
-.60	4.2587	4.5692	4.6767	4.6944	4.7054	4.7181	4.7252	4.7379	4.7442	4.7459	4.7467
-.62	4.0280	4.3703	4.4909	4.5111	4.5238	4.5387	4.5470	4.5623	4.5703	4.5726	4.5736
-.64	3.8065	4.1816	4.3157	4.3385	4.3529	4.3700	4.3798	4.3979	4.4077	4.4107	4.4121
-.66	3.5932	4.0022	4.1500	4.1755	4.1918	4.2112	4.2223	4.2435	4.2553	4.2590	4.2609
-.68	3.3871	3.8312	3.9930	4.0212	4.0394	4.0612	4.0738	4.0981	4.1121	4.1168	4.1191
-.70	3.1873	3.6678	3.8438	3.8749	3.8949	3.9192	3.9334	3.9610	3.9774	3.9830	3.9859
-.72	2.9929	3.5113	3.7017	3.7357	3.7577	3.7845	3.8003	3.8314	3.8503	3.8569	3.8605
-.74	2.8031	3.3611	3.5662	3.6031	3.6271	3.6565	3.6739	3.7086	3.7301	3.7379	3.7422
-.76	2.6170	3.2168	3.4367	3.4765	3.5025	3.5345	3.5536	3.5920	3.6163	3.6253	3.6304
-.78	2.4336	3.0778	3.3127	3.3555	3.3835	3.4182	3.4389	3.4811	3.5084	3.5186	3.5245
-.80	2.2519	2.9437	3.1937	3.2396	3.2697	3.3070	3.3295	3.3755	3.4057	3.4174	3.4241
-.82	2.0707	2.8140	3.0795	3.1284	3.1605	3.2006	3.2247	3.2747	3.3080	3.3210	3.3287
-.84	1.8886	2.6884	2.9696	3.0215	3.0557	3.0985	3.1244	3.1783	3.2148	3.2293	3.2380
-.86	1.7035	2.5666	2.8636	2.9186	2.9550	3.0005	3.0282	3.0861	3.1257	3.1418	3.1515
-.88	1.5126	2.4482	2.7614	2.8195	2.8580	2.9063	2.9357	2.9976	3.0406	3.0582	3.0690
-.90	1.3117	2.3329	2.6627	2.7239	2.7645	2.8155	2.8467	2.9127	2.9590	2.9782	2.9901
-.92	1.0929	2.2205	2.5671	2.6315	2.6742	2.7280	2.7610	2.8311	2.8807	2.9016	2.9146
-.94	.8384	2.1107	2.4746	2.5420	2.5869	2.6435	2.6783	2.7525	2.8055	2.8280	2.8423
-.96	.4812	2.0032	2.3848	2.4555	2.5025	2.5619	2.5985	2.6768	2.7332	2.7574	2.7729
-.98		1.8978	2.2976	2.3715	2.4207	2.4829	2.5213	2.6038	2.6636	2.6896	2.7063
-1.00		1.7942	2.2128	2.2899	2.3413	2.4064	2.4466	2.5332	2.5965	2.6242	2.6422
-1.02		1.6922	2.1303	2.2107	2.2643	2.3322	2.3742	2.4650	2.5318	2.5612	2.5805
-1.04		1.5915	2.0499	2.1336	2.1894	2.2602	2.3040	2.3990	2.4693	2.5004	2.5210
-1.06		1.4920	1.9714	2.0586	2.1166	2.1903	2.2359	2.3350	2.4088	2.4417	2.4636
-1.08		1.3932	1.8948	1.9854	2.0457	2.1222	2.1697	2.2730	2.3503	2.3850	2.4082
-1.10		1.2949	1.8199	1.9139	1.9765	2.0560	2.1053	2.2128	2.2936	2.3301	2.3546
-1.12		1.1969	1.7466	1.8442	1.9091	1.9915	2.0426	2.1543	2.2386	2.2769	2.3028
-1.14		1.0988	1.6748	1.7760	1.8432	1.9285	1.9815	2.0975	2.1853	2.2254	2.2526
-1.16		1.0002	1.6044	1.7092	1.7788	1.8671	1.9220	2.0421	2.1335	2.1754	2.2039
-1.18		.9007	1.5352	1.6438	1.7158	1.8072	1.8639	1.9883	2.0831	2.1268	2.1567
-1.20		.7998	1.4672	1.5797	1.6541	1.7485	1.8071	1.9358	2.0341	2.0797	2.1109
-1.22		.6967	1.4003	1.5167	1.5937	1.6912	1.7517	1.8846	1.9865	2.0338	2.0664
-1.24		.5908	1.3344	1.4549	1.5344	1.6350	1.6975	1.8346	1.9400	1.9892	2.0231
-1.26		.4811	1.2694	1.3941	1.4762	1.5800	1.6444	1.7858	1.8948	1.9457	1.9810
-1.28		.3660	1.2053	1.3343	1.4191	1.5261	1.5924	1.7382	1.8506	1.9034	1.9401
-1.30		.2436	1.1419	1.2754	1.3629	1.4732	1.5414	1.6915	1.8075	1.8621	1.9001
-1.32		.1108	1.0791	1.2173	1.3076	1.4212	1.4915	1.6459	1.7654	1.8218	1.8612
-1.34			1.0170	1.1599	1.2531	1.3701	1.4424	1.6013	1.7243	1.7825	1.8233
-1.36			.9553	1.1033	1.1994	1.3199	1.3943	1.5575	1.6841	1.7441	1.7862
-1.38			.8940	1.0473	1.1465	1.2705	1.3469	1.5147	1.6448	1.7066	1.7501
-1.40			.8331	.9919	1.0942	1.2219	1.3004	1.4726	1.6063	1.6699	1.7147
-1.42			.7725	.9370	1.0426	1.1740	1.2546	1.4313	1.5686	1.6340	1.6802
-1.44			.7120	.8826	.9915	1.1267	1.2096	1.3908	1.5316	1.5988	1.6464
-1.46			.6516	.8285	.9410	1.0801	1.1651	1.3510	1.4954	1.5644	1.6134
-1.48			.5912	.7748	.8910	1.0340	1.1214	1.3119	1.4599	1.5307	1.5810

TABLE 12.5 Continued

z_1 \ n	5	10	20	25	30	40	50	100	250	500	1000
-1.50			.5306	.7215	.8414	.9886	1.0782	1.2735	1.4250	1.4976	1.5493
-1.52			.4699	.6683	.7921	.9436	1.0356	1.2357	1.3908	1.4652	1.5182
-1.54			.4088	.6153	.7433	.8991	.9936	1.1984	1.3572	1.4334	1.4878
-1.56			.3473	.5624	.6947	.8551	.9520	1.1618	1.3242	1.4022	1.4579
-1.58			.2852	.5096	.6464	.8115	.9109	1.1257	1.2917	1.3715	1.4286
-1.60			.2224	.4567	.5983	.7683	.8703	1.0901	1.2598	1.3414	1.3998
-1.62			.1588	.4038	.5504	.7254	.8301	1.0550	1.2284	1.3118	1.3716
-1.64			.0940	.3507	.5027	.6829	.7902	1.0203	1.1975	1.2827	1.3439
-1.66				.2974	.4550	.6406	.7508	.9862	1.1671	1.2541	1.3166
-1.68				.2439	.4073	.5986	.7117	.9525	1.1371	1.2260	1.2898
-1.70				.1899	.3596	.5569	.6729	.9191	1.1076	1.1983	1.2634
-1.72				.1354	.3119	.5153	.6344	.8862	1.0785	1.1710	1.2375
-1.74				.0804	.2640	.4739	.5962	.8537	1.0498	1.1441	1.2120
-1.76					.2160	.4327	.5582	.8215	1.0215	1.1177	1.1869
-1.78					.1678	.3916	.5204	.7897	.9936	1.0916	1.1621
-1.80					.1193	.3506	.4829	.7582	.9661	1.0659	1.1378
-1.82					.0705	.3096	.4455	.7270	.9389	1.0405	1.1138
-1.84						.2686	.4083	.6961	.9120	1.0155	1.0901
-1.86						.2277	.3713	.6655	.8855	.9909	1.0668
-1.88						.1867	.3343	.6352	.8593	.9665	1.0438
-1.90						.1457	.2975	.6051	.8334	.9425	1.0211
-1.92						.1045	.2607	.5753	.8077	.9187	.9988
-1.94						.0633	.2239	.5457	.7824	.8953	.9767
-1.96							.1873	.5164	.7573	.8721	.9549
-1.98							.1506	.4872	.7325	.8492	.9333
-2.00							.1139	.4582	.7080	.8266	.9121
-2.02							.0772	.4295	.6837	.8042	.8911
-2.04								.4009	.6596	.7821	.8703
-2.06								.3725	.6357	.7602	.8498
-2.08								.3442	.6121	.7385	.8295
-2.10								.3161	.5887	.7171	.8095
-2.12								.2881	.5655	.6959	.7896
-2.14								.2602	.5424	.6748	.7700
-2.16								.2325	.5196	.6540	.7506
-2.18								.2048	.4970	.6334	.7314
-2.20								.1773	.4745	.6130	.7124
-2.22								.1499	.4522	.5928	.6936
-2.24								.1225	.4300	.5727	.6749
-2.26								.0952	.4081	.5528	.6565
-2.28								.0679	.3862	.5331	.6382
-2.30									.3645	.5135	.6201
-2.32									.3430	.4941	.6021
-2.34									.3216	.4749	.5843
-2.36									.3003	.4558	.5667
-2.38									.2791	.4368	.5492
-2.40									.2581	.4180	.5319
-2.42									.2372	.3993	.5147
-2.44									.2164	.3808	.4976
-2.46									.1956	.3624	.4807
-2.48									.1750	.3441	.4639
-2.50									.1545	.3259	.4473
-2.52									.1341	.3078	.4307
-2.54									.1138	.2899	.4143
-2.56									.0935	.2721	.3980
-2.58									.0734	.2543	.3819
-2.60									.0533	.2367	.3658
-2.62										.2192	.3498
-2.64										.2017	.3340
-2.66										.1844	.3183
-2.68										.1671	.3026

TABLE 12.5 Continued

z_1 \ n	5	10	20	25	30	40	50	100	250	500	1000
-2.70										.1500	.2871
-2.72										.1329	.2716
-2.74										.1159	.2563
-2.76										.0990	.2410
-2.78										.0821	.2258
-2.80										.0654	.2107

$\delta = 1.0$

z_1 \ n	5	10	20	25	30	40	50	100	250	500	1000
-.36	5.5555										
-.38	5.2631										
-.40	4.9997										
-.42	4.7612	4.7619									
-.44	4.5440	4.5454									
-.46	4.3450	4.3477									
-.48	4.1615	4.1663	4.1666	4.1667							
-.50	3.9914	3.9993	3.9999	4.0000	4.0000						
-.52	3.8327	3.8448	3.8460	3.8461	3.8461	3.8461					
-.54	3.6837	3.7013	3.7034	3.7036	3.7036	3.7037	3.7037				
-.56	3.5428	3.5675	3.5709	3.5712	3.5713	3.5714	3.5714				
-.58	3.4087	3.4423	3.4474	3.4478	3.4480	3.4482	3.4482	3.4483			
-.60	3.2804	3.3245	3.3319	3.3325	3.3328	3.3331	3.3332	3.3333			
-.62	3.1567	3.2132	3.2235	3.2245	3.2250	3.2254	3.2256	3.2258			
-.64	3.0367	3.1077	3.1216	3.1230	3.1237	3.1243	3.1246	3.1249	3.1250		
-.66	2.9195	3.0072	3.0253	3.0273	3.0283	3.0293	3.0297	3.0302	3.0303		
-.68	2.8044	2.9111	2.9342	2.9368	2.9382	2.9396	2.9402	2.9409	2.9411	2.9412	
-.70	2.6904	2.8188	2.8477	2.8511	2.8529	2.8548	2.8556	2.8568	2.8571	2.8571	
-.72	2.5770	2.7299	2.7652	2.7696	2.7720	2.7744	2.7756	2.7772	2.7777	2.7778	2.7778
-.74	2.4632	2.6438	2.6864	2.6919	2.6949	2.6981	2.6996	2.7018	2.7025	2.7027	2.7027
-.76	2.3482	2.5602	2.6109	2.6176	2.6214	2.6254	2.6274	2.6303	2.6313	2.6315	2.6316
-.78	2.2309	2.4787	2.5383	2.5464	2.5510	2.5560	2.5585	2.5623	2.5637	2.5640	2.5641
-.80	2.1104	2.3990	2.4684	2.4779	2.4835	2.4896	2.4927	2.4975	2.4994	2.4998	2.4999
-.82	1.9852	2.3208	2.4008	2.4120	2.4186	2.4259	2.4296	2.4357	2.4382	2.4387	2.4389
-.84	1.8535	2.2438	2.3353	2.3483	2.3560	2.3646	2.3692	2.3766	2.3798	2.3805	2.3808
-.86	1.7130	2.1677	2.2716	2.2866	2.2956	2.3056	2.3110	2.3200	2.3240	2.3250	2.3253
-.88	1.5602	2.0924	2.2096	2.2268	2.2371	2.2487	2.2549	2.2656	2.2706	2.2719	2.2724
-.90	1.3897	2.0175	2.1492	2.1685	2.1803	2.1936	2.2008	2.2134	2.2194	2.2210	2.2217
-.92	1.1919	1.9429	2.0900	2.1118	2.1250	2.1401	2.1484	2.1631	2.1703	2.1723	2.1732
-.94	.9453	1.8683	2.0320	2.0563	2.0712	2.0883	2.0977	2.1146	2.1231	2.1256	2.1267

TABLE 12.5 Continued

$\delta = 0.5$

z_1 \ n	5	10	20	25	30	40	50	100	250	500	1000
-.96	.5713	1.7936	1.9751	2.0021	2.0187	2.0378	2.0484	2.0677	2.0777	2.0807	2.0821
-.98		1.7185	1.9191	1.9489	1.9673	1.9886	2.0005	2.0223	2.0339	2.0375	2.0392
-1.00		1.6429	1.8638	1.8967	1.9169	1.9405	1.9538	1.9783	1.9917	1.9959	1.9980
-1.02		1.5664	1.8093	1.8453	1.8675	1.8935	1.9082	1.9356	1.9508	1.9558	1.9582
-1.04		1.4890	1.7554	1.7947	1.8190	1.8475	1.8636	1.8940	1.9113	1.9170	1.9199
-1.06		1.4103	1.7019	1.7447	1.7712	1.8023	1.8200	1.8536	1.8730	1.8795	1.8829
-1.08		1.3302	1.6489	1.6954	1.7241	1.7580	1.7773	1.8142	1.8358	1.8432	1.8472
-1.10		1.2483	1.5962	1.6465	1.6777	1.7144	1.7354	1.7757	1.7996	1.8080	1.8126
-1.12		1.1645	1.5438	1.5981	1.6318	1.6714	1.6942	1.7381	1.7645	1.7739	1.7791
-1.14		1.0783	1.4916	1.5501	1.5863	1.6291	1.6537	1.7012	1.7302	1.7407	1.7466
-1.16		.9895	1.4396	1.5024	1.5413	1.5873	1.6137	1.6652	1.6968	1.7085	1.7150
-1.18		.8976	1.3876	1.4550	1.4967	1.5460	1.5744	1.6298	1.6643	1.6771	1.6844
-1.20		.8022	1.3356	1.4078	1.4524	1.5052	1.5355	1.5951	1.6324	1.6465	1.6546
-1.22		.7028	1.2836	1.3607	1.4084	1.4647	1.4972	1.5610	1.6013	1.6166	1.6256
-1.24		.5985	1.2315	1.3139	1.3647	1.4246	1.4592	1.5274	1.5708	1.5875	1.5973
-1.26		.4887	1.1793	1.2670	1.3211	1.3849	1.4217	1.4943	1.5409	1.5590	1.5697
-1.28		.3720	1.1268	1.2203	1.2777	1.3454	1.3845	1.4618	1.5116	1.5311	1.5428
-1.30		.2471	1.0742	1.1735	1.2344	1.3062	1.3476	1.4297	1.4828	1.5038	1.5165
-1.32		.1117	1.0212	1.1267	1.1913	1.2672	1.3111	1.3980	1.4545	1.4771	1.4908
-1.34			.9679	1.0798	1.1482	1.2285	1.2748	1.3667	1.4268	1.4508	1.4656
-1.36			.9142	1.0329	1.1051	1.1899	1.2387	1.3357	1.3994	1.4251	1.4409
-1.38			.8601	.9858	1.0621	1.1514	1.2029	1.3051	1.3725	1.3998	1.4168
-1.40			.8055	.9385	1.0190	1.1131	1.1673	1.2749	1.3460	1.3750	1.3931
-1.42			.7504	.8911	.9759	1.0749	1.1318	1.2449	1.3198	1.3505	1.3698
-1.44			.6947	.8434	.9327	1.0367	1.0965	1.2152	1.2940	1.3265	1.3469
-1.46			.6384	.7954	.8895	.9987	1.0613	1.1857	1.2685	1.3028	1.3245
-1.48			.5814	.7472	.8461	.9606	1.0262	1.1565	1.2434	1.2795	1.3024
-1.50			.5237	.6987	.8026	.9226	.9913	1.1275	1.2185	1.2565	1.2807
-1.52			.4652	.6498	.7589	.8846	.9564	1.0987	1.1940	1.2338	1.2593
-1.54			.4059	.6006	.7150	.8466	.9216	1.0701	1.1696	1.2114	1.2382
-1.56			.3456	.5509	.6710	.8086	.8868	1.0417	1.1456	1.1893	1.2174
-1.58			.2844	.5008	.6267	.7705	.8521	1.0135	1.1217	1.1674	1.1969
-1.60			.2221	.4503	.5822	.7324	.8174	.9854	1.0981	1.1458	1.1767
-1.62			.1587	.3992	.5374	.6942	.7827	.9575	1.0748	1.1245	1.1568
-1.64			.0941	.3476	.4923	.6559	.7481	.9297	1.0516	1.1033	1.1371
-1.66				.2955	.4470	.6175	.7134	.9020	1.0286	1.0824	1.1176
-1.68				.2427	.4013	.5790	.6787	.8744	1.0058	1.0617	1.0984
-1.70				.1893	.3552	.5404	.6440	.8469	.9831	1.0412	1.0793
-1.72				.1352	.3088	.5017	.6092	.8196	.9606	1.0209	1.0605
-1.74				.0804	.2620	.4628	.5744	.7923	.9383	1.0008	1.0419
-1.76					.2148	.4237	.5395	.7651	.9161	.9808	1.0235
-1.78					.1672	.3845	.5046	.7380	.8941	.9610	1.0052
-1.80					.1190	.3451	.4696	.7109	.8722	.9414	.9872
-1.82					.0704	.3055	.4345	.6839	.8504	.9219	.9693
-1.84						.2657	.3993	.6570	.8287	.9025	.9515
-1.86						.2257	.3640	.6301	.8072	.8833	.9339
-1.88						.1854	.3286	.6032	.7857	.8642	.9165
-1.90						.1449	.2931	.5764	.7644	.8453	.8992
-1.92						.1042	.2574	.5496	.7431	.8264	.8820
-1.94						.0631	.2216	.5228	.7220	.8077	.8649
-1.96							.1857	.4961	.7009	.7891	.8480
-1.98							.1496	.4694	.6799	.7706	.8312
-2.00							.1134	.4427	.6590	.7522	.8145
-2.02							.0769	.4160	.6382	.7339	.7979
-2.04								.3893	.6174	.7157	.7814

TABLE 12.5 Continued

$\delta = 1.0$

z_1 \ n	5	10	20	25	30	40	50	100	250	500	1000
-2.06								.3625	.5967	.6975	.7651
-2.08								.3358	.5761	.6795	.7488
-2.10								.3091	.5555	.6615	.7326
-2.12								.2824	.5350	.6436	.7165
-2.14								.2556	.5146	.6258	.7005
-2.16								.2289	.4942	.6081	.6846
-2.18								.2021	.4738	.5904	.6688
-2.20								.1753	.4535	.5728	.6530
-2.22								.1484	.4332	.5553	.6373
-2.24								.1216	.4130	.5378	.6217
-2.26								.0946	.3928	.5204	.6062
-2.28								.0677	.3726	.5030	.5907
-2.30									.3525	.4857	.5753
-2.32									.3324	.4684	.5599
-2.34									.3123	.4512	.5446
-2.36									.2923	.4341	.5294
-2.38									.2722	.4169	.5142
-2.40									.2522	.3999	.4991
-2.42									.2323	.3828	.4840
-2.44									.2123	.3658	.4690
-2.46									.1924	.3489	.4541
-2.48									.1724	.3319	.4392
-2.50									.1525	.3151	.4243
-2.52									.1326	.2982	.4095
-2.54									.1127	.2814	.3947
-2.56									.0928	.2646	.3799
-2.58									.0729	.2478	.3652
-2.60									.0531	.2311	.3506
-2.62										.2144	.3359
-2.64										.1977	.3214
-2.66										.1810	.3068
-2.68										.1644	.2923
-2.70										.1478	.2778
-2.72										.1312	.2633
-2.74										.1146	.2489
-2.76										.0980	.2345
-2.78										.0815	.2201
-2.80										.0650	.2058

TABLE 12.5 Continued

z_1 \ n	5	10	20	25	30	40	50	100	250	500	1000
						δ = 1.5					
-.42	3.9952										
-.44	3.8095										
-.46	3.6398										
-.48	3.4840	3.4844									
-.50	3.3404	3.3412									
-.52	3.2075	3.2089									
-.54	3.0838	3.0863	3.0865								
-.56	2.9683	2.9724	2.9727	2.9727							
-.58	2.8597	2.8661	2.8666	2.8666	2.8666						
-.60	2.7573	2.7667	2.7676	2.7676	2.7677	2.7677					
-.62	2.6601	2.6735	2.6749	2.6750	2.6750	2.6750	2.6750				
-.64	2.5673	2.5857	2.5879	2.5880	2.5881	2.5881	2.5882				
-.66	2.4783	2.5028	2.5061	2.5063	2.5064	2.5065	2.5065				
-.68	2.3923	2.4243	2.4289	2.4293	2.4295	2.4296	2.4296	2.4297			
-.70	2.3088	2.3497	2.3561	2.3566	2.3568	2.3571	2.3571	2.3572			
-.72	2.2271	2.2785	2.2871	2.2878	2.2882	2.2885	2.2886	2.2888			
-.74	2.1466	2.2104	2.2215	2.2226	2.2231	2.2236	2.2238	2.2240	2.2240		
-.76	2.0667	2.1449	2.1592	2.1606	2.1613	2.1620	2.1623	2.1626	2.1627		
-.78	1.9867	2.0818	2.0997	2.1016	2.1026	2.1035	2.1039	2.1044	2.1045	2.1045	
-.80	1.9058	2.0208	2.0429	2.0453	2.0466	2.0478	2.0483	2.0490	2.0492	2.0492	
-.82	1.8232	1.9615	1.9884	1.9914	1.9931	1.9947	1.9954	1.9963	1.9966	1.9966	
-.84	1.7376	1.9037	1.9360	1.9398	1.9418	1.9439	1.9448	1.9461	1.9465	1.9465	1.9466
-.86	1.6475	1.8471	1.8856	1.8902	1.8927	1.8953	1.8965	1.8982	1.8987	1.8988	1.8988
-.88	1.5505	1.7916	1.8369	1.8424	1.8455	1.8487	1.8502	1.8524	1.8531	1.8532	1.8533
-.90	1.4429	1.7368	1.7897	1.7963	1.8000	1.8039	1.8058	1.8086	1.8095	1.8097	1.8098
-.92	1.3175	1.6826	1.7439	1.7517	1.7561	1.7608	1.7631	1.7666	1.7679	1.7681	1.7682
-.94	1.1578	1.6287	1.6994	1.7085	1.7137	1.7192	1.7220	1.7263	1.7279	1.282	1.7283
-.96	.9003	1.5749	1.6560	1.6665	1.6725	1.6790	1.6824	1.6876	1.6896	1.6900	1.6902
-.98		1.5209	1.6135	1.6256	1.6326	1.6402	1.6441	1.6503	1.6528	1.6534	1.6536
-1.00		1.4665	1.5720	1.5857	1.5937	1.6024	1.6070	1.6144	1.6175	1.6182	1.6185
-1.02		1.4115	1.5311	1.5467	1.5558	1.5658	1.5710	1.5797	1.5835	1.5844	1.5847
-1.04		1.3554	1.4909	1.5084	1.5187	1.5301	1.5361	1.5462	1.5507	1.5519	1.5523
-1.06		1.2981	1.4512	1.4709	1.4825	1.4953	1.5022	1.5138	1.5191	1.5205	1.5211
-1.08		1.2389	1.4119	1.4339	1.4469	1.4614	1.4691	1.4824	1.4886	1.4903	1.4910
-1.10		1.1776	1.3729	1.3975	1.4120	1.4281	1.4368	1.4519	1.4592	1.4612	1.4620
-1.12		1.1134	1.3342	1.3614	1.3775	1.3955	1.4053	1.4222	1.4306	1.4330	1.4341
-1.14		1.0459	1.2957	1.3258	1.3436	1.3636	1.3744	1.3934	1.4030	1.4058	1.4071
-1.16		.9740	1.2571	1.2904	1.3101	1.3321	1.3441	1.3653	1.3762	1.3794	1.3810
-1.18		.8969	1.2186	1.2552	1.2768	1.3011	1.3143	1.3379	1.3502	1.3539	1.3557
-1.20		.8133	1.1799	1.2202	1.2439	1.2705	1.2851	1.3111	1.3249	1.3292	1.3313
-1.22		.7220	1.1409	1.1852	1.2112	1.2403	1.2562	1.2849	1.3003	1.3052	1.3076
-1.24		.6215	1.1017	1.1502	1.1786	1.2104	1.2278	1.2593	1.2763	1.2818	1.2846
-1.26		.5107	1.0620	1.1151	1.1461	1.1808	1.1997	1.2341	1.2530	1.2592	1.2623
-1.28		.3890	1.0218	1.0799	1.1137	1.1514	1.1720	1.2094	1.2302	1.2371	1.2406
-1.30		.2564	.9809	1.0445	1.0812	1.1221	1.1445	1.1851	1.2079	1.2156	1.2196
-1.32		.1140	.9393	1.0088	1.0487	1.0930	1.1172	1.1613	1.1861	1.1946	1.1991
-1.34			.8968	.9727	1.0160	1.0640	1.0901	1.1377	1.1648	1.1741	1.1791
-1.36			.8534	.9362	.9832	1.0350	1.0631	1.1145	1.1439	1.1541	1.1597
-1.38			.8089	.8992	.9501	1.0060	1.0363	1.0916	1.1234	1.1345	1.1407
-1.40			.7632	.8616	.9167	.9770	1.0095	1.0690	1.1033	1.1154	1.1221
-1.42			.7162	.8233	.8830	.9478	.9828	1.0466	1.0835	1.0967	1.1040
-1.44			.6677	.7844	.8488	.9186	.9561	1.0244	1.0640	1.0783	1.0863
-1.46			.6178	.7446	.8142	.8893	.9294	1.0024	1.0449	1.0603	1.0690
-1.48			.5662	.7040	.7792	.8597	.9026	.9806	1.0260	1.0425	1.0520
-1.50			.5131	.6625	.7435	.8299	.8758	.9589	1.0074	1.0251	1.0354
-1.52			.4582	.6200	.7072	.7998	.8489	.9374	.9890	1.0080	1.0191
-1.54			.4016	.5765	.6703	.7695	.8218	.9160	.9709	.9912	1.0030
-1.56			.3434	.5319	.6327	.7388	.7945	.8946	.9529	.9746	.9873
-1.58			.2834	.4862	.5944	.7078	.7671	.8733	.9352	.9583	.9718
-1.60			.2219	.4393	.5553	.6763	.7395	.8521	.9176	.9421	.9566

TABLE 12.5 Continued

$\delta = 1.5$

z_1 \ n	5	10	20	25	30	40	50	100	250	500	1000
-1.62			.1588	.3914	.5154	.6445	.7116	.8309	.9002	.9262	.9416
-1.64			.0942	.3423	.4746	.6122	.6834	.8097	.8830	.9105	.9269
-1.66			.0793	.2920	.4331	.5794	.6550	.7885	.8658	.8950	.9123
-1.68				.2407	.3907	.5461	.6262	.7673	.8488	.8796	.8980
-1.70				.1883	.3474	.5123	.5971	.7461	.8319	.8644	.8838
-1.72				.1348	.3033	.4780	.5677	.7248	.8151	.8493	.8699
-1.74				.0803	.2583	.4431	.5379	.7035	.7984	.8344	.8561
-1.76				.0793	.2125	.4076	.5077	.6821	.7818	.8196	.8424
-1.78					.1659	.3716	.4771	.6606	.7652	.8049	.8289
-1.80					.1184	.3350	.4462	.6390	.7487	.7903	.8155
-1.82					.0793	.2979	.4147	.6172	.7322	.7758	.8023
-1.84					.0793	.2601	.3829	.5954	.7158	.7614	.7892
-1.86						.2218	.3506	.5734	.6994	.7471	.7762
-1.88						.1829	.3179	.5513	.6830	.7328	.7633
-1.90						.1435	.2847	.5290	.6666	.7187	.7505
-1.92						.1034	.2511	.5066	.6502	.7045	.7378
-1.94						.0793	.2171	.4840	.6338	.6905	.7252
-1.96						.0793	.1826	.4612	.6174	.6765	.7126
-1.98							.1476	.4382	.6010	.6625	.7001
-2.00							.1123	.4150	.5846	.6485	.6877
-2.02							.0793	.3916	.5681	.6346	.6754
-2.04							.0793	.3681	.5516	.6207	.6631
-2.06							.0793	.3442	.5350	.6068	.6508
-2.08								.3202	.5184	.5929	.6386
-2.10								.2960	.5017	.5790	.6265
-2.12								.2715	.4849	.5651	.6143
-2.14								.2468	.4681	.5512	.6022
-2.16								.2218	.4512	.5373	.5901
-2.18								.1966	.4342	.5234	.5781
-2.20								.1712	.4172	.5094	.5661
-2.22								.1455	.4000	.4954	.5540
-2.24								.1196	.3827	.4814	.5420
-2.26								.0935	.3654	4674	.5300
-2.28								.0793	.3480	.4533	.5180
-2.30								.0793	.3304	.4392	.5060
-2.32								.0793	.3127	.4250	.4939
-2.34									.2950	.4108	.4819
-2.36									.2771	.3965	.4699
-2.38									.2591	.3822	.4578
-2.40									.2409	.3678	.4457
-2.42									.2227	.3534	.4336
-2.44									.2043	.3389	.4215
-2.46									.1858	.3243	.4093
-2.48									.1672	.3097	.3971
-2.50									.1484	.2950	.3849
-2.52									.1295	.2802	.3727
-2.54									.1105	.2653	.3604
-2.56									.0913	.2503	.3480
-2.58									.0793	.2353	.3357
-2.60									.0793	.2202	.3232
-2.62									.0793	.2050	.3108
-2.64									.0793	.1897	.2983
-2.66										.1743	.2857
-2.68										.1589	.2731
-2.70										.1433	.2604
-2.72										.1277	.2477
-2.74										.1119	.2349
-2.76										.0961	.2220
-2.78										.0801	.2091
-2.80										.0793	.1962

TABLE 12.5 Continued

$\delta = 2.0$

z_1 \ n	5	10	20	25	30	40	50	100	250	500	1000
-.44	3.4699										
-.46	3.3108										
-.48	3.1647										
-.50	3.0300										
-.52	2.9053	2.9057									
-.54	2.7895	2.7903									
-.56	2.6816	2.6829									
-.58	2.5805	2.5828	2.5829								
-.60	2.4855	2.4891	2.4893	2.4894							
-.62	2.3959	2.4013	2.4017	2.4017	2.4017						
-.64	2.3108	2.3187	2.3194	2.3194	2.3194						
-.66	2.2299	2.2409	2.2420	2.2420	2.2421	2.2421	2.2421				
-.68	2.1523	2.1673	2.1690	2.1691	2.1691	2.1691	2.1692				
-.70	2.0777	2.0976	2.1000	2.1002	2.1003	2.1003	2.1003				
-.72	2.0056	2.0314	2.0348	2.0350	2.0351	2.0352	2.0352	2.0353			
-.74	1.9353	1.9682	1.9729	1.9733	1.9734	1.9736	1.9736	1.9737			
-.76	1.8666	1.9078	1.9141	1.9146	1.9149	1.9151	1.9152	1.9152			
-.78	1.7989	1.8500	1.8582	1.8589	1.8592	1.8595	1.8596	1.8598			
-.80	1.7317	1.7943	1.8048	1.8058	1.8062	1.8067	1.8068	1.8070	1.8070		
-.82	1.6645	1.7406	1.7538	1.7551	1.7557	1.7563	1.7565	1.7568	1.7569		
-.84	1.5967	1.6886	1.7049	1.7066	1.7074	1.7082	1.7085	1.7089	1.7090		
-.86	1.5274	1.6382	1.6581	1.6601	1.6612	1.6623	1.6627	1.6633	1.6634	1.6634	
-.88	1.4558	1.5890	1.6130	1.6156	1.6170	1.6183	1.6189	1.6197	1.6198	1.6199	
-.90	1.3804	1.5410	1.5695	1.5728	1.5745	1.5762	1.5769	1.5779	1.5782	1.5782	1.5782
-.92	1.2991	1.4939	1.5276	1.5315	1.5336	1.5357	1.5367	1.5380	1.5384	1.5384	1.5384
-.94	1.2080	1.4476	1.4870	1.4917	1.4942	1.4968	1.4980	1.4997	1.5002	1.5003	1.5003
-.96	1.0989	1,4018	1.4477	1.4532	1.4562	1.4593	1.4608	1.4629	1.4636	1.4637	1.4638
-.98		1.3564	1.4094	1.4159	1.4195	1.4232	1.4250	1.4276	1.4285	1.4287	1.4287
-1.00		1.3112	1.3722	1.3797	1.3839	1.3883	1.3905	1.3937	1.3948	1.3950	1.3951
-1.02		1.2660	1.3359	1.3445	1.3494	1.3545	1.3571	1.3610	1.3624	1.3627	1.3628
-1.04		1.2205	1.3004	1.3102	1.3158	1.3218	1.3248	1.3295	1.3313	1.3316	1.3318
-1.06		1.1747	1.2655	1.2768	1.2832	1.2901	1.2936	1.2991	1.3013	1.3017	1.3019
-1.08		1.1281	1.2313	1.2440	1.2514	1.2592	1.2633	1.2697	1.2724	1.2729	1.2732
-1.10		1.0804	1.1977	1.2120	1.2203	1.2292	1.2338	1.2413	1.2445	1.2452	1.2455
-1.12		1.0313	1.1645	1.1805	1.1898	1.2000	1.2052	1.2138	1.2176	1.2184	1.2188
-1.14		.9802	1.1317	1.1496	1.1600	1.1714	1.1773	1.1872	1.1915	1.1926	1.1930
-1.16		.9265	1.0991	1.1191	1.1307	1.1434	1.1501	1.1613	1.1664	1.1677	1.1682
-1.18		.8693	1.0668	1.0891	1.1020	1.1161	1.1236	1.1362	1.1420	1.1436	1.1442
-1.20		.8074	1.0347	1.0593	1.0736	1.0893	1.0976	1.1118	1.1185	1.1203	1.1210
-1.22		.7387	1.0026	1.0298	1.0456	1.0629	1.0722	1.0880	1.0956	1.0977	1.0986
-1.24		.6601	.9704	1.0005	1.0179	1.0370	1.0472	1.0648	1.0734	1.0758	1.0769
-1.26		.5664	.9382	.9714	.9905	1.0115	1.0227	1.0422	1.0519	1.0547	1.0559
-1.28		.4483	.9057	.9424	.9633	.9864	.9987	1.0201	1.0309	1.0341	1.0356
-1.30		.2952	.8730	.9133	.9363	.9615	.9750	.9985	1.0105	1.0141	1.0159
-1.32		.1223	.8398	.8842	.9094	.9369	.9516	.9774	.9907	.9948	.9967
-1.34			.8061	.8550	.8826	.9125	.9285	.9566	.9713	.9759	.9781
-1.36			.7716	.8256	.8557	.8884	.9057	.9363	.9525	.9576	.9601
-1.38			.7364	.7960	.8289	.8643	.8832	.9164	.9341	.9397	.9425
-1.40			.7001	.7659	.8019	.8404	.8608	.8968	.9161	.9223	.9255
-1.42			.6626	.7355	.7747	.8166	.8386	.8775	.8985	.9053	.9089
-1.44			.6237	.7044	.7474	.7927	.8166	.8585	.8812	.8888	.8927
-1.46			.5830	.6728	.7197	.7689	.7946	.8398	.8644	.8726	.8769
-1.48			.5402	.6403	.6917	.7450	.7728	.8214	.8479	.8568	.8615
-1.50			.4951	.6069	.6633	.7211	.7509	.8031	.8316	.8413	.8465
-1.52			.4472	.5724	.6343	.6970	.7291	.7851	.8157	.8262	.8319
-1.54			.3961	.5367	.6047	.6727	.7073	.7673	.8001	.8114	.8176
-1.56			.3417	.4996	.5744	.6482	.6854	.7496	.7847	.7969	.8036
-1.58			.2840	.4608	.5433	.6235	.6635	.7321	.7696	.7827	.7899
-1.60			.2232	.4203	.5112	.5984	.6414	.7147	.7547	.7687	.7765
-1.62			.1598	.3777	.4781	.5729	.6192	.6975	.7400	.7550	.7633

TABLE 12.5 Continued

$\delta = 2.0$

z_1 \ n	5	10	20	25	30	40	50	100	250	500	1000
-1.64			.0946	.3331	.4438	.5469	.5968	.6803	.7256	.7415	.7505
-1.66				.2863	.4082	.5205	.5741	.6632	.7113	.7283	.7379
-1.68				.2374	.3712	.4935	.5512	.6462	.6972	.7152	.7255
-1.70				.1866	.3327	.4658	.5279	.6292	.6832	.7024	.7133
-1.72				.1340	.2926	.4374	.5043	.6122	.6694	.6898	.7014
-1.74				.0800	.2510	.4083	.4803	.5953	.6558	.6773	.6897
-1.76					.2078	.3783	.4558	.5783	.6423	.6650	.6781
-1.78					.1631	.3473	.4308	.5613	.6289	.6529	.6668
-1.80					.1171	.3154	.4053	.5443	.6156	.6409	.6556
-1.82					.0698	.2824	.3792	.5272	.6024	.6291	.6446
-1.84						.2484	.3524	.5100	.5893	.6174	.6337
-1.86						.2134	.3248	.4927	.5763	.6058	.6230
-1.88						.1772	.2966	.4753	.5633	.5944	.6124
-1.90						.1400	.2675	.4578	.5504	.5830	.6020
-1.92						.1016	.2377	.4400	.5376	.5718	.5917
-1.94						.0622	.2070	.4221	.5248	.5606	.5815
-1.96							.1754	.4040	.5120	.5495	.5714
-1.98							.1429	.3857	.4993	.5385	.5615
-2.00							.1095	.3671	.4865	.5276	.5516
-2.02							.0752	.3482	.4738	.5168	.5418
-2.04								.3290	.4610	.5059	.5322
-2.06								.3095	.4482	.4952	.5226
-2.08								.2897	.4354	.4845	.5130
-2.10								.2694	.4226	.4738	.5036
-2.12								.2488	.4097	.4632	.4942
-2.14								.2277	.3967	.4525	.4849
-2.16								.2062	.3837	.4419	.4756
-2.18								.1841	.3706	.4313	.4664
-2.20								.1616	.3573	.4207	.4572
-2.22								.1385	.3440	.4101	.4480
-2.24								.1148	.3306	.3995	.4389
-2.26								.0905	.3170	.3888	.4298
-2.28								.0655	.3033	.3782	.4208
-2.30									.2894	.3675	.4117
-2.32									.2754	.3567	.4027
-2.34									.2612	.3459	.3937
-2.36									.2467	.3351	.3847
-2.38									.2321	.3242	.3756
-2.40									.2172	.3132	.3666
-2.42									.2020	.3021	.3576
-2.44									.1866	.2909	.3485
-2.46									.1709	.2797	.3394
-2.48									.1549	.2683	.3303
-2.50									.1386	.2568	.3212
-2.52									.1219	.2452	.3120
-2.54									.1049	.2334	.3028
-2.56									.0874	.2215	.2935
-2.58									.0696	.2094	.2841
-2.60									.0513	.1971	.2747
-2.62										.1847	.2652
-2.64										.1720	.2557
-2.66										.1592	.2460
-2.68										.1461	.2363
-2.70										.1327	.2264
-2.72										.1191	.2165
-2.74										.1053	.2064
-2.76										.0911	.1962
-2.78										.0766	.1858
-2.80										.0618	.1754

TABLE 12.5 Continued

	$\delta = 2.5$										
z_1 \ n	5	10	20	25	30	40	50	100	250	500	1000
-.46	3.1273										
-.48	2.9860										
-.50	2.8555										
-.52	2.7347										
-.54	2.6225	2.6229									
-.56	2.5179	2.5186									
-.58	2.4200	2.4212									
-.60	2.3281	2.3301									
-.62	2.2415	2.2446	2.2448								
-.64	2.1596	2.1642	2.1645	2.1646							
-.66	2.0817	2.0884	2.0890	2.0890	2.0890						
-.68	2.0075	2.0168	2.0177	2.0177	2.0177	2.0177					
-.70	1.9363	1.9489	1.9502	1.9503	1.9503	1.9504	1.9504				
-.72	1.8678	1.8845	1.8864	1.8865	1.8866	1.8866	1.8866				
-.74	1.8015	1.8232	1.8259	1.8260	1.8261	1.8262	1.8262				
-.76	1.7370	1.7646	1.7683	1.7686	1.7687	1.7688	1.7688	1.7689			
-.78	1.6739	1.7086	1.7136	1.7139	1.7141	1.7143	1.7143	1.7144			
-.80	1.6119	1.6548	1.6613	1.6619	1.6621	1.6623	1.6624	1.6625			
-.82	1.5506	1.6031	1.6114	1.6122	1.6125	1.6128	1.6129	1.6131			
-.84	1.4894	1.5532	1.5637	1.5647	1.5651	1.5656	1.5657	1.5659	1.5659		
-.86	1.4279	1.5049	1.5179	1.5192	1.5198	1.5204	1.5206	1.5209	1.5209		
-.88	1.3655	1.4580	1.4740	1.4756	1.4764	1.4771	1.4775	1.4778	1.4779		
-.90	1.3016	1.4125	1.4317	1.4337	1.4347	1.4357	1.4361	1.4366	1.4368	1.4368	
-.92	1.2349	1.3679	1.3909	1.3934	1.3947	1.3959	1.3965	1.3972	1.3973	1.3974	
-.94	1.1641	1.3244	1.3516	1.3546	1.3562	1.3577	1.3584	1.3593	1.3596	1.3596	1.3596
-.96	1.0867	1.2816	1.3135	1.3171	1.3190	1.3209	1.3218	1.3230	1.3233	1.3234	1.3234
-.98		1.2394	1.2766	1.2808	1.2832	1.2855	1.2866	1.2881	1.2885	1.2886	1.2886
-1.00		1.1977	1.2407	1.2457	1.2485	1.2513	1.2527	1.2545	1.2551	1.2552	1.2552
-1.02		1.1563	1.2058	1.2117	1.2150	1.2183	1.2199	1.2222	1.2230	1.2231	1.2231
-1.04		1.1151	1.1719	1.1786	1.1824	1.1864	1.1883	1.1911	1.1920	1.1922	1.1923
-1.06		1.0739	1.1387	1.1465	1.1508	1.1554	1.1577	1.1611	1.1623	1.1625	1.1625
-1.08		1.0326	1.1062	1.1151	1.1201	1.1254	1.1281	1.1321	1.1335	1.1338	1.1339
-1.10		.9908	1.0744	1.0845	1.0902	1.0963	1.0993	1.1041	1.1059	1.1062	1.1063
-1.12		.9483	1.0431	1.0545	1.0610	1.0679	1.0715	1.0770	1.0791	1.0796	1.0797
-1.14		.9050	1.0124	1.0251	1.0325	1.0403	1.0444	1.0507	1.0533	1.0539	1.0541
-1.16		.8603	.9820	.9963	1.0046	1.0134	1.0180	1.0253	1.0284	1.0291	1.0293
-1.18		.8139	.9521	.9680	.9772	.9872	.9923	1.0007	1.0042	1.0051	1.0054
-1.20		.7651	.9224	.9402	.9504	.9615	.9672	.9767	.9809	.9819	.9823
-1.22		.7129	.8930	.9127	.9240	.9363	.9428	.9535	.9582	.9594	.9599
-1.24		.6561	.8638	.8856	.8981	.9117	.9189	.9308	.9363	.9377	.9383
-1.26		.5921	.8347	.8587	.8725	.8875	.8955	.9088	.9150	.9166	.9173
-1.28		.5161	.8056	.8321	.8472	.8638	.8725	.8873	.8943	.8962	.8971
-1.30		.4161	.7765	.8056	.8223	.8404	.8500	.8664	.8743	.8764	.8774
-1.32		.2300	.7472	.7793	.7976	.8174	.8279	.8460	.8548	.8572	.8583
-1.34			.7178	.7531	.7730	.7947	.8062	.8260	.8358	.8386	.8399
-1.36			.6880	.7268	.7487	.7723	.7848	.8065	.8173	.8205	.8219
-1.38			.6579	.7006	.7244	.7502	.7638	.7874	.7993	.8028	.8045
-1.40			.6272	.6743	.7003	.7282	.7430	.7686	.7817	.7857	.7876
-1.42			.5959	.6477	.6761	.7065	.7225	.7503	.7646	.7690	.7712
-1.44			.5637	.6210	.6520	.6849	.7022	.7323	.7479	.7528	.7552
-1.46			.5305	.5940	.6278	.6634	.6821	.7146	.7316	.7369	.7396
-1.48			.4959	.5665	.6034	.6421	.6622	.6972	.7156	.7215	.7244
-1.50			.4597	.5385	.5789	.6208	.6425	.6801	.7000	.7064	.7097
-1.52			.4214	.5099	.5542	.5995	.6229	.6633	.6847	.6917	.6953
-1.54			.3804	.4806	.5291	.5782	.6033	.6467	.6698	.6774	.6813
-1.56			.3360	.4503	.5037	.5569	.5839	.6304	.6551	.6633	.6676
-1.58			.2869	.4188	.4778	.5355	.5645	.6142	.6407	.6496	.6543
-1.60			.2317	.3859	.4513	.5140	.5452	.5983	.6266	.6362	.6412
-1.62			.1689	.3513	.4241	.4923	.5259	.5826	.6128	.6230	.6285
-1.64			.0991	.3145	.3961	.4705	.5065	.5670	.5992	.6101	.6161

TABLE 12.5 Continued

z_1 \ n	5	10	20	25	30	40	50	100	250	500	1000
						δ = 2.5					
-1.66			.0285	.2750	.3671	.4483	.4871	.5516	.5858	.5975	.6039
-1.68				.2322	.3369	.4259	.4675	.5363	.5727	.5852	.5920
-1.70				.1855	.3052	.4030	.4479	.5211	.5597	.5730	.5803
-1.72				.1345	.2718	.3797	.4281	.5061	.5470	.5611	.5689
-1.74				.0803	.2362	.3558	.4081	.4911	.5345	.5494	.5578
-1.76				.0247	.1983	.3313	.3878	.4763	.5221	.5380	.5468
-1.78					.1576	.3061	.3673	.4615	.5099	.5267	.5361
-1.80					.1141	.2799	.3464	.4468	.4978	.5156	.5256
-1.82					.0684	.2527	.3251	.4321	.4859	.5047	.5152
-1.84					.0210	.2244	.3033	.4175	.4742	.4939	.5051
-1.86						.1946	.2809	.4028	.4626	.4833	.4951
-1.88						.1634	.2579	.3882	.4511	.4729	.4853
-1.90						.1305	.2342	.3736	.4397	.4626	.4757
-1.92						.0959	.2097	.3589	.4285	.4525	.4662
-1.94						.0596	.1841	.3442	.4173	.4425	.4569
-1.96						.0213	.1576	.3294	.4063	.4326	.4477
-1.98							.1298	.3145	.3953	.4229	.4387
-2.00							.1008	.2995	.3844	.4132	.4298
-2.02							.0704	.2844	.3736	.4037	.4211
-2.04							.0383	.2691	.3628	.3943	.4124
-2.06								.2536	.3521	.3850	.4039
-2.08								.2379	.3415	.3758	.3955
-2.10								.2220	.3309	.3667	.3872
-2.12								.2058	.3203	.3576	.3790
-2.14								.1893	.3098	.3487	.3709
-2.16								.1724	.2993	.3398	.3629
-2.18								.1551	.2888	.3309	.3550
-2.20								.1373	.2783	.3222	.3472
-2.22								.1190	.2678	.3135	.3394
-2.24								.1000	.2573	.3048	.3318
-2.26								.0803	.2467	.2962	.3242
-2.28								.0595	.2361	.2877	.3167
-2.30								.0374	.2255	.2791	.3092
-2.32									.2148	.2707	.3019
-2.34									.2041	.2622	.2945
-2.36									.1932	.2537	.2873
-2.38									.1823	.2453	.2800
-2.40									.1712	.2369	.2729
-2.42									.1600	.2285	.2657
-2.44									.1486	.2200	.2586
-2.46									.1370	.2116	.2516
-2.48									.1252	.2032	.2446
-2.50									.1131	.1947	.2376
-2.52									.1007	.1861	.2306
-2.54									.0879	.1776	.2236
-2.56									.0746	.1689	.2167
-2.58									.0607	.1602	.2098
-2.60									.0459	.1515	.2029
-2.62									.0301	.1426	.1959
-2.64										.1336	.1890
-2.66										.1244	.1820
-2.68										.1151	.1751
-2.70										.1056	.1681
-2.72										.0959	.1611
-2.74										.0859	.1540
-2.76										.0755	.1469
-2.78										.0647	.1397
-2.80										.0535	.1324

TABLE 12.5 Continued

$\delta = 3.0$

z_1 \ n	5	10	20	25	30	40	50	100	250	500	1000
-.48	2.8744										
-.50	2.7464										
-.52	2.6278										
-.54	2.5176										
-.56	2.4148	2.4152									
-.58	2.3186	2.3194									
-.60	2.2283	2.2296									
-.62	2.1432	2.1453	2.1455								
-.64	2.0628	2.0661	2.0663	2.0663							
-.66	1.9865	1.9913	1.9916	1.9916	1.9916						
-.68	1.9138	1.9206	1.9211	1.9212	1.9212						
-.70	1.8442	1.8536	1.8545	1.8545	1.8545	1.8545					
-.72	1.7774	1.7900	1.7913	1.7913	1.7914	1.7914	1.7914				
-.74	1.7129	1.7294	1.7313	1.7314	1.7315	1.7315	1.7315				
-.76	1.6504	1.6717	1.6743	1.6745	1.6745	1.6746	1.6746				
-.78	1.5895	1.6164	1.6200	1.6202	1.6203	1.6204	1.6205	1.6205			
-.80	1.5299	1.5634	1.5682	1.5685	1.5687	1.5688	1.5689	1.5689			
-.82	1.4712	1.5125	1.5187	1.5192	1.5194	1.5196	1.5197	1.5198			
-.84	1.4131	1.4635	1.4713	1.4720	1.4723	1.4726	1.4727	1.4728			
-.86	1.3551	1.4161	1.4259	1.4268	1.4272	1.4276	1.4278	1.4279	1.4280		
-.88	1.2969	1.3702	1.3823	1.3835	1.3840	1.3845	1.3848	1.3850	1.3850		
-.90	1.2378	1.3256	1.3404	1.3418	1.3426	1.3432	1.3435	1.3439	1.3439		
-.92	1.1773	1.2821	1.3000	1.3018	1.3027	1.3036	1.3040	1.3044	1.3045	1.3045	
-.94	1.1145	1.2397	1.2610	1.2632	1.2644	1.2655	1.2660	1.2666	1.2667	1.2668	
-.96	1.0480	1.1982	1.2233	1.2260	1.2274	1.2288	1.2295	1.2302	1.2304	1.2305	
-.98		1.1574	1.1868	1.1900	1.1918	1.1935	1.1943	1.1953	1.1956	1.1956	1.1956
-1.00		1.1172	1.1514	1.1552	1.1573	1.1594	1.1604	1.1617	1.1620	1.1621	1.1621
-1.02		1.0775	1.1170	1.1215	1.1240	1.1265	1.1277	1.1293	1.1298	1.1298	1.1299
-1.04		1.0381	1.0835	1.0888	1.0917	1.0947	1.0961	1.0981	1.0987	1.0988	1.0988
-1.06		.9990	1.0508	1.0570	1.0604	1.0638	1.0655	1.0680	1.0687	1.0689	1.0689
-1.08		.9598	1.0190	1.0260	1.0299	1.0340	1.0360	1.0389	1.0399	1.0400	1.0401
-1.10		.9206	.9878	.9958	1.0003	1.0050	1.0073	1.0108	1.0120	1.0122	1.0123
-1.12		.8812	.9572	.9663	.9714	.9768	.9795	.9836	.9851	.9854	.9854
-1.14		.8412	.9272	.9374	.9432	.9494	.9525	.9572	.9590	.9594	.9595
-1.16		.8006	.8977	.9092	.9157	.9227	.9262	.9317	.9339	.9343	.9345
-1.18		.7589	.8686	.8815	.8888	.8967	.9007	.9070	.9095	.9101	.9103
-1.20		.7159	.8399	.8542	.8624	.8712	.8758	.8830	.8860	.8867	.8869
-1.22		.6711	.8116	.8275	.8366	.8464	.8515	.8597	.8631	.8640	.8643
-1.24		.6237	.7835	.8011	.8112	.8221	.8278	.8370	.8410	.8420	.8423
-1.26		.5727	.7556	.7751	.7862	.7983	.8046	.8150	.8195	.8207	.8211
-1.28		.5165	.7279	.7493	.7616	.7749	.7819	.7935	.7987	.8000	.8006
-1.30		.4518	.7003	.7239	.7374	.7520	.7597	.7725	.7784	.7800	.7806
-1.32		.3710	.6727	.6986	.7134	.7295	.7379	.7521	.7588	.7605	.7613
-1.34		.6451	.6736	.6897	.7073	.7165	.7322	.7396	.7416	.7425	
-1.36		.6174	.6486	.6663	.6855	.6955	.7127	.7210	.7233	.7243	
-1.38		.5896	.6238	.6431	.6639	.6749	.6937	.7029	.7055	.7066	
-1.40		.5614	.5990	.6200	.6426	.6546	.6751	.6852	.6881	.6894	
-1.42		.5330	.5742	.5970	.6216	.6346	.6569	.6680	.6712	.6727	
-1.44		.5041	.5493	.5742	.6008	.6148	.6390	.6512	.6548	.6565	
-1.46		.4746	.5243	.5514	.5802	.5953	.6215	.6348	.6388	.6407	
-1.48		.4444	.4991	.5286	.5598	.5761	.6043	.6187	.6232	.6253	
-1.50		.4132	.4737	.5058	.5395	.5570	.5874	.6031	.6080	.6103	
-1.52		.3809	.4480	.4829	.5193	.5382	.5709	.5878	.5931	.5958	
-1.54		.3470	.4219	.4599	.4992	.5195	.5546	.5728	.5786	.5815	
-1.56		.3112	.3952	.4367	.4792	.5010	.5385	.5582	.5645	.5677	
-1.58		.2727	.3679	.4134	.4592	.4826	.5228	.5439	.5507	.5542	
-1.60		.2305	.3398	.3897	.4392	.4643	.5072	.5298	.5372	.5410	
-1.62		.1827	.3106	.3657	.4192	.4461	.4919	.5161	.5240	.5281	
-1.64		.1257	.2803	.3412	.3992	.4279	.4768	.5025	.5111	.5155	
-1.66		.0487	.2482	.3161	.3790	.4098	.4618	.4893	.4984	.5033	

TABLE 12.5 Continued

$\delta = 3.0$

z_1 \ n	5	10	20	25	30	40	50	100	250	500	1000
-1.68				.2141	.2904	.3587	.3918	.4471	.4763	.4861	.4913
-1.70				.1771	.2638	.3383	.3737	.4325	.4635	.4740	.4796
-1.72				.1360	.2362	.3176	.3556	.4181	.4510	.4621	.4681
-1.74				.0885	.2072	.2967	.3375	.4039	.4386	.4505	.4569
-1.76				.0303	.1765	.2754	.3192	.3897	.4265	.4390	.4459
-1.78					.1435	.2538	.3009	.3757	.4146	.4278	.4351
-1.80					.1075	.2316	.2824	.3619	.4028	.4169	.4246
-1.82					.0672	.2088	.2637	.3481	.3913	.4061	.4143
-1.84					.0207	.1853	.2448	.3344	.3799	.3955	.4042
-1.86						.1609	.2257	.3208	.3686	.3851	.3943
-1.88						.1354	.2062	.3073	.3575	.3749	.3846
-1.90						.1085	.1863	.2939	.3466	.3648	.3750
-1.92						.0799	.1659	.2805	.3358	.3549	.3657
-1.94						.0491	.1449	.2671	.3252	.3452	.3565
-1.96							.1232	.2538	.3147	.3356	.3475
-1.98							.1007	.2405	.3043	.3262	.3386
-2.00							.0773	.2272	.2940	.3169	.3299
-2.02							.0526	.2140	.2839	.3078	.3214
-2.04							.0268	.2007	.2739	.2987	.3129
-2.06								.1873	.2639	.2899	.3047
-2.08								.1739	.2541	.2811	.2965
-2.10								.1605	.2444	.2725	.2886
-2.12								.1470	.2347	.2639	.2807
-2.14								.1334	.2251	.2555	.2729
-2.16								.1197	.2157	.2472	.2653
-2.18								.1058	.2063	.2390	.2578
-2.20								.0918	.1969	.2309	.2504
-2.22								.0776	.1877	.2229	.2431
-2.24								.0633	.1785	.2150	.2359
-2.26								.0487	.1693	.2072	.2288

TABLE 12.5 Continued

$\delta = 3.5$

z_1 \ n	5	10	20	25	30	40	50	100	250	500	1000
-.48	2.7995										
-.50	2.6730										
-.52	2.5558										
-.54	2.4468										
-.56	2.3451	2.3454									
-.58	2.2499	2.2505									
-.60	2.1606	2.1616									
-.62	2.0764	2.0780									
-.64	1.9968	1.9994	1.9995								
-.66	1.9214	1.9252	1.9254	1.9254							
-.68	1.8495	1.8550	1.8554	1.8554	1.8554						
-.70	1.7808	1.7885	1.7891	1.7892	1.7892	1.7892					
-.72	1.7149	1.7253	1.7263	1.7263	1.7264	1.7264					
-.74	1.6514	1.6651	1.6666	1.6667	1.6667	1.6667	1.6667				
-.76	1.5900	1.6078	1.6098	1.6099	1.6100	1.6100	1.6100				
-.78	1.5303	1.5529	1.5557	1.5559	1.5560	1.5560	1.5561				
-.80	1.4719	1.5003	1.5041	1.5043	1.5045	1.5046	1.5046	1.5046			
-.82	1.4146	1.4497	1.4547	1.4551	1.4553	1.4554	1.4555	1.4555			
-.84	1.3581	1.4010	1.4074	1.4080	1.4082	1.4084	1.4085	1.4086			
-.86	1.3020	1.3540	1.3621	1.3628	1.3632	1.3635	1.3636	1.3637			
-.88	1.2459	1.3085	1.3186	1.3195	1.3200	1.3204	1.3205	1.3207	1.3207		
-.90	1.1894	1.2644	1.2768	1.2779	1.2785	1.2790	1.2793	1.2795	1.2795		
-.92	1.1320	1.2215	1.2364	1.2379	1.2387	1.2393	1.2396	1.2400	1.2400		
-.94	1.0731	1.1796	1.1975	1.1994	1.2003	1.2012	1.2016	1.2020	1.2021	1.2021	
-.96	1.0117	1.1386	1.1599	1.1622	1.1633	1.1644	1.1649	1.1655	1.1657	1.1657	
-.98		1.0985	1.1235	1.1262	1.1276	1.1290	1.1296	1.1304	1.1306	1.1306	
-1.00		1.0590	1.0882	1.0914	1.0931	1.0948	1.0956	1.0966	1.0969	1.0969	1.0969
-1.02		1.0201	1.0539	1.0577	1.0598	1.0618	1.0628	1.0640	1.0644	1.0644	1.0644
-1.04		.9816	1.0206	1.0250	1.0275	1.0299	1.0310	1.0326	1.0331	1.0331	1.0331
-1.06		.9434	.9881	.9932	.9961	.9990	1.0003	1.0023	1.0029	1.0029	1.0030
-1.08		.9054	.9564	.9623	.9656	.9690	.9706	.9730	.9737	.9738	.9739
-1.10		.8675	.9254	.9322	.9360	.9399	.9418	.9446	.9456	.9457	.9458
-1.12		.8295	.8950	.9028	.9071	.9117	.9139	.9172	.9183	.9186	.9186
-1.14		.7912	.8653	.8740	.8790	.8842	.8868	.8906	.8920	.8923	.8924
-1.16		.7525	.8360	.8459	.8515	.8574	.8604	.8649	.8666	.8669	.8670
-1.18		.7132	.8073	.8184	.8247	.8313	.8347	.8399	.8419	.8424	.8425
-1.20		.6730	.7790	.7913	.7984	.8059	.8097	.8157	.8181	.8186	.8188
-1.22		.6315	.7510	.7648	.7726	.7810	.7853	.7922	.7949	.7955	.7958
-1.24		.5885	.7234	.7386	.7473	.7567	.7615	.7693	.7725	.7732	.7735
-1.26		.5432	.6960	.7129	.7225	.7329	.7383	.7470	.7507	.7516	.7519
-1.28		.4948	.6689	.6874	.6981	.7096	.7155	.7253	.7295	.7306	.7310
-1.30		.4420	.6419	.6623	.6740	.6867	.6933	.7042	.7090	.7102	.7107
-1.32		.3823	.6150	.6375	.6503	.6642	.6715	.6835	.6890	.6904	.6910
-1.34			.5882	.6129	.6269	.6421	.6501	.6634	.6696	.6712	.6718
-1.36			.5614	.5884	.6038	.6204	.6291	.6438	.6506	.6525	.6532
-1.38			.5346	.5641	.5809	.5990	.6085	.6246	.6322	.6343	.6352
-1.40			.5077	.5400	.5582	.5779	.5882	.6058	.6142	.6166	.6176
-1.42			.4805	.5158	.5357	.5570	.5683	.5874	.5967	.5994	.6006
-1.44			.4531	.4918	.5133	.5364	.5486	.5694	.5797	.5826	.5840
-1.46			.4254	.4677	.4910	.5161	.5292	.5518	.5630	.5663	.5678
-1.48			.3972	.4435	.4689	.4959	.5101	.5345	.5467	.5504	.5521
-1.50			.3685	.4193	.4467	.4760	.4912	.5175	.5308	.5348	.5367
-1.52			.3390	.3948	.4247	.4561	.4726	.5009	.5153	.5197	.5218
-1.54			.3086	.3702	.4025	.4365	.4541	.4845	.5001	.5049	.5073
-1.56			.2770	.3452	.3804	.4169	.4358	.4684	.4852	.4905	.4931
-1.58			.2439	.3198	.3581	.3974	.4177	.4526	.4707	.4764	.4792
-1.60			.2088	.2940	.3357	.3780	.3997	.4370	.4564	.4626	.4657
-1.62			.1710	.2676	.3131	.3587	.3819	.4217	.4425	.4492	.4525
-1.64			.1292	.2404	.2902	.3394	.3642	.4066	.4288	.4360	.4397
-1.66			.0808	.2123	.2670	.3200	.3466	.3917	.4154	.4231	.4271

TABLE 12.5 Continued

$\delta = 3.5$

z_1 \ n	5	10	20	25	30	40	50	100	250	500	1000
-1.68				.1830	.2434	.3007	.3290	.3770	.4022	.4105	.4148
-1.70				.1522	.2194	.2813	.3115	.3625	.3893	.3982	.4028
-1.72				.1192	.1947	.2617	.2941	.3482	.3766	.3861	.3911
-1.74				.0833	.1693	.2421	.2767	.3340	.3641	.3742	.3796
-1.76				.0428	.1429	.2223	.2592	.3200	.3519	.3626	.3683
-1.78					.1153	.2023	.2418	.3062	.3399	.3512	.3573
-1.80					.0861	.1820	.2243	.2925	.3280	.3401	.3466
-1.82					.0547	.1614	.2068	.2790	.3164	.3291	.3360
-1.84						.1404	.1892	.2656	.3049	.3184	.3257
-1.86						.1189	.1714	.2523	.2937	.3078	.3156
-1.88						.0969	.1535	.2391	.2826	.2974	.3057
-1.90						.0741	.1354	.2260	.2716	.2873	.2959
-1.92						.0503	.1171	.2130	.2608	.2773	.2864
-1.94						.0253	.0984	.2001	.2502	.2674	.2771
-1.96							.0795	.1873	.2397	.2578	.2679
-1.98							.0601	.1745	.2294	.2483	.2589
-2.00							.0401	.1618	.2192	.2389	.2500
-2.02								.1492	.2092	.2297	.2414
-2.04								.1366	.1992	.2207	.2328
-2.06								.1241	.1894	.2118	.2245
-2.08								.1116	.1798	.2030	.2163
-2.10								.0991	.1702	.1944	.2082
-2.12								.0867	.1608	.1859	.2002
-2.14								.0743	.1514	.1775	.1924
-2.16								.0619	.1422	.1693	.1848
-2.18								.0494	.1331	.1612	.1772
-2.20								.0370	.1241	.1532	.1698
-2.22								.0246	.1152	.1453	.1625
-2.24									.1064	.1375	.1553
-2.26									.0976	.1298	.1483
-2.28									.0890	.1222	.1413
-2.30									.0805	.1148	.1345
-2.32									.0720	.1074	.1277
-2.34									.0637	.1002	.1211
-2.36									.0554	.0930	.1146
-2.38									.0472	.0860	.1082
-2.40									.0391	.0790	.1019
-2.42									.0311	.0722	.0956
-2.44									.0232	.0654	.0895
-2.46										.0587	.0835
-2.48										.0521	.0776
-2.50										.0456	.0717
-2.52										.0392	.0660
-2.54										.0329	.0603
-2.56										.0267	.0547
-2.58										.0206	.0492
-2.60											.0438
-2.62											.0385
-2.64											.0333
-2.66											.0282
-2.68											.0231

calculations for $\alpha_3^{(2)}$. We need two values, δ_i and δ_j, in a sufficiently narrow interval such that

$$\alpha_3^{(i)} \gtrless a_3 \gtrless \alpha_3^{(j)} \tag{12.5.8}$$

We then interpolate between δ_i and δ_j for the required estimate $\hat{\delta}$. We substitute this value into the expression given in (12.1.6) for $\alpha_3(X)$, and with $\alpha_3(\hat{\rho}, \hat{\delta}) = a_3$ we solve for $\hat{\rho}$. With $\hat{\delta}$ and $\hat{\rho}$ thus determined, we subsequently calculate $\hat{\beta}$ and $\hat{\gamma}$ from the second and third equations of (12.5.7).

To facilitate the calculation of estimates as described above, we include Table 12.5, which is patterned after those given in Chapter 6 for the gamma distribution for various values of δ. These tables present α_3 as a function of z_1 and n. With an assumed value of δ, i.e., an approximation, we select the appropriate table. With z_1 and n determined from the sample data, we enter the selected table and read or interpolate to obtain a corresponding approximation to α_3. We make additional selections to δ as necessary and continue until we have two values, δ_i and δ_j, and corresponding values $\alpha_3^{(i)}$ and $\alpha_3^{(j)}$ such that the inequality of (12.5.8) is satisfied. Estimates thus calculated should be sufficiently accurate for most practical purposes. When greater precision is required, additional values of the cdf can be calculated as needed to permit interpolation for $\hat{\delta}$ over a smaller interval.

Tables of the standardized cdf, $(H(z; 0, 1, \delta, \alpha_3))$, for the generalized gamma distribution are included with the appendixes as Table A.3.5.

12.6 ILLUSTRATIVE EXAMPLES

In order to illustrate simplified computational procedures for the MME described in Section 12.5, we have chosen the following example.

Example 12.6.1. Sample data were compiled by the Electrical Testing Laboratories, New York, New York, of the monthly use of electricity in 282 medium-class homes in an unnamed eastern city. This sample was employed by Croxton and Cowden (1939, p. 294) to illustrate graduation in a lognormal distribution. Data for this sample are tabulated below.

Kilowatt Hours of Electricity Used per Month in Medium-Class Homes in an Eastern City

Kilowatt hours (mid-values)	Number of homes
10	25
14	50
18	53
22	48
26	36
30	26
34	19
38	8
42	6
46	3
50	4
54	2
58	2
Total	282

A summary of these data follows: $n = 282$, $\bar{x} = 22.86685$, $s = 9.67167$, $x_1 = 8.00$, $a_3 = 1.0926$, $a_4 = 4.2948$, and $z_1 = (x_1 - \bar{x})/s = -1.5372$. In order to calculate MME, we enter Table 12.5 ($\delta = 1$) with $z_1 = -1.5372$ and $n = 282$ and interpolate to obtain $\alpha_3(1) = 1.1783$. We repeat this procedure with $\delta = 1.5$ and interpolate to read $\alpha_3(1.5) = 0.9753$. Since $\alpha_3(1.5) < a_3 < \alpha_3(1)$, we interpolate as follows to obtain $\hat{\delta} = 1.21$:

δ	α_3
1.00	1.1783
1.21	1.0926
1.50	0.9753

With $\hat{\delta} = 1.21$ and $\alpha_3 = 1.0926$, we enter Table 12.1 and interpolate to obtain $\hat{\rho} = 1.89$.

With $\hat{\delta} = 1.21$ and $\hat{\rho} = 1.89$, we use entries for $\Gamma(\ )$ from Abramowitz and Stegun (1964) to calculate $G_1(1.89, 1.21) = 1.6330697$ and $G_2(1.89, 1.21) = 3.6368429$. We then calculate estimates of β and γ by substituting these values into the last two equations of (12.3.4). We thus obtain $\hat{\beta} = 9.82$ and $\hat{\gamma} = 6.83$. Final estimates are then $\hat{\gamma} = 6.83$, $\hat{\beta} = 9.82$, $\hat{\rho} = 1.89$, and $\hat{\delta} = 1.21$.

Moment estimates for Example 12.6.1 are calculated as described in Section 12.3. We enter Table 12.4 with $a_3 = 1.0926$ and $a_4 = 4.2948$ and interpolate to obtain $\delta^* = 1.726$. We then enter Table 12.1 with $a_3 = 1.0926$ and $\delta^* = 1.726$ and interpolate to obtain $\rho^* = 0.642$. Estimates β^* and γ^* are calculated from the last two equations of (12.3.4) as in the calculation of the MME. In this instance, $G_1(0.642, 1.726) = 0.6510397$ and $G_2(0.642, 1.726) = 0.6653091$. Final ME are $\gamma^* = 10.053$, $\beta^* = 19.683$, $\rho^* = 0.642$, and $\delta^* = 1.726$, which are to be compared with the MME 6.83, 9.82, 1.89, and 1.21. Note that $\gamma^* > x_1$; unless this is due to the presence of outliers in the sample, $\gamma^* = 10.053$ would be considered inadmissible. The occurrence of ME of γ in excess of x_1 is not uncommon. In this instance it serves to strengthen our preference for the MME over the ME.

Example 12.6.2. We have chosen simulated test results to represent the life in hours of a certain type of energy cell. A summary of these results follows: $n = 500$, $\bar{x} = 30.6728$, $s = 8.2417$, $x_1 = 20.0987$, $a_3 = 1.2543$, $a_4 = 4.5136$, and $z_1 = -1.2830$. In order to calculate MME we enter Table 12.5 ($\delta = 1$) with $z_1 = -1.283$ and $n = 500$ and interpolate to obtain $\alpha_3(1) = 1.2339$. We repeat this procedure with $\delta = 1.5$ and interpolate to obtain $\alpha_3(1.5) = 1.5270$. Interpolation between these two values yields $\delta = 1.47$, which corresponds to $a_3 = 1.2543$. With $\delta = 1.47$ and $\alpha_3 = 1.2543$, we enter Table 9.1 and interpolate to obtain $\hat{\rho} = 0.80$. With $\hat{\delta} = 1.47$ and $\hat{\rho} = 0.80$, we employ the tables of $\Gamma(\ )$ given by Abramowitz and Stegun (1964) to calculate $G_1(1.47, 0.80) = 0.76080$ and $G_2(1.47, 0.80) = 0.92683$. We then substitute these values into the last two equations of (12.3.4) to calculate $\hat{\beta} = 15.187$ and $\hat{\gamma} = 17.221$. Thus, for this example, MME are $\hat{\gamma} = 17.221$, $\hat{\beta} = 15.187$, $\hat{\rho} = 0.80$, and $\hat{\delta} = 1.47$.

Moment estimates calculated for Example 12.6.2 are $\gamma^* = 21.695$, $\beta^* = 21.164$, $\rho^* = 0.33$, and $\delta^* = 2.064$. Values of G_1 and G_2 involved in these calculations are $G_1(0.33, 2.064) = 0.4241744$ and $G_2(0.33, 2.064) = 0.331568$. For this example γ^* is slightly larger than x_1, but the magnitude of the difference is small enough that it might be attributed to lack of precision in the calculations. Nevertheless, for this example, as for Example 12.6.1, there is good reason to prefer the MME over the ME.

12.7 CENSORED SAMPLES

Since our concern in this volume is primarily with models for life-spans, our interest in censored samples is limited to those that are right censored. In the most general situation, we are dealing with a progressively censored sample consisting of a total of N randomly selected specimens from a population with a four-parameter generalized gamma distribution. Censoring occurs at times $\{T_j\}$, $j = 1, 2, \ldots, k$. For Type I censoring, the T_j are fixed and the number of survivors at these times are random variables. For Type II censoring, the T_j coincide with failure times and are thus random variables, whereas the number of survivors at these times are fixed. In both types, c_j designates the number of censored observations at time T_j. Let n_0 designate the number of failures (uncensored observations) prior to T_1. Let n_j designate the number of failures between time T_j and T_{j+1}. The final stage of censoring occurs at time T_k with c_k survivors, which are then censored. Let n designate the total number of failures (i.e., complete life-spans). Let c designate the total number of censored observations. It follows that

$$N = n + c, \quad n = \sum_{j=0}^{k-1} n_j, \quad c = \sum_{1}^{k} c_j \tag{12.7.1}$$

Of course, c_j cannot exceed the number of survivors at time T_j. Single-stage censoring then becomes a special case in which $k = 1$, $n_0 = n$, $c_1 = c$. The likelihood function of a progressively censored sample as described is

$$L = \frac{C\delta^n}{[\Gamma(\rho)]^n \theta^{n\rho}} \exp\left\{-\frac{1}{\theta}\sum_{i=1}^{n}(x_i - \gamma)^\delta\right\} \prod_{i=1}^{n}(x_i - \gamma)^{\rho\delta-1} \prod_{j=1}^{k}[1 - F(T_j)]^{c_j} \tag{12.7.2}$$

where

$$C = N! \Big/ \sum_{j=1}^{k} c_j!$$

and its logarithm is

$$\ln L = n \ln \delta - n \ln \Gamma(\rho) - n\rho \ln \theta + (\rho\delta - 1)\sum_{i=1}^{n} \ln(x_i - \gamma)$$

$$-\frac{1}{\theta}\sum_{i=1}^{n}(x_i - \gamma)^\delta + \sum_{j=1}^{k} c_j \ln(1 - F_j) + \ln C \tag{12.7.3}$$

12.7.1 Maximum Likelihood Estimators for Censored Samples

When the distribution is bell-shaped—that is, when $\rho\delta > 1$—MLE equations are obtained by differentiating (12.7.3) with respect to γ, θ, ρ, δ and equating the derivatives to zero. We thus obtain

$$\frac{\partial \ln L}{\partial \gamma} = -(\rho\delta - 1)\sum_{i=1}^{n}\frac{1}{x_i - \gamma} + \frac{\delta}{\theta}\sum_{i=1}^{n}(x_i - \gamma)^{\delta-1} - \sum_{j=1}^{k} c_j \frac{1}{1 - F_j}\frac{\partial F_j}{\partial \gamma} = 0$$

$$\frac{\partial \ln L}{\partial \theta} = -\frac{n\rho}{\theta} + \frac{1}{\theta^2}\sum_{i=1}^{n}(x_i - \gamma)^\delta - \sum_{j=1}^{k} c_j \frac{1}{1 - F_j}\frac{\partial F_j}{\partial \theta} = 0$$

$$\frac{\partial \ln L}{\partial \rho} = -n[\psi(\rho) + \ln \theta] + \delta \sum_{i=1}^{n} \ln (x_i - \gamma) - \sum_{j=1}^{k} c_j \frac{1}{1 - F_j}\frac{\partial F_j}{\partial \rho} = 0 \qquad (12.7.4)$$

$$\frac{\partial \ln L}{\partial \delta} = \frac{n}{\delta} + \rho \sum_{i=1}^{n} \ln (x_i - \gamma) - \frac{1}{\theta}\sum_{i=1}^{n}(x_i - \gamma)^\delta \ln (x_i - \gamma)$$

$$- \sum_{j=1}^{k} c_j \frac{1}{1 - F_j}\frac{\partial F_j}{\partial \delta} = 0$$

where $\psi(\rho) = d \ln \Gamma(\rho)/d\rho = \Gamma'(\rho)/\Gamma(\rho)$ is the digamma function.

Evaluation of the partial derivatives of F_j with respect to the parameters involves differentiation under an integral sign, where the integrand is a function of the parameters. Our involvement is with a function of a type defined by the definite integral

$$\phi(\alpha) = \int_a^b f(x; \alpha)\, dx \qquad (12.7.5)$$

where a and/or b may or may not be functions of α. The applicable derivative is

$$\frac{\partial \phi}{\partial \alpha} = \int_a^b \frac{\partial f(x; \alpha)}{\partial \alpha}\, dx + f(b; \alpha)\frac{\partial b}{\partial \alpha} - f(a; \alpha)\frac{\partial a}{\partial \alpha} \qquad (12.7.6)$$

In various special cases, one, and sometimes two, of the three terms in (12.7.6) will vanish. The first term vanishes when f(x) does not involve α. The second and/or third terms vanish when $f(b) = 0$ and/or $f(a) = 0$, or when b and/or a are not functions of α. The simplest case is that in which the integrand does not involve α, $f(a) = 0$, and

$$\frac{\partial \phi}{\partial \alpha} = f(b)\frac{\partial b}{\partial \alpha} \qquad (12.7.7)$$

In order to simplify evaluation of the partials of F_j, we make the transformation

$$U = \frac{(X - \gamma)^\delta}{\theta} \qquad (12.7.8)$$

With this transformation, the pdf of U follows from (12.3.1) as that of the one-parameter gamma distribution, which we write as

$$g(u;\rho) = \frac{u^{\rho-1}e^{-u}}{\Gamma(\rho)}, \qquad 0 < u < \infty$$
$$= 0 \qquad \text{elsewhere} \tag{12.7.9}$$

Let ω_j designate the transformed point of censoring corresponding to T_j. We thus have

$$\omega_j = \frac{(T_j - \gamma)^\delta}{\theta} \tag{12.7.10}$$

The expected value and the variance of U are

$$E(U) = \rho \quad \text{and} \quad V(U) = \rho \tag{12.7.11}$$

We standardize the distribution of U with the transformation $Z = [U - E(U)]/\sqrt{V(U)}$. Accordingly, it follows that

$$Z = \frac{(X-\gamma)^\delta - \rho\theta}{\theta\sqrt{\rho}} \tag{12.7.12}$$

Thus Z is distributed with a mean of zero, unit variance, and with shape parameters ρ and δ. To be consistent with the notation employed in previous chapters, we let ξ_j designate the standardized value of the jth censoring point. It follows from (12.7.12) that

$$\xi_j = \frac{(T_j - \gamma)^\delta - \rho\theta}{\theta\sqrt{\rho}} \tag{12.7.13}$$

Furthermore, the cdf at the jth point of censoring is

$$F_j = F(T_j) = I_j = I(\omega_j) = H_j = H(\xi_j) \tag{12.7.14}$$

where the pdf $f(x;\gamma, \beta, \rho, \delta)$ is given by (12.3.1), $g(u;\rho)$ by (12.7.9), and $h(z;0, 1, \rho, \delta)$ by (12.5.2). It follows that

$$F_j = \int_\gamma^{T_j} f(x;\gamma, \beta, \rho, \delta)\,dx$$

$$I_j = \int_0^{\omega_j} g(u, \rho)\, du$$

$$H_j = \int_{-E}^{\xi_j} h(z; 0, 1, \rho, \delta)\, dz \tag{12.7.15}$$

where $E = G_1/\sqrt{G_2 - G_1^2}$. From (12.5.5), an alternative expression for H_j is

$$H_j = H(\xi_j; 0, 1, \rho, \delta) = \frac{1}{\Gamma(\rho)} \int_0^{(J\xi_j + G_1)^\delta} w^{\rho-1} \exp(-w)\, dw$$

where $J = \sqrt{G_2 - G_1^2}$, and G_k is as defined by (12.1.5).

Let us define

$$Q(\xi_j) = Q_j = \frac{h(\xi_j)}{1 - H(\xi_j)}$$

$$M(\omega_j) = M_j = \frac{g(\omega_j)}{1 - I(\omega_j)} \tag{12.7.16}$$

In order to evaluate the terms in (12.7.4) which involve partials of F_j, we can replace F_j with either H_j or I_j.

If we choose to employ $H(\xi_j)$, then we need

$$\frac{\partial H_j}{\partial \gamma} = h(\xi_j) \frac{\partial \xi_j}{\partial \gamma} = \frac{-\delta}{\theta\sqrt{\rho}} (T_j - \gamma)^{\delta - 1} h(\xi_j)$$

$$\frac{\partial H_j}{\partial \theta} = h(\xi_j) \frac{\partial \xi_j}{\partial \theta} = -\frac{(T_j - \gamma)^\delta}{\theta^2 \sqrt{\rho}} h(\xi_j)$$

$$\frac{\partial H_j}{\partial \rho} = \int_{-E}^{\xi_j} \frac{\partial h(z; 0, 1, \rho, \delta)}{\partial \rho}\, dz + h(\xi_j; 0, 1, \rho, \delta) \frac{\partial \xi_j}{\partial \rho} \tag{12.7.17}$$

$$\frac{\partial H_j}{\partial \delta} = \int_{-E}^{\xi_j} \frac{\partial h(z; 0, 1, \rho, \delta)}{\partial \delta} + h(\xi_j; 0, 1, \rho, \delta) \frac{\partial \xi_j}{\partial \delta}$$

If we choose to employ $I(\omega_j)$, then we need

$$\frac{\partial I_j}{\partial \gamma} = g(\omega_j)\frac{\partial \omega_j}{\partial \gamma} = -\frac{\delta \omega_j}{T_j - \gamma} g(\omega_j)$$

$$\frac{\partial I_j}{\partial \theta} = g(\omega_j)\frac{\partial \omega_j}{\partial \theta} = -\frac{\omega_j}{\theta} g(\omega_j)$$

$$\frac{\partial I_j}{\partial \rho} = \int_0^{\omega_j} \frac{\partial g(u, \rho)}{\partial \rho} du = -\psi(\rho) I(\omega_j) + \int_0^{\omega_j} (\ln u) g(u)\, du \tag{12.7.18}$$

$$\frac{\partial I_j}{\partial \delta} = \omega_j \ln (T_j - \gamma) g(\omega_j)$$

In an effort to minimize computational difficulties, we elect to use the first two derivatives from (12.7.17) and the last two from (12.7.18). When these results are substituted into (12.7.4), the estimating equations become

$$-(\rho - 1)\sum_{i=1}^{n} (x_i - \gamma)^{-1} + \frac{\delta}{\theta}\sum_{i=1}^{n} (x_i - \gamma)^{\delta-1} + \frac{\delta}{\theta\sqrt{\rho}}\sum_{j=1}^{k} c_j (T_j - \gamma)^{\delta-1} Q_j = 0$$

$$-\frac{n\rho}{\theta} + \frac{1}{\theta^2}\sum_{i=1}^{n} (x_i - \gamma)^{\delta} + \frac{1}{\theta^2\sqrt{\rho}}\sum_{j=1}^{k} c_j (T_j - \gamma)^{\delta} Q_j = 0$$

$$-n[\psi(\rho) + \ln \theta] + \delta \sum_{i=1}^{n} \ln (x_i - \gamma) + \sum_{j=1}^{k} c_j [A_j \psi(\rho) - B_j] = 0 \tag{12.7.19}$$

$$\frac{n}{\delta} + \rho \sum_{i=1}^{n} \ln (x_i - \gamma) - \frac{1}{\theta}\sum_{i=1}^{n} (x_i - \gamma)^{\delta} \ln (x_i - \gamma)$$
$$+ \frac{1}{\theta}\sum_{j=1}^{k} c_j (T_j - \gamma)^{\delta} \ln (T_j - \gamma) M_j = 0$$

where

$$A_j = A(\omega_j, \rho) = \frac{I_j}{1 - I_j}$$

$$B_j = B(\omega_j, \rho) = \frac{1}{1 - I_j}\int_0^{\omega_j} (\ln u) g(u)\, du \tag{12.7.20}$$

TABLE 12.6 The Auxiliary Functions $I(\omega)$ and $A(\omega) = I(\omega)/[1 - I(\omega)]$

ω \ ρ	1.0	1.5	2.0	2.5	3.0	3.5	4.0	4.5
.5	.3935	.1987	.0902	.0374	.0144	.0052	.0018	.0006
	.6487	.2480	.0991	.0389	.0146	.0052	.0018	.0006
.6	.4512	.2470	.1219	.0551	.0231	.0091	.0034	.0012
	.8221	.3280	.1388	.0583	.0237	.0092	.0034	.0012
.7	.5034	.2945	.1558	.0757	.0341	.0144	.0058	.0022
	1.0138	.4174	.1846	.0819	.0353	.0146	.0058	.0022
.8	.5507	.3406	.1912	.0988	.0474	.0214	.0091	.0037
	1.2255	.5166	.2364	.1096	.0498	.0218	.0092	.0037
.9	.5934	.3851	.2275	.1239	.0629	.0299	.0135	.0058
	1.4596	.6262	.2945	.1415	.0671	.0308	.0136	.0058
1.0	.6321	.4276	.2642	.1509	.0803	.0402	.0190	.0085
	1.7183	.7470	.3591	.1777	.0873	.0418	.0194	.0086
1.1	.6671	.4681	.3010	.1792	.0996	.0521	.0257	.0121
	2.0042	.8799	.4306	.2183	.1106	.0549	.0264	.0123
1.2	.6988	.5064	.3374	.2085	.1205	.0656	.0338	.0165
	2.3201	1.0258	.5091	.2635	.1370	.0702	.0349	.0168
1.3	.7275	.5425	.3732	.2386	.1429	.0806	.0431	.0219
	2.6693	1.1858	.5953	.3134	.1667	.0877	.0450	.0224
1.4	.7534	.5765	.4082	.2692	.1665	.0971	.0537	.0283
	3.0552	1.3613	.6897	.3684	.1998	.1076	.0568	.0291
1.5	.7769	.6084	.4422	.3000	.1912	.1150	.0656	.0357
	3.4817	1.5535	.7927	.4286	.2363	.1299	.0703	.0370
1.6	.7981	.6382	.4751	.3308	.2166	.1341	.0788	.0442
	3.9530	1.7639	.9050	.4944	.2766	.1549	.0856	.0462
1.7	.8173	.6660	.5068	.3614	.2428	.1543	.0932	.0537
	4.4739	1.9943	1.0274	.5660	.3206	.1825	.1028	.0567
1.8	.8347	.6920	.5372	.3917	.2694	.1755	.1087	.0643
	5.0496	2.2465	1.1606	.6439	.3687	.2128	.1220	.0687
1.9	.8504	.7161	.5663	.4214	.2963	.1975	.1253	.0759
	5.6859	2.5225	1.3055	.7284	.4210	.2461	.1432	.0822
2.0	.8647	.7385	.5940	.4506	.3233	.2202	.1429	.0886
	6.3891	2.8246	1.4630	.8201	.4778	.2824	.1667	.0972
2.1	.8775	.7593	.6204	.4790	.3504	.2435	.1614	.1022
	7.1662	3.1552	1.6342	.9194	.5393	.3219	.1924	.1139
2.2	.8892	.7786	.6454	.5066	.3773	.2673	.1806	.1168
	8.0250	3.5170	1.8203	1.0269	.6059	.3648	.2205	.1323
2.3	.8997	.7965	.6691	.5334	.4040	.2914	.2007	.1323
	8.9742	3.9130	2.0225	1.1431	.6777	.4111	.2510	.1525
2.4	.9093	.8130	.6916	.5592	.4303	.3156	.2213	.1486
	10.0232	4.3464	2.2421	1.2687	.7553	.4612	.2842	.1746
2.5	.9179	.8282	.7127	.5841	.4562	.3400	.2424	.1657
	11.1825	4.8208	2.4807	1.4045	.8389	.5152	.3200	.1986
2.6	.9257	.8423	.7326	.6080	.4816	.3644	.2640	.1835
	12.4637	5.3402	2.7399	1.5513	.9289	.5734	.3587	.2247
2.7	.9328	.8553	.7513	.6310	.5064	.3887	.2859	.2019
	13.8797	5.9088	3.0215	1.7098	1.0258	.6359	.4004	.2529
2.8	.9392	.8672	.7689	.6529	.5305	.4128	.3081	.2208
	15.4446	6.5313	3.3275	1.8810	1.1301	.7031	.4452	.2834

Top line of each entry is $I(\omega)$ and bottom line is $A(\omega)$.

$$I(\omega) = \int_0^{\omega} [\Gamma(\rho)]^{-1} u^{\rho-1} e^{-u} \, du$$

TABLE 12.6 Continued

ω \ ρ	1.0	1.5	2.0	2.5	3.0	3.5	4.0	4.5
2.9	.9450	.8782	.7854	.6738	.5540	.4367	.3304	.2402
	17.1741	7.2131	3.6600	2.0659	1.2423	.7753	.4934	.3162
3.0	.9502	.8884	.8009	.6938	.5768	.4603	.3528	.2601
	19.0855	7.9598	4.0214	2.2656	1.3630	.8527	.5450	.3515
3.1	.9550	.8977	.8153	.7128	.5988	.4834	.3752	.2803
	21.1980	8.7776	4.4141	2.4814	1.4928	.9357	.6004	.3894
3.2	.9592	.9063	.8288	.7308	.6201	.5061	.3975	.3007
	23.5325	9.6734	4.8411	2.7145	1.6322	1.0247	.6597	.4300
3.3	.9631	.9142	.8414	.7479	.6406	.5283	.4197	.3213
	26.1126	10.6549	5.3053	2.9662	1.7822	1.1201	.7231	.4734
3.4	.9666	.9214	.8532	.7641	.6603	.5500	.4416	.3421
	28.9641	11.7302	5.8100	3.2383	1.9434	1.2222	.7910	.5199
3.5	.9698	.9281	.8641	.7794	.6792	.5711	.4634	.3629
	32.1155	12.9086	6.3590	3.5323	2.1167	1.3317	.8635	.5696
3.6	.9727	.9342	.8743	.7938	.6973	.5916	.4848	.3837
	35.5982	14.2001	6.9561	3.8500	2.3031	1.4488	.9409	.6226
3.7	.9753	.9398	.8838	.8074	.7146	.6115	.5058	.4045
	39.4473	15.6156	7.6058	4.1934	2.5034	1.5743	1.0237	.6791
3.8	.9776	.9450	.8926	.8203	.7311	.6308	.5265	.4251
	43.7012	17.1673	8.3127	4.5648	2.7189	1.7087	1.1120	.7394
3.9	.9798	.9497	.9008	.8324	.7469	.6494	.5468	.4456
	48.4024	18.8684	9.0821	4.9663	2.9506	1.8526	1.2063	.8037
4.0	.9817	.9540	.9084	.8438	.7619	.6674	.5665	.4659
	53.5982	20.7336	9.9196	5.4006	3.1999	2.0067	1.3070	.8721
4.1	.9834	.9579	.9155	.8544	.7762	.6847	.5858	.4859
	59.3403	22.7788	10.8314	5.8704	3.4680	2.1717	1.4144	.9451
4.2	.9850	.9616	.9220	.8645	.7898	.7014	.6046	.5056
	65.6863	25.0218	11.8243	6.3787	3.7565	2.3484	1.5291	1.0227
4.3	.9864	.9649	.9281	.8739	.8026	.7173	.6228	.5250
	72.6998	27.4818	12.9056	6.9288	4.0670	2.5378	1.6514	1.1053
4.4	.9877	.9679	.9337	.8827	.8149	.7327	.6406	.5441
	80.4509	30.1802	14.0835	7.5243	4.4013	2.7406	1.7820	1.1933
4.5	.9889	.9707	.9389	.8909	.8264	.7473	.6577	.5627
	89.0171	33.1403	15.3668	8.1689	4.7611	2.9580	1.9214	1.2869
4.6	.9899	.9733	.9437	.8987	.8374	.7614	.6743	.5810
	98.4843	36.3879	16.7651	8.8670	5.1486	3.1909	2.0703	1.3865
4.7	.9909	.9756	.9482	.9059	.8477	.7748	.6903	.5988
	108.9472	39.9512	18.2890	9.6231	5.5660	3.4405	2.2291	1.4925
4.8	.9918	.9777	.9523	.9126	.8575	.7876	.7058	.6162
	120.5104	43.8612	19.9501	10.4422	6.0156	3.7082	2.3987	1.6053
4.9	.9926	.9797	.9561	.9189	.8667	.7998	.7207	.6331
	133.2898	48.1521	21.7610	11.3298	6.5001	3.9952	2.5798	1.7254
5.0	.9933	.9814	.9596	.9248	.8753	.8114	.7350	.6495
	147.4132	52.8615	23.7355	12.2916	7.0223	4.3030	2.7732	1.8532
5.1	.9939	.9831	.9628	.9302	.8835	.8225	.7487	.6655
	163.0219	58.0306	25.8888	13.3343	7.5853	4.6332	2.9798	1.9892
5.2	.9945	.9845	.9658	.9353	.8912	.8330	.7619	.6809
	180.2722	63.7047	28.2375	14.4648	8.1923	4.9874	3.2005	2.1340

TABLE 12.6 Continued

ω \ ρ	1.0	1.5	2.0	2.5	3.0	3.5	4.0	4.5
5.3	.9950	.9859	.9686	.9401	.8984	.8430	.7746	.6959
	199.3368	69.9338	30.7995	15.6907	8.8470	5.3677	3.4364	2.2881
5.4	.9955	.9871	.9711	.9445	.9052	.8524	.7867	.7103
	220.4064	76.7726	33.5948	17.0204	9.5532	5.7758	3.6884	2.4522
5.5	.9959	.9883	.9734	.9486	.9116	.8614	.7983	.7243
	243.6919	84.2815	36.6449	18.4628	10.3152	6.2140	3.9579	2.6270
5.6	.9963	.9893	.9756	.9524	.9176	.8699	.8094	.7378
	269.4264	92.5267	39.9737	20.0280	11.1376	6.6846	4.2460	2.8132
5.7	.9967	.9903	.9776	.9560	.9232	.8779	.8200	.7507
	297.8674	101.5813	43.6071	21.7266	12.0254	7.1901	4.5541	3.0115
5.8	.9970	.9911	.9794	.9593	.9285	.8855	.8300	.7632
	329.2996	111.5253	47.5735	23.5704	12.9839	7.7333	4.8836	3.2228
5.9	.9973	.9919	.9811	.9624	.9334	.8927	.8396	.7752
	364.0375	122.4470	51.9040	25.5721	14.0190	8.3170	5.2362	3.4480
6.0	.9975	.9926	.9826	.9652	.9380	.8994	.8488	.7867
	402.4288	134.4434	56.6327	27.7457	15.1372	8.9444	5.6136	3.6880
6.1	.9978	.9933	.9841	.9679	.9423	.9058	.8575	.7977
	444.8578	147.6210	61.7969	30.1064	16.3452	9.6190	6.0175	3.9439
6.2	.9980	.9939	.9854	.9703	.9464	.9119	.8658	.8083
	491.7490	162.0974	67.4374	32.6706	17.6506	10.3445	6.4499	4.2168
6.3	.9982	.9944	.9866	.9726	.9502	.9175	.8736	.8184
	543.5719	178.0015	73.5989	35.4566	19.0616	11.1249	6.9130	4.5079
6.4	.9983	.9949	.9877	.9747	.9537	.9229	.8811	.8281
	600.8450	195.4753	80.3304	38.4838	20.5870	11.9645	7.4091	4.8185
6.5	.9985	.9954	.9887	.9766	.9570	.9279	.8882	.8374
	664.1416	214.6750	87.6856	41.7739	22.2364	12.8680	7.9406	5.1498
6.6	.9986	.9958	.9897	.9784	.9600	.9326	.8948	.8462
	734.0952	235.7725	95.7231	45.3501	24.0203	13.8406	8.5101	5.5035
6.7	.9988	.9962	.9905	.9801	.9629	.9371	.9012	.8547
	811.4058	258.9567	104.5073	49.2381	25.9499	14.8877	9.1206	5.8811
6.8	.9989	.9965	.9913	.9816	.9656	.9412	.9072	.8627
	896.8473	284.4357	114.1086	53.4656	28.0378	16.0153	9.7752	6.2843
6.9	.9990	.9968	.9920	.9831	.9680	.9451	.9129	.8704
	991.2747	312.4382	124.6044	58.0632	30.2971	17.2299	10.4772	6.7149
7.0	.9991	.9971	.9927	.9844	.9704	.9488	.9182	.8777
	1095.6332	343.2160	136.0791	63.0639	32.7426	18.5384	11.2301	7.1749
7.1	.9992	.9974	.9933	.9856	.9725	.9523	.9233	.8846
	1210.9671	377.0461	148.6256	68.5039	35.3899	19.9484	12.0380	7.6665
7.2	.9993	.9976	.9939	.9867	.9745	.9555	.9281	.8912
	1338.4308	414.2334	162.3452	74.4228	38.2565	21.4681	12.9049	8.1919
7.3	.9993	.9978	.9944	.9878	.9764	.9585	.9326	.8975
	1479.2999	455.1132	177.3494	80.8635	41.3609	23.1065	13.8355	8.7537
7.4	.9994	.9980	.9949	.9887	.9781	.9613	.9368	.9034
	1634.9844	500.0547	193.7601	87.8733	44.7234	24.8730	14.8346	9.3543
7.5	.9994	.9982	.9953	.9896	.9797	.9640	.9409	.9091
	1807.0424	549.4641	211.7109	95.5033	48.3663	26.7782	15.9075	9.9967
7.6	.9995	.9983	.9957	.9905	.9812	.9665	.9446	.9144
	1997.1959	603.7883	231.3484	103.8098	52.3137	28.8334	17.0599	10.6840

TABLE 12.6 Continued

ω \ p	1.0	1.5	2.0	2.5	3.0	3.5	4.0	4.5
7.7	.9995	.9985	.9961	.9912	.9826	.9688	.9482	.9195
	2207.3480	663.5194	252.8331	112.8539	56.5916	31.0508	18.2980	11.4195
7.8	.9996	.9986	.9964	.9919	.9839	.9710	.9515	.9243
	2439.6020	729.1989	276.3411	122.7025	61.2285	33.4437	19.6285	12.2067
7.9	.9996	.9988	.9967	.9926	.9851	.9730	.9547	.9288
	2696.2823	801.4226	302.0654	133.4285	66.2555	36.0264	21.0586	13.0494
8.0	.9997	.9989	.9970	.9932	.9862	.9749	.9576	.9331
	2979.9580	880.8464	330.2176	145.1118	71.7063	38.8147	22.5960	13.9518
8.1	.9997	.9990	.9972	.9937	.9873	.9766	.9604	.9372
	3293.4681	968.1925	361.0295	157.8394	77.6175	41.8254	24.2491	14.9183
8.2	.9997	.9991	.9975	.9942	.9882	.9783	.9630	.9410
	3639.9503	1064.2554	394.7555	171.7064	84.0292	45.0769	26.0271	15.9537
8.3	.9998	.9991	.9977	.9947	.9891	.9798	.9654	.9446
	4022.8724	1169.9099	431.6745	186.8168	90.9847	48.5891	27.9398	17.0632
8.4	.9998	.9992	.9979	.9951	.9900	.9813	.9677	.9481
	4446.0667	1286.1190	472.0922	203.2841	98.5315	52.3836	29.9978	18.2524
8.5	.9998	.9993	.9981	.9955	.9907	.9826	.9699	.9513
	4913.7688	1413.9427	516.3441	221.2322	106.7210	56.4839	32.2126	19.5271
8.6	.9998	.9994	.9982	.9959	.9914	.9838	.9719	.9543
	5430.6596	1554.5480	564.7979	240.7969	115.6093	60.9154	34.5966	20.8940
8.7	.9998	.9994	.9984	.9962	.9921	.9850	.9738	.9572
	6001.9122	1709.2196	617.8569	262.1261	125.2575	65.7057	37.1633	22.3600
8.8	.9998	.9995	.9985	.9965	.9927	.9861	.9756	.9599
	6633.2440	1879.3719	675.9637	285.3818	135.7322	70.8848	39.9272	23.9325
8.9	.9999	.9995	.9986	.9968	.9932	.9871	.9772	.9624
	7330.9735	2066.5619	739.6034	310.7409	147.1057	76.4852	42.9041	25.6198
9.0	.9999	.9996	.9988	.9971	.9938	.9880	.9788	.9648
	8102.0839	2272.5041	809.3084	338.3967	159.4571	82.5422	46.1110	27.4305
9.1	.9999	.9996	.9989	.9973	.9942	.9889	.9802	.9671
	8954.2927	2499.0861	885.6626	368.5606	172.8723	89.0941	49.5663	29.3741
9.2	.9999	.9996	.9990	.9975	.9947	.9897	.9816	.9692
	9896.1291	2748.3862	969.3068	401.4637	187.4450	96.1827	53.2899	31.4608
9.3	.9999	.9997	.9991	.9977	.9951	.9905	.9828	.9712
	10937.0192	3022.6928	1060.9436	437.3586	203.2771	103.8530	57.3036	33.7015
9.4	.9999	.9997	.9991	.9979	.9955	.9912	.9840	.9731
	12087.3807	3324.5257	1161.3443	476.5214	220.4800	112.1543	61.6306	36.1083
9.5	.9999	.9997	.9992	.9981	.9958	.9918	.9851	.9748
	13358.7268	3656.6594	1271.3549	519.2543	239.1749	121.1398	66.2963	38.6937
9.6	.9999	.9998	.9993	.9982	.9962	.9924	.9862	.9765
	14763.7816	4022.1493	1391.9039	565.8875	259.4937	130.8674	71.3282	41.4718
9.7	.9999	.9998	.9993	.9984	.9965	.9930	.9871	.9780
	16316.6072	4424.3597	1524.0100	616.7820	281.5804	141.4002	76.7558	44.4574
9.8	.9999	.9998	.9994	.9985	.9967	.9935	.9880	.9795
	18032.7449	4866.9959	1668.7912	672.3330	305.5921	152.8065	82.6114	47.6667
9.9	.9999	.9998	.9995	.9986	.9970	.9940	.9889	.9808
	19929.3704	5354.1378	1827.4744	732.9723	331.6996	165.1607	88.9298	51.1170
10.0	1.0000	.9998	.9995	.9988	.9972	.9944	.9897	.9821

TABLE 12.6 Continued

ω \ ρ	5.0	5.5	6.0	6.5	7.0	7.5	8.0	8.5
.5	.0002	.0001	.0000	.0000	.0000	.0000	.0000	.0000
	.0002	.0001	.0000	.0000	.0000	.0000	.0000	.0000
.6	.0004	.0001	.0000	.0000	.0000	.0000	.0000	.0000
	.0004	.0001	.0000	.0000	.0000	.0000	.0000	.0000
.7	.0008	.0003	.0001	.0000	.0000	.0000	.0000	.0000
	.0008	.0003	.0001	.0000	.0000	.0000	.0000	.0000
.8	.0014	.0005	.0002	.0001	.0000	.0000	.0000	.0000
	.0014	.0005	.0002	.0001	.0000	.0000	.0000	.0000
.9	.0023	.0009	.0003	.0001	.0000	.0000	.0000	.0000
	.0023	.0009	.0003	.0001	.0000	.0000	.0000	.0000
1.0	.0037	.0015	.0006	.0002	.0001	.0000	.0000	.0000
	.0037	.0015	.0006	.0002	.0001	.0000	.0000	.0000
1.1	.0054	.0023	.0010	.0004	.0001	.0001	.0000	.0000
	.0055	.0023	.0010	.0004	.0001	.0001	.0000	.0000
1.2	.0077	.0035	.0015	.0006	.0003	.0001	.0000	.0000
	.0078	.0035	.0015	.0006	.0003	.0001	.0000	.0000
1.3	.0107	.0050	.0022	.0010	.0004	.0002	.0001	.0000
	.0108	.0050	.0022	.0010	.0004	.0002	.0001	.0000
1.4	.0143	.0069	.0032	.0014	.0006	.0003	.0001	.0000
	.0145	.0069	.0032	.0014	.0006	.0003	.0001	.0000
1.5	.0186	.0093	.0045	.0021	.0009	.0004	.0002	.0001
	.0189	.0094	.0045	.0021	.0009	.0004	.0002	.0001
1.6	.0237	.0122	.0060	.0029	.0013	.0006	.0003	.0001
	.0243	.0123	.0061	.0029	.0013	.0006	.0003	.0001
1.7	.0296	.0157	.0080	.0039	.0019	.0009	.0004	.0002
	.0305	.0159	.0081	.0040	.0019	.0009	.0004	.0002
1.8	.0364	.0198	.0104	.0052	.0026	.0012	.0006	.0003
	.0378	.0202	.0105	.0053	.0026	.0012	.0006	.0003
1.9	.0441	.0246	.0132	.0069	.0034	.0017	.0008	.0004
	.0461	.0252	.0134	.0069	.0035	.0017	.0008	.0004
2.0	.0527	.0301	.0166	.0088	.0045	.0023	.0011	.0005
	.0556	.0310	.0168	.0089	.0046	.0023	.0011	.0005
2.1	.0621	.0363	.0204	.0111	.0059	.0030	.0015	.0007
	.0662	.0377	.0209	.0113	.0059	.0030	.0015	.0007
2.2	.0725	.0433	.0249	.0139	.0075	.0039	.0020	.0010
	.0782	.0452	.0255	.0140	.0075	.0039	.0020	.0010
2.3	.0838	.0510	.0300	.0170	.0094	.0050	.0026	.0013
	.0914	.0538	.0309	.0173	.0095	.0050	.0026	.0013
2.4	.0959	.0595	.0357	.0207	.0116	.0063	.0033	.0017
	.1060	.0633	.0370	.0211	.0117	.0064	.0033	.0017
2.5	.1088	.0688	.0420	.0248	.0142	.0079	.0042	.0022
	.1221	.0739	.0439	.0254	.0144	.0079	.0043	.0022
2.6	.1226	.0789	.0490	.0295	.0172	.0097	.0053	.0029
	.1397	.0857	.0516	.0304	.0175	.0098	.0054	.0029
2.7	.1371	.0897	.0567	.0347	.0206	.0118	.0066	.0036
	.1589	.0986	.0601	.0359	.0210	.0120	.0067	.0036
2.8	.1523	.1013	.0651	.0405	.0244	.0143	.0081	.0045
	.1797	.1127	.0696	.0422	.0250	.0145	.0082	.0045

TABLE 12.6 Continued

ω \ ρ	5.0	5.5	6.0	6.5	7.0	7.5	8.0	8.5
2.9	.1682	.1136	.0742	.0469	.0287	.0171	.0099	.0056
	.2022	.1282	.0801	.0492	.0296	.0174	.0100	.0056
3.0	.1847	.1266	.0839	.0538	.0335	.0203	.0119	.0068
	.2266	.1450	.0916	.0569	.0347	.0207	.0120	.0069
3.1	.2018	.1403	.0943	.0614	.0388	.0238	.0142	.0083
	.2528	.1632	.1042	.0655	.0404	.0244	.0144	.0083
3.2	.2194	.1546	.1054	.0696	.0446	.0278	.0168	.0099
	.2810	.1829	.1178	.0748	.0467	.0286	.0171	.0100
3.3	.2374	.1695	.1171	.0784	.0510	.0322	.0198	.0118
	.3113	.2041	.1327	.0851	.0537	.0333	.0202	.0120
3.4	.2558	.1850	.1295	.0878	.0579	.0370	.0231	.0140
	.3438	.2269	.1487	.0963	.0614	.0385	.0236	.0142
3.5	.2746	.2009	.1424	.0978	.0653	.0424	.0267	.0165
	.3785	.2514	.1660	.1085	.0698	.0442	.0275	.0167
3.6	.2936	.2173	.1559	.1084	.0733	.0481	.0308	.0192
	.4156	.2777	.1847	.1216	.0791	.0506	.0318	.0196
3.7	.3128	.2342	.1699	.1196	.0818	.0544	.0352	.0222
	.4551	.3058	.2047	.1359	.0891	.0575	.0365	.0228
3.8	.3322	.2514	.1844	.1314	.0909	.0612	.0401	.0256
	.4974	.3358	.2262	.1512	.1000	.0652	.0418	.0263
3.9	.3516	.2689	.1994	.1436	.1005	.0684	.0454	.0294
	.5423	.3678	.2491	.1677	.1117	.0735	.0476	.0303
4.0	.3712	.2867	.2149	.1564	.1107	.0762	.0511	.0335
	.5902	.4019	.2737	.1854	.1244	.0825	.0539	.0346
4.1	.3907	.3047	.2307	.1697	.1214	.0845	.0573	.0379
	.6412	.4383	.2999	.2043	.1381	.0923	.0608	.0394
4.2	.4102	.3229	.2469	.1834	.1325	.0933	.0639	.0428
	.6954	.4769	.3278	.2246	.1528	.1028	.0683	.0447
4.3	.4296	.3412	.2633	.1976	.1442	.1025	.0710	.0480
	.7530	.5180	.3575	.2462	.1685	.1142	.0765	.0504
4.4	.4488	.3597	.2801	.2121	.1564	1123	.0786	.0537
	.8143	.5617	.3891	.2692	.1853	.1265	.0853	.0567
4.5	.4679	.3781	.2971	.2271	.1689	.1225	.0866	.0597
	.8793	.6080	.4226	.2938	.2033	.1396	.0948	.0635
4.6	.4868	.3966	.3142	.2423	.1820	.1332	.0951	.0662
	.9484	.6572	.4582	.3198	.2225	.1536	.1050	.0709
4.7	.5054	.4150	.3316	.2579	.1954	.1443	.1040	.0731
	1.0218	.7093	.4960	.3475	.2428	.1686	.1160	.0789
4.8	.5237	.4333	.3490	.2737	.2092	.1559	.1133	.0805
	1.0997	.7646	.5361	.3769	.2645	.1847	.1278	.0875
4.9	.5418	.4515	.3665	.2898	.2233	.1679	.1231	.0882
	1.1824	.8233	.5785	.4081	.2876	.2017	.1404	.0967
5.0	.5595	.4696	.3840	.3061	.2378	.1803	.1334	.0964
	1.2702	.8854	.6235	.4411	.3120	.2199	.1539	.1067
5.1	.5769	.4875	.4016	.3225	.2526	.1930	.1440	.1050
	1.3634	.9513	.6711	.4760	.3379	.2392	.1683	.1173
5.2	.5939	.5052	.4191	.3391	.2676	.2061	.1551	.1140

TABLE 12.6 Continued

ω \ ρ	5.0	5.5	6.0	6.5	7.0	7.5	8.0	8.5
5.3	.6105	.5226	.4365	.3557	.2829	.2196	.1665	.1234
	1.5673	1.0949	.7747	.5521	.3944	.2814	.1998	.1408
5.4	.6267	.5398	.4539	.3724	.2983	.2334	.1783	.1332
	1.6787	1.1731	.8311	.5935	.4252	.3044	.2171	.1537
5.5	.6425	.5567	.4711	.3892	.3140	.2474	.1905	.1434
	1.7971	1.2560	.8906	.6371	.4576	.3287	.2354	.1675
5.6	.6578	.5733	.4881	.4059	.3297	.2617	.2030	.1540
	1.9227	1.3438	.9537	.6833	.4920	.3545	.2547	.1821
5.7	.6728	.5896	.5050	.4227	.3456	.2762	.2159	.1650
	2.0561	1.4367	1.0203	.7321	.5282	.3817	.2753	.1976
5.8	.6873	.6056	.5217	.4393	.3616	.2910	.2290	.1763
	2.1978	1.5352	1.0907	.7835	.5664	.4104	.2970	.2140
5.9	.7013	.6211	.5381	.4559	.3776	.3059	.2424	.1879
	2.3482	1.6395	1.1651	.8379	.6068	.4407	.3199	.2314
6.0	.7149	.6364	.5543	.4724	.3937	.3210	.2560	.1999
	2.5081	1.7499	1.2438	.8952	.6493	.4727	.3441	.2498
6.1	.7281	.6512	.5702	.4887	.4098	.3362	.2699	.2121
	2.6779	1.8670	1.3269	.9557	.6942	.5064	.3697	.2692
6.2	.7408	.6657	.5859	.5049	.4258	.3515	.2840	.2247
	2.8584	1.9910	1.4148	1.0196	.7415	.5419	.3966	.2898
6.3	.7531	.6797	.6012	.5208	.4418	.3668	.2983	.2375
	3.0502	2.1224	1.5077	1.0870	.7914	.5794	.4250	.3114
6.4	.7649	.6934	.6163	.5366	.4577	.3823	.3127	.2505
	3.2541	2.2616	1.6059	1.1581	.8439	.6188	.4549	.3343
6.5	.7763	.7067	.6310	.5522	.4735	.3977	.3272	.2638
	3.4708	2.4092	1.7097	1.2331	.8993	.6603	.4864	.3584
6.6	.7873	.7195	.6453	.5675	.4892	.4131	.3419	.2773
	3.7014	2.5656	1.8195	1.3122	.9576	.7040	.5196	.3837
6.7	.7978	.7320	.6594	.5826	.5047	.4286	.3567	.2910
	3.9466	2.7315	1.9356	1.3957	1.0190	.7500	.5544	.4104
6.8	.8080	.7441	.6730	.5974	.5201	.4439	.3715	.3048
	4.2075	2.9074	2.0583	1.4838	1.0837	.7984	.5911	.4385
6.9	.8177	.7557	.6863	.6119	.5353	.4593	.3864	.3188
	4.4851	3.0940	2.1881	1.5768	1.1519	.8493	.6297	.4680
7.0	.8270	.7670	.6993	.6262	.5503	.4745	.4013	.3329
	4.7806	3.2920	2.3255	1.6749	1.2237	.9028	.6702	.4990
7.1	.8359	.7779	.7119	.6401	.5651	.4896	.4162	.3471
	5.0952	3.5020	2.4708	1.7785	1.2993	.9592	.7129	.5316
7.2	.8445	.7884	.7241	.6537	.5796	.5046	.4311	.3614
	5.4302	3.7249	2.6245	1.8878	1.3789	1.0184	.7577	.5659
7.3	.8527	.7984	.7360	.6670	.5940	.5194	.4459	.3757
	5.7870	3.9615	2.7873	2.0031	1.4629	1.0808	.8047	.6019
7.4	.8605	.8082	.7474	.6800	.6080	.5341	.4607	.3901
	6.1672	4.2127	2.9595	2.1250	1.5513	1.1463	.8541	.6397
7.5	.8679	.8175	.7586	.6926	.6218	.5486	.4754	.4045
	6.5722	4.4795	3.1419	2.2536	1.6444	1.2152	.9061	.6793
7.6	.8751	.8265	.7693	.7050	.6354	.5629	.4900	.4189

TABLE 12.6 Continued

ω \ ρ	5.0	5.5	6.0	6.5	7.0	7.5	8.0	8.5
7.7	.8819	.8351	.7797	.7169	.6486	.5770	.5044	.4333
	7.4642	5.0640	3.5395	2.5329	1.8460	1.3640	1.0179	.7646
7.8	.8883	.8434	.7897	.7286	.6616	.5909	.5188	.4477
	7.9550	5.3840	3.7562	2.6845	1.9550	1.4443	1.0781	.8105
7.9	.8945	.8513	.7994	.7399	.6743	.6045	.5330	.4619
	8.4784	5.7241	3.9858	2.8447	2.0699	1.5287	1.1413	.8585
8.0	.9004	.8589	.8088	.7509	.6866	.6179	.5470	.4762
	9.0369	6.0856	4.2291	3.0140	2.1911	1.6174	1.2077	.9090
8.1	.9060	.8661	.8178	.7615	.6987	.6311	.5609	.4903
	9.6328	6.4701	4.4871	3.1929	2.3188	1.7109	1.2774	.9619
8.2	.9113	.8731	.8264	.7718	.7104	.6440	.5746	.5043
	10.2688	6.8791	4.7606	3.3821	2.4535	1.8092	1.3507	1.0174
8.3	.9163	.8797	.8347	.7818	.7219	.6567	.5881	.5182
	10.9479	7.3142	5.0506	3.5821	2.5956	1.9126	1.4276	1.0757
8.4	.9211	.8861	.8427	.7914	.7330	.6690	.6013	.5320
	11.6729	7.7771	5.3582	3.7937	2.7454	2.0215	1.5085	1.1368
8.5	.9256	.8921	.8504	.8007	.7438	.6811	.6144	.5456
	12.4474	8.2699	5.6846	4.0175	2.9035	2.1361	1.5934	1.2009
8.6	.9299	.8979	.8578	.8097	.7543	.6930	.6272	.5591
	13.2747	8.7946	6.0310	4.2543	3.0704	2.2568	1.6826	1.2681
8.7	.9340	.9034	.8648	.8183	.7645	.7045	.6398	.5724
	14.1588	9.3533	6.3986	4.5049	3.2465	2.3839	1.7764	1.3386
8.8	.9379	.9087	.8716	.8267	.7744	.7157	.6522	.5855
	15.1036	9.9483	6.7890	4.7701	3.4324	2.5177	1.8749	1.4126
8.9	.9416	.9137	.8781	.8347	.7840	.7267	.6643	.5984
	16.1138	10.5823	7.2035	5.0510	3.6287	2.6586	1.9785	1.4903
9.0	.9450	.9184	.8843	.8425	.7932	.7373	.6761	.6112
	17.1938	11.2578	7.6438	5.3484	3.8360	2.8071	2.0874	1.5717
9.1	.9483	.9229	.8902	.8499	.8022	.7477	.6877	.6237
	18.3490	11.9778	8.1115	5.6635	4.0550	2.9636	2.2019	1.6572
9.2	.9514	.9272	.8959	.8571	.8108	.7578	.6990	.6360
	19.5848	12.7455	8.6086	5.9972	4.2864	3.1285	2.3223	1.7470
9.3	.9544	.9313	.9014	.8640	.8192	.7676	.7100	.6480
	20.9070	13.5640	9.1369	6.3509	4.5309	3.3023	2.4489	1.8412
9.4	.9571	.9352	.9065	.8706	.8273	.7771	.7208	.6599
	22.3221	14.4370	9.6985	6.7258	4.7893	3.4855	2.5820	1.9400
9.5	.9597	.9389	.9115	.8769	.8351	.7863	.7313	.6715
	23.8369	15.3684	10.2958	7.1232	5.0625	3.6787	2.7221	2.0438
9.6	.9622	.9424	.9162	.8830	.8426	.7952	.7416	.6828
	25.4587	16.3621	10.9311	7.5447	5.3513	3.8825	2.8695	2.1528
9.7	.9645	.9457	.9207	.8888	.8498	.8038	.7515	.6939
	27.1956	17.4228	11.6069	7.9917	5.6569	4.0974	3.0247	2.2673
9.8	.9667	.9489	.9250	.8944	.8567	.8122	.7612	.7048
	29.0560	18.5551	12.3260	8.4659	5.9801	4.3242	3.1880	2.3876
9.9	.9688	.9518	.9290	.8997	.8634	.8203	.7706	.7154
	31.0491	19.7641	13.0914	8.9690	6.3220	4.5636	3.3599	2.5139
10.0	.9707	.9547	.9329	.9048	.8699	.8281	.7798	.7258

TABLE 12.7 The Auxiliary Function $B(\omega, \rho)$

ρ \ ω	1.50	2.00	2.50	3.00	3.50	4.00	4.50	5.00	5.50	6.00	6.50	7.00	7.50	8.00	8.50
1.50	-.82225	-.95934	-1.04232	-1.06407	-1.00609	-.83448	-.49229	.11341	1.13251	2.80588	5.52004	9.89517	16.92689	28.21441	46.32828
2.00	-.22685	-.10560	.18286	.70883	1.58499	2.98908	5.19978	8.65328	14.03296	22.41083	35.47168	55.86817	87.78427	137.83056	216.46846
2.50	-.05888	.07696	.36040	.85002	1.63515	2.85240	4.71030	7.52732	11.79063	18.24715	28.04494	42.95318	65.70506	100.53323	154.00729
3.00	-.00895	.09829	.31968	.69816	1.29572	2.20417	3.56023	5.56802	8.53299	12.91354	19.39971	29.03345	43.39299	64.87590	97.13479
3.50	.00426	.08059	.24226	.52106	.95982	1.61958	2.58897	3.99742	6.03467	8.97986	13.24504	19.44056	28.47374	41.69797	61.13856
4.00	.00598	.05741	.17165	.37326	.69281	1.17182	1.86881	2.86736	4.28759	6.30292	9.16469	13.23855	19.05850	27.40727	39.43648
4.50	.00464	.03787	.11660	.26061	.49276	.84235	1.34919	2.06872	3.07906	4.49097	6.46221	9.21842	13.08366	18.52515	26.21923
5.00	.00297	.02363	.07656	.17808	.34612	.60229	.97467	1.50119	2.23414	3.24635	4.63981	6.55810	9.20415	12.86605	17.95440
5.50	.00171	.01407	.04870	.11911	.23988	.42772	.70323	1.09326	1.63388	2.37437	3.38260	4.75264	6.61547	9.15430	12.62646
6.00	.00091	.00804	.03004	.07790	.16374	.30098	.50541	.79674	1.20065	1.75145	2.49555	3.49632	4.84072	6.64867	9.08628
6.50	.00046	.00441	.01797	.04976	.10984	.20933	.36082	.57933	.88378	1.29875	1.85686	2.60191	3.59316	4.91123	6.66634
7.00	.00022	.00234	.01043	.03101	.07229	.14356	.25519	.41910	.64970	.96515	1.38900	1.95231	2.69646	3.67701	4.96892
7.50	.00010	.00120	.00587	.01884	.04660	.09688	.17838	.30086	.47570	.71676	1.04152	1.47256	2.03947	2.78143	3.75064
8.00	.00004	.00059	.00321	.01116	.02939	.06423	.12298	.21383	.34604	.53057	.78076	1.11346	1.55033	2.11962	2.85848
8.50	.00002	.00028	.00171	.00644	.01812	.04178	.08348	.15015	.24955	.39056	.58371	.84192	1.18143	1.62305	2.19378

$$B(\omega,\rho) = \frac{1}{1-I(\omega,\rho)} \int_0^\omega \frac{\ell n u}{\Gamma(\rho)} u^{\rho-1} e^{-u}\, du \qquad I(\omega,\rho) = \int_0^\omega \frac{1}{\Gamma(\rho)} u^{\rho-1} e^{-u}\, du$$

$$\omega_j = \frac{(T_j - \gamma)^\delta}{\theta}$$

and where Q_j and M_j are defined by (12.7.16).

When there is no censoring, $c_j = 0$ for all j and the last term of each of the equations of (12.7.19) vanishes. The resulting equations are identical to the MLE given previously for complete samples.

When $\delta = 1$, the first three equations of (12.7.19) become the applicable estimating equations for censored samples from the three-parameter gamma distribution as given in Chapter 9. When $\rho = 1$, the first, second, and fourth equations of (12.7.19) can be reduced to the applicable estimating equations for censored samples from the three-parameter Weibull distribution as given in Chapter 7.

In the general case where all four of the parameters γ, θ, ρ, δ must be estimated, it is necessary that the four equations of (12.7.19) be solved simultaneously. This can be accomplished by following the same procedures as those previously described for complete samples. In the special case where γ is known, it is only necessary to obtain a simultaneous solution of the last three equations of (12.7.19).

Tables of $A(\omega)$ and $B(\omega)$, which are included as Tables 12.6 and 12.7, respectively, will be helpful in evaluating the third equation of (12.7.19).

Modified Maximum Likelihood Estimators. Estimating equations for MMLE are $\partial \ln L/\partial\theta = 0$, $\partial \ln L/\partial\rho = 0$, $\partial \ln L/\partial\delta = 0$, and $E[F(X_1)] = F(x_1)$. The partial derivative with respect to γ employed for the MLE has been replaced by the equation involving the first-order statistic. Accordingly, with $N = n + \Sigma_{j=1}^{k} c_j$, we have

$$F(x_1; \gamma, \theta, \rho, \delta) = \frac{1}{N+1} \qquad (12.7.21)$$

plus the last three equations of (12.7.19).

The technique for obtaining a simultaneous solution of this system of estimating equations is the same as that employed in calculating MLE. Tables of standardized cdf with mean zero and unit variance for various values of α_3 and δ are included in Appendix A.3 as Table A.3.5. This table should be useful in calculating modified estimates.

12.8 ASYMPTOTIC VARIANCES AND COVARIANCES OF MAXIMUM LIKELIHOOD ESTIMATORS

The asymptotic variance-covariance matrix for MLE when the distribution is bell-shaped ($\rho\delta > 1$) can be obtained by inverting the Fisher information matrix as was done in Section 12.4 for complete samples. Elements of the information matrix are negatives of expected values of second partial derivatives of ln L with respect to the parameters. These expected values are given in Section 12.4 for complete samples. Unfortunately they become quite complex for censored samples, and in this case it is suggested that elements of the information matrix be based on approximations obtained by substituting estimated values of the parameters directly into the second partials. When samples are reasonably large ($n > 50$), these approximations should be satisfactory for most practical applications. As previously noted, these results are valid only if $\rho\delta > 2$ and are not likely to be reliable unless $\rho\delta > 4$. In the special case in which γ is known, the asymptotic 3×3 variance-covariance matrix of the MLE $(\hat{\theta}, \hat{\rho}, \hat{\delta})$ is valid if $\rho\delta > 1$. Again, however, calculations might be unreliable unless $\rho\delta > 2.5$.

The second partial derivatives are as follows:

$$\frac{\partial^2 \ln L}{\partial \gamma^2} = -(\rho\delta - 1) \sum_{i=1}^{n} \frac{1}{(x_i - \gamma)^2} - \frac{\delta}{\theta}(\delta - 1) \sum_{i=1}^{n} (x_i - \gamma)^{\delta-2}$$
$$- \sum_{j=1}^{k} c_j \frac{1}{1 - F_j}\left[\frac{\partial^2 F_j}{\partial \gamma^2} + \frac{1}{1 - F_j}\left(\frac{\partial F_j}{\partial \gamma}\right)^2\right]$$

$$\frac{\partial^2 \ln L}{\partial \theta^2} = \frac{n\rho}{\theta^2} - \frac{2}{\theta^3} \sum_{i=1}^{n} (x_i - \gamma)^{\delta} - \sum_{j=1}^{k} c_j \frac{1}{1 - F_j}\left[\frac{\partial^2 F_j}{\partial \theta^2} + \frac{1}{1 - F_j}\left(\frac{\partial F_j}{\partial \theta}\right)^2\right]$$

$$\frac{\partial^2 \ln L}{\partial \rho^2} = -n\psi'(\rho) - \sum_{j=1}^{k} c_j \frac{1}{1 - F_j}\left[\frac{\partial^2 F_j}{\partial \rho^2} + \frac{1}{1 - F_j}\left(\frac{\partial F_j}{\partial \rho}\right)^2\right]$$

$$\frac{\partial^2 \ln L}{\partial \delta^2} = -\frac{n}{\delta^2} - \frac{1}{\theta} \sum_{i=1}^{n} (x_i - \gamma)^{\delta}[\ln(x_i - \gamma)]^2$$
$$- \sum_{j=1}^{k} c_j \frac{1}{1 - F_j}\left[\frac{\partial^2 F_j}{\partial \delta^2} + \frac{1}{1 - F_j}\left(\frac{\partial F_j}{\partial \delta}\right)^2\right] \tag{12.8.1}$$

$$\frac{\partial^2 \ln L}{\partial \gamma \partial \theta} = \frac{\partial^2 \ln L}{\partial \theta \partial \gamma}$$
$$= -\frac{\delta}{\theta^2} \sum_{i=1}^{n} (x_i - \gamma)^{\delta-1} - \sum_{j=1}^{k} c_j \frac{1}{1 - F_j}\left[\frac{\partial^2 F_j}{\partial \gamma \partial \theta} + \frac{1}{1 - F_j} \frac{\partial F_j}{\partial \gamma} \frac{\partial F_j}{\partial \theta}\right]$$

$$\frac{\partial^2 \ln L}{\partial\gamma\partial\rho} = \frac{\partial^2 \ln L}{\partial\rho\,\partial\gamma}$$

$$= -\delta \sum_{i=1}^{n} \frac{1}{x_i - \gamma} - \sum_{j=1}^{k} c_j \frac{1}{1 - F_j}\left[\frac{\partial^2 F_j}{\partial\rho\,\partial\gamma} + \frac{1}{1 - F_j}\,\frac{\partial F_j}{\partial\rho}\,\frac{\partial F_j}{\partial\gamma}\right]$$

$$\frac{\partial^2 \ln L}{\partial\gamma\partial\delta} = \frac{\partial^2 \ln L}{\partial\delta\,\partial\gamma}$$

$$= -\rho \sum_{i=1}^{n} \frac{1}{x_i - \gamma} + \frac{1}{\theta}\sum_{i=1}^{n} (x_i - \gamma)^{\delta-1} + \frac{\delta}{\theta}\sum_{i=1}^{n} (x_i - \gamma)^{\delta-1} \ln (x_i - \gamma)$$

$$- \sum_{j=1}^{k} c_j \frac{1}{1 - F_j}\left[\frac{\partial^2 F_j}{\partial\delta\,\partial\gamma} + \frac{1}{1 - F_j}\,\frac{\partial F_j}{\partial\delta}\,\frac{\partial F_j}{\partial\gamma}\right]$$

$$\frac{\partial^2 \ln L}{\partial\theta\partial\rho} = \frac{\partial^2 \ln L}{\partial\rho\,\partial\theta} = -\frac{n}{\theta} - \sum_{j=1}^{k} c_j \frac{1}{1 - F_j}\left[\frac{\partial^2 F_j}{\partial\rho\,\partial\theta} + \frac{1}{1 - F_j}\,\frac{\partial F_j}{\partial\rho}\,\frac{\partial F_j}{\partial\theta}\right]$$

$$\frac{\partial^2 \ln L}{\partial\theta\partial\delta} = \frac{\partial^2 \ln L}{\partial\delta\,\partial\theta} = \frac{1}{\theta^2}\sum_{i=1}^{n} (x_i - \gamma)^{\delta} \ln (x_i - \gamma)$$

$$- \sum_{j=1}^{k} c_j \left(\frac{1}{1 - F_j}\right) \frac{\partial^2 F_j}{\partial\delta\,\partial\theta} + \left(\frac{1}{1 - F_j}\right)\left(\frac{\partial F_j}{\partial\delta}\right)\left(\frac{\partial F_j}{\partial\theta}\right)$$

$$\frac{\partial^2 \ln L}{\partial\rho\,\partial\delta} = \frac{\partial^2 \ln L}{\partial\delta\,\partial\rho}$$

$$= \sum_{i=1}^{n} \ln (x_i - \gamma) - \sum_{j=1}^{k} c_j \left(\frac{1}{1 - F_j}\right) \frac{\partial^2 F_j}{\partial\rho\,\partial\delta} + \left(\frac{1}{1 - F_j}\right)\left(\frac{\partial F_j}{\partial\rho}\right)\left(\frac{\partial F_j}{\partial\delta}\right)$$

where $\psi'(\rho) = d\psi(\rho)/d\rho$, and the first partials of F_j (where $F_j = I_j = H_j$) are given in (12.7.17) and (12.7.18).

Since $F(T_j) = I(\omega_j)$, where $\omega_j = (T_j - \gamma)^{\delta}/\theta$, and since $I(\omega_j) = \int_0^{\omega_j} g(u)\,du$, where $g(u) = [\Gamma(\rho)]^{-1}u^{\rho-1}e^{-u}$, it is convenient to express the second partials of F as follows:

$$\frac{\partial^2 F_j}{\partial \gamma^2} = \frac{\delta \omega_j}{(T_j - \gamma)^2} (\delta \omega_j - \rho\delta + 1) g(\omega_j)$$

$$\frac{\partial^2 F_j}{\partial \theta^2} = \frac{\omega_j}{\theta^2} (\rho + 1 - \omega_j) g(\omega_j)$$

$$\frac{\partial^2 F_j}{\partial \rho^2} = I(\omega_j) \left\{-\psi'(\rho) + [\psi(\rho)]^2\right\} - 2\psi(\rho) \int_0^{\omega_j} (\ln u) g(u)\, du + \int_0^{\omega_j} (\ln u)^2 g(u)\, du$$

$$\frac{\partial^2 F_j}{\partial \delta^2} = \omega_j (\rho - \omega_j)[\ln (T_j - \gamma)]^2 g(\omega_j)$$

$$\frac{\partial^2 F_j}{\partial \gamma \partial \theta} = \omega_j (\rho - \omega_j) \frac{\delta}{(T_j - \gamma)} g(\omega_j)$$

$$\frac{\partial^2 F_j}{\partial \gamma \partial \rho} = [-\psi(\rho) + \ln \omega_j] \frac{-\delta \omega_j}{T_j - \gamma} g(\omega_j) \qquad (12.8.2)$$

$$\frac{\partial^2 F_j}{\partial \gamma \partial \delta} = \frac{\omega_j}{T_j - \gamma} [\delta(\omega_j - \rho) \ln (T_j - \gamma) - 1] g(\omega_j)$$

$$\frac{\partial^2 F_j}{\partial \theta \partial \rho} = \frac{-\omega_j}{\theta} [-\psi(\rho) + \ln \omega_j] g(\omega_j)$$

$$\frac{\partial^2 F_j}{\partial \theta \partial \delta} = -\omega_j (\rho - \omega_j) \frac{\ln (T_j - \gamma)}{\theta} g(\omega_j)$$

$$\frac{\partial^2 F_j}{\partial \rho \partial \delta} = \omega_j \ln (T_j - \gamma)[-\psi(\rho) + \ln \omega_j] g(\omega_j)$$

Although the asymptotic variances and covariances of MLE are not strictly applicable to MMLE, various simulation studies have indicated that the MLE variances and covariances closely approximate corresponding variances and covariances of the MMLE.

12.9 AN ILLUSTRATIVE EXAMPLE

A computer-generated progressively censored sample of a life test, originally employed by Hsieh (1977), has been selected to illustrate the practical application of MLE and MMLE. This sample is from a four-parameter population in which $\gamma = 100$, $\rho = 2$, $\delta = 2$, and $\theta = 500$. It represents a life test conducted on 100 units of an electronic device which were placed "on test" at the same time. At time $T_1 = 125.64$, five randomly selected survivors were removed from the test (i.e., censored). At time $T_2 = 140.87$ the test was terminated with 12 survivors. Complete life-spans of the 83 units which failed during the test are tabulated below.

107.63	109.73	110.19	114.17	115.05
115.45	115.98	116.24	117.06	117.31
117.70	117.79	117.97	118.59	119.74
120.63	120.64	120.71	121.40	121.45
122.01	122.71	122.73	122.80	122.87
123.11	123.35	123.59	123.79	123.83
125.17	125.64	125.76	125.86	125.89
126.14	126.86	127.43	127.51	127.67
127.70	127.81	127.99	128.83	128.88
129.16	129.25	129.42	130.42	130.70
130.99	131.10	131.22	131.30	131.36
131.48	131.54	131.67	131.92	132.41
132.51	132.63	132.80	133.04	133.32
133.81	134.19	134.56	134.72	137.21
135.27	135.69	136.33	136.64	137.21
137.27	138.04	138.12	138.88	139.54
140.15	140.16	140.87		

In summary, we have $N = 100$, $n = 83$, $c_1 = 5$, $c_2 = 12$, $T_1 = 125.64$, $T_2 = 140.87$, $x_1 = 107.63$, $\Sigma_{i=1}^{83} x_i = 10,566.04$, and $\bar{x}_{83} = 127.302$. Estimates calculated as described herein are tabulated below.

	Parameter estimates								
Estimators	γ	θ	$\beta = \theta^{1/\delta}$	ρ	δ	E(X)	V(X)	α_3	α_4
MLE	102.67	498.94	23.20	1.731	1.976	131.18	127.38	0.4673	3.097
MMLE	100.63	498.95	22.98	1.992	1.982	131.17	125.56	0.4157	3.071
Population values	100	500	22.36	2	2	129.725	116.43	0.4658	3.0593

Approximate variances and covariances calculated by Hsieh (1977), as described in Section 12.8, are as follows:

$$\text{var}(\hat{\rho}) = 0.1096, \quad \text{var}(\hat{\delta}) = 0.0026, \quad \text{var}(\hat{\gamma}) = 11.488, \quad \text{var}(\hat{\theta}) = 6708.63$$

$$\text{cov}(\hat{\rho}, \hat{\delta}) = 0.0019, \quad \text{cov}(\hat{\rho}, \hat{\gamma}) = 0.9595, \quad \text{cov}(\hat{\rho}, \hat{\theta}) = -2.955$$

$$\text{cov}(\hat{\delta}, \hat{\gamma}) = 0.0468, \quad \text{cov}(\hat{\delta}, \hat{\theta}) = 3.254, \quad \text{cov}(\hat{\gamma}, \hat{\theta}) = -3.837$$

12.10 SOME COMMENTS AND RECOMMENDATIONS

The generalized gamma distribution is not recommended as a model for the analysis of sample data unless the sample size is large enough to justify grouping of data into a frequency table. As a rule of thumb, 100 observations might be considered to be a minimum sample for this purpose. Even larger samples are desired if maximum advantage is to be realized from employing a four-parameter model.

Calculation of parameter estimates for this distribution is not a simple task. Even with the tabular and graphical aids provided here, computations are laborious. The solution of MLE equations is particularly complicated because of regularity considerations and the complex structure of the estimating equations. Moment estimators, though somewhat easier to calculate than MLE, often produce estimates of the threshold parameter γ that exceed one or more of the smaller sample observations and, therefore, are inadmissible. Modified moment estimators possess advantages over the MLE and the ME for estimation in the four-parameter distribution. They are unbiased with respect to mean and variance. They are relatively easy to calculate with the aid of tables and charts provided here. Furthermore they are free from regularity and existence problems.

When samples include as many as 500-1000 observations, the threshold parameter γ might appropriately be estimated as $\gamma^* = x_1$, where x_1 is the smallest sample observation. The remaining parameters ρ, δ, and β can then be estimated by employing three-parameter estimators. With sample data arrayed into a frequency table, expected frequencies can be calculated on the basis of parameter estimates. Either the chi-square or the Kolmogorov-Smirnov goodness-of-fit test can be used to assess agreement of observed and expected frequencies.

Appendix

A.1 SOME CONCLUDING REMARKS

Simulation studies by Cohen and Whitten (1980, 1981, 1982, 1985, 1986), Cohen et al. (1984, 1985), and Chan et al. (1984), which included comparisons between maximum likelihood (MLE), moment (ME), and modified moment estimators (MME), have confirmed advantages of the MME with respect to both bias and estimate variances for each of the distributions considered here when α_3 is large ($\alpha_3 > 1$). As pointed out throughout this presentation, the MME are unbiased with respect to distribution means and variances. Furthermore, bias is small for other parameters of interest. Use of the first-order statistic, which contains more information about the threshold parameter γ than any of the other order statistics, and often more than all of the other sample observations combined, accounts for major reductions in both bias and variance for estimates of this parameter. The MME are not subject to restrictions which limit use of MLE. They are applicable over the entire parameter space, and with the aid of tables and charts provided here they are quite easy to calculate from sample data.

As previously noted, asymptotic variances and covariances of the MLE are not valid for some parameter values. However, when they are applicable, they closely approximate simulated variances and covariances of both MLE and MME.

When α_3 is small ($\alpha_3 < 0.5$), advantages of the MME begin to diminish and the MLE might become a better choice. As $\alpha_3 \to 0$, it follows that $\gamma \to -\infty$ and estimate variances for this parameter become quite large. When α_3 is near zero, then the normal distribution might be a better choice as a model, and in that case the MLE and the ME would be identical.

As a consequence of the above considerations, it is therefore recommended that the MME be employed when $a_3 > 1$ and that the MLE be employed when $a_3 < 0.5$. When $0.5 \leq a_3 \leq 1$, either the MLE or the MME would usually be satisfactory. These cutoff points were chosen by a somewhat arbitrary selection process and might be adjusted to meet specific needs as they arise. It is recognized that when α_3 is near zero that a_3 in small samples might even be negative, although $\alpha_3 > 0$. In these circumstances, the normal distribution should at least be considered for the model.

A.2 FURTHER COMPARISONS

In comparisons involving the Weibull, lognormal, inverse Gaussian, and gamma distributions, the lognormal is usually perceived as being a "long-tailed" distribution. Both the lognormal and the IG always exhibit a discernible mode greater than the threshold value. These distributions never become reverse J-shaped as do the Weibull and gamma distributions. The exponential distribution, which is reverse J-shaped, can be obtained as a special case from both the Weibull and gamma distributions. Furthermore, both of these distributions remain reverse J-shaped for all values of the third standard moment $\alpha_3 \geq 2.0$. Unlike the lognormal, IG, and gamma distributions, which are always positively skewed, the Weibull becomes negatively skewed in a subregion of its parameter space. The normal distribution emerges as a limiting distribution from the lognormal, IG, and gamma distributions, but not from the Weibull distribution. However, the Weibull with $\alpha_3 = 0$ might be described as being "near normal" with a finite left terminal at 3.2431 standard units below the mean.

Frequency curves of the lognormal, IG, and gamma distributions (when $\alpha_3 < 2.0$) rise somewhat slowly from the threshold to a maximum (mode) before tailing off to the right. The Weibull curve rises rather sharply to present a somewhat stubby appearance on the left before reaching its modal value and tailing off to infinity on the right.

These comments might be further clarified by reference to tables for standardized distributions with zero mean, unit variance, and skewness α_3, which were included in Chapter 1.

Additional comparisons are provided by the parameter estimates obtained from illustrative examples 3.11.1 and 3.11.2 for the four distributions. These results are summarized in Tables A.2.1 and A.2.2.

It is well known that the sample moment a_3 tends to underestimate the distribution shape parameter α_3. In small samples such as we have here, the extent of underestimation might be rather large. For this reason, it appears likely that the larger estimates of α_3 given by the MME and the MLE are more plausible than ME of 1.0673 in Example 3.11.1 and 1.8636 in Example 3.11.2.

TABLE A.2.1 Estimate Summary for Example 3.11.1:
$n = 20$, $\bar{x} = 0.423125$, $x_1 = 0.265$, $s = 0.1252789$, $a_3 = 1.0673243$

	MME			ME			MLE		
Distribution	$\hat{\gamma}$	$\hat{\alpha}_3$	$\sqrt{V(\hat{X})}$	$\hat{\gamma}$	$\hat{\alpha}_3$	$\sqrt{V(\hat{X})}$	$\hat{\gamma}$	$\hat{\alpha}_3$	$\sqrt{V(\hat{X})}$
Weibull	0.2410	1.0966	0.1253	0.2382	1.0673	0.1253	0.2611	1.4393	0.1302
Lognormal	0.1713	1.6159	0.1253	0.0572	1.0673	0.1253	0.1850	1.7843	0.1294
IG	0.1216	1.2468	0.1253	0.0710	1.0673	0.1253	0.1782	1.5533	0.1268
Gamma	0.2096	1.1736	0.1253	0.1882	1.0673	0.1253	0.2627	1.8303	0.1468

TABLE A.2.2 Estimate Summary for Example 3.11.2:
$n = 10$, $\bar{x} = 220.48$, $x_1 = 152.7$, $s = 78.405638$, $a_3 = 1.8635835$

	MME			ME			MLE		
Distribution	$\hat{\gamma}$	$\hat{\alpha}_3$	$\sqrt{V(\hat{X})}$	$\hat{\gamma}$	$\hat{\alpha}_3$	$\sqrt{V(\hat{X})}$	$\hat{\gamma}$	$\hat{\alpha}_3$	$\sqrt{V(\hat{X})}$
Weibull	146.14	2.1653	78.4056	138.31	1.8636	78.4056			
Lognormal	132.38	3.3749	78.4056	80.97	1.8636	78.4056			
IG	124.28	2.4451	78.4056	94.26	1.8636	78.4056	139.66	2.8714	77.3562
Gamma	147.57	2.1508	78.4056	136.34	1.8636	78.4056			
Exponential[a]	145.17	2	75.3111						

[a]MME are also BLUE and MVUE.

The first-order statistic provides an upper bound on estimates of the threshold parameter. Accordingly, $\gamma < 0.265$ in Example 3.11.1 and $\gamma < 152.7$ in Example 3.11.2. For these examples, estimates of γ obtained for the lognormal and IG distributions are smaller than those obtained for the Weibull and gamma distributions.

The exponential distribution might be better as a model for the data of Example 3.11.2. In this case the estimators are MME, BLUE, and MVUE. Accordingly, even though the sample is small and thus subject to larger sampling errors, estimates are both unbiased and of minimum variance.

When the choice of a model must be based solely on sample data, without regard for the process which produced the data, comparisons such as those presented here will often be helpful in reaching the correct decision.

A.3 TABLES OF CUMULATIVE DISTRIBUTION FUNCTIONS

TABLE A.3.1 Cumulative Distribution Function of Standardized Weibull Distribution

(α_3; 0, 1)

z \ α_3	0.0	0.1	0.2	0.3	0.4	0.5	0.6
-3.0	.000061						
-2.9	.000210	.000000					
-2.8	.000528	.000033					
-2.7	.001099	.000202					
-2.6	.002020	.000636	.000021				
-2.5	.003397	.001480	.000252				
-2.4	.005349	.002886	.000949	.000028			
-2.3	.007999	.005016	.002348	.000476			
-2.2	.011480	.008031	.004675	.001807	.000127		
-2.1	.015930	.012099	.008146	.004361	.001296		
-2.0	.021487	.017384	.012966	.008421	.004127	.000848	
-1.9	.028296	.024049	.019326	.014228	.008981	.004055	.000501
-1.8	.036496	.032250	.027404	.021991	.016115	.010027	.004283
-1.7	.046226	.042133	.037359	.031882	.025722	.018975	.011890
-1.6	.057615	.053835	.049330	.044044	.037938	.031011	.023329
-1.5	.070786	.067474	.063431	.058580	.052848	.046176	.038529
-1.4	.085847	.083151	.079751	.075559	.070486	.064446	.057355
-1.3	.102888	.100944	.098348	.095011	.090843	.085746	.079620
-1.2	.121983	.120906	.119249	.116930	.113859	.109946	.105091
-1.1	.143179	.143060	.142447	.141265	.139435	.136874	.133495
-1.0	.166497	.167398	.167898	.167931	.167427	.166317	.164525
-.9	.191928	.193879	.195522	.196800	.197654	.198024	.197851
-.8	.219429	.222425	.225201	.227709	.229898	.231717	.233118
-.7	.248922	.252922	.256782	.260460	.263911	.267092	.269963
-.6	.280294	.285220	.290077	.294823	.299420	.303829	.308014
-.5	.313393	.319136	.324864	.330541	.336130	.341595	.346902
-.4	.348031	.354449	.360895	.367335	.373732	.380052	.386263
-.3	.383986	.390911	.397897	.404909	.411910	.418867	.425748
-.2	.421003	.428248	.435577	.442954	.450344	.457711	.465024
-.1	.458797	.466164	.473630	.481160	.488719	.496271	.503782
.0	.497064	.504347	.511743	.519215	.526729	.534249	.541742
.1	.535480	.542479	.549604	.556818	.564086	.571374	.578648
.2	.573712	.580239	.586906	.593678	.600521	.607400	.614281
.3	.611426	.617313	.623358	.629528	.635790	.642109	.648452
.4	.648295	.653399	.658685	.664122	.669677	.675317	.681008
.5	.684008	.688219	.692641	.697246	.702000	.706871	.711825
.6	.718274	.721518	.725008	.728714	.732606	.736651	.740817
.7	.750836	.753078	.755601	.758378	.761379	.764572	.767925
.8	.781473	.782716	.784274	.786123	.788235	.790578	.793120
.9	.810008	.810289	.810918	.811871	.813123	.814643	.816399
1.0	.836308	.835700	.835463	.835579	.836024	.836770	.837785
1.1	.860291	.858891	.857878	.857237	.856948	.856985	.857319
1.2	.881920	.879848	.878166	.876865	.875929	.875336	.875062
1.3	.901208	.898598	.896367	.894513	.893027	.891892	.891088
1.4	.918210	.915200	.912546	.910253	.908319	.906735	.905483

TABLE A.3.1 Continued

(α_3; 0, 1)

z \ α_3	0.0	0.1	0.2	0.3	0.4	0.5	0.6	
1.5	.933020	.929748	.926796	.924178	.921900	.919959	.918344	1.5
1.6	.945762	.942361	.939231	.936398	.933875	.931668	.929772	1.6
1.7	.956589	.953177	.949981	.947034	.944360	.941972	.939873	1.7
1.8	.965670	.962349	.959184	.956215	.953474	.950983	.948752	1.8
1.9	.973187	.970041	.966987	.964074	.961341	.958816	.956518	1.9
2.0	.979325	.976416	.973538	.970745	.968082	.965583	.963273	2.0
2.1	.984267	.981638	.978984	.976360	.973816	.971393	.969120	2.1
2.2	.988190	.985864	.983464	.981046	.978660	.976351	.974153	2.2
2.3	.991259	.989243	.987113	.984923	.982722	.980556	.978464	2.3
2.4	.993623	.991910	.990055	.988103	.986103	.984101	.982137	2.4
2.5	.995416	.993990	.992400	.990688	.988897	.987071	.985251	2.5
2.6	.996755	.995590	.994251	.992771	.991188	.989544	.987878	2.6
2.7	.997739	.996805	.995695	.994435	.993055	.991591	.990082	2.7
2.8	.998450	.997715	.996810	.995752	.994563	.993275	.991923	2.8
2.9	.998955	.998387	.997662	.996785	.995773	.994652	.993451	2.9
3.0	.999307	.998877	.998304	.997587	.996737	.995771	.994715	3.0
3.1	.999549	.999229	.998783	.998205	.997498	.996675	.995755	3.1
3.2	.999711	.999478	.999137	.998677	.998096	.997400	.996605	3.2
3.3	.999819	.999651	.999395	.999033	.998561	.997979	.997298	3.3
3.4	.999888	.999771	.999580	.999300	.998920	.998438	.997859	3.4
3.5	.999932	.999851	.999712	.999498	.999196	.998800	.998311	3.5
3.6	.999960	.999905	.999805	.999643	.999405	.999083	.998674	3.6
3.7	.999977	.999940	.999870	.999749	.999563	.999303	.998964	3.7
3.8	.999987	.999963	.999914	.999825	.999682	.999474	.999194	3.8
3.9	.999993	.999978	.999944	.999879	.999770	.999605	.999376	3.9
4.0	.999996	.999987	.999964	.999917	.999835	.999705	.999519	4.0
4.1	.999998	.999992	.999977	.999944	.999882	.999781	.999631	4.1
4.2	.999999	.999995	.999985	.999962	.999917	.999839	.999718	4.2
4.3	.999999	.999997	.999991	.999975	.999942	.999882	.999786	4.3
4.4	1.000000	.999999	.999994	.999984	.999959	.999914	.999838	4.4
4.5	1.000000	.999999	.999997	.999989	.999972	.999937	.999878	4.5
4.6	1.000000	1.000000	.999998	.999993	.999981	.999955	.999909	4.6
4.7	1.000000	1.000000	.999999	.999996	.999987	.999968	.999932	4.7
4.8	1.000000	1.000000	.999999	.999997	.999991	.999977	.999949	4.8
4.9	1.000000	1.000000	1.000000	.999998	.999994	.999984	.999963	4.9
5.0	1.000000	1.000000	1.000000	.999999	.999996	.999989	.999972	5.0
5.1	1.000000	1.000000	1.000000	.999999	.999997	.999992	.999980	5.1
5.2	1.000000	1.000000	1.000000	1.000000	.999998	.999994	.999985	5.2
5.3	1.000000	1.000000	1.000000	1.000000	.999999	.999996	.999989	5.3
5.4	1.000000	1.000000	1.000000	1.000000	.999999	.999997	.999992	5.4
5.5	1.000000	1.000000	1.000000	1.000000	1.000000	.999998	.999994	5.5
5.6	1.000000	1.000000	1.000000	1.000000	1.000000	.999999	.999996	5.6
5.7	1.000000	1.000000	1.000000	1.000000	1.000000	.999999	.999997	5.7
5.8	1.000000	1.000000	1.000000	1.000000	1.000000	.999999	.999998	5.8
5.9	1.000000	1.000000	1.000000	1.000000	1.000000	1.000000	.999999	5.9

TABLE A.3.1 Continued

$(\alpha_3;\ 0,\ 1)$

z \ α_3	0.7	0.8	0.9	1.0	1.1	1.2	1.3	
-3.0								-3.0
-2.9								-2.9
-2.8								-2.8
-2.7								-2.7
-2.6								-2.6
-2.5								-2.5
-2.4								-2.4
-2.3								-2.3
-2.2								-2.2
-2.1								-2.1
-2.0								-2.0
-1.9								-1.9
-1.8	.000284							-1.8
-1.7	.005068	.000226						-1.7
-1.6	.015105	.006899	.000485					-1.6
-1.5	.029927	.020502	.010673	.001843				-1.5
-1.4	.049143	.039769	.029260	.017819	.006227			-1.4
-1.3	.072365	.063878	.054062	.042843	.030214	.016390	.002651	-1.3
-1.2	.099191	.092134	.083800	.074059	.062769	.049787	.034993	-1.2
-1.1	.129205	.123905	.117488	.109834	.100807	.090248	.077968	-1.1
-1.0	.161977	.158593	.154287	.148968	.142533	.134865	.125826	-1.0
-.9	.197073	.195630	.193458	.190492	.186664	.181896	.176105	-.9
-.8	.234053	.234475	.234338	.233598	.232209	.230126	.227300	-.8
-.7	.272484	.274619	.276333	.277595	.278377	.278649	.278385	-.7
-.6	.311943	.315583	.318909	.321895	.324522	.326772	.328629	-.6
-.5	.352021	.356925	.361589	.365994	.370121	.373959	.377497	-.5
-.4	.392334	.398240	.403956	.409465	.414750	.419799	.424604	-.4
-.3	.432521	.439161	.445645	.451952	.458068	.463980	.469680	-.3
-.2	.472250	.479364	.486342	.493162	.499810	.506271	.512537	-.2
-.1	.511223	.518565	.525784	.532858	.539771	.546509	.553061	-.1
.0	.549175	.556521	.563756	.570856	.577804	.584587	.591193	.0
.1	.585876	.593030	.600084	.607017	.613809	.620445	.626916	.1
.2	.621134	.627928	.634639	.641244	.647724	.654063	.660249	.2
.3	.654789	.661089	.667326	.673478	.679524	.685450	.691240	.3
.4	.686719	.692420	.698085	.703690	.709214	.714641	.719956	.4
.5	.716832	.721862	.726886	.731880	.736822	.741695	.746482	.5
.6	.745071	.749383	.753725	.758070	.762396	.766682	.770913	.6
.7	.771405	.774980	.778622	.782303	.785999	.789689	.793353	.7
.8	.795828	.798670	.801616	.804637	.807709	.810807	.813912	.8
.9	.818359	.820491	.822762	.825145	.827611	.830137	.832700	.9
1.0	.839038	.840497	.842130	.843907	.845799	.847782	.849831	1.0
1.1	.857921	.858757	.859798	.861011	.862370	.863846	.865417	1.1
1.2	.875076	.875350	.875852	.876552	.877422	.878435	.879566	1.2
1.3	.890587	.890363	.890386	.890627	.891058	.891652	.892385	1.3
1.4	.904543	.903888	.903494	.903333	.903376	.903599	.903977	1.4

TABLE A.3.1 Continued

$(\alpha_3; 0, 1)$

z \ α_3	0.7	0.8	0.9	1.0	1.1	1.2	1.3	
1.5	.917039	.916024	.915275	.914767	.914475	.914373	.914439	1.5
1.6	.928178	.926869	.925825	.925026	.924449	.924069	.923865	1.6
1.7	.938059	.936520	.935241	.934204	.933390	.932776	.932342	1.7
1.8	.946785	.945076	.943617	.942392	.941385	.940578	.939953	1.8
1.9	.954456	.952632	.951041	.949674	.948518	.947557	.946775	1.9
2.0	.961170	.959279	.957603	.956135	.954867	.953787	.952881	2.0
2.1	.967019	.965105	.963382	.961850	.960505	.959338	.958337	2.1
2.2	.972094	.970191	.968457	.966894	.965501	.964275	.963206	2.2
2.3	.976476	.974617	.972899	.971332	.969919	.968658	.967543	2.3
2.4	.980246	.978452	.976776	.975229	.973817	.972542	.971402	2.4
2.5	.983473	.981765	.980149	.978641	.977249	.975979	.974831	2.5
2.6	.986225	.984616	.983076	.981621	.980265	.979014	.977873	2.6
2.7	.988562	.987062	.985607	.984218	.982910	.981691	.980568	2.7
2.8	.990538	.989152	.987791	.986476	.985225	.984048	.982953	2.8
2.9	.992201	.990931	.989668	.988434	.987247	.986120	.985062	2.9
3.0	.993596	.992442	.991279	.990129	.989011	.987938	.986923	3.0
3.1	.994761	.993720	.992657	.991592	.990545	.989532	.988564	3.1
3.2	.995730	.994798	.993832	.992852	.991879	.990927	.990009	3.2
3.3	.996533	.995704	.994831	.993935	.993035	.992145	.991280	3.3
3.4	.997195	.996462	.995679	.994864	.994036	.993209	.992397	3.4
3.5	.997739	.997095	.996396	.995659	.994900	.994135	.993377	3.5
3.6	.998184	.997622	.997002	.996338	.995646	.994941	.994236	3.6
3.7	.998547	.998059	.997511	.996916	.996289	.995642	.994988	3.7
3.8	.998842	.998420	.997939	.997408	.996841	.996250	.995647	3.8
3.9	.999080	.998718	.998297	.997826	.997315	.996777	.996222	3.9
4.0	.999272	.998963	.998596	.998179	.997721	.997233	.996724	4.0
4.1	.999426	.999163	.998846	.998478	.998069	.997627	.997162	4.1
4.2	.999549	.999327	.999053	.998730	.998366	.997967	.997543	4.2
4.3	.999647	.999460	.999225	.998943	.998619	.998261	.997876	4.3
4.4	.999725	.999568	.999367	.999121	.998835	.998514	.998164	4.4
4.5	.999786	.999656	.999484	.999270	.999018	.998731	.998415	4.5
4.6	.999834	.999726	.999580	.999396	.999174	.998918	.998633	4.6
4.7	.999872	.999783	.999659	.999500	.999305	.999078	.998822	4.7
4.8	.999902	.999828	.999724	.999587	.999417	.999215	.998985	4.8
4.9	.999925	.999864	.999777	.999660	.999511	.999333	.999127	4.9
5.0	.999942	.999893	.999820	.999720	.999591	.999433	.999249	5.0
5.1	.999956	.999916	.999856	.999770	.999658	.999519	.999355	5.1
5.2	.999967	.999935	.999884	.999812	.999715	.999593	.999446	5.2
5.3	.999975	.999949	.999907	.999846	.999762	.999655	.999525	5.3
5.4	.999981	.999960	.999926	.999874	.999802	.999708	.999593	5.4
5.5	.999986	.999969	.999941	.999897	.999835	.999753	.999651	5.5
5.6	.999989	.999976	.999953	.999916	.999863	.999792	.999702	5.6
5.7	.999992	.999982	.999963	.999932	.999886	.999824	.999745	5.7
5.8	.999994	.999986	.999970	.999945	.999906	.999852	.999782	5.8
5.9	.999996	.999989	.999977	.999955	.999922	.999876	.999814	5.9

TABLE A.3.1 Continued

	$(\alpha_3;\ 0,\ 1)$							
$z \backslash \alpha_3$	1.4	1.5	1.6	1.7	1.8	1.9	2.0	
-1.5								-1.5
-1.4								-1.4
-1.3								-1.3
-1.2	.018423	.001341						-1.2
-1.1	.063738	.047278	.028267	.006598				-1.1
-1.0	.115251	.102933	.088600	.071890	.052275	.028907		-1.0
-.9	.169192	.161043	.151518	.140439	.127580	.112632	.095163	-.9
-.8	.223679	.219206	.213815	.207429	.199956	.191283	.181269	-.8
-.7	.277561	.276148	.274119	.271445	.268090	.264018	.259182	-.7
-.6	.330080	.331114	.331721	.331891	.331614	.330881	.329680	-.6
-.5	.380726	.383643	.386244	.388527	.390492	.392139	.393469	-.5
-.4	.429160	.433461	.437508	.441302	.444844	.448138	.451188	-.4
-.3	.475160	.480419	.485456	.490270	.494866	.499245	.503415	-.3
-.2	.518601	.524459	.530110	.535553	.540792	.545830	.550671	-.2
-.1	.559421	.565584	.571548	.577312	.582878	.588250	.593430	-.1
.0	.597614	.603845	.609883	.615729	.621381	.626844	.632121	.0
.1	.633211	.639325	.645255	.650999	.656558	.661934	.667129	.1
.2	.666273	.672129	.677813	.683322	.688656	.693816	.698806	.2
.3	.696886	.702380	.707718	.712896	.717913	.722770	.727468	.3
.4	.725149	.730210	.735133	.739916	.744555	.749051	.753403	.4
.5	.751172	.755755	.760222	.764570	.768795	.772895	.776870	.5
.6	.775074	.779155	.783145	.787039	.790832	.794521	.798103	.6
.7	.796978	.800549	.804057	.807493	.810851	.814127	.817316	.7
.8	.817006	.820076	.823109	.826094	.829025	.831896	.834701	.8
.9	.835283	.837868	.840442	.842994	.845514	.847995	.850431	.9
1.0	.851928	.854054	.856193	.858334	.860465	.862578	.864665	1.0
1.1	.867060	.868756	.870490	.872247	.874015	.875783	.877544	1.1
1.2	.880792	.882093	.883453	.884855	.886287	.887738	.889197	1.2
1.3	.893233	.894175	.895194	.896273	.897398	.898558	.899741	1.3
1.4	.904486	.905106	.905818	.906605	.907453	.908349	.909282	1.4
1.5	.914649	.914983	.915421	.915948	.916548	.917207	.917915	1.5
1.6	.923814	.923897	.924095	.924391	.924771	.925220	.925726	1.6
1.7	.932068	.931934	.931923	.932017	.932203	.932466	.932794	1.7
1.8	.939490	.939172	.938980	.938900	.938917	.939018	.939190	1.8
1.9	.946156	.945683	.945339	.945110	.944981	.944941	.944977	1.9
2.0	.952135	.951534	.951063	.950708	.950456	.950295	.950213	2.0
2.1	.957492	.956788	.956213	.955753	.955397	.955133	.954951	2.1
2.2	.962285	.961501	.960842	.960297	.959855	.959506	.959238	2.2
2.3	.966568	.965724	.965000	.964388	.963876	.963456	.963117	2.3
2.4	.970392	.969505	.968734	.968069	.967502	.967024	.966627	2.4
2.5	.973802	.972888	.972083	.971380	.970770	.970247	.969803	2.5
2.6	.976839	.975912	.975086	.974356	.973715	.973158	.972676	2.6
2.7	.979542	.978612	.977777	.977030	.976369	.975786	.975276	2.7
2.8	.981944	.981022	.980186	.979433	.978759	.978159	.977629	2.8
2.9	.984078	.983172	.982343	.981590	.980911	.980302	.979758	2.9

TABLE A.3.1 Continued

(α_3; 0, 1)

z \ α_3	1.4	1.5	1.6	1.7	1.8	1.9	2.0	
3.0	.985971	.985087	.984272	.983526	.982848	.982235	.981684	3.0
3.1	.987648	.986792	.985996	.985263	.984592	.983981	.983427	3.1
3.2	.989134	.988309	.987537	.986821	.986161	.985556	.985004	3.2
3.3	.990448	.989658	.988913	.988218	.987573	.986977	.986431	3.3
3.4	.991610	.990856	.990142	.989470	.988842	.988260	.987723	3.4
3.5	.992636	.991921	.991238	.990591	.989984	.989417	.988891	3.5
3.6	.993541	.992865	.992215	.991596	.991010	.990461	.989948	3.6
3.7	.994339	.993702	.993086	.992495	.991933	.991402	.990905	3.7
3.8	.995042	.994444	.993862	.993300	.992762	.992252	.991770	3.8
3.9	.995661	.995102	.994553	.994020	.993507	.993017	.992553	3.9
4.0	.996205	.995683	.995168	.994664	.994176	.993708	.993262	4.0
4.1	.996683	.996198	.995715	.995240	.994777	.994331	.993903	4.1
4.2	.997103	.996653	.996202	.995755	.995317	.994892	.994483	4.2
4.3	.997471	.997055	.996634	.996215	.995801	.995398	.995008	4.3
4.4	.997794	.997410	.997019	.996626	.996236	.995855	.995483	4.4
4.5	.998077	.997723	.997360	.996993	.996627	.996266	.995913	4.5
4.6	.998325	.998000	.997663	.997321	.996977	.996637	.996302	4.6
4.7	.998542	.998243	.997932	.997613	.997292	.996971	.996654	4.7
4.8	.998731	.998458	.998171	.997875	.997574	.997272	.996972	4.8
4.9	.998897	.998647	.998383	.998108	.997826	.997543	.997261	4.9
5.0	.999041	.998813	.998570	.998315	.998053	.997788	.997521	5.0
5.1	.999167	.998960	.998736	.998501	.998257	.998008	.997757	5.1
5.2	.999277	.999089	.998884	.998666	.998439	.998206	.997971	5.2
5.3	.999373	.999202	.999014	.998813	.998602	.998385	.998164	5.3
5.4	.999456	.999301	.999130	.998944	.998749	.998546	.998338	5.4
5.5	.999529	.999389	..999232	.999061	.998880	.998691	.998497	5.5
5.6	.999592	.999465	.999322	.999166	.998998	.998822	.998640	5.6
5.7	.999647	.999532	.999402	.999258	.999103	.998939	.998769	5.7
5.8	.999695	.999591	.999473	.999341	.999198	.999045	.998886	5.8
5.9	.999736	.999643	.999535	.999414	.999282	.999141	.998992	5.9
6.0	.999772	.999688	.999590	.999480	.999358	.999227	.999088	6.0
6.1	.999803	.999728	.999639	.999538	.999426	.999304	.999175	6.1
6.2	.999830	.999762	.999682	.999590	.999486	.999374	.999253	6.2
6.3	.999853	.999793	.999720	.999636	.999541	.999436	.999324	6.3
6.4	.999874	.999819	.999754	.999677	.999589	.999493	.999389	6.4
6.5	.999891	.999843	.999783	.999713	.999633	.999544	.999447	6.5
6.6	.999906	.999863	.999809	.999745	.999672	.999590	.999500	6.6
6.7	.999919	.999881	.999832	.999774	.999707	.999631	.999547	6.7
6.8	.999930	.999896	.999852	.999800	.999738	.999668	.999590	6.8
6.9	.999940	.999909	.999870	.999822	.999766	.999701	.999629	6.9
7.0	.999949	.999921	.999886	.999842	.999791	.999731	.999665	7.0
7.1	.999956	.999932	.999900	.999860	.999813	.999758	.999696	7.1
7.2	.999962	.999940	.999912	.999876	.999833	.999782	.999725	7.2
7.3	.999967	.999948	.999923	.999890	.999851	.999804	.999751	7.3
7.4	.999972	.999955	.999932	.999903	.999867	.999824	.999775	7.4

TABLE A.3.1 Continued

z \ α_3	2.1	2.2	2.3	2.5	3.0	4.0	5.0	
				$(\alpha_3; 0, 1)$				
-1.5								-1.5
-1.4								-1.4
-1.3								-1.3
-1.2								-1.2
-1.1								-1.1
-1.0								-1.0
-.9	.074506	.049502	.017379					-.9
-.8	.169733	.156436	.141048	.101845				-.8
-.7	.253529	.246996	.239502	.221221	.147016			-.7
-.6	.328000	.325826	.323141	.316159	.287790	.139295		-.6
-.5	.394484	.395183	.395569	.395401	.389439	.350496	.247418	-.5
-.4	.454000	.456578	.458928	.462967	.469462	.468717	.450570	-.4
-.3	.507379	.511144	.514716	.521309	.534883	.552022	.559306	-.3
-.2	.555320	.559783	.564067	.572121	.589569	.615450	.633006	-.2
-.1	.598425	.603238	.607877	.616654	.635975	.665812	.687555	-.1
.0	.637215	.642132	.646877	.655876	.675789	.706898	.729950	.0
.1	.672148	.676995	.681675	.690558	.710241	.741067	.763966	.1
.2	.703628	.708285	.712784	.721324	.740255	.769896	.791883	.2
.3	.732011	.736400	.740641	.748694	.766552	.794501	.815192	.3
.4	.757614	.761686	.765622	.773101	.789702	.815695	.834918	.4
.5	.780720	.784446	.788052	.794911	.810164	.834092	.851794	.5
.6	.801580	.804950	.808215	.814436	.828315	.850167	.866366	.6
.7	.820419	.823432	.826358	.831946	.844467	.864292	.879044	.7
.8	.837438	.840104	.842699	.847672	.858879	.876765	.890149	.8
.9	.852818	.855151	.857430	.861815	.871773	.887827	.899932	.9
1.0	.866720	.868739	.870719	.874550	.883334	.897676	.908595	1.0
1.1	.879290	.881015	.882716	.886030	.893721	.906477	.916301	1.1
1.2	.890657	.892111	.893555	.896391	.903072	.914364	.923183	1.2
1.3	.900939	.902145	.903352	.905751	.911504	.921454	.929353	1.3
1.4	.910242	.911222	.912214	.914213	.919121	.927844	.934902	1.4
1.5	.918661	.919437	.920234	.921872	.926011	.933617	.939908	1.5
1.6	.926281	.926873	.927496	.928808	.932252	.938844	.944437	1.6
1.7	.933178	.933608	.934075	.935095	.937914	.943587	.948546	1.7
1.8	.939423	.939708	.940037	.940798	.943056	.947899	.952281	1.8
1.9	.945079	.945237	.945444	.945974	.947732	.951826	.955685	1.9
2.0	.950200	.950248	.950348	.950675	.951988	.955408	.958793	2.0
2.1	.954840	.954792	.954798	.954947	.955866	.958682	.961637	2.1
2.2	.959043	.958912	.958838	.958833	.959403	.961678	.964244	2.2
2.3	.962851	.962650	.962507	.962368	.962633	.964423	.966637	2.3
2.4	.966302	.966042	.965840	.965586	.965584	.966942	.968837	2.4
2.5	.969429	.969121	.968870	.968517	.968283	.969257	.970864	2.5
2.6	.972264	.971915	.971623	.971188	.970753	.971386	.972734	2.6
2.7	.974834	.974453	.974127	.973623	.973016	.973347	.974461	2.7
2.8	.977164	.976757	.976405	.975845	.975091	.975155	.976058	2.8
2.9	.979276	.978851	.978478	.977872	.976994	.976824	.977536	2.9

TABLE A.3.1 Continued

$z \backslash \alpha_3$	2.1	2.2	2.3	2.5	3.0	4.0	5.0	
	$(\alpha_3;\ 0,\ 1)$							
3.0	.981191	.980753	.980364	.979722	.978741	.978366	.978907	3.0
3.1	.982928	.982481	.982081	.981412	.980346	.979791	.980179	3.1
3.2	.984504	.984052	.983645	.982956	.981821	.981111	.981361	3.2
3.3	.985933	.985480	.985069	.984367	.983178	.982333	.982461	3.3
3.4	.987229	.986778	.986367	.985657	.984426	.983467	.983484	3.4
3.5	.988405	.987959	.987550	.986837	.985576	.984519	.984437	3.5
3.6	.989472	.989032	.988628	.987917	.986636	.985495	.985327	3.6
3.7	.990440	.990009	.989610	.988905	.987612	.986403	.986157	3.7
3.8	.991319	.990898	.990507	.989810	.988513	.987248	.986932	3.8
3.9	.992116	.991707	.991324	.990639	.989344	.988033	.987658	3.9
4.0	.992840	.992442	.992070	.991398	.990111	.988765	.988336	4.0
4.1	.993497	.993112	.992750	.992094	.990820	.989447	.988972	4.1
4.2	.994093	.993722	.993371	.992732	.991475	.990083	.989567	4.2
4.3	.994634	.994277	.993938	.993317	.992081	.990677	.990125	4.3
4.4	.995125	.994783	.994456	.993853	.992641	.991231	.990649	4.4
4.5	.995572	.995243	.994928	.994345	.993159	.991748	.991142	4.5
4.6	.995977	.995662	.995360	.994797	.993638	.992232	.991604	4.6
4.7	.996344	.996044	.995754	.995211	.994082	.992685	.992039	4.7
4.8	.996678	.996392	.996115	.995592	.994494	.993108	.992447	4.8
4.9	.996982	.996709	.996444	.995941	.994875	.993504	.992832	4.9
5.0	.997257	.996998	.996745	.996263	.995229	.993876	.993195	5.0
5.1	.997507	.997261	.997020	.996558	.995557	.994224	.993536	5.1
5.2	.997735	.997501	.997271	.996829	.995861	.994550	.993859	5.2
5.3	.997941	.997720	.997501	.997078	.996143	.994856	.994162	5.3
5.4	.998129	.997919	.997711	.997307	.996405	.995143	.994449	5.4
5.5	.998299	.998101	.997904	.997518	.996649	.995413	.994720	5.5
5.6	.998454	.998267	.998079	.997712	.996875	.995666	.994976	5.6
5.7	.998595	.998418	.998240	.997890	.997085	.995904	.995217	5.7
5.8	.998722	.998556	.998388	.998054	.997280	.996127	.995446	5.8
5.9	.998838	.998681	.998522	.998205	.997462	.996338	.995662	5.9
6.0	.998944	.998796	.998646	.998344	.997631	.996535	.995867	6.0
6.1	.999040	.998901	.998759	.998472	.997788	.996722	.996061	6.1
6.2	.999127	.998996	.998862	.998590	.997934	.996897	.996244	6.2
6.3	.999206	.999083	.998957	.998699	.998070	.997062	.996418	6.3
6.4	.999278	.999163	.999043	.998799	.998197	.997218	.996583	6.4
6.5	.999344	.999235	.999123	.998891	.998315	.997364	.996739	6.5
6.6	.999403	.999301	.999195	.998976	.998425	.997502	.996887	6.6
6.7	.999457	.999362	.999262	.999055	.998527	.997633	.997028	6.7
6.8	.999506	.999417	.999323	.999127	.998623	.997756	.997162	6.8
6.9	.999551	.999467	.999379	.999193	.998712	.997872	.997288	6.9
7.0	.999592	.999513	.999430	.999255	.998795	.997981	.997409	7.0
7.1	.999629	.999555	.999477	.999311	.998873	.998085	.997524	7.1
7.2	.999662	.999593	.999520	.999364	.998945	.998182	.997632	7.2
7.3	.999693	.999628	.999560	.999412	.999012	.998275	.997736	7.3
7.4	.999720	.999660	.999596	.999456	.999075	.998362	.997834	7.4

TABLE A.3.2 Cumulative Distribution Function of Standardized Lognormal Distribution

(α_3; 0, 1)

z \ α_3	0.0	0.1	0.2	0.3	0.4	0.5	0.6	
-3.0	.001350	.000831	.000455	.000213	.000080	.000022	.000004	-3.0
-2.9	.001866	.001207	.000705	.000360	.000153	.000050	.000011	-2.9
-2.8	.002555	.001731	.001074	.000595	.000282	.000107	.000029	-2.8
-2.7	.003467	.002451	.001608	.000957	.000500	.000218	.000073	-2.7
-2.6	.004661	.003429	.002366	.001503	.000857	.000422	.000169	-2.6
-2.5	.006210	.004738	.003423	.002308	.001422	.000777	.000360	-2.5
-2.4	.008198	.006469	.004875	.003466	.002288	.001371	.000721	-2.4
-2.3	.010724	.008729	.006833	.005097	.003578	.002323	.001360	-2.3
-2.2	.013903	.011641	.009433	.007345	.005442	.003788	.002431	-2.2
-2.1	.017864	.015346	.012830	.010379	.008066	.005963	.004138	-2.1
-2.0	.022750	.020003	.017198	.014395	.011665	.009086	.006738	-2.0
-1.9	.028717	.025785	.022731	.019607	.016481	.013429	.010534	-1.9
-1.8	.035930	.032877	.029635	.026250	.022779	.019292	.015870	-1.8
-1.7	.044565	.041470	.038125	.034564	.030833	.026991	.023109	-1.7
-1.6	.054799	.051759	.048416	.044791	.040918	.036841	.032618	-1.6
-1.5	.066807	.063935	.060717	.057163	.053294	.049139	.044740	-1.5
-1.4	.080757	.078176	.075221	.071890	.068193	.064145	.059774	-1.4
-1.3	.096800	.094640	.092093	.089148	.085804	.082066	.077947	-1.3
-1.2	.115070	.113459	.111464	.109067	.106259	.103037	.099403	-1.2
-1.1	.135666	.134727	.133420	.131724	.129625	.127113	.124182	-1.1
-1.0	.158655	.158494	.157994	.157133	.155892	.154258	.152220	-1.0
-.9	.184060	.184761	.185162	.185239	.184972	.184346	.183348	-.9
-.8	.211855	.213474	.214836	.215919	.216701	.217165	.217298	-.8
-.7	.241964	.244521	.246868	.248980	.250837	.252422	.253719	-.7
-.6	.274253	.277732	.281045	.284166	.287076	.289755	.292190	-.6
-.5	.308538	.312883	.317099	.321164	.325054	.328752	.332242	-.5
-.4	.344578	.349694	.354714	.359614	.364370	.368964	.373378	-.4
-.3	.382089	.387846	.393532	.399123	.404593	.409923	.415095	-.3
-.2	.420740	.426980	.433166	.439275	.445280	.451162	.456900	-.2
-.1	.460172	.466712	.473212	.479647	.485993	.492225	.498325	-.1
.0	.500000	.506644	.513261	.519824	.526307	.532688	.538945	.0
.1	.539828	.546379	.552914	.559406	.565830	.572162	.578382	.1
.2	.579260	.585525	.591788	.598024	.604205	.610310	.616315	.2
.3	.617911	.623714	.629534	.635346	.641124	.646844	.652484	.3
.4	.655422	.660608	.665839	.671087	.676326	.681533	.686685	.4
.5	.691462	.695910	.700435	.705008	.709606	.714202	.718774	.5
.6	.725747	.729368	.733103	.736927	.740811	.744730	.748660	.6
.7	.758036	.760778	.763678	.766708	.769839	.773046	.776302	.7
.8	.788145	.789993	.792044	.794269	.796638	.799124	.801699	.8
.9	.815940	.816915	.818136	.819574	.821198	.822980	.824890	.9
1.0	.841345	.841497	.841935	.842628	.843548	.844662	.845942	1.0
1.1	.864334	.863740	.863463	.863476	.863748	.864248	.864946	1.1
1.2	.884930	.883685	.882781	.882191	.881886	.881835	.882010	1.2
1.3	.903200	.901412	.899978	.898872	.898069	.897541	.897257	1.3
1.4	.919243	.917029	.915167	.913639	.912422	.911489	.910815	1.4

TABLE A.3.2 Continued

(α_3; 0, 1)

z \ α_3	0.0	0.1	0.2	0.3	0.4	0.5	0.6	
1.5	.933193	.930667	.928482	.926624	.925076	.923815	.922818	1.5
1.6	.945201	.942475	.940066	.937968	.936169	.934652	.933399	1.6
1.7	.955435	.952611	.950070	.947815	.945841	.944137	.942689	1.7
1.8	.964070	.961237	.958647	.956310	.954229	.952400	.950813	1.8
1.9	.971283	.968518	.965949	.963595	.961468	.959569	.957893	1.9
2.0	.977250	.974612	.972122	.969807	.967684	.965761	.964042	2.0
2.1	.982136	.979671	.977305	.975072	.972996	.971090	.969363	2.1
2.2	.986097	.983837	.981628	.979511	.977515	.975659	.973955	2.2
2.3	.989276	.987239	.985210	.983233	.981342	.979561	.977906	2.3
2.4	.991802	.989995	.988159	.986338	.984570	.982882	.981296	2.4
2.5	.993790	.992211	.990571	.988915	.987281	.985700	.984196	2.5
2.6	.995339	.993978	.992531	.991043	.989549	.988083	.986671	2.6
2.7	.996533	.995376	.994116	.992791	.991439	.990092	.988778	2.7
2.8	.997445	.996474	.995388	.994221	.993008	.991781	.990568	2.8
2.9	.998134	.997330	.996403	.995385	.994306	.993197	.992085	2.9
3.0	.998650	.997991	.997210	.996329	.995377	.994381	.993368	3.0
3.1	.999032	.998499	.997846	.997090	.996256	.995368	.994451	3.1
3.2	.999313	.998886	.998345	.997702	.996976	.996189	.995364	3.2
3.3	.999517	.999179	.998735	.998192	.997564	.996871	.996132	3.3
3.4	.999663	.999399	.999038	.998582	.998042	.997435	.996777	3.4
3.5	.999767	.999563	.999271	.998892	.998431	.997901	.997317	3.5
3.6	.999841	.999684	.999451	.999137	.998745	.998285	.997769	3.6
3.7	.999892	.999773	.999588	.999330	.998999	.998601	.998148	3.7
3.8	.999928	.999838	.999692	.999481	.999203	.998861	.998463	3.8
3.9	.999952	.999885	.999771	.999600	.999367	.999074	.998726	3.9
4.0	.999968	.999919	.999831	.999692	.999498	.999248	.998945	4.0
4.1	.999979	.999943	.999875	.999764	.999603	.999390	.999127	4.1
4.2	.999987	.999961	.999908	.999819	.999686	.999506	.999279	4.2
4.3	.999991	.999973	.999933	.999862	.999753	.999600	.999404	4.3
4.4	.999995	.999981	.999951	.999895	.999805	.999677	.999508	4.4
4.5	.999997	.999987	.999965	.999921	.999847	.999739	.999594	4.5
4.6	.999998	.999991	.999974	.999940	.999880	.999790	.999666	4.6
4.7	.999999	.999994	.999982	.999955	.999906	.999831	.999725	4.7
4.8	.999999	.999996	.999987	.999966	.999927	.999864	.999773	4.8
4.9	1.000000	.999997	.999991	.999974	.999943	.999891	.999813	4.9
5.0	1.000000	.999998	.999993	.999981	.999956	.999912	.999847	5.0
5.1	1.000000	.999999	.999995	.999986	.999965	.999930	.999874	5.1
5.2	1.000000	.999999	.999997	.999989	.999973	.999944	.999896	5.2
5.3	1.000000	1.000000	.999998	.999992	.999979	.999955	.999915	5.3
5.4	1.000000	1.000000	.999998	.999994	.999984	.999964	.999930	5.4
5.5	1.000000	1.000000	.999999	.999996	.999988	.999971	.999943	5.5
5.6	1.000000	1.000000	.999999	.999997	.999990	.999977	.999953	5.6
5.7	1.000000	1.000000	.999999	.999998	.999993	.999982	.999961	5.7
5.8	1.000000	1.000000	1.000000	.999998	.999994	.999985	.999968	5.8
5.9	1.000000	1.000000	1.000000	.999999	.999996	.999988	.999974	5.9

TABLE A.3.2 Continued

(α_3; 0, 1)

z \ α_3	0.7	0.8	0.9	1.0	1.1	1.2	1.3	
-3.0	.000000	.000000	.000000	.000000				-3.0
-2.9	.000001	.000000	.000000	.000000				-2.9
-2.8	.000005	.000000	.000000	.000000	.000000			-2.8
-2.7	.000017	.000002	.000000	.000000	.000000			-2.7
-2.6	.000050	.000009	.000001	.000000	.000000	.000000		-2.6
-2.5	.000132	.000034	.000005	.000000	.000000	.000000		-2.5
-2.4	.000316	.000106	.000023	.000003	.000000	.000000	.000000	-2.4
-2.3	.000692	.000289	.000090	.000017	.000001	.000000	.000000	-2.3
-2.2	.001402	.000697	.000281	.000082	.000014	.000001	.000000	-2.2
-2.1	.002646	.001517	.000748	.000296	.000084	.000014	.000001	-2.1
-2.0	.004696	.003018	.001742	.000866	.000347	.000099	.000016	-2.0
-1.9	.007884	.005560	.003631	.002140	.001095	.000456	.000138	-1.9
-1.8	.012602	.009580	.006894	.004623	.002824	.001518	.000678	-1.8
-1.7	.019270	.015566	.012094	.008951	.006228	.003996	.002297	-1.7
-1.6	.028319	.024026	.019830	.015830	.012129	.008825	.006008	-1.6
-1.5	.040152	.035440	.030680	.025959	.021374	.017026	.013019	-1.5
-1.4	.055115	.050217	.045138	.039943	.034713	.029533	.024499	-1.4
-1.3	.073469	.068662	.063565	.058224	.052696	.047046	.041346	-1.3
-1.2	.095366	.090942	.086154	.081031	.075611	.069936	.064054	-1.2
-1.1	.120833	.117072	.112911	.108365	.103457	.098212	.092660	-1.1
-1.0	.149774	.146919	.143660	.140004	.135962	.131550	.126785	-1.0
-.9	.181969	.180206	.178059	.175530	.172625	.169354	.165726	-.9
-.8	.217090	.216535	.215630	.214376	.212776	.210836	.208564	-.8
-.7	.254718	.255412	.255798	.255873	.255641	.255104	.254270	-.7
-.6	.294367	.296281	.297924	.299296	.300398	.301231	.301802	-.6
-.5	.335510	.338548	.341350	.343912	.346235	.348320	.350171	-.5
-.4	.377599	.381616	.385423	.389014	.392388	.395546	.398490	-.4
-.3	.420094	.424908	.429529	.433950	.438169	.442185	.446000	-.3
-.2	.462478	.467885	.473109	.478143	.482985	.487632	.492084	-.2
-.1	.504275	.510062	.515674	.521104	.526346	.531398	.536259	-.1
.0	.545060	.551021	.556813	.562430	.567864	.573113	.578175	.0
.1	.584470	.590412	.596195	.601810	.607251	.612513	.617594	.1
.2	.622203	.627958	.633567	.639019	.644308	.649429	.654379	.2
.3	.658025	.663451	.668748	.673906	.678915	.683772	.688471	.3
.4	.691762	.696748	.701627	.706387	.711021	.715520	.719881	.4
.5	.723301	.727764	.732148	.736439	.740628	.744706	.748668	.5
.6	.752579	.756467	.760307	.764084	.767786	.771403	.774929	.6
.7	.779583	.782869	.786140	.789380	.792576	.795715	.798789	.7
.8	.804338	.807018	.809718	.812419	.815107	.817767	.820390	.8
.9	.826902	.828991	.831134	.833311	.835504	.837700	.839884	.9
1.0	.847359	.848887	.850501	.852180	.853905	.855659	.857427	1.0
1.1	.865812	.866821	.867945	.869162	.870452	.871795	.873176	1.1
1.2	.882381	.882919	.883598	.884395	.885287	.886256	.887284	1.2
1.3	.897190	.897312	.897596	.898018	.898555	.899188	.899898	1.3
1.4	.910373	.910134	.910074	.910167	.910392	.910728	.911157	1.4

TABLE A.3.2 Continued

z \ α_3	0.7	0.8	0.9	1.0	1.1	1.2	1.3	
	(α_3; 0, 1)							
1.5	.922061	.921518	.921164	.920976	.920931	.921009	.921192	1.5
1.6	.932387	.931593	.930995	.930569	.930295	.930153	.930125	1.6
1.7	.941478	.940484	.939687	.939066	.938602	.938275	.938067	1.7
1.8	.949454	.948308	.947355	.946578	.945959	.945479	.945123	1.8
1.9	.956433	.955175	.954105	.953207	.952465	.951863	.951385	1.9
2.0	.962520	.961188	.960035	.959048	.958212	.957514	.956939	2.0
2.1	.967815	.966442	.965236	.964187	.963283	.962513	.961863	2.1
2.2	.972410	.971023	.969789	.968702	.967753	.966930	.966224	2.2
2.3	.976388	.975009	.973770	.972665	.971689	.970833	.970087	2.3
2.4	.979823	.978472	.977244	.976139	.975152	.974277	.973507	2.4
2.5	.982783	.981474	.980273	.979182	.978198	.977317	.976534	2.5
2.6	.985330	.984074	.982911	.981844	.980874	.979998	.979212	2.6
2.7	.987516	.986322	.985205	.984172	.983224	.982362	.981582	2.7
2.8	.989389	.988262	.987198	.986206	.985288	.984446	.983679	2.8
2.9	.990992	.989935	.988929	.987981	.987098	.986282	.985534	2.9
3.0	.992360	.991376	.990430	.989531	.988687	.987901	.987175	3.0
3.1	.993528	.992616	.991730	.990882	.990079	.989327	.988627	3.1
3.2	.994522	.993681	.992857	.992061	.991301	.990583	.989912	3.2
3.3	.995368	.994596	.993832	.993087	.992371	.991690	.991049	3.3
3.4	.996087	.995381	.994676	.993982	.993309	.992665	.992055	3.4
3.5	.996696	.996054	.995405	.994761	.994132	.993525	.992946	3.5
3.6	.997213	.996630	.996035	.995439	.994852	.994282	.993735	3.6
3.7	.997651	.997124	.996579	.996029	.995484	.994950	.994433	3.7
3.8	.998021	.997546	.997049	.996543	.996037	.995538	.995052	3.8
3.9	.998334	.997906	.997455	.996991	.996522	.996056	.995601	3.9
4.0	.998598	.998215	.997805	.997380	.996947	.996514	.996087	4.0
4.1	.998821	.998478	.998107	.997718	.997319	.996917	.996518	4.1
4.2	.999009	.998703	.998368	.998013	.997646	.997273	.996900	4.2
4.3	.999167	.998894	.998593	.998269	.997932	.997586	.997240	4.3
4.4	.999300	.999058	.998786	.998492	.998183	.997864	.997541	4.4
4.5	.999413	.999197	.998953	.998686	.998403	.998108	.997808	4.5
4.6	.999507	.999316	.999097	.998855	.998596	.998324	.998045	4.6
4.7	.999586	.999418	.999222	.999002	.998765	.998515	.998256	4.7
4.8	.999653	.999504	.999329	.999130	.998914	.998683	.998444	4.8
4.9	.999709	.999578	.999421	.999242	.999044	.998832	.998611	4.9
5.0	.999756	.999640	.999500	.999339	.999158	.998964	.998759	5.0
5.1	.999796	.999694	.999569	.999423	.999259	.999081	.998891	5.1
5.2	.999829	.999739	.999628	.999496	.999347	.999184	.999009	5.2
5.3	.999857	.999778	.999679	.999560	.999425	.999275	.999113	5.3
5.4	.999880	.999811	.999723	.999616	.999493	.999356	.999207	5.4
5.5	.999899	.999839	.999761	.999665	.999553	.999427	.999290	5.5
5.6	.999916	.999863	.999793	.999707	.999606	.999491	.999364	5.6
5.7	.999929	.999883	.999821	.999744	.999652	.999547	.999430	5.7
5.8	.999941	.999900	.999846	.999776	.999693	.999597	.999489	5.8
5.9	.999950	.999915	.999867	.999804	.999729	.999641	.999542	5.9

TABLE A.3.2 Continued

	$(\alpha_3;\ 0,\ 1)$						
$z \backslash \alpha_3$	1.4	1.5	1.6	1.7	1.8	1.9	2.0
-1.5	.009455	.006424	.003992	.002192	.001005	.000347	.000073
-1.4	.019711	.015273	.011288	.007850	.005035	.002889	.001410
-1.3	.035678	.030129	.024793	.019771	.015164	.011069	.007576
-1.2	.058022	.051902	.045763	.039679	.033732	.028010	.022605
-1.1	.086836	.080779	.074531	.068142	.061662	.055148	.048663
-1.0	.121689	.116283	.110595	.104651	.098481	.092118	.085596
-.9	.161757	.157458	.152847	.147940	.142755	.137310	.131623
-.8	.205968	.203060	.199849	.196349	.192571	.188528	.184232
-.7	.253146	.251740	.250061	.248120	.245926	.243491	.240823
-.6	.302116	.302182	.302007	.301600	.300971	.300130	.299084
-.5	.351793	.353194	.354379	.355359	.356141	.356733	.357146
-.4	.401225	.403755	.406089	.408233	.410194	.411982	.413603
-.3	.449617	.453041	.456276	.459330	.462210	.464922	.467475
-.2	.496344	.500415	.504302	.508011	.511548	.514920	.518133
-.1	.540932	.545418	.549721	.553848	.557803	.561594	.565225
.0	.583052	.587744	.592257	.596594	.600761	.604763	.608608
.1	.622496	.627218	.631765	.636140	.640348	.644395	.648286
.2	.659157	.663766	.668207	.672483	.676599	.680559	.684370
.3	.693013	.697397	.701624	.705698	.709622	.713400	.717036
.4	.724101	.728180	.732117	.735915	.739576	.743104	.746502
.5	.752509	.756229	.759825	.763299	.766652	.769886	.773005
.6	.778358	.781687	.784912	.788035	.791054	.793972	.796789
.7	.801791	.804716	.807559	.810320	.812996	.815588	.818096
.8	.822966	.825488	.827951	.830352	.832688	.834958	.837160
.9	.842046	.844178	.846274	.848327	.850334	.852292	.854199
1.0	.859197	.860960	.862706	.864430	.866126	.867789	.869417
1.1	.874582	.876001	.877422	.878840	.880246	.881635	.883004
1.2	.888356	.889460	.890585	.891721	.892862	.894000	.895131
1.3	.900670	.901489	.902345	.903228	.904127	.905038	.905953
1.4	.911662	.912229	.912845	.913500	.914184	.914890	.915612
1.5	.921463	.921807	.922212	.922666	.923160	.923685	.924233
1.6	.930194	.930345	.930565	.930843	.931170	.931535	.931931
1.7	.937964	.937949	.938011	.938137	.938317	.938543	.938807
1.8	.944874	.944720	.944646	.944642	.944697	.944803	.944951
1.9	.951017	.950745	.950558	.950443	.950392	.950394	.950443
2.0	.956474	.956107	.955825	.955618	.955476	.955392	.955356
2.1	.961321	.960876	.960517	.960234	.960018	.959860	.959752
2.2	.965624	.965119	.964698	.964353	.964075	.963856	.963690
2.3	.969444	.968892	.968424	.968030	.967702	.967434	.967217
2.4	.972834	.972249	.971744	.971312	.970946	.970637	.970381
2.5	.975842	.975235	.974704	.974244	.973847	.973507	.973219
2.6	.978512	.977891	.977344	.976864	.976445	.976081	.975767
2.7	.980882	.980255	.979698	.979205	.978771	.978389	.978056
2.8	.982985	.982359	.981799	.981299	.980855	.980462	.980115
2.9	.984851	.984233	.983674	.983173	.982724	.982323	.981967

TABLE A.3.2. Continued

$z \backslash \alpha_3$	1.4	1.5	1.6	1.7	1.8	1.9	2.0
	(α_3; 0, 1)						
3.0	.986509	.985901	.985349	.984850	.984400	.983996	.983635
3.1	.987980	.987387	.986845	.986352	.985905	.985501	.985138
3.2	.989288	.988711	.988182	.987698	.987256	.986856	.986493
3.3	.990449	.989892	.989377	.988904	.988471	.988076	.987716
3.4	.991481	.990945	.990447	.989987	.989563	.989175	.988820
3.5	.992398	.991884	.991404	.990958	.990546	.990166	.989818
3.6	.993214	.992722	.992261	.991830	.991431	.991061	.990720
3.7	.993939	.993470	.993029	.992614	.992228	.991869	.991537
3.8	.994585	.994139	.993717	.993319	.992947	.992599	.992276
3.9	.995160	.994736	.994334	.993953	.993595	.993259	.992946
4.0	.995671	.995271	.994888	.994524	.994180	.993857	.993554
4.1	.996127	.995748	.995385	.995037	.994708	.994398	.994106
4.2	.996533	.996176	.995831	.995501	.995186	.994888	.994607
4.3	.996896	.996559	.996232	.995918	.995618	.995332	.995062
4.4	.997219	.996902	.996593	.996295	.996008	.995735	.995476
4.5	.997507	.997209	.996918	.996635	.996362	.996101	.995853
4.6	.997764	.997485	.997210	.996942	.996683	.996434	.996196
4.7	.997994	.997732	.997473	.997219	.996973	.996736	.996508
4.8	.998200	.997954	.997710	.997470	.997237	.997010	.996793
4.9	.998383	.998153	.997924	.997697	.997476	.997260	.997053
5.0	.998547	.998332	.998117	.997903	.997693	.997488	.997290
5.1	.998694	.998493	.998291	.998089	.997890	.997696	.997507
5.2	.998826	.998638	.998448	.998258	.998069	.997885	.997705
5.3	.998943	.998768	.998590	.998410	.998233	.998057	.997886
5.4	.999049	.998885	.998718	.998549	.998381	.998215	.998052
5.5	.999144	.998991	.998834	.998675	.998516	.998359	.998204
5.6	.999228	.999086	.998939	.998790	.998640	.998490	.998343
5.7	.999304	.999172	.999034	.998894	.998752	.998611	.998471
5.8	.999373	.999249	.999120	.998988	.998854	.998720	.998588
5.9	.999434	.999319	.999198	.999074	.998948	.998821	.998695
6.0	.999489	.999382	.999269	.999152	.999033	.998913	.998794
6.1	.999539	.999439	.999333	.999224	.999111	.998998	.998884
6.2	.999583	.999490	.999391	.999289	.999183	.999075	.998967
6.3	.999623	.999537	.999444	.999348	.999248	.999146	.999044
6.4	.999659	.999579	.999492	.999402	.999308	.999212	.999115
6.5	.999692	.999617	.999536	.999451	.999362	.999272	.999179
6.6	.999721	.999651	.999576	.999496	.999412	.999327	.999239
6.7	.999747	.999683	.999612	.999537	.999458	.999377	.999294
6.8	.999771	.999711	.999645	.999575	.999500	.999424	.999345
6.9	.999793	.999737	.999675	.999609	.999539	.999466	.999392
7.0	.999812	.999760	.999702	.999640	.999574	.999506	.999435
7.1	.999829	.999781	.999727	.999669	.999607	.999542	.999475
7.2	.999845	.999800	.999750	.999695	.999637	.999576	.999512
7.3	.999859	.999818	.999771	.999719	.999664	.999606	.999546
7.4	.999872	.999833	.999790	.999741	.999690	.999635	.999578

TABLE A.3.2 Continued

$(\alpha_3;\ 0,\ 1)$

$z \backslash \alpha_3$	2.1	2.2	2.3	2.5	3.0	4.0	5.0	
-1.5	.000006	.000000						-1.5
-1.4	.000536	.000133	.000014	.000000				-1.4
-1.3	.004754	.002642	.001225	.000088				-1.3
-1.2	.017611	.013123	.009232	.003520	.000000			-1.2
-1.1	.042274	.036054	.030080	.019209	.002164			-1.1
-1.0	.078952	.072227	.065465	.052022	.021633	.000000		-1.0
-.9	.125716	.119608	.113322	.100303	.066395	.009402		-.9
-.8	.179695	.174931	.169950	.159389	.130051	.064652	.009878	-.8
-.7	.237933	.234831	.231526	.224342	.203405	.151542	.091297	-.7
-.6	.297845	.296419	.294817	.291115	.279286	.246893	.205695	-.6
-.5	.357386	.357464	.357387	.356799	.353125	.338604	.317354	-.5
-.4	.415067	.416380	.417552	.419499	.422371	.421802	.415710	-.4
-.3	.469876	.472132	.474252	.478108	.485818	.495160	.499287	-.3
-.2	.521196	.524115	.526897	.532077	.543087	.558929	.569419	-.2
-.1	.568705	.572040	.575237	.581243	.594277	.613987	.628101	-.1
.0	.612300	.615847	.619256	.625682	.639747	.661396	.677278	.0
.1	.652028	.655626	.659087	.665623	.679975	.702206	.718639	.1
.2	.688036	.691564	.694959	.701374	.715480	.737373	.753582	.2
.3	.720537	.723906	.727150	.733283	.746777	.767735	.783248	.3
.4	.749775	.752927	.755963	.761705	.774353	.794012	.808558	.4
.5	.776011	.778909	.781702	.786989	.798654	.816814	.830258	.5
.6	.799508	.802132	.804664	.809465	.820080	.836657	.848950	.6
.7	.820522	.822867	.825132	.829436	.838987	.853973	.865123	.7
.8	.839295	.841363	.843366	.847181	.855690	.869129	.879177	.8
.9	.856054	.857857	.859607	.862953	.870463	.882430	.891438	.9
1.0	.871008	.872560	.874072	.876976	.883548	.894137	.902177	1.0
1.1	.884349	.885668	.886959	.889452	.895153	.904467	.911616	1.1
1.2	.896250	.897355	.898442	.900558	.905460	.913608	.919941	1.2
1.3	.906868	.907778	.908681	.910454	.914630	.921715	.927308	1.3
1.4	.916342	.917078	.917815	.919279	.922799	.928924	.933847	1.4
1.5	.924800	.925380	.925968	.927156	.930087	.935348	.939668	1.5
1.6	.932353	.932795	.933251	.934194	.936600	.941087	.944864	1.6
1.7	.939102	.939422	.939763	.940489	.942429	.946224	.949514	1.7
1.8	.945135	.945349	.945588	.946124	.947653	.950832	.953687	1.8
1.9	.950532	.950654	.950805	.951174	.952341	.954974	.957439	1.9
2.0	.955363	.955406	.955480	.955705	.956555	.958704	.960822	2.0
2.1	.959690	.959665	.959674	.959774	.960348	.962070	.963877	2.1
2.2	.963568	.963487	.963440	.963433	.963767	.965112	.966643	2.2
2.3	.967047	.966917	.966823	.966725	.966852	.967867	.969152	2.3
2.4	.970170	.970000	.969867	.969692	.969640	.970365	.971431	2.4
2.5	.972976	.972773	.972607	.972367	.972163	.972634	.973507	2.5
2.6	.975498	.975269	.975076	.974782	.974449	.974699	.975399	2.6
2.7	.977767	.977517	.977303	.976965	.976523	.976580	.977128	2.7
2.8	.979811	.979544	.979313	.978940	.978406	.978297	.978710	2.8
2.9	.981652	.981374	.981129	.980728	.980119	.979865	.980159	2.9

TABLE A.3.2 Continued

$(\alpha_3;\ 0,\ 1)$

z \ α_3	2.1	2.2	2.3	2.5	3.0	4.0	5.0	
3.0	.983313	.983026	.982771	.982348	.981677	.981301	.981489	3.0
3.1	.984811	.984519	.984258	.983819	.983098	.982616	.982711	3.1
3.2	.986165	.985870	.985605	.985154	.984394	.983822	.983836	3.2
3.3	.987389	.987093	.986826	.986368	.985578	.984930	.984872	3.3
3.4	.988496	.988201	.987934	.987472	.986660	.985949	.985828	3.4
3.5	.989498	.989206	.988940	.988478	.987650	.986887	.986712	3.5
3.6	.990406	.990119	.989855	.989394	.988558	.987752	.987528	3.6
3.7	.991230	.990947	.990687	.990230	.989390	.988550	.988285	3.7
3.8	.991977	.991700	.991445	.990994	.990153	.989287	.988986	3.8
3.9	.992655	.992385	.992135	.991691	.990855	.989968	.989637	3.9
4.0	.993272	.993009	.992765	.992329	.991500	.990598	.990241	4.0
4.1	.993832	.993577	.993339	.992913	.992094	.991183	.990803	4.1
4.2	.994343	.994095	.993864	.993448	.992641	.991724	.991326	4.2
4.3	.994807	.994568	.994344	.993938	.993145	.992227	.991814	4.3
4.4	.995231	.995000	.994782	.994388	.993610	.992694	.992268	4.4
4.5	.995617	.995394	.995184	.994801	.994039	.993128	.992692	4.5
4.6	.995969	.995754	.995552	.995180	.994436	.993532	.993089	4.6
4.7	.996291	.996084	.995889	.995529	.994804	.993908	.993459	4.7
4.8	.996585	.996386	.996198	.995850	.995143	.994259	.993806	4.8
4.9	.996854	.996663	.996482	.996146	.995458	.994586	.994131	4.9
5.0	.997100	.996917	.996742	.996418	.995750	.994892	.994435	5.0
5.1	.997325	.997150	.996982	.996670	.996020	.995177	.994720	5.1
5.2	.997531	.997363	.997202	.996901	.996272	.995443	.994988	5.2
5.3	.997720	.997559	.997405	.997115	.996505	.995693	.995240	5.3
5.4	.997894	.997740	.997592	.997313	.996722	.995927	.995476	5.4
5.5	.998053	.997906	.997764	.997496	.996924	.996145	.995699	5.5
5.6	.998199	.998058	.997922	.997665	.997111	.996350	.995908	5.6
5.7	.998333	.998199	.998069	.997821	.997286	.996543	.996106	5.7
5.8	.998457	.998329	.998204	.997966	.997449	.996723	.996291	5.8
5.9	.998570	.998448	.998329	.998100	.997601	.996893	.996467	5.9
6.0	.998675	.998558	.998444	.998225	.997743	.997052	.996632	6.0
6.1	.998771	.998660	.998551	.998341	.997875	.997202	.996788	6.1
6.2	.998860	.998754	.998649	.998448	.997999	.997343	.996936	6.2
6.3	.998942	.998840	.998741	.998547	.998114	.997476	.997075	6.3
6.4	.999017	.998921	.998825	.998640	.998222	.997601	.997207	6.4
6.5	.999087	.998995	.998904	.998726	.998323	.997719	.997332	6.5
6.6	.999151	.999063	.998976	.998806	.998418	.997830	.997450	6.6
6.7	.999211	.999127	.999044	.998881	.998506	.997935	.997562	6.7
6.8	.999266	.999186	.999106	.998950	.998589	.998034	.997668	6.8
6.9	.999316	.999240	.999164	.999015	.998667	.998127	.997769	6.9
7.0	.999363	.999291	.999218	.999075	.998740	.998216	.997864	7.0
7.1	.999407	.999338	.999269	.999131	.998809	.998299	.997955	7.1
7.2	.999447	.999381	.999315	.999184	.998873	.998378	.998041	7.2
7.3	.999485	.999422	.999359	.999233	.998934	.998453	.998123	7.3
7.4	.999519	.999459	.999399	.999278	.998990	.998523	.998200	7.4

TABLE A.3.3 Cumulative Distribution Function of Standardized Inverse Gaussian Distribution (α_3; 0, 1)

z \ α_3	0.0	0.1	0.2	0.3	0.4	0.5	0.6	
-3.0	.001350	.000827	.000446	.000200	.000068	.000015	.000002	-3.0
-2.9	.001866	.001203	.000693	.000342	.000134	.000037	.000005	-2.9
-2.8	.002555	.001726	.001059	.000569	.000252	.000083	.000017	-2.8
-2.7	.003467	.002446	.001589	.000922	.000456	.000178	.000048	-2.7
-2.6	.004661	.003423	.002342	.001458	.000795	.000358	.000120	-2.6
-2.5	.006210	.004731	.003396	.002252	.001339	.000682	.000275	-2.5
-2.4	.008198	.006462	.004844	.003398	.002181	.001236	.000585	-2.4
-2.3	.010724	.008721	.006799	.005018	.003445	.002141	.001156	-2.3
-2.2	.013903	.011633	.009396	.007256	.005284	.003556	.002145	-2.2
-2.1	.017864	.015338	.012792	.010284	.007886	.005683	.003764	-2.1
-2.0	.022750	.019996	.017161	.014297	.011470	.008764	.006277	-2.0
-1.9	.028717	.025780	.022698	.019513	.016281	.013079	.010001	-1.9
-1.8	.035930	.032873	.029608	.026165	.022586	.018935	.015292	-1.8
-1.7	.044565	.041468	.038106	.034495	.030663	.026652	.022527	-1.7
-1.6	.054799	.051760	.048409	.044747	.040786	.036550	.032079	-1.6
-1.5	.066807	.063939	.060724	.057150	.053216	.048925	.044297	-1.5
-1.4	.080757	.078184	.075243	.071914	.068182	.064037	.059476	-1.4
-1.3	.096800	.094652	.092131	.089212	.085871	.082085	.077838	-1.3
-1.2	.115070	.113475	.111520	.109174	.106410	.103199	.099513	-1.2
-1.1	.135666	.134746	.133492	.131873	.129860	.127421	.124526	-1.1
-1.0	.158655	.158516	.158081	.157321	.156207	.154708	.152794	-1.0
-.9	.184060	.184786	.185261	.185460	.185356	.184923	.184132	-.9
-.8	.211855	.213500	.214944	.216164	.217139	.217844	.218258	-.8
-.7	.241964	.244548	.246979	.249240	.251310	.253172	.254806	-.7
-.6	.274253	.277759	.281156	.284429	.287563	.290541	.293350	-.6
-.5	.308538	.312907	.317205	.321418	.325533	.329537	.333418	-.5
-.4	.344578	.349716	.354810	.359849	.364819	.369712	.374514	-.4
-.3	.382089	.387864	.393614	.399327	.404993	.410601	.416141	-.3
-.2	.420740	.426993	.433230	.439441	.445615	.451742	.457812	-.2
-.1	.460172	.466720	.473256	.479769	.486250	.492688	.499073	-.1
.0	.500000	.506647	.513283	.519898	.526479	.533019	.539507	.0
.1	.539828	.546376	.552914	.559430	.565914	.572357	.578748	.1
.2	.579260	.585517	.591767	.597999	.604203	.610369	.616487	.2
.3	.617911	.623701	.629493	.635277	.641041	.646776	.652471	.3
.4	.655422	.660591	.665781	.670978	.676171	.681350	.686504	.4
.5	.691462	.695890	.700362	.704867	.709390	.713920	.718447	.5
.6	.725747	.729345	.733021	.736759	.740547	.744369	.748213	.6
.7	.758036	.760755	.763589	.766523	.769540	.772625	.775762	.7
.8	.788145	.789969	.791953	.794075	.796318	.798663	.801094	.8
.9	.815940	.816891	.818045	.819378	.820869	.822498	.824247	.9
1.0	.841345	.841475	.841847	.842438	.843222	.844177	.845283	1.0
1.1	.864334	.863719	.863382	.863296	.863434	.863774	.864293	1.1
1.2	.884930	.883667	.882708	.882026	.881593	.881386	.881382	1.2
1.3	.903200	.901397	.899915	.898726	.897804	.897125	.896666	1.3
1.4	.919243	.917017	.915115	.913513	.912188	.911115	.910273	1.4

TABLE A.3.3 Continued

$(\alpha_3; 0, 1)$

z \ α_3	0.0	0.1	0.2	0.3	0.4	0.5	0.6	
1.5	.933193	.930659	.928441	.926520	.924875	.923486	.922332	1.5
1.6	.945201	.942469	.940035	.937885	.936003	.934372	.932973	1.6
1.7	.955435	.952607	.950049	.947753	.945709	.943904	.942325	1.7
1.8	.964070	.961236	.958636	.956268	.954130	.952215	.950510	1.8
1.9	.971283	.968519	.965946	.963571	.961399	.959427	.957650	1.9
2.0	.977250	.974615	.972126	.969798	.967642	.965661	.963853	2.0
2.1	.982136	.979675	.977314	.975077	.972978	.971026	.969226	2.1
2.2	.986097	.983842	.981642	.979527	.977517	.975627	.973864	2.2
2.3	.989276	.987244	.985227	.983257	.981361	.979556	.977855	2.3
2.4	.991802	.990001	.988177	.986369	.984603	.982901	.981279	2.4
2.5	.993790	.992217	.990591	.988950	.987324	.985737	.984208	2.5
2.6	.995339	.993983	.992552	.991080	.989599	.988135	.986707	2.6
2.7	.996533	.995381	.994135	.992829	.991494	.990155	.988833	2.7
2.8	.997445	.996479	.995407	.994259	.993066	.991852	.990638	2.8
2.9	.998134	.997334	.996421	.995423	.994365	.993272	.992166	2.9
3.0	.998650	.997995	.997226	.996365	.995434	.994458	.993457	3.0
3.1	.999032	.998503	.997861	.997124	.996312	.995446	.994544	3.1
3.2	.999313	.998889	.998359	.997733	.997029	.996265	.995459	3.2
3.3	.999517	.999182	.998747	.998220	.997614	.996944	.996227	3.3
3.4	.999663	.999401	.999048	.998607	.998089	.997505	.996870	3.4
3.5	.999767	.999565	.999280	.998914	.998473	.997967	.997407	3.5
3.6	.999841	.999686	.999458	.999157	.998784	.998347	.997855	3.6
3.7	.999892	.999774	.999594	.999347	.999033	.998658	.998229	3.7
3.8	.999928	.999839	.999698	.999496	.999234	.998913	.998539	3.8
3.9	.999952	.999886	.999776	.999613	.999394	.999121	.998797	3.9
4.0	.999968	.999920	.999834	.999703	.999522	.999290	.999010	4.0
4.1	.999979	.999944	.999878	.999773	.999624	.999428	.999187	4.1
4.2	.999987	.999961	.999911	.999827	.999705	.999540	.999333	4.2
4.3	.999991	.999973	.999935	.999869	.999769	.999630	.999453	4.3
4.4	.999995	.999982	.999953	.999901	.999819	.999704	.999553	4.4
4.5	.999997	.999987	.999966	.999925	.999859	.999763	.999634	4.5
4.6	.999998	.999991	.999975	.999944	.999890	.999810	.999701	4.6
4.7	.999999	.999994	.999982	.999958	.999915	.999849	.999756	4.7
4.8	.999999	.999996	.999987	.999968	.999934	.999879	.999801	4.8
4.9	1.000000	.999997	.999991	.999976	.999949	.999904	.999838	4.9
5.0	1.000000	.999998	.999994	.999982	.999960	.999924	.999869	5.0
5.1	1.000000	.999999	.999996	.999987	.999969	.999939	.999893	5.1
5.2	1.000000	.999999	.999997	.999990	.999977	.999952	.999913	5.2
5.3	1.000000	1.000000	.999998	.999993	.999982	.999962	.999930	5.3
5.4	1.000000	1.000000	.999998	.999995	.999986	.999970	.999943	5.4
5.5	1.000000	1.000000	.999999	.999996	.999989	.999976	.999954	5.5
5.6	1.000000	1.000000	.999999	.999997	.999992	.999981	.999963	5.6
5.7	1.000000	1.000000	.999999	.999998	.999994	.999985	.999970	5.7
5.8	1.000000	1.000000	1.000000	.999998	.999995	.999988	.999976	5.8
5.9	1.000000	1.000000	1.000000	.999999	.999996	.999991	.999980	5.9

TABLE A.3.3 Continued

	(α_3; 0, 1)							
$z \backslash \alpha_3$	0.7	0.8	0.9	1.0	1.1	1.2	1.3	
-3.0	.000000	.000000	.000000					-3.0
-2.9	.000000	.000000	.000000	.000000				-2.9
-2.8	.000001	.000000	.000000	.000000				-2.8
-2.7	.000007	.000000	.000000	.000000	.000000			-2.7
-2.6	.000025	.000002	.000000	.000000	.000000			-2.6
-2.5	.000077	.000011	.000000	.000000	.000000	.000000		-2.5
-2.4	.000209	.000047	.000005	.000000	.000000	.000000		-2.4
-2.3	.000506	.000159	.000028	.000002	.000000	.000000	.000000	-2.3
-2.2	.001105	.000448	.000122	.000016	.000000	.000000	.000000	-2.2
-2.1	.002215	.001095	.000413	.000098	.000010	.000000	.000000	-2.1
-2.0	.004115	.002379	.001141	.000406	.000086	.000007	.000000	-2.0
-1.9	.007161	.004684	.002691	.001271	.000438	.000086	.000005	-1.9
-1.8	.011765	.008482	.005590	.003238	.001541	.000532	.000103	-1.8
-1.7	.018373	.014306	.010470	.007033	.004175	.002055	.000744	-1.7
-1.6	.027434	.022702	.018001	.013484	.009341	.005784	.003024	-1.6
-1.5	.039362	.034176	.028819	.023402	.018081	.013053	.008560	-1.5
-1.4	.054507	.049151	.043448	.037464	.031296	.025085	.019019	-1.4
-1.3	.073118	.067919	.062248	.056128	.049601	.042738	.035647	-1.3
-1.2	.095328	.090622	.085379	.079588	.073252	.066385	.059023	-1.2
-1.1	.121142	.117240	.112791	.107768	.102147	.095909	.089045	-1.1
-1.0	.150434	.147594	.144243	.140347	.135873	.130789	.125063	-1.0
-.9	.182955	.181363	.179324	.176808	.173783	.170215	.166069	-.9
-.8	.218354	.218110	.217500	.216499	.215078	.213212	.210872	-.8
-.7	.256193	.257314	.258150	.258681	.258887	.258750	.258247	-.7
-.6	.295973	.298397	.300608	.302590	.304331	.305818	.307036	-.6
-.5	.337163	.340762	.344203	.347478	.350575	.353488	.356206	-.5
-.4	.379218	.383812	.388290	.392642	.396863	.400945	.404883	-.4
-.3	.421604	.426982	.432267	.437454	.442535	.447505	.452362	-.3
-.2	.463818	.469750	.475602	.481367	.487041	.492618	.498095	-.2
-.1	.505397	.511653	.517832	.523930	.529940	.535859	.541681	-.1
.0	.545934	.552293	.558577	.564779	.570895	.576919	.582849	.0
.1	.585080	.591344	.597533	.603641	.609663	.615594	.621430	.1
.2	.622549	.628546	.634471	.640319	.646084	.651761	.657347	.2
.3	.658118	.663708	.669235	.674691	.680072	.685373	.690588	.3
.4	.691623	.696700	.701726	.706695	.711600	.716436	.721199	.4
.5	.722959	.727448	.731903	.736319	.740688	.745004	.749262	.5
.6	.752068	.755924	.759769	.763595	.767395	.771162	.774889	.6
.7	.778939	.782143	.785363	.788588	.791810	.795020	.798212	.7
.8	.803596	.806154	.808755	.811387	.814040	.816705	.819372	.8
.9	.826096	.828031	.830037	.832100	.834209	.836352	.838520	.9
1.0	.846521	.847872	.849320	.850849	.852448	.854103	.855804	1.0
1.1	.864970	.865785	.866721	.867762	.868893	.870100	.871372	1.1
1.2	.881557	.881893	.882370	.882970	.883680	.884483	.885368	1.2
1.3	.896405	.896320	.896393	.896606	.896943	.897388	.897929	1.3
1.4	.909641	.909197	.908922	.908799	.908812	.908945	.909185	1.4

TABLE A.3.3 Continued

$(\alpha_3; 0, 1)$

z \ α_3	0.7	0.8	0.9	1.0	1.1	1.2	1.3	
1.5	.921393	.920649	.920082	.919675	.919412	.919277	.919257	1.5
1.6	.931790	.930803	.929997	.929354	.928859	.928499	.928259	1.6
1.7	.940954	.939778	.938781	.937948	.937265	.936718	.936294	1.7
1.8	.949006	.947690	.946547	.945565	.944732	.944034	.943460	1.8
1.9	.956058	.954644	.953396	.952303	.951354	.950538	.949845	1.9
2.0	.962216	.960742	.959424	.958253	.957220	.956315	.955528	2.0
2.1	.967577	.966077	.964719	.963499	.962408	.961439	.960584	2.1
2.2	.972232	.970732	.969362	.968117	.966992	.965981	.965078	2.2
2.3	.976264	.974787	.973425	.972176	.971037	.970003	.969070	2.3
2.4	.979747	.978312	.976976	.975740	.974603	.973562	.972614	2.4
2.5	.982749	.981369	.980073	.978864	.977743	.976709	.975759	2.5
2.6	.985331	.984016	.982772	.981601	.980507	.979489	.978548	2.6
2.7	.987546	.986305	.985119	.983995	.982936	.981945	.981021	2.7
2.8	.989443	.988280	.987159	.986087	.985071	.984112	.983212	2.8
2.9	.991064	.989981	.988928	.987914	.986944	.986023	.985154	2.9
3.0	.992447	.991445	.990462	.989507	.988588	.987709	.986873	3.0
3.1	.993625	.992703	.991790	.990896	.990029	.989194	.988395	3.1
3.2	.994627	.993782	.992939	.992106	.991291	.990502	.989742	3.2
3.3	.995476	.994707	.993931	.993158	.992397	.991654	.990934	3.3
3.4	.996196	.994498	.994787	.994073	.993364	.992667	.991988	3.4
3.5	.996806	.996175	.995526	.994868	.994210	.993559	.992921	3.5
3.6	.997320	.996752	.996162	.995559	.994950	.994344	.993745	3.6
3.7	.997754	.997245	.996710	.996158	.995597	.995034	.994474	3.7
3.8	.998120	.997665	.997181	.996678	.996162	.995640	.995119	3.8
3.9	.998428	.998022	.997586	.997128	.996655	.996173	.995688	3.9
4.0	.998687	.998326	.997934	.997519	.997086	.996642	.996192	4.0
4.1	.998904	.998584	.998233	.997857	.997462	.997053	.996637	4.1
4.2	.999086	.998803	.998489	.998150	.997790	.997415	.997030	4.2
4.3	.999239	.998989	.998709	.998403	.998075	.997732	.997377	4.3
4.4	.999366	.999147	.998897	.998622	.998325	.998011	.997683	4.4
4.5	.999473	.999280	.999058	.998811	.998542	.998255	.997954	4.5
4.6	.999562	.999393	.999196	.998975	.998731	.998470	.998194	4.6
4.7	.999636	.999489	.999314	.999116	.998896	.998658	.998405	4.7
4.8	.999698	.999569	.999415	.999238	.999040	.998823	.998592	4.8
4.9	.999750	.999637	.999502	.999343	.999165	.998968	.998756	4.9
5.0	.999793	.999695	.999575	.999434	.999274	.999095	.998902	5.0
5.1	.999828	.999744	.999638	.999513	.999368	.999207	.999030	5.1
5.2	.999858	.999784	.999692	.999580	.999451	.999305	.999144	5.2
5.3	.999883	.999819	.999738	.999639	.999523	.999391	.999244	5.3
5.4	.999903	.999848	.999777	.999689	.999585	.999466	.999333	5.4
5.5	.999920	.999872	.999810	.999732	.999639	.999532	.999411	5.5
5.6	.999934	.999893	.999838	.999770	.999687	.999590	.999480	5.6
5.7	.999945	.999910	.999863	.999802	.999728	.999640	.999541	5.7
5.8	.999955	.999925	.999883	.999830	.999763	.999685	.999594	5.8
5.9	.999963	.999937	.999901	.999853	.999794	.999724	.999642	5.9

TABLE A.3.3 Continued

(α_3; 0, 1)

$z \backslash \alpha_3$	1.4	1.5	1.6	1.7	1.8	1.9	2.0	
-1.5	.004865	.002204	.000674	.000094	.000002	.000000		-1.5
-1.4	.013347	.008372	.004418	.001749	.000406	.000029	.000000	-1.4
-1.3	.028488	.021482	.014926	.009188	.004673	.001713	.000331	-1.3
-1.2	.051233	.043119	.034844	.026640	.018835	.011857	.006210	-1.2
-1.1	.081556	.073460	.064802	.055664	.046179	.036556	.027105	-1.1
-1.0	.118667	.111575	.103769	.095242	.086004	.076091	.065577	-1.0
-.9	.161309	.155899	.149802	.142979	.135396	.127020	.117823	-.9
-.8	.208025	.204642	.200686	.196123	.190912	.185015	.178385	-.8
-.7	.257358	.256059	.254328	.252138	.249463	.246274	.242537	-.7
-.6	.307972	.308612	.308943	.308950	.308617	.307928	.306865	-.6
-.5	.358722	.361028	.363116	.364980	.366612	.368004	.369149	-.5
-.4	.408673	.412309	.415788	.419106	.422260	.425248	.428065	-.4
-.3	.457100	.461717	.466211	.470579	.474821	.478935	.482920	-.3
-.2	.503468	.508735	.513894	.518944	.523884	.528713	.533432	-.2
-.1	.547405	.553028	.558548	.563964	.569276	.574482	.579583	-.1
.0	.588680	.594411	.600039	.605564	.610985	.616302	.621514	.0
.1	.627169	.632808	.638345	.643779	.649110	.654337	.659459	.1
.2	.662838	.668231	.673526	.678721	.683815	.688808	.693700	.2
.3	.695716	.700754	.705699	.710550	.715307	.719968	.724534	.3
.4	.725885	.730491	.735015	.739455	.743810	.748079	.752261	.4
.5	.753458	.757588	.761650	.765640	.769558	.773402	.777171	.5
.6	.778572	.782206	.785787	.789313	.792781	.796189	.799536	.6
.7	.801379	.804515	.807617	.810680	.813701	.816678	.819608	.7
.8	.822036	.824689	.827325	.829940	.832530	.835090	.837619	:8
.9	.840704	.842896	.845091	.847281	.849463	.851631	.853782	.9
1.0	.857540	.859303	.861086	.862882	.864684	.866487	.868287	1.0
1.1	.872698	.874067	.875472	.876905	.878360	.879829	.881308	1.1
1.2	.886323	.887336	.888400	.889505	.890644	.891811	.892999	1.2
1.3	.898553	.899250	.900008	.900820	.901677	.902572	.903499	1.3
1.4	.909519	.909936	.910425	.910977	.911584	.912238	.912933	1.4
1.5	.919340	.919514	.919768	.920093	.920480	.920923	.921413	1.5
1.6	.928127	.928092	.928144	.928272	.928469	.928727	.929038	1.6
1.7	.935982	.935771	.935650	.935610	.935643	.935741	.935897	1.7
1.8	.942999	.942641	.942375	.942193	.942086	.942047	.942070	1.8
1.9	.949263	.948784	.948398	.948097	.947873	.947719	.947628	1.9
2.0	.954852	.954276	.953793	.953394	.953072	.952821	.952634	2.0
2.1	.959834	.959183	.958622	.958145	.957744	.957413	.957146	2.1
2.2	.964275	.963567	.962946	.962407	.961942	.961546	.961213	2.2
2.3	.968231	.967483	.966817	.966230	.965715	.965267	.964881	2.3
2.4	.971754	.970979	.970282	.969660	.969108	.968619	.968191	2.4
2.5	.974890	.974100	.973384	.972738	.972158	.971640	.971179	2.5
2.6	.977681	.976886	.976160	.975500	.974902	.974362	.973878	2.6
2.7	.980164	.979373	.978645	.977979	.977370	.976817	.976315	2.7
2.8	.982373	.981592	.980870	.980203	.979591	.979030	.978519	2.8
2.9	.984337	.983573	.982861	.982201	.981590	.981027	.980511	2.9

TABLE A.3.3 Continued

(α_3; 0, 1)

z \ α_3	1.4	1.5	1.6	1.7	1.8	1.9	2.0	
3.0	.986083	.985340	.984643	.983994	.983390	.982830	.982313	3.0
3.1	.987635	.986916	.986239	.985604	.985010	.984457	.983943	3.1
3.2	.989015	.988323	.987668	.987050	.986470	.985926	.985419	3.2
3.3	.990241	.989578	.988947	.988349	.987785	.987253	.986755	3.3
3.4	.991331	.990698	.990093	.989516	.988969	.988452	.987965	3.4
3.5	.992299	.991697	.991119	.990565	.990037	.989536	.989062	3.5
3.6	.993159	.992589	.992038	.991507	.991000	.990516	.990056	3.6
3.7	.993923	.993384	.992860	.992354	.991868	.991402	.990958	3.7
3.8	.994602	.994094	.993598	.993116	.992651	.992204	.991775	3.8
3.9	.995205	.994727	.994258	.993801	.993357	.992929	.992517	3.9
4.0	.995740	.995292	.994850	.994416	.993994	.993585	.993190	4.0
4.1	.996216	.995796	.995380	.994970	.994569	.994179	.993801	4.1
4.2	.996639	.996246	.995855	.995469	.995089	.994717	.994356	4.2
4.3	.997014	.996648	.996281	.995917	.995557	.995205	.994860	4.3
4.4	.997347	.997006	.996663	.996320	.995981	.995646	.995318	4.4
4.5	.997644	.997326	.997005	.996683	.996363	.996046	.995735	4.5
4.6	.997906	.997612	.997312	.997010	.996708	.996409	.996113	4.6
4.7	.998140	.997866	.997587	.997304	.997020	.996737	.996457	4.7
4.8	.998348	.998094	.997834	.997569	.997302	.997036	.996770	4.8
4.9	.998532	.998297	.998055	.997808	.997557	.997306	.997055	4.9
5.0	.998695	.998479	.998254	.998023	.997788	.997551	.997314	5.0
5.1	.998841	.998640	.998432	.998216	.997996	.997774	.997550	5.1
5.2	.998970	.998785	.998591	.998391	.998185	.997976	.997765	5.2
5.3	.999085	.998914	.998735	.998548	.998355	.998159	.997960	5.3
5.4	.999187	.999030	.998863	.998690	.998510	.998325	.998138	5.4
5.5	.999277	.999133	.998979	.998817	.998649	.998477	.998301	5.5
5.6	.999358	.999225	.999082	.998932	.998776	.998614	.998448	5.6
5.7	.999429	.999307	.999176	.999036	.998890	.998739	.998583	5.7
5.8	.999492	.999380	.999259	.999130	.998994	.998852	.998706	5.8
5.9	.999549	.999446	.999334	.999214	.999087	.998955	.998818	5.9
6.0	.999599	.999505	.999401	.999290	.999172	.999049	.998920	6.0
6.1	.999644	.999557	.999462	.999359	.999249	.999134	.999013	6.1
6.2	.999683	.999604	.999516	.999421	.999319	.999211	.999098	6.2
6.3	.999718	.999646	.999565	.999477	.999382	.999281	.999176	6.3
6.4	.999749	.999683	.999609	.999527	.999439	.999345	.999246	6.4
6.5	.999777	.999717	.999648	.999573	.999491	.999404	.999311	6.5
6.6	.999802	.999746	.999684	.999614	.999538	.999457	.999370	6.6
6.7	.999824	.999773	.999715	.999651	.999581	.999505	.999424	6.7
6.8	.999843	.999797	.999744	.999685	.999619	.999549	.999473	6.8
6.9	.999861	.999818	.999770	.999715	.999654	.999589	.999518	6.9
7.0	.999876	.999837	.999793	.999742	.999686	.999625	.999559	7.0
7.1	.999890	.999855	.999814	.999767	.999715	.999658	.999596	7.1
7.2	.999902	.999870	.999832	.999789	.999741	.999688	.999631	7.2
7.3	.999913	.999884	.999849	.999809	.999765	.999716	.999662	7.3
7.4	.999922	.999896	.999864	.999828	.999786	.999741	.999691	7.4

TABLE A.3.3 Continued

z \ α_3	(α_3; 0, 1) 2.1	2.2	2.3	2.5	3.0	4.0	5.0	
-1.5								-1.5
-1.4	.000000							-1.4
-1.3	.000014	.000000	.000000					-1.3
-1.2	.002366	.000480	.000021					-1.2
-1.1	.018263	.010608	.004799	.000129				-1.1
-1.0	.054597	.043373	.032248	.012543	.000000			-1.0
-.9	.107792	.096930	.085272	.059978	.004076			-.9
-.8	.170980	.162750	.153649	.132662	.063754			-.8
-.7	.238220	.233284	.227688	.214331	.165727	.006350		-.7
-.6	.305411	.303545	.301245	.295244	.270614	.157885	.000000	-.6
-.5	.370038	.370664	.371018	.370872	.364976	.321474	.199098	-.5
6.4	.430711	.433183	.435478	.439527	.446384	.444418	.414582	-.4
-.3	.486777	.490506	.494105	.500920	.515738	.536137	.544306	-.3
-.2	.538041	.542540	.546931	.555390	.574724	.606335	.629925	-.2
-.1	.584579	.589472	.594262	.603537	.625023	.661482	.690804	-.1
.0	.626622	.631627	.636530	.646033	.668102	.705781	.736403	.0
.1	.664479	.669395	.674210	.683538	.705178	.742030	.771865	.1
.2	.698490	.703181	.707772	.716662	.737246	.772151	.800234	.2
.3	.729005	.733381	.737663	.745950	.765114	.797504	.823429	.3
.4	.756357	.760367	.764291	.771885	.789442	.819076	.842730	.4
.5	.780865	.784483	.788026	.794887	.810768	.837602	.859020	.5
.6	.802820	.806040	.809198	.815321	.829535	.853641	.872933	.6
.7	.822489	.825320	.828100	.833505	.846108	.867623	.884935	.7
.8	.840114	.842572	.844992	.849713	.860793	.879889	.895377	.8
.9	.855913	.858020	.860103	.864183	.873844	.890707	.904531	.9
1.0	.870081	.871864	.873633	.877122	.885475	.900294	.912606	1.0
1.1	.882792	.884278	.885761	.888709	.895869	.908828	.919771	1.1
1.2	.894203	.895419	.896643	.899100	.905179	.916454	.926160	1.2
1.3	.904452	.905426	.906417	.908433	.913536	.923293	.931883	1.3
1.4	.913662	.914420	.915203	.916824	.921055	.929446	.937030	1.4
1.5	.921944	.922511	.923110	.924380	.927832	.934998	.941676	1.5
1.6	.929396	.929795	.930231	.931192	.933952	.940022	.945885	1.6
1.7	.936104	.936357	.936650	.937340	.939487	.944579	.949708	1.7
1.8	.942147	.942273	.942442	.942894	.944502	.948721	.953191	1.8
1.9	.947593	.947610	.947673	.947918	.949053	.952496	.956372	1.9
2.0	.952505	.952428	.952399	.952466	.953188	.955941	.959284	2.0
2.1	.956937	.956781	.956674	.956588	.956950	.959093	.961956	2.1
2.2	.960938	.960717	.960544	.960327	.960378	.961980	.964412	2.2
2.3	.964552	.964276	.964048	.963721	.963504	.964629	.966675	2.3
2.4	.967819	.967498	.967225	.966806	.966358	.967064	.968762	2.4
2.5	.970773	.970416	.970106	.969610	.968968	.969305	.970692	2.5
2.6	.973445	.973060	.972720	.972163	.971355	.971370	.972478	2.6
2.7	.975863	.975457	.975095	.974488	.973542	.973276	.974133	2.7
2.8	.978054	.977632	.977252	.976606	.975547	.975037	.975670	2.8
2.9	.980038	.979606	.979214	.978539	.977387	.976665	.977098	2.9

TABLE A.3.3 Continued

$(\alpha_3; 0, 1)$

$z \backslash \alpha_3$	2.1	2.2	2.3	2.5	3.0	4.0	5.0	
3.0	.981836	.981399	.980999	.980302	.979076	.978172	.978427	3.0
3.1	.983467	.983028	.982623	.981912	.980629	.979570	.979665	3.1
3.2	.984946	.984508	.984102	.983384	.982057	.980866	.980820	3.2
3.3	.986289	.985854	.985450	.984729	.983372	.982069	.981898	3.3
3.4	.987508	.987079	.986679	.985959	.984583	.983188	.982905	3.4
3.5	.988615	.988194	.987800	.987086	.985699	.984228	.983847	3.5
3.6	.989621	.989209	.988822	.988117	.986728	.985197	.984729	3.6
3.7	.990535	.990134	.989755	.989062	.987678	.986099	.985556	3.7
3.8	.991366	.990977	.990608	.989928	.988555	.986940	.986331	3.8
3.9	.992122	.991745	.991386	.990723	.989366	.987725	.987058	3.9
4.0	.992810	.992446	.992098	.991452	.990115	.988457	.987741	4.0
4.1	.993436	.993085	.992749	.992121	.990808	.989142	.988383	4.1
4.2	.994006	.993669	.993344	.992735	.991450	.989781	.988986	4.2
4.3	.994526	.994202	.993889	.993300	.992044	.990379	.989554	4.3
4.4	.994999	.994688	.994387	.993818	.992594	.990939	.990089	4.4
4.5	.995430	.995133	.994844	.994295	.993104	.991463	.990593	4.5
4.6	.995823	.995539	.995262	.994734	.993576	.991954	.991068	4.6
4.7	.996181	.995910	.995645	.995138	.994015	.992414	.991516	4.7
4.8	.996508	.996249	.995996	.995509	.994422	.992846	.991939	4.8
4.9	.996806	.996560	.996318	.995851	.994799	.993250	.992339	4.9
5.0	.997078	.996844	.996614	.996166	.995150	.993630	.992716	5.0
5.1	.997326	.997104	.996885	.996456	.995476	.993987	.993072	5.1
5.2	.997553	.997342	.997133	.996724	.995779	.994322	.993409	5.2
5.3	.997760	.997560	.997361	.996970	.996060	.994637	.993728	5.3
5.4	.997949	.997760	.997571	.997198	.996322	.994933	.994030	5.4
5.5	.998122	.997943	.997763	.997407	.996566	.995212	.994316	5.5
5.6	.998280	.998110	.997940	.997601	.996793	.995474	.994586	5.6
5.7	.998425	.998264	.998102	.997780	.997004	.995720	.994843	5.7
5.8	.998557	.998405	.998252	.997944	.997200	.995952	.995086	5.8
5.9	.998677	.998534	.998389	.998097	.997383	.996171	.995316	5.9
6.0	.998788	.998652	.998515	.998237	.997554	.996377	.995535	6.0
6.1	.998889	.998761	.998631	.998367	.997713	.996571	.995742	6.1
6.2	.998981	.998861	.998738	.998487	.997861	.996754	.995939	6.2
6.3	.999066	.998952	.998836	.998598	.997999	.996927	.996126	6.3
6.4	.999143	.999036	.998927	.998701	.998128	.997090	.996304	6.4
6.5	.999214	.999113	.999010	.998796	.998248	.997243	.996472	6.5
6.6	.999279	.999184	.999086	.998884	.998361	.997388	.996633	6.6
6.7	.999338	.999249	.999157	.998965	.998465	.997525	.996785	6.7
6.8	.999393	.999309	.999222	.999040	.998563	.997655	.996930	6.8
6.9	.999443	.999364	.999282	.999110	.998655	.997777	.997068	6.9
7.0	.999489	.999414	.999337	.999174	.998740	.997892	.997199	7.0
7.1	.999531	.999461	.999388	.999234	.998820	.998001	.997324	7.1
7.2	.999569	.999503	.999435	.999289	.998894	.998104	.997442	7.2
7.3	.999604	.999543	.999478	.999340	.998964	.998202	.997555	7.3
7.4	.999636	.999579	.999518	.999387	.999029	.998294	.997663	7.4

TABLE A.3.4 Cumulative Distribution Function of Standardized Gamma Distribution

(α_3; 0, 1)

$z \backslash \alpha_3$	0.0	0.1	0.2	0.3	0.4	0.5	0.6	
-3.0	.001350	.000824	.000430	.000177	.000047	.000005	.000000	-3.0
-2.9	.001866	.001199	.000673	.000309	.000099	.000016	.000000	-2.9
-2.8	.002555	.001721	.001034	.000523	.000199	.000043	.000002	-2.8
-2.7	.003467	.002440	.001557	.000862	.000378	.000108	.000011	-2.7
-2.6	.004661	.003416	.002305	.001382	.000686	.000244	.000042	-2.6
-2.5	.006210	.004724	.003352	.002157	.001192	.000509	.000130	-2.5
-2.4	.008198	.006454	.004794	.003286	.001994	.000992	.000340	-2.4
-2.3	.010724	.008713	.006745	.004890	.003217	.001816	.000783	-2.3
-2.2	.013903	.011624	.009340	.007116	.005020	.003149	.001623	-2.2
-2.1	.017864	.015330	.012736	.010136	.007594	.005202	.003088	-2.1
-2.0	.022750	.019989	.017108	.014150	.011165	.008231	.005468	-2.0
-1.9	.028717	.025774	.022652	.019377	.015982	.012525	.009103	-1.9
-1.8	.035930	.032869	.029572	.026050	.022315	.018402	.014372	-1.8
-1.7	.044565	.041467	.038086	.034414	.030445	.026188	.021666	-1.7
-1.6	.054799	.051762	.048407	.044709	.040646	.036202	.031368	-1.6
-1.5	.066807	.063945	.060744	.057165	.053176	.048740	.043821	-1.5
-1.4	.080757	.078193	.075288	.071990	.068260	.064051	.059306	-1.4
-1.3	.096800	.094665	.092202	.089353	.086077	.082321	.078021	-1.3
-1.2	.115070	.113492	.111616	.109379	.106746	.103663	.100068	-1.2
-1.1	.135666	.134766	.133612	.132139	.130319	.128104	.125440	-1.1
-1.0	.158655	.158539	.158221	.157639	.156773	.155584	.154029	-1.0
-.9	.184060	.184811	.185416	.185819	.186007	.185953	.185624	-.9
-.8	.211855	.213527	.215109	.216549	.217845	.218978	.219926	-.8
-.7	.241964	.244575	.247147	.249634	.252039	.254353	.256561	-.7
-.6	.274253	.277785	.281319	.284815	.288281	.291711	.295099	-.6
-.5	.308538	.312931	.317357	.321779	.326207	.330640	.335076	-.5
-.4	.344578	.349737	.354944	.360167	.365419	.370699	.376006	-.4
-.3	.382089	.387881	.393723	.399591	.405494	.411432	.417406	-.3
-.2	.420740	.427005	.433311	.439640	.445999	.452389	.458809	-.2
-.1	.460172	.466727	.473304	.479896	.486505	.493132	.499776	-.1
.0	.500000	.506647	.513299	.519949	.526602	.533255	.539910	.0
.1	.539828	.546371	.552896	.559408	.565906	.572390	.578861	.1
.2	.579260	.585506	.591719	.597909	.604073	.610214	.616330	.2
.3	.617911	.623686	.629417	.635125	.640803	.646453	.652075	.3
.4	.655422	.660573	.665682	.670774	.675842	.680887	.685909	.4
.5	.691462	.695869	.700245	.704622	.708990	.713347	.717694	.5
.6	.725747	.729322	.732891	.736487	.740096	.743716	.747346	.6
.7	.758036	.760731	.763453	.766235	.769060	.771923	.774822	.7
.8	.788145	.789945	.791815	.793782	.795826	.797941	.800121	.8
.9	.815940	.816868	.817911	.819091	.820384	.821782	.823273	.9
1.0	.841345	.841454	.841721	.842165	.842758	.843487	.844340	1.0
1.1	.864334	.863701	.863268	.863045	.863004	.863128	.863403	1.1
1.2	.884930	.883651	.882608	.881802	.881205	.880797	.880562	1.2
1.3	.903200	.901384	.899830	.898533	.897464	.896602	.895931	1.3
1.4	.919243	.917007	.915046	.913353	.911899	.910664	.909629	1.4

TABLE A.3.4 Continued

$(\alpha_3; 0, 1)$

z \ α_3	0.0	0.1	0.2	0.3	0.4	0.5	0.6	
1.5	.933193	.930652	.928388	.926393	.924639	.923108	.921782	1.5
1.6	.945201	.942465	.939998	.937790	.935819	.934066	.932518	1.6
1.7	.955435	.952606	.950027	.947688	.945574	.943669	.941960	1.7
1.8	.964070	.961237	.958626	.956230	.954040	.952044	.950232	1.8
1.9	.971283	.968521	.965947	.963557	.961348	.959315	.957449	1.9
2.0	.977250	.974619	.972136	.969804	.967626	.965600	.963723	2.0
2.1	.982136	.979680	.977332	.975100	.972991	.971010	.969158	2.1
2.2	.986097	.983847	.981665	.979563	.977554	.975648	.973848	2.2
2.3	.989276	.987250	.985254	.983303	.981417	.979608	.977883	2.3
2.4	.991802	.990007	.988207	.986421	.984672	.982976	.981343	2.4
2.5	.993790	.992222	.990621	.989006	.987403	.985830	.984301	2.5
2.6	.995339	.993989	.992582	.991138	.989685	.988240	.986821	2.6
2.7	.996533	.995386	.994165	.992888	.991582	.990268	.988962	2.7
2.8	.997445	.996484	.995435	.994316	.993154	.991969	.990777	2.8
2.9	.998134	.997338	.996447	.995477	.994451	.993390	.992310	2.9
3.0	.998650	.998001	.997250	.996415	.995517	.994574	.993602	3.0
3.1	.999032	.998508	.997882	.997170	.996390	.995558	.994688	3.1
3.2	.999313	.998894	.998377	.997775	.997102	.996372	.995599	3.2
3.3	.999517	.999185	.998763	.998258	.997681	.997045	.996360	3.3
3.4	.999663	.999404	.999062	.998641	.998150	.997598	.996996	3.4
3.5	.999767	.999567	.999292	.998944	.998528	.998053	.997526	3.5
3.6	.999841	.999688	.999468	.999183	.998833	.998425	.997965	3.6
3.7	.999892	.999776	.999603	.999369	.999077	.998728	.998330	3.7
3.8	.999928	.999840	.999705	.999515	.999272	.998976	.998632	3.8
3.9	.999952	.999887	.999781	.999629	.999427	.999177	.998881	3.9
4.0	.999968	.999921	.999839	.999717	.999551	.999340	.999086	4.0
4.1	.999979	.999945	.999882	.999785	.999649	.999472	.999255	4.1
4.2	.999987	.999962	.999914	.999837	.999726	.999578	.999394	4.2
4.3	.999991	.999973	.999937	.999877	.999787	.999664	.999507	4.3
4.4	.999995	.999982	.999955	.999907	.999834	.999733	.999600	4.4
4.5	.999997	.999988	.999967	.999930	.999872	.999788	.999676	4.5
4.6	.999998	.999992	.999977	.999948	.999901	.999832	.999738	4.6
4.7	.999999	.999994	.999983	.999961	.999924	.999867	.999788	4.7
4.8	.999999	.999996	.999988	.999971	.999941	.999895	.999829	4.8
4.9	1.000000	.999998	.999992	.999979	.999955	.999917	.999862	4.9
5.0	1.000000	.999998	.999994	.999984	.999965	.999935	.999889	5.0
5.1	1.000000	.999999	.999996	.999988	.999974	.999949	.999911	5.1
5.2	1.000000	.999999	.999997	.999991	.999980	.999960	.999928	5.2
5.3	1.000000	1.000000	.999998	.999994	.999985	.999968	.999943	5.3
5.4	1.000000	1.000000	.999999	.999995	.999988	.999975	.999954	5.4
5.5	1.000000	1.000000	.999999	.999997	.999991	.999981	.999963	5.5
5.6	1.000000	1.000000	.999999	.999998	.999993	.999985	.999971	5.6
5.7	1.000000	1.000000	1.000000	.999998	.999995	.999988	.999977	5.7
5.8	1.000000	1.000000	1.000000	.999999	.999996	.999991	.999981	5.8
5.9	1.000000	1.000000	1.000000	.999999	.999997	.999993	.999985	5.9

TABLE A.3.4 Continued

(α_3; 0, 1)

z \ α_3	0.7	0.8	0.9	1.0	1.1	1.2	1.3	
-3.0								-3.0
-2.9								-2.9
-2.8	.000000							-2.8
-2.7	.000000							-2.7
-2.6	.000001							-2.6
-2.5	.000008	.000000						-2.5
-2.4	.000049	.000000						-2.4
-2.3	.000191	.000007						-2.3
-2.2	.000571	.000075	.000000					-2.2
-2.1	.001417	.000368	.000012					-2.1
-2.0	.003047	.001203	.000189	.000000				-2.0
-1.9	.005866	.003047	.000986	.000057				-1.9
-1.8	.010335	.006480	.003126	.000776	.000001			-1.8
-1.7	.016941	.012130	.007461	.003358	.000589			-1.7
-1.6	.026153	.020602	.014827	.009080	.003899	.000451		-1.6
-1.5	.038387	.032418	.025926	.018988	.011839	.005089	.000414	-1.5
-1.4	.053967	.047970	.041252	.033769	.025523	.016655	.007706	-1.4
-1.3	.073105	.067486	.061064	.053725	.045348	.035820	.025095	-1.3
-1.2	.095885	.091026	.085380	.078813	.071164	.062229	.051767	-1.2
-1.1	.122259	.118480	.114005	.108708	.102438	.094997	.086132	-1.1
-1.0	.152053	.149592	.146565	.142877	.138403	.132993	.126446	-1.0
-.9	.184981	.183978	.182556	.180648	.178168	.175012	.171048	-.9
-.8	.220664	.221160	.221379	.221277	.220800	.219885	.218454	-.8
-.7	.258649	.260599	.262391	.263998	.265392	.266536	.267388	-.7
-.6	.298439	.301720	.304932	.308063	.311095	.314012	.316792	-.6
-.5	.339511	.343942	.348363	.352768	.357150	.361500	.365807	-.5
-.4	.381340	.386698	.392079	.397480	.402897	.408324	.413755	-.4
-.3	.423415	.429459	.435535	.441643	.447779	.453939	.460119	-.3
-.2	.465260	.471740	.478249	.484784	.491343	.497922	.504517	-.2
-.1	.506438	.513116	.519809	.526515	.533232	.539956	.546683	-.1
.0	.546566	.553222	.559878	.566530	.573177	.579816	.586444	.0
.1	.585317	.591759	.598186	.604597	.610988	.617359	.623706	.1
.2	.622422	.628490	.634534	.640552	.646544	.652506	.658437	.2
.3	.657671	.663239	.668780	.674294	.679778	.685232	.690654	.3
.4	.690908	.695885	.700839	.705770	.710677	.715558	.720411	.4
.5	.722031	.726358	.730672	.734974	.739262	.743535	.747790	.5
.6	.750985	.754630	.758281	.761935	.765590	.769244	.772895	.6
.7	.777754	.780714	.783700	.786709	.789738	.792783	.795841	.7
.8	.802359	.804651	.806992	.809378	.811803	.814263	.816755	.8
.9	.824852	.826510	.828240	.830037	.831894	.833806	.835767	.9
1.0	.845306	.846377	.847543	.848796	.850129	.851535	.853008	1.0
1.1	.863815	.864354	.865009	.865771	.866630	.867578	.868609	1.1
1.2	.880486	.880554	.880756	.881081	.881519	.882060	.882697	1.2
1.3	.895433	.895095	.894904	.894849	.894919	.895103	.895394	1.3
1.4	.908778	.908098	.907574	.907194	.906949	.906827	.906819	1.4

TABLE A.3.4 Continued

$(\alpha_3;\ 0,\ 1)$

z \ α_3	0.7	0.8	0.9	1.0	1.1	1.2	1.3	
1.5	.920646	.919684	.918884	.918235	.917724	.917343	.917081	1.5
1.6	.931158	.929973	.928952	.928083	.927355	.926758	.926284	1.6
1.7	.940435	.939082	.937889	.936847	.935945	.935175	.934527	1.7
1.8	.948594	.947121	.945802	.944629	.943593	.942685	.941899	1.8
1.9	.955745	.954194	.952790	.951523	.950389	.949378	.948484	1.9
2.0	.961992	.960402	.958946	.957620	.956417	.955332	.954360	2.0
2.1	.967433	.965834	.964358	.963000	.961757	.960623	.959595	2.1
2.2	.972158	.970577	.969105	.967740	.966479	.965319	.964257	2.2
2.3	.976249	.974708	.973260	.971907	.970648	.969480	.968402	2.3
2.4	.979782	.978296	.976890	.975566	.974324	.973164	.972085	2.4
2.5	.982824	.981407	.980056	.978774	.977562	.976422	.975355	2.5
2.6	.985437	.984098	.982811	.981580	.980409	.979300	.978255	2.6
2.7	.987676	.986422	.985205	.984033	.982911	.981840	.980825	2.7
2.8	.989591	.988423	.987282	.986174	.985105	.984080	.983100	2.8
2.9	.991224	.990144	.989080	.988040	.987029	.986052	.985114	2.9
3.0	.992614	.991622	.990636	.989664	.988713	.987788	.986895	3.0
3.1	.993794	.992888	.991979	.991076	.990186	.989315	.988468	3.1
3.2	.994794	.993970	.993137	.992302	.991473	.990657	.989857	3.2
3.3	.995640	.994895	.994134	.993365	.992597	.991834	.991083	3.3
3.4	.996355	.995683	.994991	.994287	.993577	.992867	.992165	3.4
3.5	.996957	.996354	.995728	.995084	.994431	.993773	.993118	3.5
3.6	.997463	.996925	.996359	.995774	.995174	.994567	.993958	3.6
3.7	.997888	.997409	.996901	.996369	.995821	.995262	.994697	3.7
3.8	.998244	.997820	.997364	.996883	.996383	.995870	.995348	3.8
3.9	.998543	.998167	.997760	.997326	.996872	.996402	.995920	3.9
4.0	.998792	.998461	.998098	.997708	.997296	.996866	.996424	4.0
4.1	.999000	.998709	.998387	.998037	.997664	.997272	.996866	4.1
4.2	.999173	.998918	.998632	.998319	.997983	.997627	.997254	4.2
4.3	.999317	.999094	.998842	.998562	.998259	.997936	.997596	4.3
4.4	.999436	.999242	.999020	.998771	.998498	.998205	.997895	4.4
4.5	.999536	.999367	.999171	.998950	.998705	.998440	.998158	4.5
4.6	.999618	.999472	.999300	.999103	.998884	.998645	.998388	4.6
4.7	.999686	.999559	.999408	.999235	.999039	.998823	.998590	4.7
4.8	.999742	.999633	.999501	.999347	.999173	.998979	.998767	4.8
4.9	.999788	.999694	.999579	.999443	.999288	.999114	.998922	4.9
5.0	.999827	.999746	.999645	.999526	.999387	.999231	.999058	5.0
5.1	.999858	.999788	.999701	.999596	.999473	.999333	.999177	5.1
5.2	.999884	.999824	.999749	.999656	.999547	.999422	.999281	5.2
5.3	.999905	.999854	.999789	.999708	.999611	.999499	.999372	5.3
5.4	.999923	.999879	.999822	.999751	.999666	.999566	.999452	5.4
5.5	.999937	.999900	.999851	.999789	.999713	.999624	.999522	5.5
5.6	.999949	.999917	.999875	.999820	.999754	.999675	.999583	5.6
5.7	.999958	.999931	.999895	.999848	.999789	.999718	.999636	5.7
5.8	.999966	.999943	.999912	.999871	.999819	.999756	.999682	5.8
5.9	.999972	.999953	.999926	.999890	.999845	.999789	.999723	5.9

TABLE A.3.4 Continued

z \ α_3	1.4	1.5	1.6	1.7	1.8	1.9	2.0	
-1.5								-1.5
-1.4	.000685							-1.4
-1.3	.013409	.002331						-1.3
-1.2	.039499	.025184	.009098					-1.2
-1.1	.075506	.062662	.046967	.027549	.003895			-1.1
-1.0	.118505	.108815	.096882	.081970	.062896	.037476	.000000	-1.0
-.9	.166109	.159975	.152353	.142839	.130847	.115471	.095163	-.9
-.8	.216408	.213628	.209961	.205208	.199108	.191299	.181269	-.8
-.7	.267897	.268002	.267630	.266688	.265067	.262625	.259182	-.7
-.6	.319410	.321836	.324036	.325972	.327596	.328853	.329680	-.6
-.5	.370058	.374240	.378336	.382328	.386197	.389919	.393469	-.5
-.4	.419184	.424602	.430001	.435370	.440699	.445976	.451188	-.4
-.3	.466314	.472518	.478724	.484924	.491112	.497278	.503415	-.3
-.2	.511124	.517738	.524351	.530958	.537552	.544125	.550671	-.2
-.1	.553409	.560129	.566837	.573528	.580195	.586831	.593430	-.1
.0	.593056	.599648	.606215	.612751	.619251	.625710	.632121	.0
.1	.630026	.636314	.642567	.648780	.654948	.661066	.667129	.1
.2	.664335	.670195	.676015	.681790	.687516	.693190	.698806	.2
.3	.696042	.701392	.706702	.711968	.717187	.722355	.727468	.3
.4	.725235	.730027	.734784	.739504	.744182	.748817	.753403	.4
.5	.752026	.756239	.760428	.764588	.768718	.772813	.776870	.5
.6	.776539	.780175	.783798	.787406	.790995	.794562	.798103	.6
.7	.798909	.801983	.805060	.808135	.811206	.814267	.817316	.7
.8	.819273	.821814	.824373	.826945	.829527	.832113	.834701	.8
.9	.837771	.839814	.841891	.843995	.846124	.848270	.850431	.9
1.0	.854540	.856126	.857760	.859437	.861150	.862894	.864665	1.0
1.1	.869714	.870886	.872120	.873408	.874746	.876126	.877544	1.1
1.2	.883420	.884224	.885100	.886041	.887042	.888096	.889197	1.2
1.3	.895783	.896260	.896820	.897455	.898157	.898922	.899741	1.3
1.4	.906916	.907111	.907395	.907760	.908201	.908711	.909282	1.4
1.5	.916929	.916881	.916927	.917060	.917273	.917561	.917915	1.5
1.6	.925924	.925669	.925513	.925447	.925465	.925560	.925726	1.6
1.7	.933994	.933568	.933241	.933007	.932858	.932790	.932794	1.7
1.8	.941226	.940660	.940192	.939818	.939530	.939323	.939190	1.8
1.9	.947701	.947022	.946441	.945952	.945548	.945225	.944977	1.9
2.0	.953493	.952727	.952055	.951474	.950976	.950557	.950213	2.0
2.1	.958668	.957837	.957096	.956442	.955870	.955374	.954951	2.1
2.2	.963289	.962411	.961620	.960911	.960281	.959724	.959238	2.2
2.3	.967411	.966504	.965678	.964930	.964256	.963653	.963117	2.3
2.4	.971086	.970164	.969317	.968542	.967838	.967200	.966627	2.4
2.5	.974359	.973434	.972577	.971788	.971064	.970403	.969803	2.5
2.6	.977273	.976354	.975498	.974704	.973971	.973295	.972676	2.6
2.7	.979865	.978961	.978114	.977323	.976588	.975906	.975276	2.7
2.8	.982169	.981287	.980456	.979675	.978944	.978262	.977629	2.8
2.9	.984216	.983362	.982551	.981786	.981065	.980389	.979758	2.9

TABLE A.3.4 Continued

z \ α_3	1.4	1.5	1.6	1.7	1.8	1.9	2.0	
3.0	.986034	.985211	.984426	.983680	.982974	.982309	.981684	3.0
3.1	.987648	.986859	.986102	.985379	.984692	.984042	.983427	3.1
3.2	.989079	.988326	.987600	.986904	.986239	.985605	.985004	3.2
3.3	.990348	.989633	.988940	.988271	.987630	.987016	.986431	3.3
3.4	.991473	.990795	.990136	.989497	.988881	.988289	.987723	3.4
3.5	.992469	.991830	.991205	.990597	.990007	.989438	.988891	3.5
3.6	.993351	.992750	.992159	.991582	.991019	.990474	.989948	3.6
3.7	.994131	.993568	.993012	.992465	.991930	.991409	.990905	3.7
3.8	.994821	.994295	.993772	.993256	.992749	.992253	.991770	3.8
3.9	.995432	.994941	.994451	.993965	.993485	.993014	.992553	3.9
4.0	.995972	.995515	.995057	.994600	.994147	.993700	.993262	4.0
4.1	.996449	.996025	.995597	.995168	.994742	.994319	.993903	4.1
4.2	.996870	.996477	.996078	.995677	.995276	.994878	.994483	4.2
4.3	.997242	.996879	.996508	.996133	.995757	.995381	.995008	4.3
4.4	.997571	.997235	.996891	.996542	.996189	.995836	.995483	4.4
4.5	.997860	.997551	.997232	.996907	.996577	.996245	.995913	4.5
4.6	.998116	.997831	.997537	.997234	.996926	.996615	.996302	4.6
4.7	.998342	.998080	.997808	.997527	.997239	.996948	.996654	4.7
4.8	.998540	.998300	.998049	.997789	.997521	.997248	.996972	4.8
4.9	.998716	.998496	.998264	.998023	.997774	.997519	.997261	4.9
5.0	.998870	.998669	.998456	.998233	.998001	.997764	.997521	5.0
5.1	.999006	.998822	.998626	.998420	.998206	.997984	.997757	5.1
5.2	.999126	.998958	.998778	.998588	.998389	.998183	.997971	5.2
5.3	.999232	.999078	.998913	.998738	.998554	.998362	.998164	5.3
5.4	.999325	.999185	.999034	.998872	.998702	.998523	.998338	5.4
5.5	.999407	.999279	.999141	.998992	.998834	.998669	.998497	5.5
5.6	.999479	.999363	.999236	.999099	.998954	.998800	.998640	5.6
5.7	.999542	.999437	.999321	.999195	.999061	.998918	.998769	5.7
5.8	.999598	.999502	.999396	.999281	.999157	.999025	.998886	5.8
5.9	.999647	.999560	.999463	.999358	.999243	.999121	.998992	5.9
6.0	.999690	.999611	.999523	.999426	.999321	.999208	.999088	6.0
6.1	.999728	.999656	.999576	.999487	.999391	.999286	.999175	6.1
6.2	.999761	.999696	.999623	.999542	.999453	.999357	.999253	6.2
6.3	.999790	.999732	.999665	.999591	.999509	.999420	.999324	6.3
6.4	.999816	.999763	.999703	.999635	.999560	.999477	.999389	6.4
6.5	.999838	.999791	.999736	.999674	.999605	.999529	.999447	6.5
6.6	.999858	.999815	.999765	.999709	.999645	.999576	.999500	6.6
6.7	.999876	.999837	.999792	.999740	.999682	.999617	.999547	6.7
6.8	.999891	.999856	.999815	.999768	.999714	.999655	.999590	6.8
6.9	.999904	.999873	.999836	.999793	.999744	.999689	.999629	6.9
7.0	.999916	.999888	.999854	.999815	.999770	.999720	.999665	7.0
7.1	.999926	.999901	.999870	.999835	.999794	.999748	.999696	7.1
7.2	.999935	.999913	.999885	.999852	.999815	.999773	.999725	7.2
7.3	.999943	.999923	.999898	.999868	.999834	.999795	.999751	7.3
7.4	.999950	.999932	.999909	.999882	.999851	.999815	.999775	7.4

TABLE A.3.4 Continued

z \ α_3	2.1	2.2	2.3	2.5	3.0	4.0	5.0	
-1.5								-1.5
-1.4								-1.4
-1.3								-1.3
-1.2								-1.2
-1.1								-1.1
-1.0								-1.0
-.9	.066775	.020163						-.9
-.8	.168238	.150925	.126921	.000000				-.8
-.7	.254500	.248258	.239997	.214268				-.7
-.6	.329998	.329714	.328711	.323915	.279143			-.6
-.5	.396821	.399943	.402800	.407558	.411126	.000000		-.5
-.4	.456324	.461371	.466313	.475830	.496757	.516555	.000000	-.4
-.3	.509513	.515565	.521561	.533355	.561337	.608339	.639089	-.3
-.2	.557182	.563651	.570070	.582735	.613116	.666828	.710210	-.2
-.1	.599986	.606491	.612941	.625647	.656061	.709851	.753804	-.1
.0	.638478	.644776	.651010	.663264	.692458	.743678	.785204	.0
.1	.673132	.679071	.684941	.696455	.723770	.771331	.809575	.1
.2	.704361	.709850	.715271	.725889	.751006	.794523	.829335	.2
.3	.732524	.737518	.742446	.752097	.774898	.814332	.845823	.3
.4	.757939	.762420	.766844	.775509	.795998	.831482	.859865	.4
.5	.780886	.784858	.788783	.796481	.814732	.846486	.872005	.5
.6	.801616	.805096	.808540	.815311	.831442	.859725	.882627	.6
.7	.820349	.823363	.826353	.832252	.846402	.871487	.892008	.7
.8	.837286	.839863	.842430	.847519	.859839	.881996	.900358	.8
.9	.852602	.854778	.856955	.861298	.871942	.891431	.907836	.9
1.0	.866457	.868266	.870087	.873751	.882869	.899937	.914571	1.0
1.1	.878993	.880470	.881970	.885018	.892756	.907633	.920665	1.1
1.2	.890339	.891518	.892728	.895224	.901718	.914618	.926200	1.2
1.3	.900610	.901523	.902476	.904477	.909857	.920975	.931246	1.3
1.4	.909910	.910588	.911311	.912874	.917258	.926775	.935860	1.4
1.5	.918331	.918802	.919324	.920498	.923998	.932079	.940091	1.5
1.6	.925958	.926249	.926594	.927428	.930144	.936938	.943980	1.6
1.7	.932867	.933002	.933194	.933729	.935754	.941398	.947563	1.7
1.8	.939126	.939127	.939186	.939463	.940881	.945498	.950870	1.8
1.9	.944798	.944684	.944630	.944683	.945570	.949273	.953928	1.9
2.0	.949938	.949727	.949576	.949437	.949863	.952753	.956761	2.0
2.1	.954596	.954304	.954072	.953770	.953795	.955966	.959390	2.1
2.2	.958818	.958460	.958160	.957721	.957402	.958935	.961831	2.2
2.3	.962645	.962233	.961878	.961324	.960710	.961682	.964103	2.3
2.4	.966114	.965659	.965260	.964612	.963748	.964226	.966218	2.4
2.5	.969260	.968772	.968337	.967613	.966539	.966584	.968191	2.5
2.6	.972112	.971600	.971138	.970354	.969105	.968771	.970032	2.6
2.7	.974698	.974169	.973687	.972857	.971464	.970802	.971752	2.7
2.8	.977043	.976504	.976008	.975143	.973636	.972689	.973360	2.8
2.9	.979170	.978625	.978122	.977234	.975635	.974444	.974865	2.9

TABLE A.3.4 Continued

z \ α_3	2.1	2.2	2.3	2.5	3.0	4.0	5.0	
3.0	.981100	.980554	.980047	.979144	.977476	.976077	.976275	3.0
3.1	.982849	.982307	.981801	.980891	.979172	.977597	.977596	3.1
3.2	.984436	.983901	.983399	.982489	.980736	.979014	.978836	3.2
3.3	.985876	.985351	.984855	.983951	.982178	.980334	.979999	3.3
3.4	.987182	.986669	.986182	.985289	.983508	.981565	.981092	3.4
3.5	.988367	.987867	.987391	.986513	.984736	.982714	.982119	3.5
3.6	.989442	.988957	.988494	.987634	.985869	.983787	.983085	3.6
3.7	.990418	.989949	.989499	.988660	.986916	.984788	.983993	3.7
3.8	.991303	.990851	.990416	.989600	.987882	.985724	.984848	3.8
3.9	.992106	.991671	.991252	.990461	.988775	.986599	.985653	3.9
4.0	.992834	.992418	.992014	.991249	.989601	.987417	.986412	4.0
4.1	.993495	.993097	.992710	.991971	.990364	.988183	.987128	4.1
4.2	.994095	.993715	.993344	.992633	.991069	.988899	.987802	4.2
4.3	.994640	.994278	.993923	.993240	.991722	.989570	.988438	4.3
4.4	.995134	.994790	.994451	.993796	.992325	.990198	.989039	4.4
4.5	.995583	.995256	.994933	.994306	.992884	.990786	.989606	4.5
4.6	.995990	.995680	.995373	.994774	.993401	.991338	.990141	4.6
4.7	.996359	.996066	.995774	.995202	.993879	.991854	.990647	4.7
4.8	.996695	.996417	.996140	.995595	.994322	.992339	.991125	4.8
4.9	.996999	.996737	.996475	.995956	.994732	.992793	.991577	4.9
5.0	.997276	.997028	.996780	.996287	.995112	.993219	.992005	5.0
5.1	.997526	.997293	.997058	.996590	.995465	.993619	.992409	5.1
5.2	.997754	.997534	.997313	.996868	.995791	.993995	.992792	5.2
5.3	.997961	.997754	.997545	.997124	.996093	.994347	.993154	5.3
5.4	.998148	.997954	.997757	.997358	.996373	.994678	.993497	5.4
5.5	.998319	.998136	.997951	.997573	.996633	.994989	.993822	5.5
5.6	.998473	.998302	.998128	.997771	.996874	.995281	.994129	5.6
5.7	.998614	.998453	.998289	.997952	.997097	.995555	.994421	5.7
5.8	.998741	.998591	.998437	.998118	.997304	.995813	.994697	5.8
5.9	.998857	.998716	.998571	.998271	.997496	.996056	.994958	5.9
6.0	.998962	.998831	.998694	.998411	.997674	.996284	.995207	6.0
6.1	.999057	.998935	.998807	.998540	.997839	.996498	.995442	6.1
6.2	.999144	.999029	.998910	.998659	.997993	.996699	.995665	6.2
6.3	.999223	.999115	.999003	.998767	.998135	.996889	.995877	6.3
6.4	.999294	.999194	.999089	.998867	.998267	.997067	.996078	6.4
6.5	.999359	.999266	.999167	.998959	.998389	.997235	.996268	6.5
6.6	.999418	.999331	.999239	.999043	.998503	.997393	.996449	6.6
6.7	.999471	.999390	.999304	.999120	.998609	.997541	.996621	6.7
6.8	.999520	.999444	.999364	.999191	.998707	.997681	.996784	6.8
6.9	.999564	.999494	.999419	.999256	.998798	.997813	.996939	6.9
7.0	.999604	.999539	.999469	.999316	.998883	.997937	.997086	7.0
7.1	.999640	.999579	.999514	.999371	.998961	.998054	.997226	7.1
7.2	.999673	.999617	.999556	.999422	.999034	.998164	.997358	7.2
7.3	.999703	.999651	.999594	.999469	.999102	.998268	.997485	7.3
7.4	.999731	.999682	.999629	.999511	.999165	.998365	.997604	7.4

TABLE A.3.5 Cumulative Distribution Function of Standardized Four-Parameter Generalized Gamma Distribution (α_3; 0, 1)

$\delta = 0.5$

z \ α_3	.20	.30	.40	.50	.60	.70	.80	.90	1.00	1.50	2.00	2.50	3.00	4.00	5.00	α_3 / z
-3.9	0	0	0	0	0	0	0	0	0	0	0	0	0	0	0	-3.9
-3.8	1	0	0	0	0	0	0	0	0	0	0	0	0	0	0	-3.8
-3.7	1	0	0	0	0	0	0	0	0	0	0	0	0	0	0	-3.7
-3.6	2	0	0	0	0	0	0	0	0	0	0	0	0	0	0	-3.6
-3.5	4	1	0	0	0	0	0	0	0	0	0	0	0	0	0	-3.5
-3.4	6	2	0	0	0	0	0	0	0	0	0	0	0	0	0	-3.4
-3.3	11	3	1	0	0	0	0	0	0	0	0	0	0	0	0	-3.3
-3.2	17	6	2	0	0	0	0	0	0	0	0	0	0	0	0	-3.2
-3.1	28	11	3	1	0	0	0	0	0	0	0	0	0	0	0	-3.1
-3.0	45	20	7	1	0	0	0	0	0	0	0	0	0	0	0	-3.0
-2.9	69	34	13	4	0	0	0	0	0	0	0	0	0	0	0	-2.9
-2.8	106	57	25	8	2	0	0	0	0	0	0	0	0	0	0	-2.8
-2.7	159	92	46	18	5	1	0	0	0	0	0	0	0	0	0	-2.7
-2.6	234	146	80	36	12	2	0	0	0	0	0	0	0	0	0	-2.6
-2.5	340	226	134	68	27	7	1	0	0	0	0	0	0	0	0	-2.5
-2.4	485	340	219	124	58	20	4	0	0	0	0	0	0	0	0	-2.4
-2.3	680	503	345	215	116	50	15	2	0	0	0	0	0	0	0	-2.3
-2.2	940	727	530	357	215	110	43	11	1	0	0	0	0	0	0	-2.2
-2.1	1280	1029	790	570	378	222	108	39	8	0	0	0	0	0	0	-2.1
-2.0	1717	1431	1149	880	631	414	238	112	37	0	0	0	0	0	0	-2.0
-1.9	2270	1953	1631	1312	1005	721	472	269	123	0	0	0	0	0	0	-1.9
-1.8	2961	2618	2262	1899	1536	1185	856	564	323	0	0	0	0	0	0	-1.8
-1.7	3811	3451	3069	2671	2262	1849	1444	1060	712	0	0	0	0	0	0	-1.7
-1.6	4841	4476	4082	3661	3218	2759	2290	1823	1371	13	0	0	0	0	0	-1.6
-1.5	6073	5716	5325	4899	4441	3954	3443	2914	2378	194	0	0	0	0	0	-1.5

TABULATED VALUES MUST BE MULTIPLIED BY .00001 TO OBTAIN PROBABILITIES

δ = 0.5

z \ α_3	.20	.30	.40	.50	.60	.70	.80	.90	1.00	1.50	2.00	2.50	3.00	4.00	5.00	α_3 / z
-1.4	7524	7192	6821	6410	5959	5470	4944	4385	3798	853	0	0	0	0	0	-1.4
-1.3	9213	8922	8589	8213	7794	7330	6821	6269	5673	2250	0	0	0	0	0	-1.3
-1.2	11152	10917	10642	10323	9960	9549	9090	8581	8021	4499	582	0	0	0	0	-1.2
-1.1	13349	13187	12986	12744	12458	12126	11746	11317	10834	7583	2993	0	0	0	0	-1.1
-1.0	15807	15731	15619	15470	15281	15050	14774	14452	14080	11399	7115	1348	0	0	0	-1.0
-.9	18525	18544	18533	18489	18411	18296	18142	17947	17709	15794	12408	7034	37	0	0	-.9
-.8	21493	21614	21710	21779	21819	21829	21806	21750	21658	20602	18363	14521	8210	0	0	-.8
-.7	24697	24921	25126	25309	25470	25606	25716	25799	25854	25665	24590	22403	18653	0	0	-.7
-.6	28114	28439	28750	29044	29320	29578	29815	30031	30225	30842	30815	30056	28403	20622	0	-.6
-.5	31719	32138	32546	32942	33324	33692	34043	34379	34697	36014	36857	37208	37040	34782	27890	-.5
-.4	35480	35981	36475	36959	37432	37893	38342	38778	39200	41088	42601	43754	44570	45218	44384	-.4
-.3	39360	39929	40492	41047	41594	42130	42656	43170	43672	45990	47983	49675	51103	53293	54734	-.3
-.2	43322	43941	44555	45162	45761	46351	46931	47501	48059	50669	52971	54991	56767	59735	62109	-.2
-.1	47325	47974	48619	49257	49888	50510	51122	51724	52315	55090	57558	59744	61685	64983	67697	-.1
.0	51328	51987	52643	53291	53933	54566	55189	55802	56404	59232	61750	63981	65963	69329	72093	.0
.1	55291	55941	56587	57227	57860	58484	59099	59703	60296	63084	65563	67754	69695	72973	75643	.1
.2	59177	59799	60417	61030	61636	62235	62824	63404	63973	66646	69019	71111	72958	76060	78566	.2
.3	62949	63527	64102	64673	65238	65796	66347	66888	67421	69922	72143	74098	75820	78699	81010	.3
.4	66578	67098	67617	68133	68645	69152	69653	70146	70632	72921	74960	76755	78335	80971	83078	.4
.5	70037	70488	70940	71392	71842	72290	72734	73173	73605	75657	77496	79121	80552	82941	84846	.5
.6	73303	73677	74057	74439	74822	75206	75588	75967	76344	78145	79777	81228	82511	84657	86371	.6
.7	76360	76654	76957	77266	77580	77897	78215	78534	78853	80400	81826	83106	84245	86160	87695	.7
.8	79196	79410	79635	79871	80115	80366	80622	80881	81141	82439	83665	84782	85785	87483	88852	.8
.9	81806	81940	82091	82256	82433	82619	82814	83015	83220	84279	85316	86279	87154	88651	89869	.9
1.0	84186	84246	84327	84425	84538	84664	84801	84948	85102	85937	86796	87616	88373	89687	90768	1.0

TABULATED VALUES MUST BE MULTIPLIED BY .00001 TO OBTAIN PROBABILITIES

TABLE A.3.5 Continued

$\delta = 0.5$

z \ α_3	.20	.30	.40	.50	.60	.70	.80	.90	1.00	1.50	2.00	2.50	3.00	4.00	5.00	α_3 / z
1.1	86339	86332	86348	86385	86440	86511	86595	86691	86798	87427	88124	88813	89463	90609	91565	1.1
1.2	88272	88205	88164	88146	88149	88170	88208	88259	88322	88766	89315	89885	90437	91432	92275	1.2
1.3	89992	89875	89785	89720	89678	89656	89651	89663	89689	89966	90383	90846	91309	92168	92911	1.3
1.4	91512	91354	91223	91119	91038	90979	90939	90917	90909	91042	91341	91708	92093	92829	93481	1.4
1.5	92845	92654	92492	92356	92244	92154	92085	92033	91998	92004	92200	92483	92797	93424	93994	1.5
1.6	94004	93790	93604	93443	93307	93193	93099	93024	92965	92865	92971	93179	93431	93960	94457	1.6
1.7	95005	94777	94574	94396	94241	94109	93996	93901	93824	93634	93663	93806	94002	94444	94876	1.7
1.8	95864	95628	95416	95226	95059	94913	94786	94677	94584	94322	94285	94370	94518	94883	95256	1.8
1.9	96595	96358	96142	95947	95772	95617	95480	95360	95256	94935	94843	94879	94985	95281	95601	1.9
2.0	97213	96980	96766	96569	96391	96231	96088	95961	95849	95483	95344	95339	95407	95642	95916	2.0
2.1	97731	97508	97299	97105	96928	96766	96620	96489	96372	95971	95795	95754	95790	95972	96203	2.1
2.2	98164	97953	97752	97565	97390	97230	97084	96951	96831	96407	96200	96129	96137	96272	96466	2.2
2.3	98523	98326	98136	97957	97789	97632	97487	97355	97235	96795	96565	96468	96453	96546	96707	2.3
2.4	98818	98637	98460	98291	98130	97979	97838	97708	97589	97142	96893	96775	96740	96797	96928	2.4
2.5	99059	98895	98732	98574	98422	98278	98143	98016	97899	97450	97188	97054	97001	97026	97132	2.5
2.6	99255	99108	98960	98813	98671	98535	98406	98284	98171	97725	97455	97307	97239	97237	97319	2.6
2.7	99413	99282	99149	99015	98884	98756	98634	98518	98408	97971	97694	97536	97456	97430	97491	2.7
2.8	99540	99425	99306	99184	99064	98945	98830	98720	98616	98189	97911	97744	97654	97608	97650	2.8
2.9	99642	99542	99436	99326	99216	99106	98999	98896	98797	98384	98106	97934	97836	97772	97798	2.9
3.0	99722	99636	99543	99445	99345	99244	99145	99048	98955	98557	98282	98106	98002	97922	97934	3.0
3.1	99786	99712	99630	99543	99453	99362	99270	99180	99092	98712	98441	98263	98153	98061	98060	3.1
3.2	99836	99773	99702	99625	99545	99461	99377	99294	99212	98850	98584	98406	98293	98190	98177	3.2
3.3	99875	99822	99761	99693	99621	99546	99469	99393	99316	98973	98714	98536	98420	98308	98286	3.3
3.4	99905	99860	99808	99749	99685	99618	99548	99478	99407	99083	98832	98655	98537	98418	98387	3.4
3.5	99928	99891	99847	99796	99739	99679	99616	99551	99486	99180	98938	98764	98645	98520	98481	3.5

TABULATED VALUES MUST BE MULTIPLIED BY .00001 TO OBTAIN PROBABILITIES

$\delta = 1.0$

z \ α_3	.20	.30	.40	.50	.60	.70	.80	.90	1.00	1.50	2.00	2.50	3.00	4.00	5.00	α_3 / z
-1.4	7529	7199	6826	6405	5931	5397	4797	4125	3377	0	0	0	0	0	0	-1.4
-1.3	9220	8935	8608	8232	7802	7311	6749	6106	5373	233	0	0	0	0	0	-1.3
-1.2	11162	10938	10675	10366	10007	9589	9103	8538	7881	2518	0	0	0	0	0	-1.2
-1.1	13361	13214	13032	12810	12544	12226	11848	11400	10871	6266	0	0	0	0	0	-1.1
-1.0	15822	15764	15677	15558	15403	15205	14959	14657	14288	10882	0	0	0	0	0	-1.0
-.9	18542	18582	18601	18595	18562	18498	18398	18256	18065	15997	9516	0	0	0	0	-.9
-.8	21511	21655	21784	21898	21993	22066	22116	22138	22128	21363	18127	0	0	0	0	-.8
-.7	24715	24963	25204	25435	25656	25865	26060	26239	26400	26800	25918	21427	0	0	0	-.7
-.6	28132	28481	28828	29171	29510	29844	30172	30493	30806	32184	32968	32391	27914	0	0	-.6
-.5	31736	32178	32621	33064	33508	33951	34394	34836	35277	37424	39347	40756	41113	0	0	-.5
-.4	35494	36017	36542	37070	37601	38134	38670	39208	39748	42460	45119	47583	49676	51656	0	-.4
-.3	39372	39959	40549	41143	41741	42342	42946	43554	44164	47252	50341	53336	56134	60834	63909	-.3
-.2	43331	43964	44600	45239	45881	46526	47174	47825	48478	51774	55067	58274	61312	66683	71021	-.2
-.1	47330	47990	48650	49313	49978	50644	51312	51981	52652	56013	59343	62565	65606	70985	75380	-.1
.0	51330	51995	52660	53326	53991	54657	55322	55988	56653	59965	63212	66326	69246	74368	78520	.0
.1	55290	55941	56591	57239	57886	58532	59176	59819	60460	63631	66713	69645	72377	77133	80957	.1
.2	59172	59791	60407	61021	61633	62242	62849	63453	64055	67020	69881	72589	75101	79452	82934	.2
.3	62942	63512	64080	64645	65208	65767	66324	66878	67429	70139	72747	75210	77490	81433	84582	.3
.4	66568	67077	67584	68089	68591	69091	69589	70084	70577	73003	75340	77551	79600	83148	85986	.4
.5	70025	70462	70899	71335	71769	72203	72636	73067	73497	75624	77687	79648	81473	84649	87201	.5
.6	73289	73649	74010	74372	74735	75098	75463	75828	76193	78017	79810	81531	83144	85973	88263	.6
.7	76345	76623	76906	77192	77482	77775	78071	78370	78671	80198	81732	83225	84640	87149	89201	.7
.8	79181	79378	79583	79794	80012	80236	80465	80699	80938	82181	83470	84752	85984	88200	90036	.8
.9	81791	81909	82038	82178	82327	82485	82651	82824	83004	83981	85043	86130	87194	89143	90784	.9
1.0	84172	84217	84276	84349	84434	84531	84638	84754	84880	85613	86466	87375	88287	89994	91457	1.0

TABULATED VALUES MUST BE MULTIPLIED BY .00001 TO OBTAIN PROBABILITIES

TABLE A.3.5 Continued

$\delta = 1.0$

z \ α_3	.20	.30	.40	.50	.60	.70	.80	.90	1.00	1.50	2.00	2.50	3.00	4.00	5.00	α_3 / z
1.1	86327	86305	86300	86313	86340	86382	86435	86501	86577	87089	87754	88502	89276	90763	92066	1.1
1.2	88261	88180	88120	88080	88056	88049	88055	88076	88108	88422	88920	89522	90172	91462	92620	1.2
1.3	89983	89853	89746	89660	89593	89543	89510	89490	89485	89626	89974	90448	90986	92097	93125	1.3
1.4	91505	91335	91190	91066	90963	90878	90810	90757	90719	90711	90928	91287	91726	92678	93586	1.4
1.5	92839	92639	92464	92311	92178	92065	91968	91888	91823	91688	91792	92050	92400	93208	94009	1.5
1.6	94000	93779	93582	93407	93252	93116	92997	92895	92808	92567	92573	92743	93014	93694	94398	1.6
1.7	95003	94769	94557	94367	94196	94044	93908	93789	93685	93357	93279	93373	93575	94140	94756	1.7
1.8	95863	95623	95404	95204	95023	94859	94712	94580	94463	94066	93919	93946	94088	94550	95087	1.8
1.9	96595	96356	96135	95931	95745	95574	95419	95279	95152	94702	94498	94468	94557	94927	95393	1.9
2.0	97214	96980	96763	96560	96372	96199	96040	95895	95762	95273	95021	94944	94986	95275	95676	2.0
2.1	97733	97510	97299	97101	96916	96743	96583	96436	96300	95784	95495	95377	95380	95597	95939	2.1
2.2	98167	97956	97755	97565	97385	97216	97058	96911	96774	96241	95924	95772	95740	95894	96183	2.2
2.3	98525	98330	98142	97961	97788	97625	97471	97326	97191	96650	96312	96132	96071	96168	96410	2.3
2.4	98821	98642	98467	98298	98134	97978	97830	97689	97557	97016	96663	96461	96375	96423	96622	2.4
2.5	99062	98901	98740	98583	98430	98282	98141	98006	97877	97343	96980	96761	96654	96658	96819	2.5
2.6	99258	99114	98968	98824	98682	98544	98410	98281	98158	97635	97268	97035	96910	96877	97003	2.6
2.7	99416	99289	99158	99027	98896	98768	98642	98521	98403	97896	97528	97286	97146	97080	97175	2.7
2.8	99543	99432	99315	99197	99078	98959	98842	98728	98617	98129	97763	97514	97364	97269	97336	2.8
2.9	99645	99548	99445	99339	99231	99122	99014	98908	98804	98336	97976	97723	97563	97444	97487	2.9
3.0	99725	99642	99552	99457	99360	99261	99162	99064	98966	98521	98168	97914	97748	97608	97627	3.0
3.1	99788	99717	99639	99556	99469	99379	99289	99198	99108	98686	98343	98089	97917	97760	97760	3.1
3.2	99838	99778	99710	99637	99560	99479	99397	99314	99230	98833	98500	98249	98074	97901	97884	3.2
3.3	99876	99826	99768	99704	99636	99564	99489	99413	99337	98963	98643	98395	98218	98033	98000	3.3
3.4	99906	99864	99815	99760	99700	99635	99568	99499	99429	99080	98772	98529	98351	98157	98109	3.4
3.5	99929	99894	99853	99805	99753	99696	99635	99573	99508	99183	98889	98651	98474	98271	98212	3.5

TABULATED VALUES MUST BE MULTIPLIED BY .00001 TO OBTAIN PROBABILITIES

δ = 1.5

z \ α_3	.20	.30	.40	.50	.60	.70	.80	.90	1.00	1.50	2.00	2.50	3.00	4.00	5.00	α_3 / z
-1.3	9237	8970	8659	8292	7848	7298	6601	5700	4506	0	0	0	0	0	0	-1.3
-1.2	11185	10988	10757	10481	10143	9719	9175	8462	7508	0	0	0	0	0	0	-1.2
-1.1	13390	13277	13143	12976	12765	12492	12130	11643	10979	0	0	0	0	0	0	-1.1
-1.0	15855	15839	15812	15768	15697	15588	15421	15173	14808	8037	0	0	0	0	0	-1.0
-.9	18578	18666	18754	18838	18913	18971	19000	18983	18900	15923	0	0	0	0	0	-.9
-.8	21549	21745	21949	22162	22380	22598	22810	23007	23174	22778	11340	0	0	0	0	-.8
-.7	24754	25055	25373	25707	26058	26423	26799	27179	27556	28977	27257	0	0	0	0	-.7
-.6	28170	28570	28993	29438	29906	30397	30910	31441	31985	34668	36302	33544	0	0	0	-.6
-.5	31771	32260	32774	33313	33879	34472	35093	35738	36406	39926	43195	45277	43972	0	0	-.5
-.4	35525	36089	36677	37290	37930	38599	39297	40022	40774	44801	48881	52505	55245	53246	0	-.4
-.3	39397	40018	40660	41326	42017	42734	43479	44251	45049	49327	53751	57916	61585	67042	69060	-.3
-.2	43349	44008	44683	45378	46095	46834	47598	48387	49198	53530	58015	62285	66137	72392	76916	-.2
-.1	47341	48016	48704	49406	50125	50863	51620	52399	53198	57433	61799	65957	69717	75876	80444	-.1
.0	51333	52004	52683	53371	54071	54785	55515	56263	57027	61055	65189	69121	72672	78480	82778	.0
.1	55285	55933	56584	57239	57901	58574	59259	59958	60670	64412	68246	71892	75185	80562	84530	.1
.2	59160	59768	60373	60979	61589	62206	62832	63469	64118	67521	71016	74349	77364	82295	85933	.2
.3	62923	63475	64022	64566	65112	65662	66219	66785	67362	70396	73534	76545	79282	83775	87101	.3
.4	66545	67029	67506	67979	68452	68928	69410	69901	70401	73051	75828	78521	80987	85064	88099	.4
.5	69997	70405	70805	71201	71597	71995	72399	72812	73234	75500	77924	80309	82516	86202	88969	.5
.6	73259	73585	73905	74221	74538	74857	75184	75518	75863	77756	79839	81933	83896	87217	89738	.6
.7	76313	76556	76795	77032	77270	77513	77763	78023	78294	79830	81593	83412	85147	88130	90426	.7
.8	79150	79310	79470	79630	79794	79963	80142	80331	80532	81734	83199	84762	86287	88958	91045	.8
.9	81760	81843	81928	82016	82110	82212	82324	82447	82584	83480	84670	85998	87328	89711	91608	.9
1.0	84143	84154	84170	84193	84224	84265	84316	84381	84460	85079	86018	87131	88282	90400	92122	1.0
1.1	86300	86247	86203	86168	86143	86129	86128	86141	86168	86540	87253	88171	89158	91034	92593	1.1

TABULATED VALUES MUST BE MULTIPLIED BY .00001 TO OBTAIN PROBABILITIES

TABLE A.3.5 Continued

$\delta = 1.5$

z \ α_3	.20	.30	.40	.50	.60	.70	.80	.90	1.00	1.50	2.00	2.50	3.00	4.00	5.00	α_3 / z
1.2	88238	88130	88033	87948	87875	87815	87768	87736	87719	87873	88385	89126	89965	91617	93028	1.2
1.3	89964	89810	89670	89544	89431	89332	89247	89177	89122	89089	89423	90004	90708	92156	93430	1.3
1.4	91489	91300	91126	90967	90822	90691	90575	90473	90387	90195	90373	90812	91393	92655	93803	1.4
1.5	92827	92611	92412	92228	92059	91904	91763	91636	91525	91200	91242	91556	92027	93119	94150	1.5
1.6	93992	93758	93542	93341	93154	92981	92821	92676	92544	92112	92038	92241	92613	93550	94474	1.6
1.7	94998	94755	94528	94317	94119	93934	93761	93602	93456	92938	92767	92872	93155	93951	94777	1.7
1.8	95861	95615	95385	95169	94965	94773	94593	94425	94269	93686	93432	93453	93658	94325	95060	1.8
1.9	96595	96353	96125	95909	95704	95510	95327	95154	94992	94361	94041	93988	94123	94675	95325	1.9
2.0	97216	96983	96761	96549	96348	96155	95972	95797	95633	94970	94596	94481	94555	95001	95574	2.0
2.1	97738	97516	97304	97100	96905	96717	96536	96364	96199	95519	95104	94936	94955	95306	95809	2.1
2.2	98172	97965	97766	97572	97385	97205	97030	96861	96699	96013	95566	95354	95326	95592	96029	2.2
2.3	98532	98341	98156	97975	97799	97627	97459	97297	97139	96456	95988	95740	95671	95860	96237	2.3
2.4	98828	98655	98484	98317	98152	97991	97832	97677	97526	96854	96372	96095	95990	96111	96433	2.4
2.5	99069	98914	98760	98606	98454	98303	98155	98008	97864	97211	96721	96421	96287	96347	96618	2.5
2.6	99265	99128	98989	98850	98711	98572	98433	98296	98160	97530	97039	96722	96562	96568	96793	2.6
2.7	99423	99302	99179	99054	98928	98800	98673	98545	98418	97815	97328	96999	96818	96775	96958	2.7
2.8	99550	99445	99336	99225	99111	98995	98878	98760	98642	98070	97591	97253	97055	96970	97114	2.8
2.9	99651	99560	99465	99367	99265	99161	99054	98946	98837	98296	97829	97487	97275	97153	97262	2.9
3.0	99730	99653	99571	99484	99394	99301	99204	99106	99005	98498	98045	97702	97480	97325	97401	3.0
3.1	99793	99728	99657	99582	99502	99419	99332	99243	99151	98678	98241	97899	97669	97487	97534	3.1
3.2	99842	99787	99727	99662	99592	99518	99441	99360	99277	98837	98418	98081	97845	97639	97659	3.2
3.3	99880	99834	99783	99727	99666	99602	99533	99460	99385	98978	98579	98247	98008	97782	97778	3.3
3.4	99909	99872	99829	99781	99728	99671	99610	99546	99478	99103	98724	98400	98159	97916	97891	3.4
3.5	99932	99901	99865	99824	99779	99729	99676	99618	99557	99214	98855	98540	98300	98043	97998	3.5
3.6	99949	99924	99894	99860	99821	99778	99731	99680	99626	99312	98973	98669	98430	98162	98100	3.6

TABULATED VALUES MUST BE MULTIPLIED BY .00001 TO OBTAIN PROBABILITIES

$\delta = 2.0$

z \ α_3	.20	.30	.40	.50	.60	.70	.80	.90	1.00	1.50	2.00	2.50	3.00	4.00	5.00	α_3 / z
-1.2	11254	11136	10994	10790	10471	9964	9165	7898	5813	0	0	0	0	0	0	-1.2
-1.1	13473	13460	13447	13401	13285	13045	12609	11870	10650	0	0	0	0	0	0	-1.1
-1.0	15950	16049	16169	16285	16368	16382	16278	15993	15432	0	0	0	0	0	0	-1.0
-.9	18681	18894	19145	19417	19688	19931	20117	20206	20148	12241	0	0	0	0	0	-.9
-.8	21656	21981	22357	22771	23208	23651	24078	24466	24785	23549	0	0	0	0	0	-.8
-.7	24859	25288	25776	26314	26893	27498	28117	28733	29330	31192	24325	0	0	0	0	-.7
-.6	28269	28790	29373	30013	30703	31434	32196	32978	33768	37371	38704	0	0	0	0	-.6
-.5	31860	32457	33115	33831	34602	35421	36281	37173	38088	42679	46476	48057	35915	0	0	-.5
-.4	35600	36255	36964	37730	38551	39422	40339	41293	42278	47378	52174	56087	58612	0	0	-.4
-.3	39455	40146	40884	41674	42515	43406	44342	45319	46328	51615	56768	61337	65149	70303	68931	-.3
-.2	43388	44093	44837	45624	46459	47340	48266	49230	50228	55477	60652	65327	69358	75578	79832	-.2
-.1	47359	48058	48784	49547	50352	51199	52087	53013	53970	59024	64031	68575	72514	78664	83016	-.1
.0	51330	52002	52692	53410	54164	54956	55787	56653	57549	62298	67025	71326	75056	80878	84991	.0
.1	55262	55889	56525	57182	57869	58592	59349	60139	60959	65329	69712	73714	77191	82614	86434	.1
.2	59119	59686	60254	60837	61446	62086	62759	63463	64197	68142	72145	75825	79032	84045	87576	.2
.3	62867	63362	63852	64352	64874	65425	66007	66619	67259	70756	74363	77714	80651	85263	88520	.3
.4	66475	66890	67295	67707	68139	68597	69084	69601	70146	73186	76396	79420	82094	86323	89326	.4
.5	69918	70247	70565	70887	71227	71592	71985	72407	72858	75445	78265	80972	83393	87260	90029	.5
.6	73174	73416	73646	73881	74131	74405	74706	75036	75395	77545	79989	82392	84572	88100	90652	.6
.7	76227	76384	76529	76679	76844	77031	77246	77489	77761	79495	81583	83696	85650	88860	91211	.7
.8	79064	79140	79206	79278	79364	79472	79606	79768	79960	81305	83057	84899	86640	89552	91717	.8
.9	81678	81680	81675	81675	81690	81726	81788	81877	81995	82983	84423	86011	87552	90188	92179	.9
1.0	84068	84003	83935	83873	83826	83799	83796	83820	83873	84537	85688	87042	88397	90773	92604	1.0
1.1	86233	86112	85991	85876	85776	85694	85636	85603	85598	85973	86862	87998	89180	91316	92997	1.1
1.2	88180	88013	87847	87689	87545	87419	87314	87233	87178	87299	87950	88886	89909	91820	93361	1.2

TABULATED VALUES MUST BE MULTIPLIED BY .00001 TO OBTAIN PROBABILITIES

TABLE A.3.5 Continued

$\delta = 2.0$

z \ α_3	.20	.30	.40	.50	.60	.70	.80	.90	1.00	1.50	2.00	2.50	3.00	4.00	5.00	α_3 / z
1.3	89916	89713	89513	89322	89143	88980	88837	88717	88621	88521	88959	89712	90587	92290	93700	1.3
1.4	91452	91222	90998	90782	90577	90386	90214	90061	89932	89645	89894	90481	91220	92729	94018	1.4
1.5	92800	92553	92313	92080	91857	91647	91453	91276	91119	90676	90760	91197	91812	93141	94315	1.5
1.6	93975	93718	93470	93228	92995	92772	92562	92368	92192	91622	91561	91864	92365	93527	94595	1.6
1.7	94989	94732	94482	94238	94000	93771	93552	93346	93156	92487	92302	92485	92882	93890	94858	1.7
1.8	95860	95607	95361	95120	94884	94653	94431	94219	94020	93276	92986	93064	93366	94232	95107	1.8
1.9	96601	96358	96121	95887	95656	95429	95208	94995	94792	93994	93618	93602	93820	94555	95342	1.9
2.0	97227	96997	96773	96550	96328	96108	95892	95681	95478	94647	94201	94104	94245	94859	95565	2.0
2.1	97752	97539	97329	97120	96910	96700	96492	96287	96087	95239	94737	94570	94643	95146	95776	2.1
2.2	98189	97994	97801	97607	97411	97213	97015	96818	96624	95775	95231	95004	95016	95417	95976	2.2
2.3	98551	98374	98199	98021	97840	97656	97470	97283	97098	96258	95684	95407	95365	95674	96166	2.3
2.4	98847	98690	98532	98371	98206	98037	97864	97688	97513	96694	96099	95782	95693	95916	96347	2.4
2.5	99089	98950	98810	98666	98517	98362	98203	98040	97875	97085	96480	96129	95999	96146	96519	2.5
2.6	99284	99163	99040	98912	98779	98639	98494	98344	98191	97435	96829	96452	96286	96364	96683	2.6
2.7	99441	99337	99230	99118	98999	98874	98743	98606	98464	97749	97147	96751	96555	96570	96839	2.7
2.8	99567	99478	99385	99287	99183	99072	98954	98830	98701	98028	97437	97027	96806	96765	96988	2.8
2.9	99666	99591	99512	99427	99336	99238	99133	99022	98905	98277	97701	97283	97041	96950	97130	2.9
3.0	99744	99681	99614	99542	99463	99377	99285	99185	99080	98498	97941	97520	97261	97125	97265	3.0
3.1	99805	99753	99697	99635	99568	99493	99412	99324	99229	98694	98159	97738	97466	97291	97395	3.1
3.2	99853	99809	99763	99711	99653	99589	99518	99441	99357	98866	98357	97939	97658	97448	97518	3.2
3.3	99889	99854	99815	99772	99723	99668	99607	99539	99465	99018	98535	98125	97836	97596	97636	3.3
3.4	99917	99889	99857	99821	99780	99733	99680	99621	99556	99152	98697	98296	98003	97737	97748	3.4
3.5	99938	99916	99890	99860	99826	99786	99741	99690	99633	99269	98842	98453	98158	97871	97856	3.5
3.6	99955	99936	99916	99891	99863	99830	99791	99747	99698	99372	98973	98597	98303	97997	97959	3.6
3.7	99967	99952	99936	99916	99892	99865	99832	99795	99752	99461	99091	98729	98437	98117	98057	3.7
3.8	99976	99964	99951	99935	99916	99893	99866	99834	99797	99539	99196	98850	98562	98230	98151	3.8

TABULATED VALUES MUST BE MULTIPLIED BY .00001 TO OBTAIN PROBABILITIES

$\delta = 2.5$

z \ α_3	.20	.30	.40	.50	.60	.70	.80	.90	1.00	1.50	2.00	2.50	3.00	4.00	5.00	z
-.7	25180	25838	26553	27309	28087	28871	29643	30382	31068	32596	0	0	0	0	0	-.7
-.6	28562	29293	30092	30945	31839	32760	33695	34629	35550	39419	39700	0	0	0	0	-.6
-.5	32112	32892	33744	34659	35627	36635	37673	38728	39790	44852	48650	49205	0	0	0	-.5
-.4	35801	36604	37481	38425	39428	40479	41568	42685	43819	49457	54467	58324	60413	0	0	-.4
-.3	39596	40398	41272	42214	43218	44274	45373	46504	47658	53495	58916	63553	67308	72024	0	-.3
-.2	43465	44242	45087	46000	46975	48004	49078	50187	51322	57111	62568	67338	71353	77394	81405	-.2
-.1	47371	48101	48896	49756	50677	51653	52675	53733	54819	60394	65689	70341	74278	80281	84435	-.1
.0	51278	51944	52669	53457	54305	55207	56155	57141	58156	63405	68424	72846	76591	82293	86230	.0
.1	55152	55739	56380	57080	57839	58652	59510	60408	61337	66184	70864	75003	78512	83848	87519	.1
.2	58957	59455	60001	60604	61264	61976	62734	63533	64364	68763	73068	76900	80158	85118	88528	.2
.3	62662	63064	63510	64009	64562	65168	65820	66514	67242	71165	75077	78595	81601	86195	89358	.3
.4	66239	66541	66884	67277	67723	68219	68762	69348	69969	73407	76922	80126	82886	87129	90065	.4
.5	69661	69866	70107	70395	70733	71121	71556	72034	72550	75502	78624	81521	84043	87956	90680	.5
.6	72907	73018	73162	73349	73584	73867	74197	74571	74984	77463	80202	82802	85097	88696	91225	.6
.7	75960	75984	76038	76131	76269	76454	76685	76959	77274	79299	81668	83984	86062	89367	91714	.7
.8	78808	78754	78725	78734	78783	78878	79016	79198	79421	81017	83035	85080	86952	89979	92158	.8
.9	81440	81319	81220	81153	81124	81137	81192	81289	81428	82623	84310	86099	87777	90543	92563	.9
1.0	83854	83676	83517	83387	83290	83231	83213	83235	83297	84125	85502	87049	88544	91065	92937	1.0
1.1	86049	85826	85619	85436	85282	85163	85081	85037	85032	85527	86617	87938	89259	91549	93283	1.1
1.2	88027	87772	87528	87304	87105	86935	86799	86699	86636	86834	87659	88769	89929	92002	93606	1.2
1.3	89796	89519	89250	88995	88761	88552	88373	88226	88113	88050	88635	89549	90557	92426	93907	1.3
1.4	91364	91075	90791	90516	90257	90018	89806	89621	89468	89180	89547	90280	91147	92824	94190	1.4
1.5	92743	92451	92160	91875	91600	91341	91104	90891	90706	90227	90400	90967	91702	93199	94456	1.5
1.6	93945	93658	93368	93080	92798	92528	92274	92041	91832	91196	91196	91612	92225	93552	94707	1.6
1.7	94984	94708	94426	94142	93860	93586	93323	93077	92851	92090	91940	92218	92718	93887	94944	1.7

TABULATED VALUES MUST BE MULTIPLIED BY .00001 TO OBTAIN PROBABILITIES

TABLE A.3.5 Continued

$\delta = 2.5$

z \ α_3	.20	.30	.40	.50	.60	.70	.80	.90	1.00	1.50	2.00	2.50	3.00	4.00	5.00	α_3 / z
1.8	95875	95614	95346	95072	94796	94523	94259	94006	93770	92913	92633	92787	93183	94204	95169	1.8
1.9	96633	96391	96140	95880	95615	95350	95088	94835	94594	93668	93278	93321	93622	94504	95383	1.9
2.0	97272	97051	96819	96577	96327	96073	95820	95571	95330	94359	93878	93823	94036	94789	95587	2.0
2.1	97806	97608	97397	97174	96942	96703	96461	96220	95985	94990	94436	94294	94427	95060	95781	2.1
2.2	98248	98073	97884	97683	97469	97247	97020	96791	96563	95564	94952	94735	94797	95318	95965	2.2
2.3	98612	98459	98292	98112	97919	97715	97504	97289	97073	96085	95430	95148	95146	95563	96142	2.3
2.4	98909	98777	98631	98472	98299	98115	97922	97722	97518	96555	95871	95535	95475	95797	96310	2.4
2.5	99149	99036	98911	98772	98619	98454	98279	98095	97907	96980	96278	95897	95785	96019	96472	2.5
2.6	99341	99246	99140	99019	98886	98740	98583	98416	98243	97361	96652	96235	96078	96231	96626	2.6
2.7	99494	99415	99325	99223	99107	98979	98840	98690	98533	97701	96996	96551	96354	96433	96774	2.7
2.8	99615	99550	99474	99388	99289	99178	99056	98923	98781	98005	97310	96845	96615	96625	96916	2.8
2.9	99709	99656	99594	99522	99438	99343	99236	99119	98992	98275	97598	97118	96859	96809	97052	2.9
3.0	99782	99739	99688	99629	99559	99478	99386	99284	99172	98514	97860	97372	97090	96984	97182	3.0
3.1	99837	99804	99763	99714	99656	99588	99509	99421	99323	98725	98098	97607	97306	97151	97308	3.1
3.2	99880	99853	99821	99781	99733	99677	99611	99535	99449	98909	98314	97826	97509	97310	97428	3.2
3.3	99912	99891	99866	99834	99795	99748	99693	99628	99555	99071	98509	98028	97700	97461	97543	3.3
3.4	99936	99920	99900	99875	99843	99805	99759	99705	99642	99212	98685	98214	97879	97605	97654	3.4
3.5	99954	99942	99926	99906	99881	99850	99812	99767	99714	99334	98844	98385	98046	97743	97760	3.5
3.6	99967	99958	99946	99931	99911	99886	99855	99817	99772	99440	98986	98543	98202	97874	97862	3.6
3.7	99977	99970	99961	99949	99933	99913	99888	99858	99820	99530	99113	98688	98348	97998	97961	3.7
3.8	99984	99979	99972	99963	99951	99935	99915	99890	99859	99608	99227	98821	98485	98117	98055	3.8
3.9	99989	99985	99980	99973	99964	99951	99935	99915	99890	99674	99328	98943	98611	98230	98145	3.9
4.0	99992	99989	99986	99981	99973	99964	99951	99935	99914	99730	99417	99054	98730	98337	98232	4.0
4.1	99995	99993	99990	99986	99981	99973	99964	99951	99934	99778	99496	99155	98839	98439	98316	4.1
4.2	99996	99995	99993	99990	99986	99981	99973	99963	99949	99818	99566	99246	98941	98535	98396	4.2

TABULATED VALUES MUST BE MULTIPLIED BY .00001 TO OBTAIN PROBABILITIES

δ = 3.0

z \ α_3	.20	.30	.40	.50	.60	.70	.80	.90	1.00	1.50	2.00	2.50	3.00	4.00	5.00	α_3 / z
-2.6	0	0	0	0	0	0	0	0	0	0	0	0	0	0	0	-2.6
-2.5	4	0	0	0	0	0	0	0	0	0	0	0	0	0	0	-2.5
-2.4	46	0	0	0	0	0	0	0	0	0	0	0	0	0	0	-2.4
-2.3	162	0	0	0	0	0	0	0	0	0	0	0	0	0	0	-2.3
-2.2	380	0	0	0	0	0	0	0	0	0	0	0	0	0	0	-2.2
-2.1	722	42	0	0	0	0	0	0	0	0	0	0	0	0	0	-2.1
-2.0	1208	345	0	0	0	0	0	0	0	0	0	0	0	0	0	-2.0
-1.9	1858	924	0	0	0	0	0	0	0	0	0	0	0	0	0	-1.9
-1.8	2686	1770	440	0	0	0	0	0	0	0	0	0	0	0	0	-1.8
-1.7	3708	2877	1555	0	0	0	0	0	0	0	0	0	0	0	0	-1.7
-1.6	4935	4241	3078	1168	0	0	0	0	0	0	0	0	0	0	0	-1.6
-1.5	6375	5853	4926	3346	522	0	0	0	0	0	0	0	0	0	0	-1.5
-1.4	8037	7709	7052	5882	3818	0	0	0	0	0	0	0	0	0	0	-1.4
-1.3	9924	9800	9424	8667	7297	4772	0	0	0	0	0	0	0	0	0	-1.3
-1.2	12038	12117	12015	11645	10866	9415	6656	0	0	0	0	0	0	0	0	-1.2
-1.1	14375	14648	14802	14779	14493	13808	12474	9906	3266	0	0	0	0	0	0	-1.1
-1.0	16932	17383	17766	18042	18160	18046	17586	16583	14619	0	0	0	0	0	0	-1.0
-.9	19699	20306	20886	21413	21854	22168	22300	22168	21643	0	0	0	0	0	0	-.9
-.8	22665	23401	24142	24870	25564	26197	26740	27156	27394	19505	0	0	0	0	0	-.8
-.7	25815	26650	27516	28397	29278	30141	30969	31740	32433	33289	0	0	0	0	0	-.7
-.6	29130	30034	30987	31976	32987	34008	35025	36024	36992	40815	39723	0	0	0	0	-.6
-.5	32590	33531	34535	35588	36680	37798	38930	40066	41195	46408	50041	49360	0	0	0	-.5
-.4	36170	37119	38139	39218	40346	41510	42700	43905	45116	50979	56015	59739	61378	0	0	-.4
-.3	39844	40772	41777	42849	43974	45144	46345	47568	48803	54897	60394	65004	68670	72996	0	-.3
-.2	43585	44466	45429	46462	47554	48693	49869	51071	52289	58353	63908	68675	72634	78512	82345	-.2

TABULATED VALUES MUST BE MULTIPLIED BY .00001 TO OBTAIN PROBABILITIES

TABLE A.3.5 Continued

$\delta = 3.0$

z \ α_3	.20	.30	.40	.50	.60	.70	.80	.90	1.00	1.50	2.00	2.50	3.00	4.00	5.00	α_3 / z
-.1	47363	48175	49072	50041	51073	52154	53276	54426	55596	61460	66869	71532	75427	81289	85297	-.1
.0	51149	51874	52684	53570	54519	55522	56566	57643	58741	64290	69441	73889	77604	83185	86990	.0
.1	54911	55535	56245	57031	57883	58789	59740	60726	61737	66893	71721	75904	79397	84634	88190	.1
.2	58621	59135	59735	60410	61151	61950	62795	63679	64591	69304	73774	77668	80926	85812	89123	.2
.3	62250	62649	63132	63690	64315	64998	65731	66504	67310	71550	75643	79239	82261	86805	89889	.3
.4	65771	66055	66419	66857	67362	67926	68543	69202	69898	73649	77358	80657	83448	87666	90538	.4
.5	69159	69331	69579	69899	70284	70730	71229	71774	72358	75618	78943	81949	84517	88426	91102	.5
.6	72393	72460	72597	72802	73072	73402	73786	74219	74694	77467	80416	83137	85489	89106	91602	.6
.7	75453	75424	75460	75558	75719	75938	76213	76538	76907	79207	81788	84235	86381	89722	92050	.7
.8	78326	78212	78156	78158	78217	78335	78506	78729	78997	80845	83072	85255	87205	90285	92456	.8
.9	80999	80814	80678	80594	80563	80587	80665	80792	80968	82387	84277	86206	87969	90804	92828	.9
1.0	83464	83222	83019	82862	82753	82694	82687	82729	82818	83838	85408	87097	88683	91284	93170	1.0
1.1	85719	85432	85177	84959	84784	84654	84573	84538	84551	85202	86471	87933	89350	91731	93488	1.1
1.2	87761	87444	87150	86885	86656	86467	86322	86222	86167	86485	87473	88720	89977	92150	93784	1.2
1.3	89595	89260	88940	88641	88370	88135	87938	87782	87668	87687	88415	89461	90567	92542	94061	1.3
1.4	91227	90886	90551	90229	89930	89658	89420	89219	89056	88814	89303	90160	91124	92912	94321	1.4
1.5	92664	92327	91988	91656	91338	91042	90774	90537	90335	89867	90139	90820	91650	93261	94567	1.5
1.6	93919	93594	93260	92926	92601	92290	92001	91739	91507	90849	90926	91444	92148	93592	94799	1.6
1.7	95004	94697	94376	94049	93724	93409	93108	92829	92575	91762	91665	92034	92620	93905	95019	1.7
1.8	95932	95648	95346	95033	94717	94403	94099	93811	93544	92609	92360	92591	93067	94203	95228	1.8
1.9	96718	96460	96182	95888	95586	95281	94981	94692	94417	93392	93012	93118	93492	94487	95427	1.9
2.0	97377	97148	96895	96624	96340	96051	95760	95475	95201	94114	93622	93616	93895	94757	95617	2.0
2.1	97924	97723	97497	97252	96991	96720	96444	96168	95899	94777	94193	94086	94278	95015	95798	2.1
2.2	98374	98200	98001	97782	97546	97296	97038	96777	96517	95384	94726	94530	94641	95261	95972	2.2
2.3	98738	98591	98419	98227	98016	97790	97552	97308	97061	95937	95222	94948	94987	95496	96138	2.3

TABULATED VALUES MUST BE MULTIPLIED BY .00001 TO OBTAIN PROBABILITIES

$\delta = 3.0$

z \ α_3	.20	.30	.40	.50	.60	.70	.80	.90	1.00	1.50	2.00	2.50	3.00	4.00	5.00	α_3 / z
2.4	99031	98908	98762	98595	98410	98208	97992	97767	97537	96440	95683	95343	95315	95721	96297	2.4
2.5	99264	99162	99040	98898	98737	98559	98366	98162	97949	96894	96110	95714	95626	95936	96450	2.5
2.6	99447	99364	99263	99144	99006	98852	98682	98498	98305	97302	96506	96062	95921	96142	96597	2.6
2.7	99589	99523	99440	99341	99226	99093	98945	98783	98609	97668	96871	96390	96201	96340	96738	2.7
2.8	99698	99646	99579	99499	99402	99290	99163	99021	98867	97994	97206	96696	96466	96529	96874	2.8
2.9	99780	99740	99687	99622	99543	99450	99342	99219	99084	98284	97514	96983	96717	96710	97005	2.9
3.0	99842	99811	99770	99719	99655	99578	99487	99383	99265	98539	97795	97251	96955	96883	97130	3.0
3.1	99888	99865	99833	99793	99742	99679	99604	99516	99416	98763	98051	97501	97179	97049	97252	3.1
3.2	99921	99904	99880	99849	99809	99759	99697	99624	99539	98958	98283	97733	97391	97208	97368	3.2
3.3	99946	99933	99915	99891	99860	99820	99771	99711	99639	99128	98493	97949	97590	97361	97481	3.3
3.4	99963	99954	99940	99922	99899	99868	99828	99779	99721	99274	98683	98148	97778	97506	97589	3.4
3.5	99975	99968	99959	99945	99927	99904	99873	99833	99785	99399	98853	98332	97955	97646	97694	3.5
3.6	99983	99979	99972	99962	99949	99931	99906	99875	99837	99506	99005	98502	98121	97779	97795	3.6
3.7	99989	99986	99981	99974	99964	99951	99932	99908	99877	99596	99141	98659	98276	97907	97892	3.7
3.8	99993	99991	99987	99982	99975	99965	99951	99932	99908	99672	99261	98802	98422	98029	97986	3.8
3.9	99995	99994	99992	99988	99983	99976	99965	99951	99932	99736	99367	98933	98558	98145	98076	3.9
4.0	99997	99996	99995	99992	99989	99983	99976	99965	99950	99788	99460	99052	98685	98257	98164	4.0
4.1	99998	99998	99997	99995	99992	99989	99983	99975	99964	99832	99542	99161	98803	98363	98248	4.1
4.2	99999	99998	99998	99997	99995	99992	99988	99983	99974	99867	99613	99259	98912	98464	98329	4.2
4.3	99999	99999	99999	99998	99997	99995	99992	99988	99982	99896	99675	99348	99014	98560	98407	4.3
4.4	100000	99999	99999	99999	99998	99997	99995	99992	99987	99919	99728	99428	99108	98652	98482	4.4
4.5	100000	100000	100000	99999	99999	99998	99997	99994	99991	99937	99774	99500	99195	98739	98555	4.5
4.6	100000	100000	100000	100000	99999	99999	99998	99996	99994	99952	99812	99565	99275	98822	98625	4.6
4.7	100000	100000	100000	100000	100000	99999	99999	99998	99996	99964	99845	99622	99349	98901	98692	4.7
4.8	100000	100000	100000	100000	100000	99999	99999	99998	99997	99973	99873	99673	99416	98975	98756	4.8

TABULATED VALUES MUST BE MULTIPLIED BY .00001 TO OBTAIN PROBABILITIES

TABLE A.3.5 Continued

$\delta = 3.5$

z \ α_3	.20	.30	.40	.50	.60	.70	.80	.90	1.00	1.50	2.00	2.50	3.00	4.00	5.00	α_3 / z
-2.2	0	0	0	0	0	0	0	0	0	0	0	0	0	0	0	-2.2
-2.1	157	0	0	0	0	0	0	0	0	0	0	0	0	0	0	-2.1
-2.0	600	0	0	0	0	0	0	0	0	0	0	0	0	0	0	-2.0
-1.9	1301	59	0	0	0	0	0	0	0	0	0	0	0	0	0	-1.9
-1.8	2247	869	0	0	0	0	0	0	0	0	0	0	0	0	0	-1.8
-1.7	3427	2151	159	0	0	0	0	0	0	0	0	0	0	0	0	-1.7
-1.6	4835	3765	1955	0	0	0	0	0	0	0	0	0	0	0	0	-1.6
-1.5	6464	5651	4225	1660	0	0	0	0	0	0	0	0	0	0	0	-1.5
-1.4	8308	7773	6762	4951	1339	0	0	0	0	0	0	0	0	0	0	-1.4
-1.3	10359	10103	9495	8336	6200	1398	0	0	0	0	0	0	0	0	0	-1.3
-1.2	12611	12623	12384	11775	10582	8324	3069	0	0	0	0	0	0	0	0	-1.2
-1.1	15056	15313	15402	15252	14748	13687	11633	7121	0	0	0	0	0	0	0	-1.1
-1.0	17684	18159	18531	18756	18772	18483	17729	16191	13030	0	0	0	0	0	0	-1.0
-.9	20485	21146	21754	22280	22691	22937	22950	22621	21763	0	0	0	0	0	0	-.9
-.8	23448	24260	25056	25818	26523	27147	27656	28005	28134	9257	0	0	0	0	0	-.8
-.7	26560	27486	28424	29362	30282	31170	32006	32770	33435	33533	0	0	0	0	0	-.7
-.6	29805	30809	31847	32906	33974	35039	36087	37107	38085	41761	39119	0	0	0	0	-.6
-.5	33167	34214	35309	36443	37602	38776	39953	41125	42281	47515	50954	48902	0	0	0	-.5
-.4	36628	37684	38800	39965	41167	42395	43639	44889	46136	52082	57084	60675	61898	0	0	-.4
-.3	40168	41202	42305	43465	44670	45908	47169	48442	49720	55927	61431	65992	69577	73580	0	-.3
-.2	43767	44750	45811	46934	48107	49319	50559	51815	53080	59280	64860	69595	73500	79248	82953	-.2
-.1	47401	48311	49302	50362	51476	52633	53821	55030	56251	62269	67718	72361	76209	81958	85861	-.1
.0	51048	51864	52766	53739	54771	55850	56964	58102	59255	64976	70183	74623	78300	83780	87490	.0
.1	54682	55389	56185	57055	57988	58971	59992	61042	62110	67457	72358	76546	80012	85164	88635	.1
.2	58280	58867	59545	60300	61120	61993	62909	63858	64829	69750	74310	78223	81466	86283	89521	.2

TABULATED VALUES MUST BE MULTIPLIED BY .00001 TO OBTAIN PROBABILITIES

$\delta = 3.5$

z \ α_3	.20	.30	.40	.50	.60	.70	.80	.90	1.00	1.50	2.00	2.50	3.00	4.00	5.00	α_3 / z
.3	61817	62278	62830	63461	64160	64916	65718	66556	67422	71883	76083	79713	82732	87224	90245	.3
.4	65268	65602	66026	66529	67101	67734	68417	69140	69896	73878	77708	81056	83855	88038	90858	.4
.5	68612	68820	69116	69491	69937	70444	71006	71613	72255	75750	79211	82278	84865	88755	91391	.5
.6	71825	71915	72088	72338	72658	73043	73484	73974	74505	77514	80607	83402	85783	89397	91861	.6
.7	74889	74869	74927	75059	75260	75526	75850	76226	76647	79177	81911	84441	86626	89978	92283	.7
.8	77786	77669	77622	77644	77734	77887	78100	78367	78682	80749	83133	85407	87404	90508	92666	.8
.9	80501	80301	80162	80087	80074	80124	80234	80398	80613	82235	84282	86310	88128	90997	93015	.9
1.0	83024	82756	82539	82379	82276	82233	82247	82317	82439	83640	85365	87157	88803	91450	93338	1.0
1.1	85346	85026	84746	84515	84335	84210	84140	84125	84161	84969	86387	87954	89436	91872	93637	1.1
1.2	87463	87106	86779	86491	86247	86053	85910	85819	85779	86225	87353	88706	90031	92267	93915	1.2
1.3	89373	88995	88635	88305	88012	87761	87557	87401	87294	87411	88267	89416	90593	92638	94176	1.3
1.4	91079	90693	90315	89957	89628	89333	89079	88870	88706	88529	89132	90089	91125	92988	94422	1.4
1.5	92587	92205	91822	91449	91096	90770	90479	90226	90015	89581	89950	90726	91628	93319	94653	1.5
1.6	93904	93536	93159	92784	92419	92075	91757	91472	91224	90569	90725	91331	92106	93633	94873	1.6
1.7	95042	94696	94334	93966	93601	93249	92916	92609	92333	91493	91457	91906	92561	93931	95081	1.7
1.8	96013	95695	95356	95004	94648	94297	93958	93639	93345	92357	92149	92451	92993	94215	95278	1.8
1.9	96831	96546	96235	95905	95566	95225	94889	94566	94262	93161	92802	92969	93405	94485	95467	1.9
2.0	97512	97261	96981	96680	96363	96038	95713	95394	95089	93907	93418	93461	93798	94744	95647	2.0
2.1	98072	97855	97608	97337	97047	96745	96436	96128	95827	94597	93997	93929	94172	94991	95820	2.1
2.2	98525	98341	98128	97889	97629	97352	97065	96773	96482	95231	94541	94372	94529	95228	95985	2.2
2.3	98887	98734	98553	98347	98118	97869	97606	97334	97059	95812	95050	94792	94870	95455	96143	2.3
2.4	99172	99047	98897	98721	98523	98304	98067	97818	97562	96342	95527	95190	95195	95673	96295	2.4
2.5	99393	99293	99170	99024	98855	98665	98456	98232	97997	96822	95971	95567	95505	95882	96442	2.5
2.6	99561	99483	99385	99265	99124	98962	98780	98582	98370	97256	96383	95923	95800	96082	96582	2.6
2.7	99688	99628	99550	99454	99339	99203	99048	98875	98686	97644	96766	96259	96081	96275	96718	2.7

TABULATED VALUES MUST BE MULTIPLIED BY .00001 TO OBTAIN PROBABILITIES

TABLE A.3.5 Continued

$\delta = 3.5$

z \ α_3	.20	.30	.40	.50	.60	.70	.80	.90	1.00	1.50	2.00	2.50	3.00	4.00	5.00	α_3 / z
2.8	99782	99736	99676	99601	99507	99396	99265	99117	98952	97991	97119	96575	96349	96460	96849	2.8
2.9	99850	99816	99771	99712	99638	99548	99440	99315	99173	98298	97444	96872	96604	96638	96975	2.9
3.0	99898	99874	99840	99796	99738	99667	99579	99475	99354	98568	97742	97151	96846	96809	97097	3.0
3.1	99932	99915	99891	99857	99814	99758	99687	99602	99502	98804	98013	97412	97076	96974	97215	3.1
3.2	99956	99944	99926	99902	99869	99826	99771	99703	99620	99008	98261	97656	97294	97132	97328	3.2
3.3	99972	99963	99951	99934	99910	99878	99835	99781	99714	99184	98484	97883	97500	97284	97438	3.3
3.4	99982	99977	99968	99956	99939	99915	99883	99841	99787	99334	98686	98094	97696	97430	97544	3.4
3.5	99989	99985	99980	99972	99959	99942	99918	99886	99844	99461	98867	98289	97880	97571	97647	3.5
3.6	99993	99991	99987	99982	99973	99961	99944	99919	99887	99568	99028	98470	98053	97706	97746	3.6
3.7	99996	99995	99992	99989	99983	99974	99962	99944	99919	99657	99170	98636	98217	97835	97842	3.7
3.8	99998	99997	99995	99993	99989	99983	99975	99962	99943	99730	99296	98789	98370	97959	97935	3.8
3.9	99999	99998	99997	99996	99993	99989	99983	99974	99961	99789	99407	98928	98514	98078	98025	3.9
4.0	99999	99999	99998	99998	99996	99993	99989	99983	99973	99837	99503	99055	98648	98192	98112	4.0
4.1	100000	99999	99999	99999	99998	99996	99993	99989	99982	99876	99586	99170	98774	98302	98196	4.1
4.2	100000	100000	100000	99999	99999	99998	99996	99993	99988	99906	99658	99275	98890	98406	98277	4.2
4.3	100000	100000	100000	100000	99999	99999	99997	99996	99992	99930	99719	99368	98998	98506	98356	4.3
4.4	100000	100000	100000	100000	100000	99999	99999	99997	99995	99948	99771	99453	99099	98601	98432	4.4
4.5	100000	100000	100000	100000	100000	100000	99999	99998	99997	99962	99814	99528	99191	98692	98505	4.5
4.6	100000	100000	100000	100000	100000	100000	100000	99999	99998	99973	99851	99594	99277	98779	98576	4.6
4.7	100000	100000	100000	100000	100000	100000	100000	99999	99999	99981	99881	99653	99355	98861	98645	4.7
4.8	100000	100000	100000	100000	100000	100000	100000	100000	99999	99986	99906	99705	99426	98940	98711	4.8
4.9	100000	100000	100000	100000	100000	100000	100000	100000	100000	99990	99926	99751	99492	99014	98775	4.9
5.0	100000	100000	100000	100000	100000	100000	100000	100000	100000	99993	99943	99790	99551	99085	98836	5.0
5.1	100000	100000	100000	100000	100000	100000	100000	100000	100000	99996	99956	99824	99605	99152	98896	5.1
5.2	100000	100000	100000	100000	100000	100000	100000	100000	100000	99997	99966	99854	99654	99215	98953	5.2

TABULATED VALUES MUST BE MULTIPLIED BY .00001 TO OBTAIN PROBABILITIES

A.4 COMPUTER PROGRAMS

PROGRAM A.4.1 Estimates of Weibull Distribution Parameters

```
      SUBROUTINE MOMEST(AMOM,V,ERROR)
      DIMENSION  AMOM(3),DG(3),V(6)
      LOGICAL ERROR
C*********************************************************************
C THIS SUBROUTINE CALCULATES THE MOMENT ESTIMATES OF THE THREE
C PARAMETER WEIBULL DISTRIBUTION.
C
C  INPUT
C              AMOM   = ARRAY OF MOMENTS
C                       AMOM(1)  = MEAN
C                       AMOM(2)  = STANDARD DEVIATION
C                       AMOM(3)  = THIRD STANDARD MOMENT
C
C  OUTPUT
C              V      = ARRAY OF ESTIMATES OF THE PARAMETERS
C                       V(1)  = T TERMINUS ( AMOM(1) -V(3)DG(1))
C                       V(2)  = D (THE SHAPE PARAMETER)
C                       V(3)  = B (THE SCALE PARAMETER)
C
C              ERROR = .TRUE IF NO ESTIMATES ARE FOUND
C
C FUNCTIONS
C              GAMMA(X)   = AN IMSL FUNCTION THAT EVALUATES
C                                THE GAMMA FUNCTION
C*********************************************************************
C
C  SET UPPER AND LOWER BOUNDS ON D.  USE A BINARY SEARCH OF FIND
C  SOLUTIONS TO THE ESTIMATING EQUATIONS.
C
C  DEFINE THE FUNCTION TO BE EVALUATED
      F(A,B,C,D1)=(C-3.0*B*A+2.0*A**3)/(B-A**2)**1.5-D1
      ERROR = .FALSE.
      DL=.5
      DU=3.22
C
C  SET THE TOLERANCE LEVEL
      EPS=.1E-7
C
C  COMPUTE F(X) AT THE INTERVAL ENDPOINTS
      DO 13 I=1,3
      A=I
   13 DG(I)=GAMMA(1.0+A/DL)
      FL=F(DG(1),DG(2),DG(3),AMOM(3))
      DO 14 I=1,3
      A=I
   14 DG(I)=GAMMA(1.0+A/DU)
      FU=F(DG(1),DG(2),DG(3),AMOM(3))
C
C  PROVIDED A ZERO EXISTS IN THE GIVEN INTERVAL, BISECT THE INTERVAL
C  AND CHOOSE THE SUBINTERVAL WITH A SIGN CHANGE, UNTIL THE INTERVAL
C  WIDTH IS WITHIN TOLERANCE
C
      IF (FL*FU.GT.0.0) THEN
C  NO SOLUTION EXISTS, THEREFORE, SET ERROR = .TRUE.
         ERROR= .TRUE.
```

PROGRAM A.4.1 Continued

```
C      SUBROUTINE MOMEST(AMOM,V,ERROR)  CONTINUED
       ELSE
          D = (DU+DL)/2.0
  100     DO 15 I=1,3
          A=I
   15     DG(I)=GAMMA(1.0+A/D)
          FD=F(DG(1),DG(2),DG(3),AMOM(3))
          V(3)=AMOM(2)/(DG(2)-DG(1)**2)**.5
          V(1)=AMOM(1)-V(3)*DG(1)
          V(2)=D
          V(4)=V(1)+V(3)*DG(1)
          V(5)=V(3)*(DG(2)-DG(1)**2)**.5
          V(6)=(DG(3)-3*DG(2)*DG(1)+2*DG(1)**3)/(DG(2)-DG(1)**2)**1.5
          IF (ABS(DL-D).GT.EPS) THEN
             IF (FD*FL.LE.0.0) THEN
C               CHOOSE THE LEFT SUBINTERVAL
                DU=D
             ELSE
C               CHOOSE THE RIGHT SUBINTERVAL
                DL=D
                FL=FD
             ENDIF
             D =(DU+DL)/2.0
             GO TO 100
          ENDIF
       ENDIF
       RETURN
       END
```

PROGRAM A.4.1 Continued

```
      SUBROUTINE WMME(AMOM,XMIN,N,V,ERROR)
      DIMENSION AMOM(3),V(6),DG(3)
      LOGICAL ERROR
C**********************************************************************
C THIS SUBROUTINE CALCULATES THE MODIFIED MOMENT ESTIMATES OF THE
C THREE-PARAMETER WEIBULL DISTRIBUTION.
C INPUT
C             AMOM  = ARRAY OF MOMENTS
C                       AMOM(1) = MEAN
C                       AMOM(2) = STANDARD DEVIATION
C                       AMOM(3) = THIRD STANDARD MOMENT
C             XMIN  = FIRST ORDER STATISTIC (SMALLEST OBSERVATION)
C             N     = NUMBER OF OBSERVATIONS
C
C OUTPUT
C             V     = ARRAY OF ESTIMATES OF THE PARAMETERS
C                     V(1) = TERMINUS  (T)
C                     V(2) = SHAPE PARAMETER (D)
C                     V(3) = SCALE PARAMETER (B)
C          ERROR    = .TRUE IF NO ESTIMATES ARE FOUND
C
C FUNCTIONS
C          GAMMA(X)  AN IMSL FUNCTION EVALUATES THE GAMMA FUNCTION
C**********************************************************************
C
C  USE BINARY SEARCH TO FIND SOLUTIONS TO THE ESTIMATING EQUATIONS.
C
      ERROR=.FALSE.
      AN=N
C
C SET VAL=(AMOM(2)/(AMOM(1)-XMIN))**2
C
      VAL=(AMOM(2)/(AMOM(1)-XMIN))**2
      DU = 3.22
      DL = .5
C  CALCULATE GAMMA VALUES AT DU
      DO 11 I = 1,3
      A=I
   11 DG(I)=GAMMA(1.0+A/DU)
C CALCULATE FUNCTION AT UPPER BOUND, D
      FU=VAL-(DG(2)-DG(1)**2)/((1.0-AN**(-1.0/DU))*DG(1))**2
C  CALCULATE GAMMA VALUES AT DL
      DO 111 I = 1,3
      A=I
  111 DG(I)=GAMMA(1.0+A/DL)
C CALCULATE FUNCTION AT LOWER BOUND, DL
      FL=VAL-(DG(2)-DG(1)**2)/((1.0-AN**(-1.0/DL))*DG(1))**2
C  SET TOLERANCE
      EPS=.1E-7
C  PROVIDED A ZERO ESITS IN THE GIVEN INTERVAL, BISECT THE INTERVAL AND
C  CHOOSE THE SUBINTERVAL WITH A SIGN  CHANGE, UNTIL THE INTERVAL WIDTH
C  IS WITHIN TOLERANCE.
      IF(FL*FU.GT.0.0) THEN
         ERROR=.TRUE.
      ELSE
```

PROGRAM A.4.1 Continued

```
C      SUBROUTINE WMME(AMOM,XMIN,N,V,ERROR) CONTINUED
C
          D=(DU+DL)/2.0
          F=FL
  100     IF(ABS(D-DL) .GT. EPS ) THEN
C            CALCULATE GAMMA VALUES AT D
             DO 112 I = 1,3
             A=I
  112        DG(I)=GAMMA(1.0+A/D)
C            CALCULATE FUNCTION AT D
             F=VAL-(DG(2)-DG(1)**2)/((1.0-AN**(-1.0/D))*DG(1))**2
             IF(F*FL.LE.0.0) THEN
C               CHOOSE LEFT SUBINTERVAL
                DU=D
             ELSE
C               CHOOSE RIGHT SUBINTERVAL
                DL=D
                FL=F
             ENDIF
             D=(DU+DL)/2.0
             GO TO 100
          ELSE
             DO 115 I = 1,3
             A=I
  115        DG(I)=GAMMA(1.0+A/D)
             V(2)=D
             V(3)=AMOM(2)/SQRT(DG(2)-DG(1)**2)
             V(1)=AMOM(1)-V(3)*DG(1)
             V(4)=V(1)+V(3)*DG(1)
             V(5)=V(3)*(DG(2)-DG(1)**2)**.5
             V(6)=(DG(3)-3*DG(2)*DG(1)+2*DG(1)**3)/(DG(2)-DG(1)**2)**1.5
          ENDIF
       ENDIF
       RETURN
       END
```

PROGRAM A.4.1 Continued

```
      SUBROUTINE WMLE(X,N,AMOM,XMIN,ERROR,V)
      DIMENSION DG(3),X(N),AMOM(3),V(6),ML(10)
      LOGICAL ERROR
C********************************************************************
C THIS SUBROUTINE CALCULATES THE  MAXIMUM LIKELIHOOD ESTIMATES
C OF THE THREE PARAMETER WEIBULL DISTRIBUTION.
C INPUT
C             N      = NUMBER OF OBSERVATIONS
C             X(N)   = ARRAY OF OBSERVATIONS
C             AMOM   = ARRAY OF MOMENTS
C                      AMOM(1) = MEAN
C                      AMOM(2) = STANDARD DEVIATION
C                      AMOM(3) = THIRD STANDARD MOMENT
C             XMIN   = FIRST ORDER STATISTICS
C
C OUTPUT
C             V      = ARRAY OF ESTIMATES OF THE PARAMETERS
C                      V(1) = TERMINUS ( AMOM(1) -V(3)DG(1))
C                      V(2) = D (DELTA, THE SHAPE PARAMETER)
C                      V(3) = B (THE SCALE PARAMETER)
C                      V(4) = MEAN
C                      V(5) = STANDARD DEVIATION
C                      V(6) = THIRD STANDARD MOMENT
C             ERROR =  TRUE IF NO ESTIMATES ARE FOUND
C
C FUNCTION
C             GAMMA(X)  AN IMSL FUNCTION THAT EVALUATES THE
C                        GAMMA FUNCTION
C
C SUBROUTINE
C             MLDEL(X,N,T,D,ERROR,S3)   CALCULATES D  (THE  SHAPE
C                        PARAMETER) FOR GIVEN VALUES OF THE TERMINUS
C                        AND OBSERVATIONS (X) .
C********************************************************************
C
C  SET UPPER AND LOWER BOUNDS ON D.  USE A BINARY SEARCH OF FIND
C  SOLUTIONS TO THE ESTIMATING EQUATIONS.
C
      AN=N


C
C  SET UPPER  BOUND ON THE TERMINUS, T.  T CANNOT BE LESS THAN
C  BOUND = XMIN - 6*(STANDARD DEVIATION).  USE A BINARY SEARCH TO
C  FIND SOLUTIONS TO THE ESTIMATING EQUATIONS.
C
      BOUND=XMIN-6.0*AMOM(2)
      ERROR=.FALSE.
      AN=N
      STEP=AMOM(2)/50.0
C  SET THE TOLERANCE LEVEL
      EPS=.1E-7
      T= XMIN-STEP/10
C  CALL MLDEL TO SOLVE FOR (D) FOR T=XMIN-STEP/10
C  IF NO CORRESPONDING VALUE OF (D) CAN BE FOUND, ERROR=.TRUE.
   50 IF (T.GT.BOUND)THEN
```

PROGRAM A.4.1 Continued

```
C       SUBROUTINE WMLE(X,N,AMOM,XMIN,ERROR,V)  CONTINUED
           CALL MLDEL (X,N,T,D,ERROR,S3)
           IF (ERROR) THEN
              T=T-STEP
              GO TO 50
            ELSE
C             SET UPPER BOUND, TU  TO T
              TU=T
C             CALCULATE THETA AND THE VALUE OF FUNCTION AT T,
C                THE UPPER BOUND
              THETA=S3/AN
              B=THETA**(1.0/D)
              S12=S13=0.0
              DO 13 I=1,N
              S13=S13+(X(I)-TU)**(D-1.0)
   13         S12=S12+1.0/(X(I)-TU)
              FU=D/THETA*S13-(D-1.0)*S12
C             SEARCH FOR A LOWER BOUND ON T SUCH THAT FL*FU < 0
              FL=FU
  100         IF(FU*FL.GE.0.0.AND..NOT.ERROR) THEN
                 T=T-STEP
                 IF(T.GT.BOUND) THEN
                    CALL MLDEL (X,N,T,D,ERROR,S3)
                    IF (.NOT.ERROR) THEN
C                      CALCULATE THETA AND THE VALUE OF FUNCTION AT T,
C                      THE LOWER BOUND
C
                       THETA=S3/AN
                       B=THETA**(1.0/D)
                       S12=S13=0.0
                       DO 113 I=1,N
                       S13=S13+(X(I)-T)**(D-1.0)
  113                  S12=S12+1.0/(X(I)-T)
                       FL=D/THETA*S13-(D-1.0)*S12
                    ENDIF
                 ELSE
                    ERROR=.TRUE.
                    PRINT*,'ERROR.  CANNOT FIND D FOR LOWER BOUND
                 ENDIF
                 GO TO 100
             ELSE
                IF(.NOT.ERROR) THEN
                   TL=T
C END POINTS OF INTERVAL HAVE BEEN FOUND, (TL,TU)
                   T=(TL+TU)/2.0
                   FT=FU
  200              CALL MLDEL(X,N,T,D,ERROR,S3)
                   IF (ERROR) THEN
                      PRINT*,'  T NOT WITHIN REQUIRED TOLEANCE.'
                   ELSE
                      THETA=S3/AN
                      B=THETA**(1.0/D)
                      S12=S13=0.0
                      DO 114 I=1,N
                S13=S13+(X(I)-T)**(D-1.0)
```

PROGRAM A.4.1 Continued

```
C      SUBROUTINE WMLE(X,N,AMOM,XMIN,ERROR,V)  CONTINUED
  114                 S12=S12+1.0/(X(I)-T)
                      FT=D/THETA*S13-(D-1.0)*S12
                    V(1)=T
                      V(2)=D
                      V(3)=THETA**(1.0/D)
                      DO 15 I=1,3
                      A=I
   15                 DG(I)=GAMMA(1.0+A/D)
                      V(4)=V(1)+V(3)*DG(1)
                      V(5)=V(3)*(DG(2)-DG(1)**2)**.5
                      V(6)=(DG(3)-3*DG(2)*DG(1)+2*DG(1)**3)/
     *                (DG(2)-DG(1)**2)**1.5
                      IF(ABS(TU-T).GT.EPS.AND..NOT.ERROR)THEN
                         IF(FT*FU.LE.0.0) THEN
C                           CHOOSE RIGHT SUBINTERVAL
                            TL=T
                         ELSE
C                           CHOOSE LEFT SUBINTERVAL
                            TU=T
                            FU=FT
                         ENDIF
                         T= (TL+TU)/2.0
                         GO TO 200
                      ENDIF
                    ENDIF
                ENDIF
             ENDIF
          ENDIF
       ELSE
          PRINT*,'CANNOT FIND D FOR TU'
       ENDIF
       RETURN
       END
```

PROGRAM A.4.1 Continued

```
      SUBROUTINE MLDEL(X,N,T,D,ERROR,S3)
      DIMENSION X(N)
      LOGICAL ERROR
C***********************************************************************
C THIS SUBROUTINE IS CALLED BY WMLE WHICH CALCULATES THE MAXIMUM
C LIKELIHOOD ESTIMATES OF THE THREE  PARAMETER WEIBULL DISTRIBUTION.
C THIS ROUTINE CALCULATES D (THE SHAPE PARAMETER) FOR GIVEN VALUES OF
C THE TERMINUS AND OBSERVATIONS (X) .
C INPUT
C             N      = NUMBER OF OBSERVATIONS
C             X(N)   = ARRAY OF OBSERVATIONS FROM A WEIBULL
C                        DISTRIBUTION
C             T      = ESTIMATES OF THE TERMINUS
C
C OUTPUT
C             D      = VALUE OF THE SHAPE PARAMETER
C             S3     = SUM OF (X(I)-T)**D FOR I = 1,N
C             ERROR = .TRUE IF NO ESTIMATES ARE FOUND
C***********************************************************************
C
C  SET ERROR FLAG
      ERROR=.FALSE.

      AN=N
C  SET UPPER AND LOWER BOUNDS ON D
      DU = 3.0
      DL = 1.01
C CALCULATE FUNCTION AT UPPER BOUND, D
      S1=S2=S3=0.0
    6 DO 11 I = 1,N
      S3=S3+(X(I)-T)**DU
      S2=S2+(X(I)-T)**DU*ALOG(X(I)-T)
   11 S1=S1+ALOG(X(I)-T)
      FU=S2/S3-1.0/DU-S1/AN
C CALCULATE FUNCTION AT LOWER BOUND, DL
      S2=S3=0.0
      DO 12 I = 1,N
      S3=S3+(X(I)-T)**DL
   12 S2=S2+(X(I)-T)**DL*ALOG(X(I)-T)
      FL=S2/S3-1.0/DL-S1/AN
C
C  SET TOLERANCE
      EPS=.1E-7
C  PROVIDED A ZERO ESITS IN THE GIVEN INTERVAL, BISECT THE INTERVAL AND
C  CHOOSE THE SUBINTERVAL WITH A SIGN  CHANGE, UNTIL THE INTERVAL WIDTH
C  IS WITHIN TOLERANCE.
      IF(FL*FU.GT.0.0) THEN
         ERROR=.TRUE.
      ELSE
         D=(DU+DL)/2.0
         F=FL
  100    S2=S3=0.0
         DO 13 I = 1, N
         S3=S3+(X(I)-T)**D
   13    S2=S2+(X(I)-T)**D*ALOG(X(I)-T)
```

PROGRAM A.4.1 Continued

```
C      SUBROUTINE MLDEL(X,N,T,D,ERROR,S3) CONTINUED
C
          F=S2/S3-1.0/D-S1/AN
          IF(ABS(D-DL) .GT. EPS ) THEN
             IF(F*FL.LE.0.0) THEN
C               CHOOSE LEFT SUBINTERVAL
                DU=D
             ELSE
C               CHOOSE RIGHT SUBINTERVAL
                DL=D
                FL=F
             ENDIF
             D=(DU+DL)/2.0
             GO TO 100
          ENDIF
       ENDIF
       RETURN
       END
```

PROGRAM A.4.1 Continued

```
      SUBROUTINE WBE(N,X,AMOM,V)
      DIMENSION X(N),AMOM(3),V(6),DG(3)
C**********************************************************************
C THIS SUBROUTINE CALCULATES THE WYCOFF-BAIN-ENGELHARDT ESTIMATES OF
C THE THREE PARAMETER WEIBULL DISTRIBUTION.
C
C  INPUT
C            AMOM   = ARRAY OF MOMENTS
C                     AMOM(1)  = MEAN
C                     AMOM(2)  = STANDARD DEVIATION
C                     AMOM(3)  = THIRD STANDARD MOMENT
C               N   = NUMBER OF OBSERVATIONS
C             X(N)  = ARRAY OF N OBSERVATIONS
C
C  OUTPUT
C            V      = ARRAY OF ESTIMATES OF THE PARAMETERS
C                     V(1)  = T TERMINUS ( AMOM(1) -V(3)DG(1))
C                     V(2)  = D (THE SHAPE PARAMETER)
C                     V(3)  = B (THE SCALE PARAMETER)
C FUNCTIONS
C            GAMMA(X)   = AN IMSL FUNCTION THAT EVALUATES
C                              THE GAMMA FUNCTION
C**********************************************************************
      AN=N
      PI=.16731
      PK=.97366
      S=.84*N
      S1=0.0
      S2=0.0
      NPI=N*PI+1
      NPK=N*PK+1
      D0=2.989/ALOG((X(NPK)-X(1))/(X(NPI)-X(1)))
      V(1)=(X(1)-AMOM(1)/AN**(1.0/D0))/(1.0-1.0/AN**(1.0/D0))
      IS=.84*AN
      DO 13 I=1,IS
          SS1=+ALOG(X(I)-V(1))
          S1=S1+SS1
   13 CONTINUE
      IS1=.84*AN+1.0
      DO 14 I=IS1,N
          SS2= ALOG(X(I)-V(1))
          S2=S2+SS2
   14 CONTINUE
      V(2)=AN*1.4192/(-S1+IS/(AN-IS)*S2)
      V(3)=EXP((0.5772/V(2))+(S1+S2)/AN)
      DO 15 I=1,3
          DG(I)=GAMMA(1.0+I/V(2))
  15  CONTINUE
      V(4)=V(1)+V(3)*DG(1)
      V(5)=V(3)*(DG(2)-DG(1)**2)**.5
      V(6)=(DG(3)-3*DG(2)*DG(1)+2*DG(1)**3)/(DG(2)-DG(1)**2)**1.5
      RETURN
      END
```

PROGRAM A.4.1 Continued

```
      SUBROUTINE ZANA(N,X,AMOM,V)
      DIMENSION X(N),AMOM(3),V(6),DG(3)
C*****************************************************************
C THIS SUBROUTINE CALCULATES THE ZANAKIS ESTIMATES OF THE
C THREE PARAMETER WEIBULL DISTRIBUTION.
C
C  INPUT
C            AMOM  = ARRAY OF MOMENTS
C                    AMOM(1) = MEAN
C                    AMOM(2) = STANDARD DEVIATION
C                    AMOM(3) = THIRD STANDARD MOMENT
C               N  = NUMBER OF OBSERVATIONS
C             X(N) = ARRAY OF N OBSERVATIONS
C
C  OUTPUT
C            V     = ARRAY OF ESTIMATES OF THE PARAMETERS
C                    V(1) = T TERMINUS ( AMOM(1) -V(3)DG(1))
C                    V(2) = D (THE SHAPE PARAMETER)
C                    V(3) = B (THE SCALE PARAMETER)
C FUNCTIONS
C            GAMMA(X)   = AN IMSL FUNCTION THAT EVALUATES
C                                THE GAMMA FUNCTION
C*****************************************************************
      AN=N
      PI=.16731
      PK=.97366
      NPI=N*PI+1
      NPK=N*PK+1
      V(1)=(X(1)*X(N)-X(2)**2)/(X(1)+X(N)-2.0*X(2))
      V(2)=2.989/ALOG((X(NPK)-V(1))/(X(NPI)-V(1)))
      I63=.63*N+1
      V(3)=-V(1)+X(I63)
      DO 15 I=1,3
          DG(I)=GAMMA(1.0+I/V(2))
   15 CONTINUE
      V(4)=V(1)+V(3)*DG(1)
      V(5)=V(3)*(DG(2)-DG(1)**2)**.5
      V(6)=(DG(3)-3*DG(2)*DG(1)+2*DG(1)**3)/(DG(2)-DG(1)**2)**1.5
      RETURN
      END
```

PROGRAM A.4.2 Estimates of Lognormal Distribution Parameters

```
      SUBROUTINE MOMEST(AMOM,V,ERROR)
      DIMENSION AMOM(3),V(8)
      LOGICAL ERROR
C*********************************************************************
C THIS SUBROUTINE CALCULATES THE MOMENT ESTIMATES OF THE
C THREE-PARAMETER LOGNORMAL DISTRIBUTION.
C INPUT
C        AMOM  =  ARRAY OF MOMENTS
C                    AMOM(1) = MEAN
C                    AMOM(2) = STANDARD DEVIATION
C                    AMOM(3) = THIRD STANDARD MOMENT
C

C*********************************************************************
C
      F(DW)=DW**3+3.0*DW**2-4.0-AMOM(3)**2
      G(DW)=3.0*DW**2+6.0*DW
      DL(DW)=.2000*DW+1.0
      DW=DL(AMOM(3))
C
      ERROR=.FALSE.
      KOUNT=1
  100 IF (ABS(F(DW)) .GE. 1.0E-9 .AND.KOUNT .LT. 500) THEN
         DW=DW-F(DW)/G(DW)
         KOUNT=KOUNT+1
         GO TO 100
      ELSE
         IF(KOUNT.LT. 500) THEN
            V(1)=AMOM(1)-AMOM(2)/SQRT(DW-1.0)
            IF  (V(1).LE.AMOM(1)-50.0*AMOM(2)) THEN
               ERROR = .TRUE.
               PRINT*, 'NO MOMENT ESTIMATION HAVE BEEN FOUND'
            ELSE
               BETA=AMOM(2)/SQRT(DW*(DW-1.0))
               V(2)=ALOG(BETA)
               V(3)=SQRT(ALOG(DW))
               V(4)=DW
               V(5)=BETA
               V(6)=(DW+2.0)*SQRT(DW-1.0)
               V(7)=V(1)+BETA*SQRT(DW)
               V(8)=BETA *SQRT(DW*(DW-1.0))
            ENDIF
         ELSE
            ERROR=.TRUE.
            PRINT*, 'NO MOMENT ESTIMATION HAVE BEEN FOUND'
         ENDIF
      ENDIF
      RETURN
      END
```

PROGRAM A.4.2 Continued

```
      SUBROUTINE MME(AMOM,XMIN,SI,V,ERROR)
      DIMENSION AMOM(3),V(8)
      LOGICAL ERROR
C*********************************************************************
C THIS SUBROUTINE CALCULATES THE MODIFIED MOMENT ESTIMATES OF THE
C THREE-PARAMETER LOGNORMAL DISTRIBUTION.
C INPUT
C        AMOM  =  ARRAY OF MOMENTS
C                     AMOM(1)  = MEAN
C                     AMOM(2)  = STANDARD DEVIATION
C                     AMOM(3)  = THIRD STANDARD MOMENT
C        XMIN  = SMALLEST OBSERVED VALUE (FIRST ORDER STATISTIC)
C           SI = EXPECTED VALUE OF THE FIRST ORDER STATISTIC
C
C OUTPUT
C           V  =  ARRAY OF ESTIMATES OF THE PARAMETERS
C                     V(1)  =  TERMINUS = AMOM(1)-AMOM(2)/SQRT(DW-1.0)
C                     V(2)  =  SCALE PARAMETER = ALOG(BETA) = MU
C                     V(3)  =  SHAPE PARAMETER = SQRT(ALOG(DW)) =
C                                SIGMA
C                     V(4)  =  DW
C                     V(5)  =  BETA
C                     V(6)  =  A3 = DW+2.0)*SQRT(DW-1.0
C                     V(7)  =  E(X)  = V(1)+BETA*SQRT(DW)
C                     V(8)  =  STANDARD DEVIATION =
C                                   BETA*SQRT(DW*(DW-1.0))
C         ERROR = .TRUE. IF NO ESTIMATES ARE FOUND
C*********************************************************************
C
      ERROR=.FALSE.
C  SET LOWER BOUND FOR DW (DWL=1.0+.1E-10), CALCULATE FUNCTION AT DWL
      DWL=1.00004
      FL=DWL*(DWL-1.0)-((SQRT(DWL)-EXP(SI*SQRT(ALOG(DWL))))*AMOM(2)/
     *(AMOM(1)-XMIN))**2
C SET UPPER BOUND OF DW (DWU < 1.0 +(AMOM(2)/(AMOM(1)-XMIN))**2)
      DWU=1.0+(AMOM(2)/(AMOM(1)-XMIN))**2 -0.1E-7
      FU=DWU*(DWU-1.0)-((SQRT(DWU)-EXP(SI*SQRT(ALOG(DWU))))*AMOM(2)/
     *(AMOM(1)-XMIN))**2
C SET THE TOLERANCE LEVEL
      EPS=.1E-10
      IF(FL*FU.GT.0.0) THEN
         ERROR=.TRUE.
         PRINT*, ' NO LOGNORMAL MODIFIED MOMENT ESTIMATES FOUND '
      ELSE
         DW=(DWL+DWU)/2.0
         F=FL
  100    IF(ABS(DW-DWL).GT.EPS) THEN
            F=DW*(DW-1.0)-((SQRT(DW)-EXP(SI*SQRT(ALOG(DW))))*AMOM(2)/
     *      (AMOM(1)-XMIN))**2
            IF (F*FL.LT.0.0) THEN
               DWU=DW
            ELSE
               DWL=DW
               FL=F
            ENDIF
```

PROGRAM A.4.2 Continued

```
C      SUBROUTINE MME(AMOM,XMIN,SI,V,ERROR) CONTINUED
             DW=(DWL+DWU)/2.0
             GO TO 100
          ELSE
             V(1)=AMOM(1)-AMOM(2)/SQRT(DW-1.0)
             BETA=AMOM(2)/SQRT(DW*(DW-1.0))
             V(2)=ALOG(BETA)
             V(3)=SQRT(ALOG(DW))
             V(4)=DW
             V(5)=BETA
             V(6)=(DW+2.0)*SQRT(DW-1.0)
             V(7)=V(1)+BETA*SQRT(DW)
             V(8)=BETA *SQRT(DW*(DW-1.0))
          ENDIF
       ENDIF
       RETURN
       END
```

PROGRAM A.4.2 Continued

```
      SUBROUTINE MLE(N,X,AMOM,V,ERROR)
      DIMENSION X(N),AMOM(3),V(8)
      LOGICAL ERROR
C**********************************************************************
C THIS SUBROUTINE CALCULATES THE LOCAL MAXIMUM LIKELIHOOD ESTIMATES OF
C THE THREE-PARAMETER LOGNORMAL DISTRIBUTION.
C INPUT
C            N  =  NUMBER OF OBSERVATIONS
C         X(N)  =  ARRAY OF OBSERVATIONS
C         AMOM  =  ARRAY OF MOMENTS
C                        AMOM(1) = MEAN
C                        AMOM(2) = STANDARD DEVIATION
C                        AMOM(3) = THIRD STANDARD MOMENT
C
C OUTPUT
C            V  =  ARRAY OF ESTIMATES OF THE PARAMETERS
C                        V(1) =  TERMINUS = AMOM(1)-AMOM(2)/SQRT(DW-1.0)
C                        V(2) =  SCALE PARAMETER = ALOG(BETA) = MU
C                        V(3) =  SHAPE PARAMETER = SQRT(ALOG(DW)) =
C                                   SIGMA
C                        V(4) =  DW
C                        V(5) =  BETA
C                        V(6) =  A3 = DW+2.0)*SQRT(DW-1.0
C                        V(7) =  E(X) = V(1)+BETA*SQRT(DW)
C                        V(8) =  STANDARD DEVIATION =
C                                   BETA*SQRT(DW*(DW-1.0))
C           ERROR = .TRUE. IF NO ESTIMATES ARE FOUND
C**********************************************************************
C
      ERROR=.FALSE.
C SET UPPER BOUND ON THE TERMINUS, G, TO GU.  G CANNOT BE LESS THAN
C BOUND = MEAN - 60*STANDARD DEVIATION.  USE A BINARY SERCH TO FIND
C SOLUTIONS TO THE ESTIMATING EQUATIONS.
C
      BOUND=AMOM(1)-60.*AMOM(2)
      AN=N
      S1=0.0
      S2=0.0
      S3=0.0
      S4=0.0
C  SET THE UPPER BOUND ON THE TERMINUS
      GU=X(1)-1.0E-6
      DO 11 IJ = 1,N
      S1=S1+ALOG(X(IJ)-GU)
      S2=S2+ALOG(X(IJ)-GU)**2
      S3=S3+1.0/(X(IJ)-GU)
      S4=S4+ALOG(X(IJ)-GU)/(X(IJ)-GU)
   11 CONTINUE
      S11=S1/AN
      S22=S2/AN-S11**2
C CALCULATE FUNCTION AT THE UPPER BOUND
      FU=S3*(S11-S22)-S4
C SEARCH FOR A LOWER BOUND, GL, SUCH THAT THE FUNCTION IS ZERO IN THE
C INTERVAL (GL,GU)
      GL=X(1)
```

PROGRAM A.4.2 Continued

```
C      SUBROUTINE MLE(N,X,AMOM,V,ERROR) CONTINUED
       FL=FU
  100 IF (GL.GT.BOUND.AND.(FL*FU).GT.0.0)THEN
          S1=0.0
          S2=0.0
          S3=0.0
          S4=0.0
          GL=GL-1.0
          DO 12 IJ = 1,N
          S1=S1+ALOG(X(IJ)-GL)
          S2=S2+ALOG(X(IJ)-GL)**2
          S3=S3+1.0/(X(IJ)-GL)
          S4=S4+ALOG(X(IJ)-GL)/(X(IJ)-GL)
   12     CONTINUE
          S11=S1/AN
          S22=S2/AN-S11**2
          FL=S3*(S11-S22)-S4
          GO TO 100
       ELSE
          IF (GL.LT.BOUND ) THEN
             ERROR=.TRUE.
             PRINT*, 'NO LOCAL MAXIMUM LIKELIHOOD ESTIMATE CAN BE
     *        FOUND FOR THE TERMINUS'
          ELSE
C            USE A BINARY SEARCH TO LOCATE THE SOLUTIONS TO THE
C            MAXIMUM LIKELIHOOD EQUATIONS.  THE ESTIMATE FOR THE
C            TERMINUS IS BETWEEN (GL,GU)
             KOUNT=0
             G=(GU+GL)/2.0
             F=FL
             EPS=.1E-7
  200        S1=0.0
             S2=0.0
             S3=0.0
             S4=0.0
             DO 111 IJ = 1,N
             S1=S1+ALOG(X(IJ)-G)
             S2=S2+ALOG(X(IJ)-G)**2
             S3=S3+1.0/(X(IJ)-G)
             S4=S4+ALOG(X(IJ)-G)/(X(IJ)-G)
  111        CONTINUE
             S11=S1/AN
             S22=S2/AN-S11**2
             F=S3*(S11-S22)-S4
             IF(ABS(G-GL).GT.EPS) THEN
                IF (F*FL.LT.0.0) THEN
                    GU=G
                ELSE
                    GL=G
                    FL=F
                ENDIF
                G=(GU+GL)/2.0
                GO TO 200
             ELSE
                BETA=EXP(S11)
```

PROGRAM A.4.2 Continued

```
C     SUBROUTINE MLE(N,X,AMOM,V,ERROR) CONTINUED
                V(1)=G
                V(2)=S11
                V(3)=SQRT(S22)
                DW=EXP(V(3)**2)
                V(4) =DW
                V(5) =BETA
                V(6) =(DW+2.0)*SQRT(DW-1.0)
                V(7) =G+BETA*SQRT(DW)
                V(8)=BETA*SQRT(DW*(DW-1.0))
             ENDIF
          ENDIF
      ENDIF
      RETURN
      END
```

PROGRAM A.4.3 Estimates of Inverse Gaussian Distribution Parameters

```
      SUBROUTINE MOMEST(AMOM,V)
      DIMENSION AMOM(3),V(5)
C*******************************************************************
C THIS SUBROUTINE CALCULATES THE MOMENT ESTIMATE FOR THE
C INVERSE GAUSSIAN DISTRIBUTION
C INPUT
C        AMOM  =  ARRAY OF MOMENTS
C                   AMOM(1) = MEAN
C                   AMOM(2) = STANDARD DEVIATION
C                   AMOM(3) = THIRD STANDARD MOMENT
C
C OUTPUT:     V  =  ARRAY OF ESTIMATES OF THE PARAMETERS
C                   V(1) = TERMINUS
C                   V(2) = MU = 3.0*AMOM(2)/AMOM(3)
C                   V(3) = SIGMA = STANDARD DEVIATON
C                   V(4) = E(X)
C                   V(5) = A3 = THIRD STANDARD MOMENT
C
C*******************************************************************
C
      V(3)=AMOM(2)
      V(2)=3.0*V(3)/AMOM(3)
      V(1)=AMOM(1)-V(2)
      V(4)=V(1)+V(2)
      V(5)=3.0*V(3)/V(2)
      RETURN
      END
```

PROGRAM A.4.3 Continued

```
      SUBROUTINE MME(AMOM,XMIN,N,V,ERROR)
C**********************************************************************
C THIS SUBROUTINE CALCULATES THE MODIFIED MOMENT ESTIMATE FOR THE
C INVERSE GAUSSIAN DISTRIBUTION
C INPUT
C        AMOM  =  ARRAY OF MOMENTS
C                   AMOM(1) = MEAN
C                   AMOM(2) = STANDARD DEVIATION
C                   AMOM(3) = THIRD STANDARD MOMENT
C        XMIN  =  THE FIRST ORDER STATISTIC, THE SMALLEST OBSERVATION
C        N     =  THE NUMBER OF OBSERVATIONS
C
C OUTPUT
C        V     = ARRAY OF ESTIMATES OF THE PARAMETERS
C                 V(1) = TERMINUS
C                 V(2) = MU = 3.0*AMOM(2)/AMOM(3)
C                 V(3) = SIGMA = STANDARD DEVIATON
C                 V(4) = E(X)
C                 V(5) = A3 = THIRD STANDARD MOMENT
C
C        ERROR = .TRUE IF NO ESTIMATES ARE FOUND
C
C SUBROUTINES
C        MDNOR =  IMSL  FUNCTION  -   CALCULATES   THE   NORMAL
C                   PROBABILITIES
C
C**********************************************************************
C
C
      DIMENSION AMOM(3),V(5)
      LOGICAL ERROR
C
C  INITIALIZE THE ERROR FLAG AND SET TOLERANCE
      ERROR = .FALSE.
      EPS = .1E-8
      AN=N
      ANN=1.0/(AN+1.0)
C
C FIND THE STANDARDIZE FIRST ORDER STATISTIC
C
      Z=(XMIN-AMOM(1))/AMOM(2)
C
C SET THE UPPER BOUND ON A3 TO 3*AMOM(2/(AMOM(1)-XMIN)
C
      A3U=3.0*AMOM(2)/(AMOM(1)-XMIN)
C
C SET THE LOWER BOUND ON A3 TO .18
C
      A3L=.18
C
C FUNCTIONAL VALUE AT UPPER BOUND, A3U, = -1.0/(1+N)
C
       FU=-ANN
C
CALCULATE FUNCTION AT LOWER BOUND, A3L
```

PROGRAM A.4.3 Continued

```
C      SUBROUTINE MME(AMOM,XMIN,N,V,ERROR) CONTINUED
       VAL=Z/SQRT(1.0+A3L*Z/3.0)
       CALL MDNOR(VAL,P1)
       VAL1=-VAL-6.0/(A3L*SQRT(1.0+A3L*Z/3.0))
       CALL MDNOR(VAL1,P2)
       FL=(P1+EXP(18./A3L**2)*P2)-ANN
C
C      DETERMINE IF THERE EXISTS AN A3 IN THE INTERVAL (A3L,A3L)
C      SATISFIES THE EQUATION
C
       IF (FL*FU.GT.0.0) THEN
          ERROR=.TRUE.
C         NOT A3 IN THE INTERVAL (A3L,A3U) SATISFIES THE EQUATION
       ELSE
C
C  USE BINARY SEARCH TO FIND THE SOLUTION TO THE MODIFIED MOMENT
C  EQUATIONS.
C
          F=FL
  100     A3=(A3U+A3L)/2.0
          IF (ABS(A3-A3L).GT.EPS)THEN
             VAL=Z/SQRT(1.0+A3*Z/3.0)
             CALL MDNOR(VAL,P1)
             VAL1=-VAL-6.0/(A3*SQRT(1.0+A3*Z/3.0))
             CALL MDNOR(VAL1,P2)
             F=(P1+EXP(18./A3**2)*P2)-ANN
             IF(F*FL.L T.0.0)THEN
                A3U=A3
             ELSE
                A3L=A3
                FL=F
             ENDIF
             A3=(A3U+A3L)/2.0
             GO TO 100
          ELSE
             V(1)=AMOM(1)-3.0*AMOM(2)/A3
             V(2)=3.0*AMOM(2)/A3
             V(3)=AMOM(2)
             V(4)=V(1)+V(2)
             V(5)=A3
          ENDIF
       ENDIF
       RETURN
       END
```

PROGRAM A.4.3 Continued

```
      SUBROUTINE MLE(X,N,AMOM,XMIN,V,ERROR)
      DIMENSION X(N),AMOM(3),V(5)
      LOGICAL ERROR
C***********************************************************************
C  THIS SUBROUTINE CALCULATES THE MAXIMUM LIKELIHOOD
C  ESTIMATES FOR THE INVERSE GAUSSIAN DISTRIBUTION
C
C
C INPUT
C          N      = THE NUMBER OF OBSERVATIONS
C          X(N)   = ARRAY OF OBSERVATIONS
C          AMOM   = ARRAY OF MOMENTS
C                     AMOM(1) = MEAN
C                     AMOM(2) = STANDARD DEVIATION
C                     AMOM(3) = THIRD STANDARD MOMENT
C          XMIN   = THE FIRST ORDER STATISTIC, THE SMALLEST OBSERVATION
C
C
C OUTPUT
C             V   = ARRAY OF ESTIMATES OF THE PARAMETERS
C                     V(1) = TERMINUS (T)
C                     V(2) = MU = 3.0*AMOM(2)/AMOM(3)
C                     V(3) = SIGMA = STANDARD DEVIATION
C                     V(4) = E(X)
C                     V(5) = A3 (THIRD STANDARD MOMENT)
C
C          ERROR - .TRUE IF NO ESTIMATES ARE FOUND
C************************************************************************
C
C  INITIALIZE THE ERROR FLAG AND SET THE TOLERANCE
C
      ERROR = .FALSE.
      EPS = .1E-8
C
C  PLACE A LOWER BOUND (BOUND) ON THE ESTIMATE OF THE TERMINUS,TU.
C  SET THE INITIAL GUESS TO (XMIN - .1E-6). THIS IS AN UPPER BOUND
C  OF THE TERMINUS
C
      BOUND =XMIN-25.0*AMOM(2)
      STEP=AMOM(2)/50.0
      AN=N
      TU=XMIN-EPS
      S1=S2=S3=0.0
      DO 11 I = 1,N
      S3=S3+(X(I)-AMOM(1))**2/(X(I)-TU)
      S2=S2+1.0/(X(I)-TU)
   11 S1=S1+1.0/(X(I)-TU)**2
      FU=3.0*S3*S2/AN+AN-(AMOM(1)-TU)**2*S1
C     LOCATE A LOWER BOUND, TL, FOR THE TERMINUS, THE ESTIMATE OF THE
C     TERMINUS SHOULD BE IN THE INTERVAL (TL,TU)
C
      TL=XMIN
      FL=FU
  100 IF (FL*FU.GT.0.0)THEN
         TL=TL-STEP
```

```
C     SUBROUTINE MLE(X,N,AMOM,XMIN,V,ERROR) CONTINUED
         IF (TL.LT.BOUND) THEN
C           CANNOT FIND AN INTERVAL (TL,TU) SUCH THAT TL > BOUND
            PRINT*,' CANNOT FIND INTERVAL (TL,TU)'
            ERROR=.TRUE.
            GO TO 300
         ELSE
C           REQUIREMENT: TL > BOUND
            S1=S2=S3=0.0
            DO 10 I = 1,N
            S3=S3+(X(I)-AMOM(1))**2/(X(I)-TL)
            S2=S2+1.0/(X(I)-TL)
   10       S1=S1+1.0/(X(I)-TL)**2
            FL=3.0*S3*S2/AN+AN-(AMOM(1)-TL)**2*S1
            GO TO 100
         ENDIF
      ELSE
C        USE A BINARY SEARCH TO LOCATE THE ESTIMATE OF THE TERMINUS IN
C        THE INTERVAL
  200    T=(TU+TL)/2.0
         S1=S2=S3=0.0
         DO 13 I = 1,N
         S3=S3+(X(I)-AMOM(1))**2/(X(I)-T)
         S2=S2+1.0/(X(I)-T)
   13    S1=S1+1.0/(X(I)-T)**2
         F=3.0*S3*S2/AN+AN-(AMOM(1)-T)**2*S1
         IF (ABS(TL-T).GT.EPS) THEN
            IF (F*FL .LT.0.0)THEN
               TU=T
            ELSE
               TL=T
               FL=F
            ENDIF
            GO TO 200
         ELSE
            V(2)=(AMOM(1)-T)
            V(3)=SQRT(V(2)/AN*S3)
            V(1)=T
            V(4)=V(2)+V(1)
            V(5)=3.0*V(3)/V(2)
         ENDIF
      ENDIF
  300 RETURN
      END
```

PROGRAM A.4.4 Estimates of Gamma Distribution Parameters

```
      SUBROUTINE MOMEST(AMOM,V)
      DIMENSION AMOM(3), V(6)
C*******************************************************************
C THIS SUBROUTINE CALCULATES THE MOMENT ESTIMATES FOR THE
C THREE-PARAMETER GAMMA DISTRIBUTION
C INPUT
C          AMOM  = ARRAY OF MOMENTS
C                    AMOM(1) = MEAN
C                    AMOM(2) = STANDARD DEVIATION
C                    AMOM(3) = THIRD STANDARD DEVIATION
C
C OUTPUT   V     = ARRAY OF ESTIMATES OF THE PARAMETERS
C                    V(1) = TERMINUS, T
C                    V(2) = SCALE PARAMETER, BETA
C                    V(3) = SHAPE PARAMETER, RHO
C                    V(4) = MEAN, (TERMINUS+RHO*BETA)
C                    V(5) = STANDARD DEVIATION,  (RHO*BETA**2)
C                    V(6) = THIRD STANDARD MOMENT, A3
C
C*******************************************************************
C
C
       V(1)=AMOM(1)-2.0*AMOM(2)/AMOM(3)
       V(2)=AMOM(2)*AMOM(3)/2.0
       V(3)=4.0/AMOM(3)**2
       V(4)=V(2)*V(3)+V(1)
       V(5)=V(2)*SQRT(V(3))
       V(6)= AMOM(3)
       RETURN
       END
```

PROGRAM A.4.4 Continued

```
      SUBROUTINE MME(AMOM,XMIN,N,V,ERROR)
C**********************************************************************
C THIS SUBROUTINE CALCULATES THE MODIFIED MOMENT ESTIMATE FOR THE
C GAMMA DISTRIBUTION
C INPUT
C          AMOM  = ARRAY OF MOMENTS
C                    AMOM(1) = MEAN,
C                    AMOM(2) = STANDARD DEVIATION
C                    AMOM(3) = THIRD STANDARD DEVIATION
C          XMIN  = THE FIRST ORDER STATISTIC, THE SMALLEST OBSERVATION
C          N     = THE NUMBER OF OBSERVATIONS
C
C OUTPUT
C          V     = ARRAY OF ESTIMATES OF THE PARAMETERS
C                    V(1) = TERMINUS
C                    V(2) = SCALE PARAMETER, BETA
C                    V(3) = SHAPE PARAMETER, RHO
C                    V(4) = MEAN = TERMINUS+RHO*BETA
C                    V(5) = STANDARD DEVIATION = RHO*BETA**2
C                    V(6) = THIRD STANDARD MOMENT, A3
C          ERROR = .TRUE. IF NO ESTIMATES ARE FOUND
C
C ROUTINES
C          MDGAM - IMSL GAMMA PROBABILITY DISTRIBUTION FUNCTION
C
C**********************************************************************
C
      DIMENSION AMOM(3),V(6)
      LOGICAL ERROR
C
C  INITIALIZE THE ERROR FLAG AND SET TOLERANCE
      ERROR = .FALSE.
      EPS = .1E-8
,     AN=N
C
C FIND THE STANDARDIZE FIRST ORDER STATISTIC
C
      Z=(XMIN-AMOM(1))/AMOM(2)
C
C SET THE UPPER BOUND ON RHO TO 1600 ( LOWER BOUND ON A3 IS 0.05)
C
      RHOU=1600
C
C SET THE LOWER BOUND ON RHO
C
      RHOL=((AMOM(1)-XMIN)/AMOM(2))**2
C
CALCULATE FUNCTION AT UPPER BOUND, A3U
C
C SCALE Z1 SO THAT Z1> 0
C
      Z1=(Z+SQRT(RHOU))*SQRT(RHOU)
      CALL MDGAM(Z1,RHOU,PROB,IER)
      IF (IER.NE.0) THEN
         ERROR=.TRUE.
```

PROGRAM A.4.4 Continued

```
C      SUBROUTINE MME(AMOM,XMIN,N,V,ERROR)  CONTINUED
       ELSE
          FU=PROB-1.0/(AN+1.)
          Z1=(Z+SQRT(RHOL))*SQRT(RHOL)
          CALL MDGAM(Z1,RHOL,PROB,IER)
          IF (IER.NE.0) THEN
             ERROR=.TRUE.
          ELSE
             FL=PROB-1.0/(AN+1)
             IF (FL*FU.GT.0.0) THEN
                ERROR=.TRUE.
             ELSE
                RHO=(RHOU+RHOL)/2.0
                F=FL
  100           IF (ABS(RHO-RHOL).GT.EPS)THEN
                   Z1=(Z+SQRT(RHO))*SQRT(RHO)
                   CALL MDGAM(Z1,RHO,PROB,IER)
                   IF(IER.EQ.129.OR.IER.EQ.130)THEN
                      ERROR=.TRUE.
                   ELSE
                      F=PROB-1.0/(AN+1.0)
                      IF(F*FL.L T.0.0)THEN
                         RHOU=RHO
                      ELSE
                         RHOL=RHO
                         FL=F
                      ENDIF
                         RHO=(RHOU+RHOL)/2.0
                        GO TO 100
                   ENDIF
                ELSE
                   V(1)=AMOM(1)-AMOM(2)*SQRT(RHO)
                   V(2)=AMOM(2)/SQRT(RHO)
                   V(3)=RHO
                   V(4)=V(3)*V(2)+V(1)
                   V(5)=V(2)*SQRT(V(3))
                   V(6)=2.0/SQRT(V(3))
                ENDIF
             ENDIF
          ENDIF
       ENDIF
       RETURN
       END
```

PROGRAM A.4.4 Continued

```
      SUBROUTINE MLE(X,N,AMOM,XMIN,V,ERROR)
C*********************************************************************
C  THIS SUBROUTINE CALCULATES THE MAXIMUM LIKELIHOOD
C  ESTIMATES FOR THE GAMMA DISTRIBUTION
C
C*********************************************************************
C INPUT
C           N      = THE NUMBER OF OBSERVATIONS
C           X(N)   = ARRAY OF OBSERVATIONS
C
C           AMOM   = ARRAY OF MOMENTS
C                      AMOM(1) = MEAN
C                      AMOM(2) = STANDARD DEVIATION
C                      AMOM(3) = THIRD STANDARD DEVIATION
C
C           XMIN   = THE FIRST ORDER STATISTIC, THE SMALLEST OBSERVATION
C
C
C OUTPUT
C           V      = ARRAY OF ESTIMATES OF THE PARAMETERS
C                      V(1) = TERMINUS
C                      V(2) = SCALE PARAMETER, BETA
C                      V(3) = SHAPE PARAMETER, RHO
C                      V(4) = MEAN = TERMINUS+RHO*BETA
C                      V(5) = STANDARD DEVIATION = RHO*BETA**2
C                      V(6) = THIRD STANDARD MOMENT, A3
C
C           ERROR - .TRUE IF NO ESTIMATES ARE FOUND
C
C FUNCTIONS
C           PSI    STATEMENT FUNCTION INCLUDED
C
C*********************************************************************
C
C
      DIMENSION AMOM(3),V(6),X(N)
      LOGICAL ERROR
C
C*********************************************************************
C                            PSI FUNCTION
      PSI(Z)=ALOG(Z)-1.0/(2.0*Z)-1.0/(12.0*Z**2)+1.0/(120.0*Z**4)
     +-1.0/(252.0*Z**6)+1.0/(240.0*Z**8)
C*********************************************************************
C
C  INITIALIZE THE ERROR FLAG AND SET THE TOLERANCE
C
         ERROR = .FALSE.
         EPS = .1E-8
C
C  PLACE A LOWER BOUND (BOUND) ON THE ESTIMATE OF THE TERMINUS,T.
C  SET THE INITIAL GUESS TO (XMIN - .1E-8). THIS IS AN UPPER BOUND
C  OF THE TERMINUS
C
      BOUND =AMOM(1)-25.0*AMOM(2)
      STEP=AMOM(2)/100.0
```

PROGRAM A.4.4 Continued

```
C       SUBROUTINE MLE(X,N,AMOM,XMIN,V,ERROR) CONTINUED
        TU=XMIN-EPS
        AN=N
        S1=0
        S2=0
        DO 11 I=1,N
        S1=S1+1.0/(X(I)-TU)
   11   S2=ALOG(X(I)-TU)+S2
        A=(AMOM(1)-TU)*S1
        RHO=-A/(AN-A)
        B=(AMOM(1)-TU)/RHO
        FU=PSI(RHO)+ALOG(B)-S2/AN
C       LOCATE A LOWER BOUND, TL, FOR THE TERMINUS, THE ESTIMATE OF THE
C       TERMINUS SHOULD BE IN THE INTERVAL (TL,TU)
C
        TL=TU
        FL=FU
  100   IF (FL/FU.GT.0.0.AND..NOT.ERROR)THEN
           TL=TL-STEP
           IF (TL.GT.BOUND) THEN
C             REQUIREMENT: TL > BOUND
              S1=0
              S2=0
              DO 111 I=1,N
              S1=S1+1.0/(X(I)-TL)
  111         S2=ALOG(X(I)-TL)+S2
              A=(AMOM(1)-TL)*S1
              RHO=-A/(AN-A)
              B=(AMOM(1)-TL)/RHO
              FL=PSI(RHO)+ALOG(B)-S2/AN
           ELSE
C             CANNOT FIND AN INTERVAL (TL,TU) SUCH THAT TL > BOUND
              PRINT*,' CANNOT FIND INTERVAL (TL,TU)'
              ERROR=.TRUE.
           ENDIF
           GO TO 100
        ELSE
           IF (TL.LT.BOUND) THEN
              ERROR=.TRUE.
           ELSE
C             USE A BINARY SEARCH TO LOCATE THE ESTIMATE OF THE TERMINUS I
C
              T=(TL+TU)/2.0
  200         S2=0
              S1=0
              DO 12 I=1,N
              S1=S1+1.0/(X(I)-T)
              S2=S2+ALOG(X(I)-T)
 12           CONTINUE
              A=(AMOM(1)-T)*S1
              RHO=-A/(AN-A)
              B=(AMOM(1)-T)/RHO
              F=PSI(RHO)+ALOG(B)-S2/AN
              V(1)=T
              V(2)=B
```

PROGRAM A.4.4 Continued

```
C       SUBROUTINE MLE(X,N,AMOM,XMIN,V,ERROR) CONTINUED
              V(3)=RHO
              V(4)=V(3)*V(2)+T
              V(5)=V(2)*SQRT(V(3))
              V(6)= 2.0/SQRT(RHO)
              IF (ABS(TL-T).GT.EPS.AND..NOT.ERROR)THEN
                 IF (F*FL .LT.0.0)THEN
                    TU=T
                 ELSE
                    TL=T
                    FL=F
                 ENDIF
                 T=(TL+TU)/2.0
                 GO TO 200
              ENDIF
           ENDIF
        ENDIF
        RETURN
        END
```

Glossary

THE GREEK ALPHABET

Α	α	alpha
Β	β	beta
Γ	γ	gamma
Δ	δ	delta
Ε	ϵ	epsilon
Ζ	ζ	zeta
Η	η	eta
Θ	θ	theta
Ι	ι	iota
Κ	κ	kappa
Λ	λ	lambda
Μ	μ	mu
Ν	ν	nu
Ξ	ξ	xi
Ο	o	omicron
Π	π	pi
Ρ	ρ	rho
Σ	σ	sigma
Τ	τ	tau
Υ	υ	upsilon
Φ	ϕ	phi
Χ	χ	chi
Ψ	ψ	psi
Ω	ω	omega

A, B, C, D, E, etc.	arbitrary constants or functions—definitions vary from chapter to chapter.
$\alpha_3 = E\{[X - E(X)]/\sigma\}^3$	population third standard moment; a measure of skewness.
$a_3 = [\sum_1^n (x_i - \bar{x})^3/n]/[\sum_1^n (x_i - \bar{x})^2/n]^{3/2}$	sample third standard moment.
$\alpha_4 = E\{[X - E(X)]/\sigma\}^4$	population fourth standard moment; a measure of kurtosis.
$a_4 = [\sum_1^n (x_i - \bar{x})^4/n]/[\sum_1^n (x_i - \bar{x})^2/n]^2$	sample fourth standard moment.
α	a parameter of the Pareto distribution; also a parameter of the extreme value distribution.
$\alpha(\xi) = \{1 - Q(\xi)[Q(\xi) - \xi]\}/[Q(\xi) - \xi]^2$	a function of the standardized terminus in truncated and censored samples from the normal distribution.
$\hat{\alpha} = \alpha(\hat{\xi}) = s_y^2/(D - \bar{y})^2$	maximum likelihood estimates of α in singly truncated and singly censored samples from a normal distribution.
β	a scale parameter in various distributions.
β_1 and β_2	Pearson's betas; $\beta_1 = \alpha_3^2$ and $\beta_2 = \alpha_4$.
c	number of censored observations in a censored sample; n_0 is sometimes used as an alternate symbol for c.
c_j	number of observations censored at point (time) T_j; $c = \sum_1^k c_j$.
cdf	abbreviation for cumulative distribution function.
CV	coefficient of variation (sometimes abbreviated as v); $CV = \sigma/[E(X) - \gamma]$.
$C(\delta)$ and $D(\delta)$	functions of the Weibull shape parameter; introduced in Chapter 4.

Arranged in alphabetical order with Greek characters inserted according to their English pronunciation.

δ	the Weibull shape parameter.
D and D_j	censoring or truncation points in samples from the normal distribution. When samples from a lognormal distribution are censored or truncated at T or T_j, then $D = \ln(T - \gamma)$.
$e = 2.7182818$	base of natural logarithms.
E( )	the expectation symbol.
f( ), g( ), and h( )	symbols for probability density functions.
F( ), G( ), and H( )	symbols for cumulative density functions.
γ	a threshold parameter in various skewed distributions.
$\Gamma(z) = \int_0^\infty x^{z-1} e^{-x}\, dx$	the gamma function.
$\Gamma_k = \Gamma(1+k/\delta)$	a function of the Weibull shape parameter.
$G_k = \Gamma(\rho+k/\delta)/\Gamma(\rho)$	a function of shape parameters in the generalized gamma distribution.
h	proportion of censored observations in censored samples; $h = c/N$ except for the Rayleigh distributions where $h = n_0/n$ (i.e., $h = c/n$) (cf. Chapter 10).
h(x)	symbol for the hazard function, $h(x) = f(x)/[1 - F(x)]$.
H(x)	symbol for the cumulative hazard function; $H(x) = \int_0^x h(t)\, dt$.
$H_3(z; h)$	Rayleigh distribution censored sample estimating function; $(h = c/n)$.
$I(\omega) = \int_0^\omega g(u; \rho)\, du$	cdf of generalized gamma distribution (cf. Eq. 12.7.5).
$J(n, \sigma)$	the lognormal estimating function.
$J_2(z)$	Rayleigh distribution truncated sample estimating function; $p = 2$.

$J_3(z)$	Rayleigh distribution truncated sample estimating function; $p = 3$.
κ_r	symbol for the rth order cumulant.
$\lambda(\alpha, h) = \lambda(\xi, h) = \Omega(\xi, h)/[\Omega(\xi, h) - \xi]$	normal distribution estimating function for singly censored samples.
$L(\ )$	symbol for the likelihood function.
$\ln L(\ )$	symbol for the loglikelihood function.
$\ln(\ )$	symbol for the natural logarithm (base e).
$\log(\ )$	symbol for common logarithm (base 10).
$Mo(X)$	mode of X.
$Me(X)$	median of X.
$M_X(t) = E(e^{tX})$	moment generating function.
μ	symbol for a distribution mean; in the three-parameter inverse Gaussian distribution, mean $= \gamma + \mu$.
μ_k	symbol for the kth central moment.
μ'_k	symbol for the kth noncentral moment.
μ_{ij}	symbol for variance-covariance factors used with truncated and censored normal samples.
n	number of complete observations in a sample.
N	total number of observations in a sample; in censored samples $N = n + c$; in truncated samples and in complete samples $N = n$.
ν_1	first moment of truncated or censored sample about left terminus; $\nu_1 = \bar{x} - T_1$.
ω	an alternate shape parameter for the lognormal distribution where $\omega = \exp(\sigma^2)$; also used as a symbol for the transformation $\omega_j = (T_j - \gamma)/\beta$ in the gamma distribution and for the transformation $\omega_j = (T_j - \gamma)^{\delta}/\theta$ in the generalized gamma distribution.

$\Omega(h, \xi) = [h/(1-h)]Q(-\xi)$	an estimating function involved in singly censored samples from the normal distribution.
$\Omega_1(a_1, \xi_1) = a_1 Q(-\xi_1)$ and $\Omega_2(a_2, \xi_2) = a_2 Q(\xi_2)$	estimating functions involved in doubly censored samples from the normal distribution; $a_1 = c_1/n$ and $a_2 = c_2/n$.
p	a dimension symbol in the multidimensional Rayleigh distribution; also a symbol for probability.
pdf	abbreviation for probability density function.
$\phi(z) = (1/\sqrt{2\pi}) \exp(-z^2/2)$	pdf of the standard normal distribution.
$\Phi(z) = \int_{-\infty}^{z} \phi(t)\, dt$	cdf of the standard normal distribution.
ϕ_{ij}	symbol for variance-covariance factors used with complete samples.
π	symbol for pi = 3.14159 . . .
$\psi(x) = \partial \ln \Gamma(x)/\partial x = \Gamma'(x)/\Gamma(x)$	the digamma function.
$\psi'(x) = \partial\psi(x)/\partial x$	the trigamma function.
$\psi(t)$	used in Chapter 3 as a symbol for the cumulant generating function.
$Q(\xi) = \phi(\xi)/[1 - \Phi(\xi)]$	involved in singly truncated normal distribution samples.
$\bar{Q}_1(\xi_1, \xi_2) = \phi(\xi_1)/[\Phi(\xi_2) - \Phi(\xi_1)]$	involved in doubly truncated normal distribution samples.
$\bar{Q}_2(\xi_1, \xi_2) = \phi(\xi_2)/[\Phi(\xi_2) - \Phi(\xi_1)]$	involved in doubly truncated normal distribution samples.
ρ	shape parameter of the gamma distribution.
σ	symbol for distribution standard deviation; a shape parameter in the log-normal distribution.
s	symbol for sample standard deviation; $s = [\sum_{1}^{n} (x_i - \bar{x}) /(n-1)]^{\frac{1}{2}}$.
T, T_j	points or times at which truncation or censoring occurs.

θ	an alternate scale parameter sometimes used in lieu of β; also an arbitrary parameter.
$\theta(\alpha) = \theta(\xi) = Q(\xi)/[Q(\xi) - \xi]$	estimating function in singly truncated normal distribution samples.
$V(\ )$	symbol for variance; $V(X) = \sigma_X^2$.
$W(n, \delta)$	Weibull distribution estimating function.
$w = T_2 - T_1$	sample range in a doubly truncated sample.
X, Y, and sometimes other capital letters	symbols for random variables.
X_1 or $X_{1:N}$	first order statistic in a random sample of size N.
x_i, y_i, etc.	sample observations.
$\bar{x}$, $\bar{y}$, etc.	sample means; $\bar{x} = \sum_1^n x_i/n$.
ξ_j	standardized points of truncation or censoring; $\xi_j = [T_j - E(X)]/\sqrt{V(X)}$.
Z	standardized random variable; $Z = [X - E(X)]/\sqrt{V(X)}$.
$Z_{1:N}$	standardized first order statistic in a random sample of size N.
z_1	observed value of standardized first order statistic; $z_1 = (x_1 - \bar{x})/s$.

Bibliography

Abramowitz, M. and Stegun, I. A. (1964) Handbook of Mathematical Functions with Formulas, Graphs, and Mathematical Tables. National Bureau of Standards Applied Mathematics Series 55. U.S. Government Printing Office, Washington, D.C.

Adams, J. D. (1962) Failure time distribution estimation. Semiconductor Reliability, 2, 41-52.

Aitchison, J. and Brown, J. A. C. (1957) The Lognormal Distribution. London, Cambridge University Press.

Antle, C. E. (1985) Lognormal distribution. In Encyclopedia of Statistical Science, vol. 5, pp. 134-136. Eds. S. Kotz, N. L. Johnson, and C. B. Read. Wiley, New York.

Archer, C. O. (1967) Some properties of Rayleigh distributed random variables and of their sums and products. Technical memo. TM-67-15, Naval Missile Center, Point Mugu, Calif.

Arora, M. S. (1973) Theory and applications of the four-parameter generalized Weibull distribution. Ph.D. Dissertation, University of Georgia, Athens, Ga.

Bain, L. J. (1978) Statistical Analysis of Reliability and Life-Testing, Dekker, New York.

Bol'shev, L. N., Prohorov, Yu. V., and Rodinov, D. A. (1963) On the logarithmic-normal law in geology. Teoriya Veroyatnostei i ee Primeneniya, 8, 114 (Abstract). (In Russian; English translation, p. 107.)

Bowman, K. O. and Shenton, L. R. (1987) Properties of Estimators for the Gamma Distribution. Dekker, New York.

Box, E. E. P. and Cox, D. R. (1964) An analysis of transformations. J. Roy. Statist. Soc. B, 26, 211-251.

Bradu, Dan and Mundlak, Yair (1970) Estimation in lognormal linear models. J. Amer. Statist. Assoc., 65, 198-211.

Burges, S. J. and Hoshi, K. (1978) Approximation of a normal distribution by a three-parameter log normal distribution. Water Resour. Res., 14(4), 620-622.

Burges, S. J., Lettenmaier, D. P., and Bates, C. L. (1975) Properties of the three-parameter log normal probability distribution. Water Resour. Res., 11(2), 229-235.

Calitz, F. (1969) Skattingsprosedures by die lognormal verdeling (Estimation procedures for the lognormal distribution). Unpublished D. Com. thesis, University of South Africa.

Calitz, F. (1973) Maximum likelihood estimation of the parameters of the three-parameter lognormal distribution—a reconsideration. Austral. J. Statist., 3, 185-190.

Chan, Micah Y. (1982) Modified moment and maximum likelihood estimators for parameters of the three-parameter inverse Gaussian distribution. Ph.D. Dissertation, University of Georgia, Athens, Ga.

Chan, M., Cohen, A. C., and Whitten, B. J. (1983) The standardized inverse Gaussian distribution: Tables of the cumulative probability function. Comm. in Statist.—Simul. and Comp. B, 12, 423-442.

Chan, M., Cohen, A. C., and Whitten, B. J. (1984) Modified maximum likelihood and modified moment estimators for the three-parameter inverse Gaussian distribution. Com. in Statist.—Simul. and Comp. B, 13, 47-68.

Charbeneau, R. J. (1978) Comparison of the two- and three-parameter lognormal distributions used in streamflow synthesis. Water Resour. Res., 14(1), 149-150.

Cheng, R. C. H. and Amin, N. A. K. (1981) Maximum likelihood estimation of parameters in the inverse Gaussian distribution with unknown origin. Technometrics, 23, 257-263.

Chhikara, R. S. and Folks, J. L. (1974) Estimation of the inverse Gaussian distribution function. J. Amer. Statist. Assoc., 69, 250-254.

Chiepa, M. and Amato, P. (1981) A new parameter estimation procedure for the three-parameter lognormal distribution. In Statistical Distributions in Scientific Work. Vol. 5. Inferential Problems and Properties, Eds. C. Taillie, G. P. Patil, and B. A. Baldessavi. NATO Advanced Studies Ser. C, vol. 79, part 2. Reidel, Dordrecht.

Cohen, A. C. (1950) Estimating the mean and variance of normal populations from singly truncated and doubly truncated samples. Ann. Math. Statist., 21, 557-569.

Cohen, A. C. (1951) Estimating parameters of logarithmic-normal distributions by maximum likelihood. J. Amer. Statist. Assoc., 46, 206-212.

Cohen, A. C. (1955) Maximum likelihood estimation of the dispersion parameter of a chi-distributed radial error from truncated and censored samples with applications to target analysis. J. Amer. Statist. Assoc., 50, 1122-1135.

Cohen, A. C. (1957) On the solution of estimating equations for truncated and censored samples from normal populations. Biometrika, 44, 225-236.

Cohen, A. C. (1959) Simplified estimators for the normal distribution when samples are singly censored or truncated. Technometrics, 1, 217-237.

Cohen, A. C. (1961) Tables for maximum likelihood estimates: singly truncated and singly censored samples. Technometrics, 3, 535-541.

Cohen, A. C. (1963) Progressively censored samples in life testing. Technometrics, 5, 237-339.

Cohen, A. C. (1965) Maximum likelihood estimation in the Weibull distribution based on complete and censored samples. Technometrics, 5, 579-588.

Cohen, A. C. (1966) Life testing and early failure. Technometrics, 8, 539-549.

Cohen, A. C. (1969) A generalization of the Weibull distribution. Marshall Space Flight Center, NASA Contractor Report No. 61293, Cont. NAS 8-11175.

Cohen, A. C. (1973) The reflected Weibull distribution. Technometrics, 15, 867-873.

Cohen, A. C. (1975) Multi-censored sampling in the three-parameter Weibull distribution. Technometrics, 17, 347-351.

Cohen, A. C. (1976) Progressively censored sampling in the three-parameter lognormal distribution. Technometrics, 18, 99-103.

Cohen, A. C. (1988) Estimation in the lognormal distributions. In Lognormal Distributions—Theory and Applications, chaps. 4, 5, Eds. E. L. Crow and K. Shimizu. Dekker, New York.

Cohen, A. C. and Helm, Russell (1973) Estimation in the exponential distribution. Technometrics, 14, 841-846.

Cohen, A. C., Helm, F. Russell, and Sugg, Merritt (1969) Tables of areas of the standardized Pearson Type III density function. NASA Contractor Report CR-61266 Contract NAS8-11175.

Cohen, A. C. and Norgaard, N. J. (1977) Progressively censored sampling in the three-parameter gamma distribution. Technometrics, 19, 333-340.

Cohen, A. C. and Whitten, B. J. (1980) Estimation in the three-parameter lognormal distribution. J. Amer. Statist. Assoc., 75, 399-404.

Cohen, A. C. and Whitten, B. J. (1981) Estimation of lognormal distributions. Amer. J. Math. Mgt. Sci., 1, 139-153.

Cohen, A. C. and Whitten, B. J. (1982) Modified moment and maximum likelihood estimators for parameters of the three-parameter Gamma distribution. Comm. in Statist.—Simul. and Comp., 11, 197-216.

Cohen, A. C. and Whitten, B. J. (1985) Modified moment estimation for the three-parameter inverse Gaussian distribution. J. Qual. Tech., 17, 147-154.

Cohen, A. C. and Whitten, B. J. (1986) Modified moment estimation for the three-parameter gamma distribution. J. Qual. Tech., 18, 53-62.

Cohen, A. C. and Woodward, J. (1953) Tables of Pearson-Lee-Fisher functions of singly truncated normal distributions. Biometrics, 9, 489-497.

Cohen, A. C., Whitten, B. J., and Ding, Y. (1984) Modified moment estimation for the three-parameter Weibull distribution. J. Qual. Tech., 16, 159-167.

Cohen, A. C., Whitten, B. J., and Ding, Y. (1985) Modified moment estimation for the three-parameter lognormal distribution. J. Qual. Tech., 17, 92-99.

Cooley, C. G. and Cohen, A. C. (1970) Tables of maximum likelihood estimating functions for singly truncated and singly censored samples from the normal distribution. NASA Contractor Report NASA CR-61330. NASA George C. Marshall Space Flight Center, Marshall Space Flight Center, Alabama.

Croxton, F. E. and Cowden, D. J. (1939) Applied General Statistics. Prentice-Hall, Inc., Newark, New Jersey.

Crow, E. L. and Shimizu, K. (1988) Lognormal Distributions—Theory and Applications. Dekker, New York.

David, H. A. (1964) Order Statistics. Wiley, New York, pp. 120-122.

Dubey, S. D. (1966) Hyper-efficient estimator of the location parameter of the Weibull laws. Nav. Res. Log. Quart., 13, 253-263.

Dumonceaux, R. and Antle, C. E. (1973) Discrimination between the lognormal and the Weibull distributions. Technometrics, 15, 923-926.

Elandt-Johnson, R. C. and Johnson, N. L. (1980) Survival Models and Data Analysis. Wiley, New York.

Elteto, O. (1965) Large-sample lognormality tests based on new inequality measures. Bul. Intntl. Statist. Inst., 41, 382-385.

Engelhardt, M., and Bain, L. J. (1977) Simplified statistical procedures for the Weibull or extreme value distribution. Technometrics, 19, 323-331.

Epstein, B. (1947) The mathematical description of certain breakage mechanisms leading to the logarithmico-normal distribution. J. Franklin Inst., 244, 471-477.

Epstein, B. (1948) Statistical aspects of fracture problems. J. Appl. Phys., 19, 140-147.

Epstein, Benjamin (1960) Estimation from life test data. Technometrics, 2, 447-454.

Epstein, Benjamin and Sobel, M. (1953) Life testing. J. Amer. Statist. Assoc., 48, 485-502.

Evans, I. G. and Shaban, S. A. (1974) A note on estimation in lognormal models. J. Amer. Statist. Assoc., 69, 779-781.

Finney, D. J. (1941) On the distribution of a variate whose logarithm is normally distributed. J. Roy. Statist. Soc. Ser. B, 7, 155-161.

Folks, J. L. and Chhikara, R. S. (1978) The inverse Gaussian distribution and its statistical application—a review. J. Roy Statist. Soc., Ser. B, 40, 263-289.

Gaddum, J. H. (1945) Lognormal distributions. Nature (London), 156, 463-466.

Gajjar, A. V. and Khatri, C. G. (1969) Progressively censored samples from log-normal and logistic distributions. Technometrics, 11, 793-803.

Giesbrecht, F. and Kempthorne, O. (1976) Maximum likelihood estimation in the three-parameter lognormal distribution. J. Roy. Statist. Soc., Ser. B, 38, 257-264.

Gross, A. J. and Clark, V. A. (1975) Survival Distributions: Reliability Applications in the Biomedical Sciences. Wiley, New York.

Gumbel, E. J. (1958) Statistics of Extremes. Columbia University Press, New York.

Gupta, A. K. (1952) Estimation of the mean and standard deviation of a normal population from a censored sample. Biometrika, 39, 260-273.

Gupta, S. S. (1962) Life test sampling plans for normal and lognormal distributions. Technometrics, 4, 151-175.

Hagstroem, K. G. (1960) Early characterization of the Pareto distribution. Skand. Aktuarietidskr., 8, 65-88.

Hald, A. (1949) Maximum likelihood estimation of the parameters of a normal distribution which is truncated at a known point. Skand. Aktuarietidskr., 32, 119-134.

Halperin, M. (1952) Estimation in the truncated normal distribution. J. Amer. Statist. Assoc., 47, 457-465.

Hamdan, M. A. (1971) The logarithm of the sum of two correlated lognormal variates. J. Amer. Statist. Assoc., 61, 842-851.

Harris, C. M. (1968) The Pareto distribution as a queue service discipline. Operations Research, 16, 307-313.

Harter, H. L. (1961) Expected values of normal order statistics. Biometrika, 48, 151-166.

Harter, H. L. (1967) Maximum likelihood estimation of the parameter of a four-parameter generalized gamma population from complete and censored samples. Technometrics, 9, 159-165.

Harter, H. L. (1969) Order Statistics and Their Use in Testing and Estimation, Vol. 2, Table C1, pp. 426-456. Aerospace Res. Lab., Wright-Patterson AFB, Ohio.

Harter, H. L. and Moore, A. H. (1965) Maximum likelihood estimation of the parameters of the Gamma and Weibull populations from censored samples. Technometrics, 7, 639-643.

Harter, H. L. and Moore, A. H. (1966) Local-maximum-likelihood estimation of the parameters of three-parameter lognormal populations from complete and censored samples. J. Amer. Statist. Assoc., 61, 842-851.

Harter, H. L. and Moore, A. H. (1967) Asymptotic variances and covariances of maximum likelihood estimators from censored samples, of parameters of Weibull and Gamma populations. Ann. Math. Statist., 38, 557-571.

Herd, G. R. (1956) Estimation of the parameters of a population from a multi-censored sample. Ph.D. Dissertation, Iowa State College.

Herd, G. R. (1957) Estimation of reliability functions. Proc. Third Nat. Symp. on Reliability and Quality Control.

Herd, G. R. (1960) Estimation of reliability from incomplete data. Proc. Sixth Nat. Symp. on Reliability and Quality Control.

Heyde, C. C. (1963) On a property of the lognormal distribution. J. Roy. Statist. Soc. Ser. B, 25, 392-393.

Hill, B. M. (1963) The three-parameter lognormal distribution and Bayesian analysis of a point-source epidemic. J. Amer. Statist. Assoc., 58, 72-84.

Hirano, Katuomi (1986) Rayleigh distribution. In Encyclopedia of Statistical Sciences, vol. 7, pp. 647-649, Eds. Samuel Kotz, Norman L. Johnson, and Campbell B. Read, Wiley, New York.

Hoel, Paul G. (1954) Introduction to Mathematical Statistics, 2d ed. Wiley, New York.

Hoshi, K., Stedinger, J. R., and Burges, J. (1984) Estimation of log-normal quantiles: Monte Carlo results and first-order approximations. J. Hydrology, 71, 1-30.

Hsieh, Paul I-Po (1977) Progressive censoring on the four-parameter Weibull distribution. Ph.D. Dissertation, University of Georgia, Athens, Ga.

Hyrenius, H. and Gustafsson, R. (1962) Tables of Normal and Log-Normal Random Deviates. I, II. Almqvist and Wiksell, Stockholm.

Itoh, M. and Sugiyama, T. (1980) Estimation of the threshold parameter of the three-parameter lognormal distribution. In Recent Developments in Statistical Inferences and Data Analysis, Ed. K. Matusita. North-Holland, Amsterdam.

Johnson, N. L. and Kotz, S. (1970) Continuous Univariate Distributions. I. Houghton Mifflin, Boston.

Kalinske, A. A. (1946) On the logarithmic probability law. Trans. Geophys. Union, 27, 709-711.

Kane, V. E. (1978) Some topics in multivariate classifications. Presentation, SREB Summer Research Conference in Statistics, DeGray State Park Lodge, Arkadelphia, Ark.

Kane, V. E. (1982) Standard and goodness-of-fit parameter estimation methods for the three-parameter lognormal distribution. Commun. Statist. A, 11(17), 1935-1957.

Kao, J. H. K. (1959) A graphical estimation of mixed Weibull parameters in life testing of electron tubes. Technometrics, 1, 389-407.

Kapteyn, J. C. (1903) Skew Frequency Curves in Biology and Statistics, Astronomical Laboratory, Noordhoff, Groningen.

Kapteyn, J. C. and van Uven, M. J. (1916) Skew Frequency Curves in Biology and Statistics. Hotsema Brothers, Groningen.

Kendall, M. G. (1948) The Advanced Theory of Statistics, vol. 1, 4th ed. Charles Griffin, London.

Kolmogorov, A. N. (1941) Uber das logarithmisch normale Verteilungsgesetz der Dimensionen der Teilchen bei Zerstuckelung. Doklady Akademii Nauk SSSR, 31, 99-101.

Koniger, Wolfgang (1981) Die Anwendung der Extremal-3-Verteilung bei der Regenauswertung und der Niedrigwasseranalyse. GWF—Wasser/Abwasser, 122, 460-466.

Koopmans, L. H., Owen, D. B., and Rosenblatt, J. I. (1964) Confidence intervals for the coefficient of variation for the normal and lognormal distributions. Biometrika, 51, 25-31.

Kotz, S. and Srinivasan, R. (1969) Ann. Inst. Statist. Math. (Tokyo), 21, 201-210.

Krumbein, W. C. (1936) Application of logarithmic moments to size frequency distributions of sediments. J. Sedimentary Petrology, 6, 35-47.

Lambert, J. A. (1964) Estimation of parameters in the three-parameter lognormal distribution. Austral. J. Statist., 6, 29-32.

Laurent, A. G. (1963) The lognormal distribution and the translation method: description and estimation problems. J. Amer. Statist. Assoc., 58, 231-235 (correction, 58, 1163).

Lawless, J. F. (1982) Statistical Models and Methods for Lifetime Data. Wiley, New York.

Lawrence, R. J. (1979) The lognormal as inter-event time distribution, unpublished manuscript.

Lawrence, R. J. (1980) The lognormal distribution of buying frequency rates. J. Marketing Res., 17, 212-220.

Lawrence, R. J. (1984) The lognormal distribution of the duration of strikes. J. Roy. Statist. Soc. A, 147, 464-483.

Lemon, Glen H. (1975) Maximum likelihood estimation for the three parameter Weibull distribution based on censored samples. Technometrics, 17, 247-254.

Malik, H. J. (1966) Exact moments of order statistics from the Pareto distribution. Skand. Aktuarietidskr., 49, 144-157.

Malik, H. J. (1967) Exact moments of order statistics for a power-function population. Skand. Aktuarietidskr., 50, 64-69.

Malik, H. J. (1970) Products of Pareto variables. Metrika, 15, 19-22.

Mandelbrodt, B. (1960) The Pareto-Levy law and the distribution of income. Int. Econ. Rev., 1, 79-106.

Mann, Nancy R. (1968) Estimation procedures for the two-parameter Weibull and extreme-value distributions. Technometrics, 10, 231-256.

Mann, Nancy R. and Fertig, Kenneth W. (1975a) A goodness-of-fit test for the two-parameter vs. three-parameter Weibull: confidence bounds for threshold. Technometrics, 17, 237-245.

Mann, Nancy R. and Fertig, Kenneth W. (1975b) Simplified efficient point and interval estimators for Weibull parameters. Technometrics, 17, 361-368.

Mann, N. R., Schafer, R. E., and Singpurwalla, N. D. (1974) Methods for Statistical Analysis of Reliability and Life Data. Wiley, New York.

McCool, J. I. (1974) Inferential techniques for Weibull populations. Aerospace Research Laboratories Report ARL TR 74-0180, Wright-Patterson AFB, Ohio.

Meeker, W. Q. and Nelson, W. B. (1974) Tables for the Weibull and smallest extreme value distributions. General Electric Co. Corporate Research and Development TIS Report 74CRD230.

Meeker, W. Q. and Nelson, W. B. (1977) Weibull variances and confidence limits by maximum likelihood for singly censored data. Technometrics, 19, 473-476.

Mendenhall, W. (1958) A bibliography on life testing and related topics. Biometrika, 45, 521-543.

Monlezum, C. J., Antle, C. E., and Klimko, L. A. (1975) Unpublished manuscript (concerning maximum likelihood estimation of parameter in the lognormal model).

Morrison, J. (1958) The lognormal distribution in quality control. Appl. Statist., 7, 160-172.

Moshman, J. (1953) Critical values of the log-normal distribution. J. Amer. Statist. Assoc., 48, 600-609.

Munro, A. H. and Wixley, R. A. J. (1970) Estimators based on order statistics of small samples from a three-parameter lognormal distribution. J. Amer. Statist. Assoc., 65, 212-225.

Naus, J. I. (1969) The distribution of the logarithm of the sum of two lognormal variates. J. Amer. Statist. Assoc., 64, 655-659.

Nelson, W. (1969) Hazard plotting for incomplete failure data. J. Qual. Tech., 1, 27-52.

Nelson, W. (1972) Theory and application of hazard plotting for censored failure data. Technometrics, 14, 945-966.

Nelson, W. (1982) Applied Life Data Analysis. Wiley, New York.

Nelson, W. and Schmee, J. (1979) Inference for (log) normal life distributions from small singly censored samples and BLUEs. Technometrics, 21, 43-54.

Norgaard, Nicholas (1975) Estimation of parameters in continuous distributions from restricted samples. Ph.D. Dissertation, University of Georgia, Athens, Ga.

Nowick, A. S. and Berry, B. S. (1961) Lognormal distribution function for describing anelastic and other relaxation processes. IBM J. Res. Dev., 5, 297-311, 312-320.

Nydell, S. (1919) The mean errors of the characteristics in logarithmic-normal distribution. Skand. Aktuarietidskr., 2, 134-144.

Oldham, P. D. (1965) On estimating the arithmetic means of lognormally-distributed populations. Biometrics, 21, 235-239.

Padgett, W. J. and Wei, L. J. (1979) Estimation for the three-parameter inverse Gaussian distribution. Comm. in Statist. A, 8, 129-137.

Pareto, V. (1897) Cours d'Economie Politique. Rouge and Cie, Lausanne and Paris.

Pearce, S. C. (1945) Lognormal distributions. Nature (London), 156, 747.

Prohorov, Yu. V. (1963) On the lognormal distribution in geo-chemistry. Teoriya Veroyatnostei i ee Primeneniya, 10, 184-187. (In Russian)

Quant,. R. E. (1966) Old and new methods of estimation and the Pareto distribution. Metrika, 10, 55-82.

Quensel, C. E. (1945) Studies of the logarithmic normal curve. Skand. Aktuarietidskr., 28, 141-153.

Raj, Des (1953) On moment estimation of the parameters of a normal population from singly and doubly truncated samples. Ganita, 4, 79-84.

Rayleigh, J. W. S. (1919) On the problem of random vibrations, and of random flights in one, two, or three dimensions. Philos. Mag., 37, 321-347.

Ringer, L. J. and Sprinkle, E. E. (1972) Estimation of the parameters of the Weibull distribution from multi-censored samples. IEEE Transactions on Reliability, R-21, 46-51.

Roberts, H. R. (1962a) Some results in life testing based on hypercensored samples from an exponential distribution. Ph.D. Dissertation, George Washington University.

Roberts, H. R. (1962b) Life test experiments with hypercensored samples. Proc. Eighteenth Annual Quality Control Conf., Rochester Soc. for Quality Control.

Rockette, H., Antle, C., and Klimko, L. A. (1974) Maximum likelihood estimation with the Weibull model. J. Amer. Statist. Assoc., 69, 246-249.

Rohn, W. B. (1959) Reliability prediction for complex systems. Proc. 5th National Symposium on Reliability and Quality Control in Electronics, pp. 381-388.

Rukhin, A. L. (1984) Improved estimation in lognormal models. T.R. 84-38 Dept. of Statist., Purdue Univ., West Lafayette, Inc.

Salvosa, Louis R. (1936) Tables of Pearson's Type III function. Ann. Math. Statist., 1, 191-198; appendix 1-125.

Sampford, M. R. (1952) The estimation of response-time distributions. Part II. Biometrics, 9, 307-369.

Sarhan, A. E. and Greenberg, B. G. (1956) Estimation of location and scale parameters by order statistics from singly and doubly censored samples. Part I. Ann. Math. Statist., 27, 427-451.

Sarhan, A. E. and Greenberg, B. G. (1957) Tables for best linear estimates by order statistics of the parameters of single exponential distributions from singly and doubly censored samples. J. Amer. Statist. Assoc., 52, 58-87.

Sarhan, A. E. and Greenberg, B. G. (1958) Estimation of location and scale parameters by order statistics from singly and doubly censored samples. Part II. Ann. Math. Statist., 29, 79-105.

Sarhan, A. E. and Greenberg, B. G. (1962) Contributions to Order Statistics. Wiley, New York, London.

Schmee, J., Gladstein, D., and Nelson, W. (1985) Confidence limits of a normal distribution from singly censored samples using maximum likelihood. Technometrics, 27, 119-128.

Schmee, J. and Nelson, W. B. (1977) Estimates and approximate confidence limits for (log) normal life distributions from singly censored samples by maximum likelihood. General Electric Company Technical Report 76CRD250, Schenectady, N.Y.

Schneider, H. (1986) Truncated and Censored Samples from Normal Populations. Dekker, New York.

Schroedinger, E. (1915) Zur Theorie der Fall—und Steigversuche an Teilchen mit Brownscher Bewegung. Physikalische Zeitschrift, 16, 289-295.

Severo, N. C. and Olds, E. G. (1956) A comparison of tests on the mean of a logarithmico-normal distribution with known variance. Ann. Math. Statist., 27, 670-686.

Shook, B. L. (1930) Synopsis of elementary mathematical statistics. Ann. Math. Statist., 1, 14-41.

Siddiqui, M. M. (1962) Some problems connected with Rayleigh distributions. J. Res., Nat. Bu. Std., 66D, 167-174.

Smoluchowsky, M. W. (1915) Notiz uber die Berechnung der Browschen Molekularbewegung bei der Ehrenhaft-Millikanschen Versuchsanordnung. Physikalische Zeitschrift, 16, 318-321.

Srivastava, M. S. (1965) A characterization of Pareto's distribution. (Abstract). Ann. Math. Statist., 36, 361-362.

Stacy, E. W. (1962) A generalization of the gamma distribution. Ann. Math. Statist., 33, 1187-1192.

Stacy, E. W. and Mirham, G. A. (1965) Parameter estimation for a generalized gamma distribution. Technometrics, 7, 349-358.

Stedinger, J. R. (1980) Fitting log normal distributions to hydrologic data. Water Resources Res., 16, 481-490.

Stevens, W. L. (1937) The truncated normal distribution. (Appendix to paper by C. I. Bliss, The calculation of the time mortality curve.) Ann. Appl. Biol., 24, 815-852.

Sundaraiyer, V. H. (1986) Estimation of parameters of the inverse Gaussian distribution from censored samples. Ph.D. Dissertation, Dept. of Statistics, University of Georgia, Athens, Ga.

Tallis, G. M. and Young, S. S. Y. (1962) Maximum likelihood estimation of parameters of the normal, the log-normal, truncated normal and bivariate normal distributions from grouped data. Austral. J. Statist., 4, 49-54.

Thompson, G. W., Friedman, M., and Garelis, E. (1954) Estimation of mean and variance of normal populations from doubly truncated samples. Report of Ethyl Corp. Res. Lab., Detroit.

Tiago de Oliveira, J. (1963) Decision results for the parameters of the extreme value (Gumbel) distribution based on the mean and the standard deviation. Trabajos de Estadistica, 14, 61-81.

Tiku, M. L. (1968) Estimating the parameters of log-normal distribution from censored samples. J. Amer. Statist. Assoc., 63, 134-140.

Tweedie, M. C. K. (1956) Some statistical properties of the inverse Gaussian distribution. Vir. J. of Sci., 7, 160-165.

Tweedie, M. C. K. (1957a) Statistical properties of the inverse Gaussian distribution. I. Ann. Math. Stat., 28, 362-377.

Tweedie, M. C. K. (1957b) Statistical properties of the inverse Gaussian distribution. II. Ann. Math. Stat., 28, 695-705.

Uven, M. J. van (1917) Logarithmic frequency distributions. Proc. Roy. Acad. Sci. Amsterdam, 19, 533-546.

Wald, A. (1944) On cumulative sums of random variables. Ann. Math. Statist., 15, 283-296.

Wasan, M. T. (1968) On an inverse Gaussian process. Skand. Aktuarietidskr., 51, 69-96.

Wasan, M. T. and Roy, L. K. (1969) Tables of inverse Gaussian percentage points. Technometrics, 11, 590-603.

Weiss, L. L. (1957) A nomogram for log-normal frequency analysis. Trans. Amer. Geophys. Union, 38, 33-37.

Whitten, B. J. and Cohen, A. C. (1981) Percentiles and other characteristics of the four-parameter generalized gamma distribution. Comm. in Statist. B, 10(2), 175-218.

Whitten, B. J., Cohen, A. C., and Sundaraiyer, V. (1988) A pseudo-complete sample technique for estimation from censored samples. Comm. in Statist., A, 17(7), 2239-2258.

Williams, C. B. (1937) The use of logarithms in the interpretation of certain entomological problems. Ann. Appl. Biol., 24, 404-414.

Wilson, Edwin B. and Worcester, Jane (1945) The normal logarithmic transform. Rev. Econ. Statist., 27, 17-22.

Wingo, Dallas R. (1973) Solution of the three-parameter Weibull equations by constrained modified quasi-linearization (progressively censored samples). IEEE Transactions on Reliability, R-22, 2, 96-102.

Wingo, D. R. (1975) The use of interior penalty functions to overcome log-normal distribution parameter estimation anomalies. J. Statist. Comp. Simul., 4, 49-61.

Wingo, D. R. (1976) Moving truncations barrier-function methods for estimation in three-parameter lognormal models. Comm. in Statist. B, 5, 65-80.

Wise, M. E. (1966a) The geometry of log-normal and related distributions and an application to tracer-dilution curves. Statistica Neerlandica, 20, 119-142.

Wise, M. E. (1966b) Tracer dilution curves in cardiology and random walk and lognormal distributions. Acta Physiologica Pharmacologica Neerlandica, 14, 175-204.

Wycoff, J., Bain, L., and Engelhardt, M. (1980) Some complete and censored sampling results for the three-parameter Weibull distribution. J. Statist. Comp. Simul., 11, 139-151.

Yuan, P. T. (1933) On the logarithmic frequency distributions and the semi-logarithmic correlation surface. Ann. Math. Statist., 4, 30-74.

Zanakis, S. H. (1977) Computational experience with some nonlinear optimization algorithms in deriving MLE for the three-parameter Weibull distribution. In Algorithmic Methods in Probability: TIMS Studies in Management Sciences, Vol. 7, Ed. M. F. Neuts. North-Holland, Amsterdam.

Zanakis, S. H. (1979a) Monte Carlo study of some simple estimators of the three-parameter Weibull distribution. J. Statist. Comp. Simul., 9, 101-116.

Zanakis, S. H. (1979b) Extended pattern search with transformations for the three-parameter Weibull MLE problem. Mgmt. Sci., 25, 1149-1161.

Zanakis, S. H. and Mann, N. R. (1981) A good simple percentile estimator of the Weibull shape parameter for use when all three parameters are unknown. Unpublished manuscript.

Index